Eduardo Jacob-Lopes, Rosangela Rodrigues Dias, Leila Queiroz Zepka (Eds.)

Microalgae-Based Systems

Also of interest

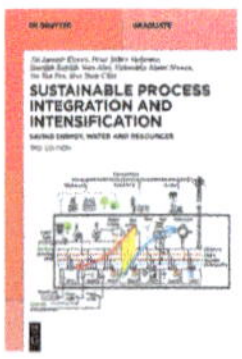

Sustainable Process Integration and Intensification.
Saving Energy, Water and Resources
3rd, completely revised and extended edition
Klemeš, Varbanow, Wan Alwi, Manan, Fan, Chin, 2023
ISBN 978-3-11-078283-7, e-ISBN (PDF) 978-3-11-078298-1,
e-ISBN (EPUB) 978-3-11-078300-1

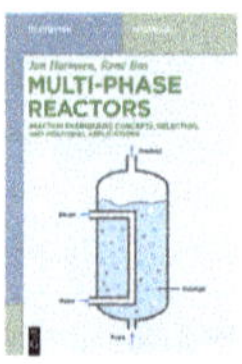

Multiphase Reactors.
Reaction Engineering Concepts, Selection, and Industrial Applications
Harmsen, Bos, 2023
ISBN 978-3-11-071376-3, e-ISBN (PDF) 978-3-11-071377-0,
e-ISBN (EPUB) 978-3-11-071384-8

Basic Process Engineering Control
2nd, completely revised and updated edition
Agachi, Cristea, Makhura, 2020
ISBN 978-3-11-064789-1, e-ISBN (PDF) 978-3-11-064793-8,
e-ISBN (EPUB) 978-3-11-064853-9

Engineering Catalysis
2nd edition
Murzin, 2020
ISBN 978-3-11-061442-8, e-ISBN (PDF) 978-3-11-061443-5,
e-ISBN (EPUB) 978-3-11-061469-5

Process Intensification.
Breakthrough in Design, Industrial Innovation Practices, and Education
Harmsen, Verkerk, 2020
ISBN 978-3-11-065734-0, e-ISBN (PDF) 978-3-11-065735-7,
e-ISBN (EPUB) 978-3-11-065752-4

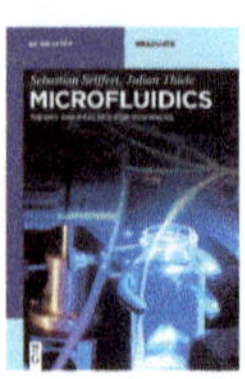

Microfluidics.
Theory and Practice for Beginners
Seiffert, Thiele, 2019
ISBN 978-3-11-048777-0, e-ISBN (PDF) 978-3-11-048770-1,
e-ISBN (EPUB) 978-3-11-048784-8

Microalgae-Based Systems

Process Integration and Process Intensification Approaches

Edited by
Eduardo Jacob-Lopes, Rosangela Rodrigues Dias
and Leila Queiroz Zepka

DE GRUYTER

Editors
Ass. Prof. Eduardo Jacob-Lopes
Campus Sede
Department of Food Technology and Science
Federal University of Santa Maria
Av. Roraima 1000
97105-900 Santa Maria
Brazil
ejacoblopes@gmail.com

Ass. Prof. Rosangela Rodrigues Dias
Campus Sede
Department of Food Technology and Science
Federal University of Santa Maria
Av. Roraima 1000
97105-900 Santa Maria
Brazil
ro.rosangelard@gmail.com

Ass. Prof. Leila Queiroz Zepka
Campus Sede
Department of Food Technology and Science
Federal University of Santa Maria
Av. Roraima 1000
97105-900 Santa Maria
Brazil
zepkaleila@yahoo.com.br

ISBN 978-3-11-078125-0
e-ISBN (PDF) 978-3-11-078126-7
e-ISBN (EPUB) 978-3-11-078136-6

Library of Congress Control Number: 2023938949

Bibliographic information published by the Deutsche Nationalbibliothek
The Deutsche Nationalbibliothek lists this publication in the Deutsche Nationalbibliografie; detailed bibliographic data are available on the internet at http://dnb.dnb.de.

Cover image: marekuliasz/iStock/Getty Images Plus; Process flow diagram by Eduardo Jacob-Lopes, Rosangela Rodrigues Dias, Leila Queiroz Zepka
Typesetting: Integra Software Services Pvt. Ltd.
Printing and binding: CPI books GmbH, Leck

www.degruyter.com

Contents

Contributing authors —— IX

Part I: **Fundamentals**

Luísa C. Schetinger, Marcele L. Nörnberg, Patrícia A. Caetano, Leila Q. Zepka
Chapter 1
A timeline on microalgal biotechnology —— 3

Melih Onay
Chapter 2
Scope of the microalgae market: a demand and supply perspective —— 19

Rosangela Rodrigues Dias, Adriane Terezinha Schneider,
Mariane Bittencourt Fagundes
Chapter 3
Challenges and opportunities for microalgae biotechnology development —— 41

Emmanuel Manirafasha, Theoneste Ndikubwimana, Hanqing Fu, Mao Lin,
Liangliang Zhang, Keju Jing
Chapter 4
Major bottlenecks in industrial microalgae-based facilities —— 55

Calvin Lo, Rene H. Wijffels, Iulian Boboescu, A. Kazbar, Michel H. M. Eppink
Chapter 5
Multimethod and multiproduct microalgae biorefineries: industrial scale feasibility: eutectic solvents as a novel extraction system for microalgae biorefinery —— 67

Samara C. Silva, Madalena M. Dias, M. Filomena Barreiro
Chapter 6
What is next in microalgae research —— 81

Diva S. Andrade, Tiago Pellini, Karla C. T. T. Rodrigues, Danilo A. Silvestre,
Heder Asdrubal Montañez Valencia, Jerusa S. Andrade,
Freddy Zambrano Gavilanes, Tiago S. Telles
Chapter 7
Microalgae supply chains —— 107

Part II: Process integration applied to microalgae-based systems

Ihana A. Severo, Diego de O. Corrêa, Wellington Balmant, Juan C. Ordonez, André B. Mariano, José V. C. Vargas
Chapter 8
Energy and heat integration applied to microalgae-based systems — 133

Jalelys Liceth Leones-Cerpa, Eduardo Luis Sánchez-Tuirán, Karina A. Ojeda-Delgado
Chapter 9
Mass integration applied to microalgae-based systems — 147

Alberto Reis, Tiago F. Lopes
Chapter 10
Water integration applied to microalgae-based systems — 165

Akhil Rautela, Shweta Rawat, Indrajeet Yadav, Agendra Gangwar, Sanjay Kumar
Chapter 11
Process integration opportunities applied to microalgae biomass production — 183

Part III: Process intensification applied to microalgae-based processes

Carlos Eduardo Guzmán-Martínez, Juan Manuel Vera-Morales, Efraín Quiroz-Pérez, Araceli Guadalupe Romero-Izquierdo, Claudia Gutiérrez-Antonio
Chapter 12
Process intensification applied to bioreactor design — 213

Priya Yadav, Parag R. Gogate
Chapter 13
Process intensification approaches applied to the downstream processing of microalgae production — 241

Part IV: **Process integration and process intensification strategies applied to microalgae-based products**

Florin Barla, Massimiliano Lega, Sandeep Kumar
Chapter 14
Process integration opportunities applied to microalgae biomass production —— 271

Júlio Cesar de Carvalho, Denisse Tatiana Molina Aulestia, Juliana Cardoso, Hissashi Iwamoto, Maria Clara Manzoki, Carlos R. Soccol
Chapter 15
Process integration opportunities applied to microalgae specialty chemicals production —— 299

Jingyan Hu, Weiqi Fu
Chapter 16
Process intensification opportunities applied to the production of microalgae specialty chemicals —— 325

Jonathan S. Harris, Anh N. Phan
Chapter 17
Process integration opportunities applied to microalgae biofuels production —— 349

Aparna Gautam, Sushil Kumar, Dipesh S. Patle
Chapter 18
Process intensification opportunities in the production of microalgal biofuels —— 377

Rafaela Basso Sartori, Eduardo Jacob-Lopes
Chapter 19
Process integration approaches applied to carbon dioxide capture and use from microalgae —— 409

Olatunde Akinbuja, Kamelia Boodhoo, Sharon Velasquez Orta
Chapter 20
Algae-based bioelectrochemical systems —— 425

Mariany Costa Deprá, Leila Queiroz Zepka, Eduardo Jacob-Lopes
Chapter 21
Process integration and process intensification approaches as enhancers of industrial sustainability in microalgae-based systems —— 441

Ahmad Farhad Talebi, Sara Kabirnataj, Elham Soleimani
Chapter 22
Patents related to process integration and process intensification applied to microalgae-based systems —— 455

Mehak Kaur, Hishita Peshwani, Mayurika Goel
Chapter 23
A brief mapping of patents in microalgae-based systems —— 479

Index —— 507

Contributing authors

Chapter 1
Luísa C. Schetinger
Department of Food Science and Technology
Federal University of Santa Maria (UFSM)
Roraima Avenue, 1000, 97105-900,
Santa Maria, RS
Brazil

Marcele L. Nörnberg
Department of Food Science and Technology
Federal University of Santa Maria (UFSM)
Roraima Avenue, 1000, 97105-900,
Santa Maria, RS
Brazil

Patrícia A. Caetano
Department of Food Science and Technology
Federal University of Santa Maria (UFSM)
Roraima Avenue, 1000, 97105-900,
Santa Maria, RS
Brazil

Leila Q. Zepka
Department of Food Science and Technology
Federal University of Santa Maria (UFSM)
Roraima Avenue, 1000, 97105-900,
Santa Maria, RS
Brazil

Chapter 2
Melih Onay
Department of Environmental Engineering
Van Yuzuncu Yil University
65080, Van
Turkey
melihonay@yyu.edu.tr

Chapter 3
Rosangela Rodrigues Dias
Bioprocess Intensification Group
Federal University of Santa Maria, UFSM
Roraima Avenue 1000, 97105-900,
Santa Maria, RS
Brazil

Adriane Terezinha Schneider
Bioprocess Intensification Group
Federal University of Santa Maria, UFSM
Roraima Avenue 1000, 97105-900,
Santa Maria, RS
Brazil

Mariane Bittencourt Fagundes
Interdisciplinary Centre of Marine and
Environmental Research, CIIMAR
Portugal

Chapter 4
Emmanuel Manirafasha
Department of Chemical and Biochemical
Engineering
College of Chemistry and Chemical Engineering,
and The Key Lab for Synthetic Biotechnology of
Xiamen City
Xiamen University
Xiamen 361005
China

Alpha Natural Resources Company Limited
(ANARECO Ltd.)
Kigali
Rwanda

Xiamen Canco Biotech Co., Ltd.
Xiamen 361000
China
meonb2003@gmail.com

Theoneste Ndikubwimana
General Higher Education Quality Standards
Department
Higher Education Council (HEC)
P.O.BOX 6311 Kigali
Rwanda
ntheo05@gmail.com,
tndikubwimana@hec.gov.rw

Hanqing Fu
Xiamen Canco Biotech Co., Ltd.
Xiamen 361000
China
1512734847@qq.com

https://doi.org/10.1515/9783110781267-203

Mao Lin
Xiamen Canco Biotech Co., Ltd.
Xiamen 361000
China

Liangliang Zhang
Academy of Advanced Carbon Conversion Technology
Huaqiao University
Xiamen 361021
China
zhangll@hqu.edu.cn

Keju Jing
Department of Chemical and Biochemical Engineering
College of Chemistry and Chemical Engineering, and The Key Lab for Synthetic Biotechnology of Xiamen City
Xiamen University
Xiamen 361005
China

Xiamen Canco Biotech Co., Ltd.
Xiamen 361000
China
jkj@xmu.edu.cn

Chapter 5

Calvin Lo
Bioprocess Engineering, AlgaePARC
Wageningen University
PO Box 16, 6700 AA Wageningen
The Netherlands

Rene H. Wijffels
Bioprocess Engineering, AlgaePARC
Wageningen University
PO Box 16, 6700 AA Wageningen
The Netherlands

Faculty of Biosciences and Aquaculture
Nord University
N-8049 Bodø
Norway

Iulian Boboescu
Bioprocess Engineering, AlgaePARC
Wageningen University
PO Box 16, 6700 AA Wageningen
The Netherlands

A. Kazbar
Bioprocess Engineering, AlgaePARC
Wageningen University
PO Box 16, 6700 AA Wageningen
The Netherlands

Michel H.M. Eppink
Bioprocess Engineering, AlgaePARC
Wageningen University
PO Box 16, 6700 AA Wageningen
The Netherlands
michel.eppink@wur.nl

Chapter 6

Samara C. Silva
Centro de Investigação de Montanha (CIMO)
Instituto Politécnico de Bragança
Campus Santa Apolónia
5300-253 Bragança
Portugal

Laboratório Associado para a Sustentabilidade e Tecnologia em Regiões de Montanha (LA SusTEC)
Instituto Politécnico de Bragança
Campus de Santa Apolónia
5300-253 Bragança
Portugal

Laboratory of Separation and Reaction Engineering – Laboratory of Catalysis and Materials (LSRE-LCM)
Faculdade de Engenharia da Universidade do Porto
R. Dr. Roberto Frias, 4200-465, Porto
Portugal

ALiCE – Associate Laboratory in Chemical Engineering, Faculty of Engineering
University of Porto
Rua Dr. Roberto Frias, 4200-465 Porto
Portugal
samaras@ipb.pt

Madalena M. Dias
Laboratory of Separation and Reaction Engineering – Laboratory of Catalysis and Materials (LSRE-LCM)
Faculdade de Engenharia da Universidade do Porto
R. Dr. Roberto Frias, 4200-465, Porto
Portugal

ALiCE – Associate Laboratory in Chemical Engineering, Faculty of Engineering
University of Porto
Rua Dr. Roberto Frias, 4200-465 Porto
Portugal
dias@fe.up.pt

M. Filomena Barreiro
Centro de Investigação de Montanha (CIMO)
Instituto Politécnico de Bragança
Campus Santa Apolónia
5300-253 Bragança
Portugal

Laboratório Associado para a Sustentabilidade e Tecnologia em Regiões de Montanha (LA SusTEC)
Instituto Politécnico de Bragança
Campus de Santa Apolónia
5300-253 Bragança
Portugal
barreiro@ipb.pt

Chapter 7

Diva S. Andrade
Instituto de Desenvolvimento Rural do Paraná – IAPAR-EMATER
86047-902, Londrina, Paraná
Brazil
diva@idr.pr.gov.br

Tiago Pellini
Instituto de Desenvolvimento Rural do Paraná – IAPAR-EMATER
86047-902, Londrina, Paraná
Brazil

Karla C. T. T. Rodrigues
Instituto de Desenvolvimento Rural do Paraná – IAPAR-EMATER
86047-902, Londrina, Paraná
Brazil

Danilo A. Silvestre
Instituto de Desenvolvimento Rural do Paraná – IAPAR-EMATER
86047-902, Londrina, Paraná
Brazil

Heder Asdrubal Montañez Valencia
Instituto de Desenvolvimento Rural do Paraná – IAPAR-EMATER
86047-902, Londrina, Paraná
Brazil

Jerusa S. Andrade
Instituto Nacional de Pesquisas da Amazônia – INPA
Manaus
Brazil

Freddy Zambrano Gavilanes
Departamento de Agronomía
Facultad de Ingeniería Agronómica
Universidad Técnica de Manabí
Portoviejo, Manabí
Ecuador

Tiago S. Telles
Instituto de Desenvolvimento Rural do Paraná – IAPAR-EMATER
86047-902, Londrina, Paraná
Brazil

Chapter 8

Ihana A. Severo
Sustainable Energy Research and Development Center (NPDEAS)
Federal University of Paraná (UFPR)
Curitiba, PR 81531-980
Brazil

Department of Mechanical Engineering
Energy and Sustainability Center and Center for Advanced Power Systems (CAPS)
Florida State University
Tallahassee, FL 32310-6046
USA
ihana.aguiar@gmail.com

Diego de O. Corrêa
Sustainable Energy Research and Development Center (NPDEAS)
Federal University of Paraná (UFPR)
Curitiba, PR 81531-980
Brazil

Wellington Balmant
Sustainable Energy Research and Development Center (NPDEAS)
Federal University of Paraná (UFPR)
Curitiba, PR 81531-980
Brazil

Juan C. Ordonez
Department of Mechanical Engineering
Energy and Sustainability Center and Center for Advanced Power Systems (CAPS)
Florida State University
Tallahassee, FL 32310-6046
USA

André B. Mariano
Sustainable Energy Research and Development Center (NPDEAS)
Federal University of Paraná (UFPR)
Curitiba, PR 81531-980
Brazil

José V. C. Vargas
Sustainable Energy Research and Development Center (NPDEAS)
Federal University of Paraná (UFPR)
Curitiba, PR 81531-980
Brazil

Chapter 9
Jalelys Liceth Leones-Cerpa
Process Design and Biomass Utilization Research Group (IDAB)
University of Cartagena, Chemical Engineering Program
Av. El Consulado Street 30 #48-150, Cartagena
Colombia
jleonesc@unicartagena.edu.co

Eduardo Luis Sánchez-Tuiránb
Process Design and Biomass Utilization Research Group (IDAB)
University of Cartagena, Chemical Engineering Program
Av. El Consulado Street 30 #48-150, Cartagena
Colombia
esanchezt2@unicartagena.edu.co

Karina A. Ojeda-Delgadoa
Process Design and Biomass Utilization Research Group (IDAB)
University of Cartagena, Chemical Engineering Program
Av. El Consulado Street 30 #48-150, Cartagena
Colombia
kojedad@unicartagena.edu.co

Chapter 10
Alberto Reis
LNEG – UBB – National Laboratory of Energy and Geology I.P.
Bioenergy and Biorefineries Unit
Estrada do Paço do Lumiar 22, 1649-038 Lisbon
Portugal

Tiago F. Lopes
LNEG – UBB – National Laboratory of Energy and Geology I.P.
Bioenergy and Biorefineries Unit
Estrada do Paço do Lumiar 22, 1649-038 Lisbon
Portugal
tiago.lopes@lneg.pt

Chapter 11
Akhil Rautela
School of Biochemical Engineering
IIT BHU, Varanasi 221005
Uttar Pradesh
India

Shweta Rawat
School of Biochemical Engineering
IIT BHU, Varanasi 221005
Uttar Pradesh
India

Indrajeet Yadav
School of Biochemical Engineering
IIT BHU, Varanasi 221005
Uttar Pradesh
India

Agendra Gangwar
School of Biochemical Engineering
IIT BHU, Varanasi 221005
Uttar Pradesh
India

Sanjay Kumar
School of Biochemical Engineering
IIT BHU, Varanasi 221005
Uttar Pradesh
India
sanjaykr.bce@iitbhu.ac.in

Chapter 12
Carlos Eduardo Guzmán-Martínez
Facultad de Ingeniería
Universidad Autónoma de Querétaro, Campus Amazcala
El Marqués-Querétaro, 76265
México

Juan Manuel Vera-Morales
Facultad de Ingeniería
Universidad Autónoma de Querétaro, Campus Amazcala
El Marqués-Querétaro, 76265
México

Efraín Quiroz-Pérez
Facultad de Ingeniería
Universidad Autónoma de Querétaro, Campus Amazcala
El Marqués-Querétaro, 76265
México

Araceli Guadalupe Romero-Izquierdo
Facultad de Ingeniería
Universidad Autónoma de Querétaro, Campus Amazcala
El Marqués-Querétaro, 76265
México

Claudia Gutiérrez-Antonio
Facultad de Ingeniería
Universidad Autónoma de Querétaro, Campus Amazcala
El Marqués-Querétaro, 76265
México
claugtez@gmail.com

Chapter 13
Priya Yadav
Chemical Engineering Department
Institute of Chemical Technology
Matunga, Mumbai 400019
Maharashtra
India

Parag R. Gogate
Chemical Engineering Department
Institute of Chemical Technology
Matunga, Mumbai 400019
Maharashtra
India
pr.gogate@ictmumbai.edu.in

Chapter 14
Florin Barla
Circ INC
Danville, VA 24540
USA

Massimiliano Lega
Dipartimento di ingegneria
Universita degli Studi di Napoli Parthenope
80143 – Napoli
Italy

Sandeep Kumar
Department of Civil and Environmental Engineering
Old Dominion University
Norfolk, VA 23529
USA
skumar@odu.edu

Chapter 15
Júlio Cesar de Carvalho
Federal University of Paraná, UFPR
Curitiba, PR
Brazil
jccarvalho@ufpr.br

Denisse Tatiana Molina Aulestia
Federal University of Paraná, UFPR
Curitiba, PR
Brazil

Juliana Cardoso
Federal University of Paraná, UFPR
Curitiba, PR
Brazil

Hissashi Iwamoto
Federal University of Paraná, UFPR
Curitiba, PR
Brazil

Maria Clara Manzoki
Federal University of Paraná, UFPR
Curitiba, PR
Brazil

Carlos R. Soccol
Federal University of Paraná, UFPR
Curitiba, PR
Brazil

Chapter 16
Jingyan Hu
Department of Marine Science
Ocean College
Zhejiang University
China

Weiqi Fu
Department of Marine Science
Ocean College
Zhejiang University
China

Center for Systems Biology and Faculty of Industrial Engineering, Mechanical Engineering and Computer Science
School of Engineering and Natural Sciences
University of Iceland
Iceland
weiqifu@zju.edu.cn

Chapter 17
Jonathan S. Harris
School of Engineering
Chemical Engineering
Newcastle University
Newcastle Upon Tyne, NE1 7RU
UK

Anh N. Phan
School of Engineering
Chemical Engineering
Newcastle University
Newcastle Upon Tyne, NE1 7RU
UK
anh.phan@newcastle.ac.uk

Chapter 18
Aparna Gautam
Department of Chemical Engineering
Motilal Nehru National Institute of Technology Allahabad
Prayagraj 211004, Uttar Pradesh
India

Sushil Kumar
Department of Chemical Engineering
Motilal Nehru National Institute of Technology Allahabad
Prayagraj 211004, Uttar Pradesh
India

Dipesh S. Patle
Department of Chemical Engineering
Motilal Nehru National Institute of Technology Allahabad
Prayagraj 211004, Uttar Pradesh
India
dipesh-patle@mnnit.ac.in

Chapter 19
Rafaela Basso Sartori
Bioprocess Intensification Group
Federal University of Santa Maria, UFSM
Roraima Avenue 1000, 97105-900,
Santa Maria, RS
Brazil

Eduardo Jacob-Lopes
Bioprocess Intensification Group
Federal University of Santa Maria, UFSM
Roraima Avenue 1000, 97105-900, Santa Maria, RS
Brazil
ejacoblopes@gmail.com

Chapter 20

Mr. Olatunde Akinbuja
School of Engineering, Merz Court
Newcastle University
Newcastle upon Tyne
United Kingdom

Prof. Kamelia Boodhoo
School of Engineering, Merz Court
Newcastle University
Newcastle upon Tyne
UK

Dr. Sharon Velasquez Orta
School of Engineering, Merz Court
Newcastle University
Newcastle upon Tyne
UK
sharon.velasquez-orta@newcastle.ac.uk

Chapter 21

Mariany Costa Deprá
Bioprocess Intensification Group
Federal University of Santa Maria, UFSM
Roraima Avenue 1000, 97105-900, Santa Maria, RS
Brazil
marianydepra@gmail.com

Leila Queiroz Zepka
Bioprocess Intensification Group
Federal University of Santa Maria, UFSM
Roraima Avenue 1000, 97105-900, Santa Maria, RS
Brazil

Eduardo Jacob-Lopes
Bioprocess Intensification Group
Federal University of Santa Maria, UFSM
Roraima Avenue 1000, 97105-900, Santa Maria, RS
Brazil

Chapter 22

Ahmad Farhad Talebi
Department of Microbial Biotechnology
Faculty of New Sciences and Technologies
Semnan University
35131-19111, Semnan
Iran
aftalebi@semnan.ac.ir

Sara Kabirnataj
Department of Microbial Biotechnology
Faculty of New Sciences and Technologies
Semnan University
35131-19111, Semnan
Iran

Elham Soleimani
Department of Microbial Biotechnology
Faculty of New Sciences and Technologies
Semnan University
35131-19111, Semnan
Iran

Chapter 23

Mehak Kaur
Sustainable Agriculture Program
The Energy and Resources Institute (TERI)
Gurugram, Haryana
India

Hishita Peshwani
Sustainable Agriculture Program
The Energy and Resources Institute (TERI)
Gurugram, Haryana
India

Mayurika Goel
Sustainable Agriculture Program
The Energy and Resources Institute (TERI)
Gurugram, Haryana
India
mayurika.goel@teri.res.in
mayurikagoel@gmail.com

Part I: **Fundamentals**

Luísa C. Schetinger, Marcele L. Nörnberg, Patrícia A. Caetano, Leila Q. Zepka

Chapter 1
A timeline on microalgal biotechnology

Abstract: Microalgae have been known for centuries, but it was only a few decades ago that advanced research on the possibilities of this biotechnology started. The applications of these microorganisms include food, animal feed, biofuels, and wastewater treatment, among others. Microalgae are a novel sustainable solution for multiple fields, being able to convert carbon dioxide into valuable nutrients and bioactive compounds. Their content of macronutrients makes some species perfect for producing alternative proteins, helping to complement the constantly growing food industry demands. However, the cost of biomass production remains high, and plenty of challenges related to sensory properties are present. To overcome that, diverse studies have been developed, and some of the most innovative methods embrace genetic engineering approaches. Strain selection, light intensity, manipulation of the cultivation environment, and stress conditions can also change the microalgal biomass profile and compounds profile. Considering the constant evolution of microalgal biotechnology, many important discoveries have been made, and challenges were already overcome. Regardless of how improvements were made, it is clear that microalgal biotechnology still requires further studies and new findings. Contemplating the past, current, and future scenarios, this chapter aims to build a broad timeline of microalgal biotechnology from some of the first discoveries until modern times. Topics covered include pioneer studies in microalgae, the development of sustainable solutions applying microalgae, the application of microalgae in food products, and challenges and future trends in microalgal biotechnology.

Keywords: biotechnology, microalgae, foods, alternative proteins, sustainability, timeline

1.1 Introduction

Microalgae can be defined as microscopic organisms, naturally found in seas, lakes, and dry terrains. This group contains an enormous diversity of species, essentially owning the same basic needs for survival: a source of carbon atoms, water, and light (Nigam and Singh, 2011). Although diverse algae varieties have been consumed for centuries, first eaten by Indigenous populations, it was only a few decades ago that

Luísa C. Schetinger, Marcele L. Nörnberg, Patrícia A. Caetano, Leila Q. Zepka, Department of Food Science and Technology, Federal University of Santa Maria (UFSM), Roraima Avenue 1000, 97105-900 Santa Maria, RS, Brazil

https://doi.org/10.1515/9783110781267-001

specific research started in the microalgae field. Also, until the 1940s, the exploration of these organisms was mostly restricted to small-scale facilities (Khan et al., 2018; Terry and Raymond, 1985). Investigations on the energetic potential of microalgae were as early as 1931 when some scientists such as Stadnichenko (1931) started publishing new ideas on the combustion of microalgae. Around the 1960s, the University of Tokyo already held a large collection of microalgae strains, demonstrating interest in future research and collaboration (Watanabe, 1960). In the late years of the 1970s, important notions on the absorption of contamination by microalgae were already known (Sakaguchi et al., 1978, 1979). By the 1980s, many countries had started investments in large-scale microalgae, especially focusing on fuels, animal feed, and food ingredients. Despite the numerous advances made so far, microalgae costs are still a challenge, inspiring new studies to overcome that (Terry and Raymond, 1985; Singh and Patidar, 2018).

Global warming and climate change, along with the pressure on oil reserves, are driving new research on sustainable energetic sources. Considering this issue, microalgae is pointed out as a promising alternative to fossil and pollutant fuels (Oliveira et al., 2022). Thus, in the last decades, many advances have been made in understanding the capabilities and limitations of microalgal biomass (Terry and Raymond, 1985; Rizwan et al., 2018). Microalgae, along with terrestrial plants are sustainable sources of renewable fuels, capable of helping to reduce carbon dioxide emissions (Williams and Laurens, 2010). Aquatic microalgal biomass is particularly upstanding for liquid fuel production. This is mostly due to their easier cultivation, quick growth, available reactor systems, and high yield of harvesting (Gallagher, 2011; Gao et al., 2012; Fon Sing et al., 2013). Around the 1980s, the discussions on renewable fuels from microalgal biomass were quite advanced. Feinberg (1984) produced a complete report on diverse species and their potential for energy generation. The biomass is essentially made of lipids, carbohydrates, and proteins; among these, lipids are the best for fuels, due to their high energetic content. Some microalgae species produce hydrocarbons, similar to petroleum compounds, which is interesting for diesel production. On the other hand, carbohydrates can be fermented to generate ethanol, and gas from this fermentation could also be used. *Botryococcus braunii* was found to be the best at lipid yield in biomass, however, at that time; Feinberg (1984) proposed the question: Would the cultivation be efficient for this utilization? This efficiency challenge has remained for decades, with new discoveries arising and advances being made.

Rodolfi et al. (2009) documented a low-cost photobioreactor, with the selection of microalgae strains and cultivate manipulation, being capable of yielding an oil production of up to 90 kg per hectare per day. Despite the huge advance and the significantly lower cost of the photobioreactor, the author still emphasized the difficulty of the technology when compared to other cultures, like conventional seeds. Other emerging technologies to increase microalgal biomass efficiency are electrocoagulation, allied to electro-assisted lipid extraction, which could be maintained through other renewable energy sources (Richardson et al., 2014; Wicker et al., 2021). However, biofuels are not the only sustainability-oriented use of microalgae. Wastewater treatment is another valuable application of microalgae, as they are flexible organisms, capable of growing in disadvantaged conditions and adapting their metabolism to photoautotrophic, mixotrophic, or heterotrophic conditions. Thus, this biomass originated from wastewater treatment could generate energy or even be applied for animal feed or food production. It is noticeable that research in the wastewater field applying microalgae has been documented since the 1950s (Wollmann et al., 2019).

Microalgae biomass is rich in lipids, proteins, and carbohydrates (Feinberg, 1984), which makes this product great for use in many food applications. Since the early years of the 1950s, *Chlorella* was already explored as an extra ingredient in foods, specially intended to increase proteins. At that time, some of the products explored were noodles, soups, and ice cream, among others (Tamiya et al., 1953). Around the late years of the 1980s, some authors already defined the food industry as the most established application of microalgal biomass and also as a source of natural food colorings (Benemann et al., 1987). The topic of microalgae applied to human consumption was of so great importance that in 1977 an international workshop was held in Neuherberg, Germany. During the meeting named "Microalgae for Food and Feed – A Status Analysis," guests discussed the current status and prospects of the field, considering research from countries such as Thailand, Peru, India, and Israel (Soeder and Binsack, 1979). Soeder (1976) also mentioned in one of their studies that microalgae safety and toxicity required further studies to become economically feasible in food applications. A few years later, Soeder (1980) published that both *Spirulina* and *Chlorella* could be securely added to food products, as they did not show any concerning toxicity. Kay and Barton (1991) defined *Chlorella* and *Spirulina*, among other reliable microalgae strains as excellent sources of nutrients and fine chemicals. Their chlorophylls, carotenoids, vitamins, and minerals were highlighted as promising health improvers and probiotics. It suggests that around the 1990s, the full potential of microalgae as a source of food was yet unknown, but much more explored than before. These more specific studies are of extreme relevance, as arable land potential is declining, oceans are at a limit of fishing capability, and freshwater is limited for human use (Draaisma et al., 2013).

The human population is estimated to reach the mark of 9.7 billion people by the year 2050. Nevertheless, this raises concern about how the existent global food supply chain will hold this demand. Agricultural and livestock ecosystems are at their limits, negatively impacting climate change and global warming (Härtel and Pearman, 2010). Considering this global challenge, studies on the potential of microalgae as a food considerably grew in quantity in the last decades. Microalgae are considered extremely valuable sources of proteins, being sustainable, and capable of thriving in inhospitable environments, such as salty and polluted waters. Their yield in some methods of culture showcases a perspective of high efficiency and rentability to producers. However, even with decades of research, some traits of microalgae, such as taste and nutrients, still need further improvements. Furthermore, some genetic engineering approaches are also presenting exciting results, opening new possibilities in the microalgae field (Torres-Tij et al., 2020). Strain selection technologies, for instance, can direct microalgae to produce specific secondary metabolites. These compounds are related to the improvement of biomass quality and yield (Laamanen et al., 2021; Grossmann et al., 2020). Although many species of microalgae are well-known and documented, it is still needed to explore and discover new lineages that might take biotechnology even further. One of the best ways to continue advances is through bioprospecting and research with new organisms (Martínez-Francés and Escudero-Oñate, 2018). In this sense, this chapter aims to create a timeline of microalgal biotechnology, providing a comprehensive discussion of the most relevant studies so far. To achieve that, four main topics are included: pioneer studies on microalgae, development of sustainable solutions applying microalgae, application of microalgae in food products, and challenges and future trends in microalgae biotechnology.

1.2 Pioneer studies in microalgae

Diverse species of algae have been utilized as food sources for centuries, first consumed by indigenous populations. Despite the early discovery of algae, it was only a few decades ago that microalgae growth and research at the industrial and scientific levels started (Khan et al., 2018). Documented studies on algae (Figure 1.1) are as old as 1877 when this kind of organism was mentioned in the *Annals of the New York Academy of Sciences* (Bold, 1877). However, regarding specific research on microalgae, it is possible to locate some of the earliest dated around 1900, such as the work executed by White (1908). This author noted the presence of microalgae in specific rocks and coals used for gas extraction. In the text, microalgae are showcased as a finding or even an issue for the formation of an ideal rock. Despite the presence in the literature, the cultivation of algae organisms was mostly restricted to laboratories until the late years of the 1940s (Terry and Raymond, 1985). To reference some studies, during a bathymetrical survey of the fresh-water lochs of Scotland, microalgae presence was reported by Murray (1901), and the quantity of this organism varied during the survey period. Robertson et al. (1907) conducted a study on humic salts and during its execution, had an important finding on microalgae growth. It was documented during the experiment that microalgae were only able to thrive in environments containing natural humic salts, which did not happen with artificial salts. The unicellular structure of unspecified microalgae at that moment was mentioned by Snider (1934) in an article discussing the origin and evolution of petroleum. According to the article, algae are an important finding in petroleum when analyzed in microscopy, including the presence of unicellular microalgae organisms. The language used in this article suggests that by this period, microalgae and the identification of unicellular structures were already established in the scientific community.

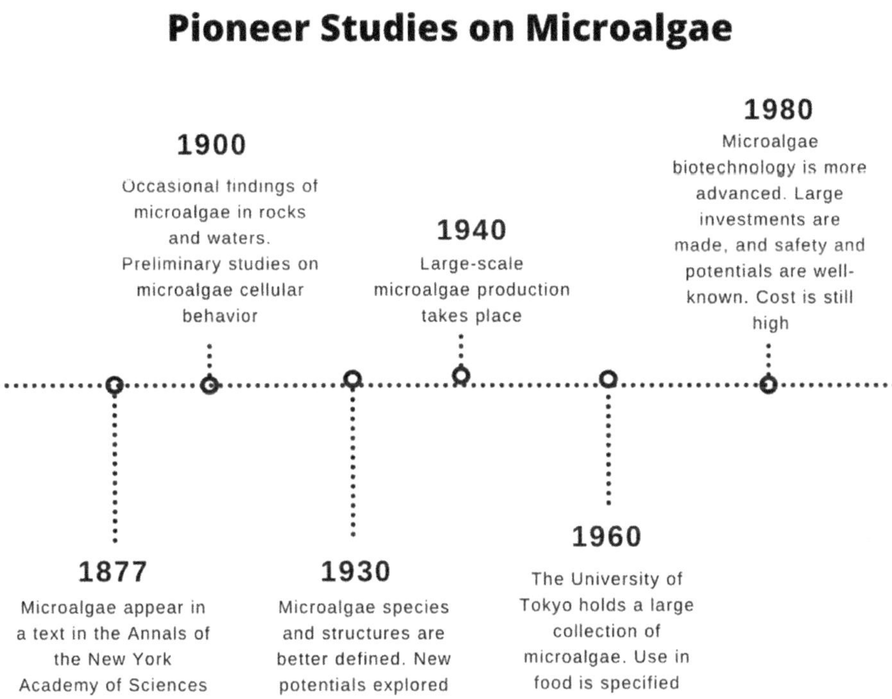

Figure 1.1: Pioneer studies on microalgae.
Source: Authors [adapted from Bold (1877), White (1908), Murray (1901), Snider (1934), Terry and Raymond (1985), and Watanabe (1960)].

Another study related to geology mentioned the presence of microalgae in *sapropelite* deposits. The author defined the found microalgae as *Botryococcus braunii*, present in Russia. The author also compared this found organism to other microalgae of the genus *Pila*, already documented in Scotland, Alaska, and the USA. Interestingly, the process of death and fermentation of microalgae is defined in the paper. With decomposition, hydrogen sulfide arises from microalgae. Additionally, the combustion of the green matter formed by these organisms was defined as a yellow solid mass (Stadnichenko, 1931). These ideas suggest that at that time, pioneer investigations on the energy potential of microalgae were made, even if the author was not completely aware of that discovery at the time. A few years ahead, around 1945, during investigations on copper-base alloy corrosion in seawater, the presence of microalgae was also documented. Microalgae were known for predominating in dark tunnels, along with mussels (Bulow, 1945). This finding already indicated the capability of microalgae to successfully grow without the presence of light, when other organisms could not. Lindeman (1941) highlighted microalgae as the most important kind of organism in a lake food cycle, being able to pass through plankton net, reaching areas of the lake where larger organisms are not present. In this 1941 article, microalgae were referred to as a Nannoplankter. Devices for large-scale cultivation of algae and microalgae were built in the Soviet Union around 1940. At that moment, many possible designs for this culture were explored, focusing on an efficient yield for food and animal feed (Terry and Raymond, 1985).

Although the years from 1930 to 1950 already presented some findings on the capabilities of microalgae, it was only around the 1960s that more specific research can be located. (Tamiya et al., 1953), for instance, explored the application of microalgae, especially *Chlorella* in foods. They suggest that soy sauce, noodles, ice cream, and soup, among others, can be enriched with it to increase protein and vitamin content. By the year 1960, the University of Tokyo already holds a collection of 136 strains of microalgae, preserved under controlled conditions. This collection was also open to other researchers that could send new strains or extract some for experiments (Watanabe, 1960). Peng et al. (1956) found that supplementing microalgae cultivates an environment with selenium-generated proteins and polysaccharides with molecules containing this metal. Due to that, the scavenging effects of the microalgae were significantly increased. Furthermore, Hughes et al. (1958) decided to investigate the specific factors associated with *Microcystis aeruginosa* toxicity, isolating the cells and doing a couple of analyses. The authors detected the presence of endotoxins when cells are broken or destroyed. Also, the stage of the culture helped to determine the grade of toxic effects. Understanding the safety of microalgae for feeding humans and different organisms is an important step in the development of new technologies. These pioneer findings doubtless opened discussions and instigated others to explore the potential of microalgae for food applications. During the 1960s and 1970s, there was an intense investment from the French Petroleum Institute in the cultivation of *Spirulina* and its carbon dioxide absorption capacity. These trial experiments extended to many countries, such as Egypt, Algeria, Mexico, and Taiwan. Germany showed interest in the applications of this field; however, the cold climate made it impossible at the time (Terry and Raymond, 1985).

Soeder (1980) defined the treatment of liquid wastes through the cultivation of microalgae as one of the most promising uses of these organisms. In the publication, it was highlighted that many species of microalgae, such as *Chlorella* and *Spirulina*, are nontoxic for humans and animals. Thus, the biomass originating from these water treatments could safely be applied in animal feed, aquaculture, and agriculture. Under the topic of safety and metabolism behavior, Sakaguchi et al. (1979) conducted an experiment on the accumulation of cadmium by microalgae. It was noted that temperature and enzymatic inhibitors did not affect how much cadmium *Chrorella* absorbed. On the other hand, divalent cations in the cultivated water and pH changes were able to significantly reduce the cadmium intake. Additionally, living cells were able to block this heavy metal absorption, while dead cells would take a much higher concentration of cadmium. On a similar line, *Chlorella regularis, Scenedesmus bijuga, Chlamydomonas angulosa, Chlamydomonas reinhardtii, Scenedesmus chloreloides*, and *Scenedesmus obliquus* demonstrated high absorption of uranium when submersed in a contaminated solution. *Chlorella regularis* absorbed less uranium when carbonate ions were present and more within decarbonated seawater. Furthermore, each species of microalgae shows a different behavior on uranium absorption (Sakaguchi et al., 1978). Uranium, for instance, can be very toxic depending on the dose, with irreversible effects on the body. Cadmium, on other hand, is known for its carcinogen effects on many organs (Keith and Faroon, 2022; Hartwig, 2013). Thus, the early understanding of microalgae's absorption of heavy metals and toxic substances plays an important role in the development of new technologies. If microalgae are intended for animal feed and human consumption, these substances could get into the food chain and cause harm.

From the year 1980 and beyond, the discussions on the potential of microalgae as a source of fuels advanced considerably. Benemann et al. (1982) prepared a report for the U.S. Department of Energy, containing all the most up-to-date information on efficient microalgae culture for energy purposes. This report also described the possible use of different light sources to enhance microalgae's qualities. Furthermore, they explored the laboratory selection of the most reliable strains for specific goals, such as lipid and protein production. Around this time, the biological and chemical structures of many microalgae species were already well known, as indicated by this report. A little bit later in time, Benemann et al. (1987) described the health food industry as the most established commercial use of microalgae production. *Chlorella* and *Spirulina* were highlighted as the major species utilized for this application. At the time, microalgae were reported to be sold as powder or pills, mostly cultivated in Japan, Taiwan, Mexico, the USA, Thailand, and Israel. The authors also mention the cultivation of *Spirulina* for the production of *phycocyanin*, a naturally blue coloring, of great value to the food industry. It is noticeable how studies on microalgae have been evolving in the last century, especially since the 1980s. New discoveries seemed to instigate researchers and push microalgae biotechnology even further.

1.3 Development of sustainable solutions applying microalgae

One of the oldest life forms on earth, microalgae cells, with their evolution and adaptation over billions of years, present diversity and complexity that allow them a range of applications (Calijuri et al., 2022). Since the beginning, humans have used microalgae as food. In the 1950s, microalgae began to be consumed as a supplement. Until today, due to their rich biochemical composition, they are considered a sustainable food source of high nutritional value, with functional potential. The microalgae strains most commonly used for consumption include *Chlorella, Spirulina, Dunaliella, Haematococcus*, and *Schizochytrium* (Moreira et al., 2022).

The first patents in the area of microalgae in the European Patent Office database appeared in the 1960s and refer to food products. This fact coincided with the first commercial developments for large-scale cultivation in Japan. This initiative was rapidly expanding in the area of climate change. Since the 1970s, research on the development of microalgae and its contribution to economic production has been carried out. In the mid-1980s, the topic of environmental issues was addressed in the literature. The current competitiveness of raw materials and the depletion of natural resources have increased the relevance of microalgae since several bioproducts can be obtained from their biomass (Severo et al., 2021; Calijuri et al., 2022).

The attention given to microalgae is mainly related to their bioaccumulation efficiency, nutrient assimilation, and biomass productivity. In the production of biomass for energy and other bioproducts (pigments, bioplastics, fatty acids, among others), microalgae have a range of characteristics that make them advantageous over conventional raw materials, such as noncompetition for agricultural land and clean water, favoring the production of food and other agricultural products. In addition, microalgae can fix atmospheric CO_2 and grow in freshwater, wastewater, or seawater. When grown in wastewater, they consume the nutrients present, favoring bioremediation and, at the same time, reducing treatment costs, making microalgae candidates for transforming waste into wealth, paving the way for a sustainable future. Due to the potential of microalgae in the production of bioactive compounds, their use of biomass as a raw material can be adapted to the concept of biorefinery, based on oil refineries where biomass can be converted into several products with high added value, contributing to different processes, production of nutraceuticals, chemicals, food, and pharmaceuticals. The by-products generated are used in several areas such as food, natural dyes, feed, nutraceuticals – human health (polyunsaturated fatty acids (PUFA), carotenoids, vitamins, phytosterols or polyphenols, biopolymers), energy (biofuels such as biogas, biodiesel, bio-oil, and biohydrogen), and organic fertilizers (Nörnberg et al., 2021; Calijuri et al., 2022; Nascimento et al., 2022; Nörnberg et al., 2022a).

Also, in addition to having a range of characteristics that make them more advantageous over conventional raw materials, microalgae play a variety of roles in agriculture, one of the most explored activities being their ability to improve the properties of plants and soil, reducing the environmental impact generated by chemical fertilizers. Microalgae polysaccharides are potential plant biostimulants to protect against biotic and abiotic stress. Agrochemicals such as pesticides and chemical fertilizers contain toxic elements and contaminants for food, soil, and water. This contamination can cause several environmental consequences on global biodiversity. Agricultural systems that aim to replace synthetic substances such as chemical fertilizers prevent environmental damage and contribute to sustainable agriculture. In this scenario, agrochemicals of biological origin stand out, such as biofertilizers, biostimulants, and biopesticides. The enrichment of soil and plants through microalgae is related to the release of bioactive substances (vitamins, amino acids, polypeptides, antibacterial or antifungal substances, phytohormones, and polysaccharides). The release of polysaccharide material collaborates to increase the germination rate and biomass accumulation in plants. Studies have demonstrated the potential of these molecules to stimulate different metabolic pathways in plants, helping their growth, and development, as well as protection against contaminants. Therefore, these compounds have a high potential to be applied to replace chemical fertilizers and pesticides, collaborating with alternatives for sustainable agriculture (Moreira et al., 2022).

In addition to the variety of roles in agriculture, microalgae biomass offers opportunities for the sustainable development of various industries, making its use important to drive global sustainable development. Microalgae can be found in any aquatic environment that contains inorganic nutrients (such as carbon, nitrogen, phosphorus, and other trace elements) and light (to carry out photosynthesis), although they can also grow heterotrophically using organic substrates. Microalgae biomass can be converted into biodiesel, bioethanol, and biogas through processes such as liquefaction, pyrolysis, transesterification, fermentation, and anaerobic digestion. In the food and pharmaceutical industries, microalgae are a proven source of carotenoids, essential amino acids, and PUFA, with antimicrobial, anticancer, and antioxidant activity, among others (Nascimento et al., 2022; Nörnberg et al., 2022b; Oliveira et al., 2022).

Microalgae are also ecological and sustainable options for effluent treatment. Most wastewater contains macronutrients such as carbon, nitrogen, and phosphorus, which are necessary for microalgae metabolism and microalgae's growth in wastewater for water treatment as well as energy production and/or other useful resources, encouraging research for the development of circular economies. Effective wastewater treatment with microalgae, while producing valuable biomass and improving water quality, can also reduce coastal eutrophication and its negative impacts on fisheries and aquaculture, tourism, and public health (Oliveira et al., 2022).

Faced with population growth, as well as rising standards of living and consumption, new sustainable sources for the production of food, feed, and raw materials have become the center of focus of development. The sustainable intensification of agriculture and the rise of high-performance forms of production such as aquaculture are becoming the focus of attention. The sustainable intensification of agriculture for food, feed, and raw materials is recurrent in the 2050 forecast for economic and social scenarios. In the same search for environmental sustainability, the development of production models that minimize waste and that exploit waste from other processes as raw material has become mandatory, in a logic of industrial symbiosis and circular economy. In all this context, the production of microalgae has been highlighted by characteristics such as high photosynthetic efficiency reflected in productivity, the ability to produce a wide range of bioactive compounds, and the possibility of using alternative resources such as land not classified as fertile soil, seawater, and reclaimed streams/wastewater (Herrera et al., 2021).

The current scenario of large-scale production of microalgae biomass is predominantly based on extensive practices, that is, it uses old, cheap, and low-productivity systems that require large volumes of water. Although the microalgae production chain is considered highly sustainable, many obstacles related to the high demand for water and high loads of organic waste, including residual microalgae biomass, still remain underexplored. To avoid the generation of waste and eliminate purification steps, the direct incorporation of microalgae biomass in products would be an even more sustainable and economical way. However, some microalgae strains, such as *Haematococcus pluvialis* and *Dunaliella salina*, have a specific market aimed at the extraction of high-value-added compounds, with components that do not exceed 10% of the biomass weight (astaxanthin and β-carotene, respectively), producing organic waste from microalgae biomass. Thus, providing more solutions for the remaining biomass is therefore considered a key step for the long-term development of the microalgae industry (Moreira et al., 2022; Oliveira et al., 2022).

1.4 Application of microalgae in food products

When discussing the exciting potential of microalgal biomasses as novel and valuable food ingredients, it is essential to consider all the variables. To start, every taxonomic group can present a particular composition of biomass as well as a unique behavior in each kind of environment. Not all the modulations of temperature, salinity, and so on will occasion the same results in all species. That said, the food industry's target markers for the nutritional use of microalgae should be protein, lipids, vitamins, and minerals (Torres-Tiji et al., 2020). Some of the most interesting species for food applications include *Chlorella* sp. and *Neochoris oleoabundan* when targeting lipids, *Isochrysis galbana, Tetraselmis chuii,* and *Skeletonema Costatum* for protein, *Haematococcus pluvialis,* and *Spirulina* sp. for high pigment and antioxidant yield, and *Chlorella* sp. for carbohydrates. One of the vitamins that can be synthesized by microalgae is vitamin B12, which means a great advantage when producing meat substitutes. In nature, most of the B12 sources remain in animal-origin ingredients. Specially *Chlorella, Spirulina,* and *Dunaliella* can be used to provide vitamin B12 content to foods, also contributing with vitamin C and pro-vitamin A. In this sense, microalgae can act as a nutritious ingredient or a sole diet supplement. Considering that, microalgae present a relevant advantage when compared to other vegan protein substitutes, as several heath-improving compounds can be added to food at the same time with the use of microalgal biomass (Souza et al., 2019).

Regarding the nonvegan conventional industry, *Chlorella vulgaris* and *Arthrospira platensis* were already utilized for cheese manufacturing and proved to increase the viability of probiotics in these products. Also, they can enhance the phenolic compounds in cheeses, which are closely related to the better radical scavenging activity of enriched cheeses (Suna and Yilmaz-Ersan, 2022). *Spirulina* is not suitable for only one kind of dairy product, as was observed in a study conducted by Barkallah et al. (2017). With supplementation of 0.25% of *Spirulina* in weight, yogurt had an increase of fiber, protein, and antioxidants, which is beneficial for both nutritional value and shelf-life. This specific yogurt did not suffer relevant impacts in sensory attributes when compared to a control group. The water holding capacity, a target physicochemical property in this kind of food, was positively affected by the *Spirulina*. Additionally, *Spirulina* is pointed as a novel ingredient for beverages as well. During a trial, it was added to nectar at a 10% proportion. In this case, microalgae did not impact panel sensory acceptance. As expected, many compounds were increased in the beverage, such as amino acids, total solids, fats, sugars, carotenoids, and chlorophyll (Aljobair et al., 2021). In a similar case, researchers developed a vegan soya drink, supplemented with lactic acid bacteria and *Chlorella vulgaris*. The microalgae were of importance to preserving the bacterial population and life cycle (Ścieszka et al., 2021). These studies suggest that microalgal biomass could represent a valuable addition to beverages and act as a natural preservative for fermented vegan products.

As new studies arise and international interest is focused on microalgae, new applications of these organisms start to be explored, both inside and outside the food field, as has been discussed in this chapter. Another remarkable use of this novel ingredient is the addition of the base "dough" of plant-based meat substitutes. A meat substitute or meat analog can be described as a food specifically designed to mimic meat. This includes meat's taste, aroma, appearance, texture, and other sensory attributes that make the meat to be recognized as it. A meat analog should deliver to the consumers the feeling of meat in all its senses as food (Boukid, 2021). Heterotrophically cultivated *Auxenochlorella protothecoides*, for instance, can assist with the production of appealing meat substitutes, considering their light yellow color. These microalgae also increase vitamin B and vitamin E contents in foods when added to them. Combined with soy and going through a high moisture extrusion, *Auxenochlorella protothecoides* bond with soy to create a fibrous structure like meat fibers (Caporgno et al., 2020). Furthermore, proteins present in microalgal biomass own essential amino acids that will help to replace meat in the consumer's diet. However, that is not the sole benefit of microalgae for the meat analog industry. The physicochemical properties of microalgae (solubility, gelation, foaming, and emulsification) are suitable for the development of many meat replacement products. In terms of protein digestibility, some microalgae species can have better performance than soy and wheat, which could attract more consumers (Fu et al., 2021). Thus, the applications of microalgae in food products are numerous and will not stop growing so soon, as new findings arise.

1.5 Challenges and future trends in microalgae biotechnology

Microalgae are excellent vegan ingredients for food manufacturing, especially for their richness in nutrients and biocompounds. Furthermore, utilizing this novel compound in products can help to turn the global food supply chain into a more sustainable and green industry, as they have a low environmental impact (Olguín et al., 2022). Despite their outstanding qualities, the mass production of microalgae biomass still poses multiple challenges. First of all, each species of this organism owns a particular form of metabolism and environmental behavior. Often, under normal conditions, the yield of nutrients in the biomass might be insufficient to serve industrial purposes. Targeted environmental modifications are required to direct biomass composition to the desired profile. In addition, these controlled changes can increase production costs, vegan product costs, and undesirable environmental impact. Producers can manipulate temperature, salinity, light intensity, radiations, and oxygen concentration to enhance cultivate efficiency, and these transformations consume energy and resources (Khan et al., 2018).

Depending on the taxonomic group and cultivation method, microalgae could still have the equivalent environmental impacts of animal meat production, which is concerning for this novel industry. Microalgae grow both in phototropic and heterotrophic conditions, adapting to their medium. Researchers suggested that in both phototropic and heterotrophic circumstances, *A. platensis* and *C. vulgaris* could present high energy and resource consumption during their growth (Smetana et al., 2017). Light intensity and wavelength are also able to modify microalgae's lipid content and carotenoid profile. However, the electricity for powering the lamps requires to be sourced, which could be resolved with the application of clean electricity sources, such as photovoltaic energy (Maltsev et al., 2021). In this sense, new technologies still need to be developed to guarantee completely clean, sustainable, and efficient microalgae biomass production.

Most explored microalgae species also present challenging sensory attributes, when considering customer acceptance. They have fishy notes, earthy odors, and a grassy feeling in the mouth. Some volatile organic substances have been found to be responsible for these undesirable characteristics. Moreover, the medium where microalgae grow is capable of altering their taste profile, which is being explored by researchers to make microalgal biomass more suitable for industrial purposes (Perez-Llorens, 2020). For instance, microalgae such as *Spirulina*, *Chlorella*, *Dunaliella*, and *Scenedesmus*, if processed with the right technique, can become attractive in flavor, turning into more valuable ingredients (Souza et al., 2019).

Furthermore, the physicochemical characteristics of microalgal biomass during industrial applications are still under constant investigation. When replacing animal and other vegetable proteins in food products, microalgae can behave in an unpredictable manner, impacting customer satisfaction and acceptance, as is the case of meat analogs, for instance. Some results suggest that microalgal biomass behaves similarly to whey and soy proteins (Fu et al., 2021). This information suggests that microalgae can be a suitable replacement for many popular alternative proteins in the food industry; however, to achieve their full potential, adjustments might be required. As an alternative approach to modifying microalgal biomass color, genetic engineering techniques can direct the conversion of carotenoids and chlorophylls into targeted pigments, by changing the enzymatic processes in their metabolism (Anila et al., 2016). Despite the advances and findings related to multiple alternative technologies, the nutritional value of microalgae and customer acceptability remain in the focus. Plenty of research is still mandatory to fully explore this novel ingredient that started to be more valued only in the last century, one of the reasons why multiple questions remain.

1.6 Final considerations

Microalgae are microscopic organisms, naturally found in waters including seas, lakes, and also some dry terrains, such as rocks. Microalgae comprise a huge number of species, essentially presenting a similar metabolism, needing a source of carbon, water, and light to survive. Despite the early consumption of algae organisms, specific research started in the microalgae field only a few decades ago. The studies on these organisms were essentially restricted to small-scale facilities until the 1940s when larger-scale research started to take place. For food applications, the most popular strains include *Chlorella*, *Spirulina*, *Dunaliella*, *Haematococcus*, and *Schizochytrium*. As researches advance and new strains are explored, the commercially viable list of microalgae species is extended. Around the 1980s many publications already recognized the environmental significance of microalgal research.

Diverse studies already indicate that microalgae biomass presents anticancer, antimicrobial, and antioxidant activities. They are also sources of essential amino acids and PUFA of great relevance to food and pharmaceutical fields. The environmental importance, high commercial value, and novel health benefits of microalgae impulsion new words attract new researchers and companies to invest in this area. Microalgal biotechnology has shown enormous advances in the past decades despite these notable signs of progress, the technology is yet lacking in many points.

References

Aljobair, M. O., Albaridi, N. A., Alkuraieef, A. N., & AlKehayez, N. M. (2021). Physicochemical properties, nutritional value, and sensory attributes of a nectar developed using date palm puree and spirulina. *International Journal of Food Properties*, *24*(1), 845–858.

Anila, N., Simon, D. P., Chandrashekar, A., Ravishankar, G. A., & Sarada, R. (2016). Metabolic engineering of *Dunaliella salina* for production of ketocarotenoids. *Photosynthesis Research*, *127*(3), 321–333.

Barkallah, M., Dammak, M., Louati, I., Hentati, F., Hadrich, B., Mechichi, T., & Abdelkafi, S. (2017). Effect of *Spirulina platensis* fortification on physicochemical, textural, antioxidant and sensory properties of yogurt during fermentation and storage. *LWT*, *84*, 323–330.

Benemann, J. R., Goebel, R. P., Weissman, J. C., & Augenstein, D. C. (1982). Microalgae as a source of liquid fuels. Report to DOE office of energy research, 1–17.

Benemann, J. R., Tillett, D. M., & Weissman, J. C. (1987). Microalgae biotechnology. *Trends in Biotechnololgy*, *5*(2), 47–53.

Bold, H. C. (1877). Soil algae. *Annals of the New York Academy of Sciences*, *1*, 601.

Boukid, F. (2021). Plant-based meat analogues: From niche to mainstream. *European Food Research and Technology (EFRT)*, *247*(2), 297–308.

Bulow, C. L. (1945). Corrosion and biofouling of copper-base alloys in sea water. *Transactions of the Electrochemical Society*, *87*(1), 127.

Calijuri, M. L., Silva, T. A., Magalhães, I. B., De paula, A. S. A., Marangon, P. B. B., de Assis, L. R., & Lorentz, J. F. (2022). Bioproducts from microalgae biomass: Technology, sustainability, challenges and opportunities. *Chemosphere*, *305*, 135508.

Caporgno, M. P., Böcker, L., Müssner, C., Stirnemann, E., Haberkorn, I., Adelmann, H., & Mathys, A. (2020). Extruded meat analogues based on yellow, heterotrophically cultivated auxenochlorella protothecoides microalgae. *Innovative Food Science and Emerging Technologies*, *59*, 102275.

Draaisma, R. B., Wijffels, R. H., Slegers, P. E., Brentner, L. B., Roy, A., & Barbosa, M. J. (2013). Food commodities from microalgae. *Current Opinion in Biotechnology*, *24*(2), 169–177.

Feinberg, D. A. (1984). Fuel Options from Microalgae with Representative Chemical Compositions (No SERI/TR-231-2427). Solar Energy Research Inst, Golden, CO (USA).

Fon Sing, S., Isdepsky, A., Borowitzka, M. A., & Moheimani, N. R. (2013). Production of biofuels from microalgae. *Mitigation and Adaptation Strategies for Global Change*, *18*(1), 47–72.

Fu, Y., Chen, T., Chen, S. H. Y., Liu, B., Sun, P., Sun, H., & Chen, F. (2021). The potentials and challenges of using microalgae as an ingredient to produce meat analogues. *Trends in Food Science & Technology*, *112*, 188–200.

Gallagher, B. J. (2011). The economics of producing biodiesel from algae. *Renewable Energy*, *36*(1), 158–162.

Gao, Y., Gregor, C., Liang, Y., Tang, D., & Tweed, C. (2012). Algae biodiesel-a feasibility report. *Chemistry Central Journal (CCJ)*, *6*(1), 1–16.

Grossmann, L., Hinrichs, J., & Weiss, J. (2020). Cultivation and downstream processing of microalgae and cyanobacteria to generate protein-based technofunctional food ingredients. *Critical Reviews in Food Science and Nutrition*, *60*(17), 2961–2989.

Härtel, C. E., & Pearman, G. I. (2010). Understanding and responding to the climate change issue: Towards a whole-of-science research agenda. *Journal of Management and Organization (JMO)*, *16*(1), 16–47.

Hartwig, A. (2013). Cadmium and cancer. *From Toxicity to Essentiality: From Toxicity to Essentiality*, *11*, 491–507.

Herrera, A., D'Imporzano, G., Acién Fernandez, F. G., & Adani, F. (2021). Sustainable production of microalgae in raceways: Nutrients and water management as key factors influencing environmental impacts. *Journal of Cleaner Production*, *287*, 125005.

Hughes, E. O., Gorham, P. R., & Zehnder, A. (1958). Toxicity of a unialgal culture of microcystis aeruginosa. *Canadian Journal of Microbiology*, *4*(3), 225–236.

Kay, R. A., & Barton, L. L. (1991). Microalgae as food and supplement. *Critical Reviews in Food Science and Nutrition*, *30*(6), 555–573.

Khan, M. I., Shin, J. H., & Kim, J. D. (2018). The promising future of microalgae: Current status, challenges, and optimization of a sustainable and renewable industry for biofuels, feed, and other products. *Microbial Cell Factories*, *17*(1), 1–21.

Keith, L. S., & Faroon, O. M. (2022). Uranium. In: Handbook on the Toxicology of Metals. Academic Press, 5, 885–936. Elsevier.

Laamanen, C. A., Desjardins, S. M., Senhorinho, G. N. A., & Scott, J. A. (2021). Harvesting microalgae for health beneficial dietary supplements. *Aquatic Life and Global Algal Research*, *54*, 102189.

Lindeman, R. L. (1941). Seasonal food-cycle dynamics in a senescent lake. *American Midland Naturalist*, *26* (3), 636–673.

Martínez-Francés, E., & Escudero-Oñate, C. (2018). Cyanobacteria and microalgae in the production of valuable bioactive compounds. *Microalgal Biotechnology*, *6*, 104–128.

Maltsev, Y., Maltseva, K., Kulikovskiy, M., & Maltseva, S. (2021). Influence of light conditions on microalgae growth and content of lipids, carotenoids, and fatty acid composition. *Biology, 10*(10), 1060.

Moreira, J. B., Vaz, B. S., Cardias, B. B., Cruz, C. G., Almeida, A. C. A. D., Costa, J. A. V., & Morais, M. G. (2022). Microalgae polysaccharides: An alternative source for food production and sustainable agriculture. *Polysaccharides, 3*(2), 441–457.

Morimura, Y., & Tamiya, N. (1954). Preliminary experiments in the use of chlorella as human food. *Food Technology, 8*(179).

Murray, J. (1901). A bathymetrical survey of the fresh-water lochs of scotland: Continued. *The Geographical Journal, 17*(3), 273–295.

Nascimento, T. C., Nass, P. P., Fernandes, A. S., Nörnberg, M. L., Zepka, Q. Z., & Jacob-Lopes, E. (2022). Microalgae carotenoids: An overview of biomedical applications. *Algal Biotechnology, 1*, 409–425.

Nigam, P. S., & Singh, A. (2011). Production of liquid biofuels from renewable resources. *Progress in energy and combustion science, 37*(1), 52–68.

Nörnberg, M. L., Nass, P. P., Nascimento, T. C., Fernandes, A. S., Jacob-Lopes, E., & Zepka, L. Q. (2021). Carotenoids profile of *desertifilum* spp. in mixotrophic conditions. *Brazilian Journal of Development (BJD), 7*(3), 33017–33029.

Nörnberg, M. L., Caetano, P. A., Nass, P. P., Vieira, K. R., Jacob-Lopes, E., & Zepka, L. Q. (2022a). Limonene production in microalgal photoautotrophic cultivation. *Brazilian Journal of Development (BJD), 8*(2), 10241–10254.

Nörnberg, M. L., Nass, P. P., Nascimento, T. C., Fernandes, A. S., Jacob-Lopes, E., & Zepka, L. Q. (2022b). Production of microalgae biocompounds in different cultivation conditions. *Brazilian Journal of Development, 8*(2), 10226–10240.

Olguín, E. J., Sánchez-Galván, G., Arias-Olguín, I. I., Melo, F. J., González-Portela, R. E., Cruz, L., & Adessi, A. (2022). Microalgae-based biorefineries: Challenges and future trends to produce carbohydrate enriched biomass, high-added value products and bioactive compounds. *Biology, 11*(8), 1146.

Oliveira, C. Y. B., Jacob, A., Nader, C., Oliveira, C. D. L., Matos, A. P., Araújo, E. S., Shabnam, N., Ashokg, B., & Gálveza, A. O. (2022). An overview on microalgae as renewable resources for meeting sustainable development goals. *Journal of Environmental Management, 320*, 115897.

Peng, Y., Huang, Z., & Guo, B. (1956). The scavenging effects of se enriched *Spirulina platensis* on oxygen free radicals. *Acta Nutrimenta Sinica, 6*.

Pérez-Lloréns, J. L. (2020). Microalgae: From staple foodstuff to avant-garde cuisine. *International Journal of Gastronomy and Food Science, 21*, 100221.

Richardson, J. W., Johnson, M. D., Lacey, R., Oyler, J., & Capareda, S. (2014). Harvesting and extraction technology contributions to algae biofuels economic viability. *Aquatic Life and Global Algal Research, 5*, 70–78.

Rizwan, M., Mujtaba, G., Memon, S. A., Lee, K., & Rashid, N. (2018). Exploring the potential of microalgae for new biotechnology applications and beyond: A review. *Renewable and Sustainable, Energy Reviews, 92*, 394–404.

Robertson, R. A., Irvine, J. C., & Dobson, M. E. (1907). A contribution to the chemistry and physiological action of the humic acids. *Biochemical Journal, 2*(10), 458.

Rodolfi, L., Chini Zittelli, G., Bassi, N., Padovani, G., Biondi, N., Bonini, G., & Tredici, M. R. (2009). Microalgae for oil: Strain selection, induction of lipid synthesis and outdoor mass cultivation in a low-cost photobioreactor. *Biotechnology & Bioengineering, 102*(1), 100–112.

Sakaguchi, T., Horikoshi, T., & Nakajima, A. (1978). Uptake of uranium from sea water by microalgae. *Journal of Fermentation Technology (JFT), 56*(6), 561–565.

Sakaguchi, T., Tsuji, T., Nakajima, A., & Horikoshi, T. (1979). Accumulation of cadmium by green microalgae. *European Journal of Applied Microbiology and Biotechnology, 8*(3), 207–215.

Ścieszka, S., Gorzkiewicz, M., & Klewicka, E. (2021). Innovative fermented soya drink with the microalgae chlorella vulgaris and the probiotic strain levilactobacillus brevis ŁOCK 0944. *LWT*, *151*, 112131.

Severo, I. A., Santos, A. M., Deprá, M. C., Barin, J. S., & Jacob-Lopes, E. (2021). Microalgae photobioreactors integrated into combustion processes: A patent-based analysis to map technological trends. *Algal Research*, *60*, 102529.

Singh, G., & Patidar, S. K. (2018). Microalgae harvesting techniques: A review. *Journal of Environmental Management*, *217*, 499–508.

Smetana, S., Sandmann, M., Rohn, S., Pleissner, D., & Heinz, V. (2017). Autotrophic and heterotrophic microalgae and cyanobacteria cultivation for food and feed: Life cycle assessment. *Bioresour Technology*, *245*, 162–170.

Snider, L. C. (1934). Foreword: Part II. Origin and Evolution of Petroleum: Group 1. Origin.

Stadnichenko, T. (1931). Some effects of metamorphism on certain debris in source rocks. *AAPG Bulletin*, *15*(2), 161–164.

Soeder, C. J. (1976). The technical production of microalgae and its prospects in marine aquaculture. In *Harvesting Polluted Waters: Waste Heat and Nutrient-Loaded Effluents in the Aquaculture* (pp. 11–26). Boston, MA: Springer US.

Soeder, C. J., & Binsack, R. (1979). Microalgae for food and feed. *Advances in Limnology*, 11, 1–300.

Soeder, C. J. (1980). Massive cultivation of microalgae: Results and prospects. *Hydrobiologia*, *72*(1), 197–209.

Souza, M. P., Hoeltz, M., Gressler, P. D., Benitez, L. B., & Schneider, R. (2019). Potential of microalgal bioproducts: General perspectives and main challenges. *Waste Biomass Valorization*, *10*(8), 2139–2156.

Suna, G., & Yilmaz-Ersan, L. (2022). Utilization of microalgae in probiotic white brined cheese. Mljekarstvo: časopis za unaprjeđenje proizvodnje i prerade mlijeka, *72*(2), 88–104.

Tamiya, H., Shibata, K., Sasa, T., Iwamura, T., & Morimura, Y. (1953). Effect of diurnally intermittent illumination on the growth and some cellular characteristics of Chlorella. *ALGAL CULTURE*, *1*, 76.

Terry, K. L., & Raymond, L. P. (1985). System design for the autotrophic production of microalgae. *Enzyme & Microbial Technology*, *7*(10), 474–487.

Torres-Tiji, Y., Fields, F. J., & Mayfield, S. P. (2020). Microalgae as a future food source. *Biotechnology Advances*, *41*, 107536.

United Nations Population Division. (2011). World Population Prospects: The 2010 Revision. File 1: Total Population (Both Sexes Combined) by Major Area, Region and Country Annually for 1950–2100 (Thousands).

Watanabe, A. (1960). List of algal strains in collection at the Institute of Applied Microbiology, University of Tokyo. *The Journal of General and Applied Microbiology (JGAM)*, *6*(4), 283–292.

Wicker, R. J., Kumar, G., Khan, E., & Bhatnagar, A. (2021). Emergent green technologies for cost-effective valorization of microalgal biomass to renewable fuel products under a biorefinery scheme. *The Full Form for Chemical Engineering Journal Is CEJ*, *415*, 128932.

White, C. D. (1908). Some problems of the formation of coal. *Economic Geology*, *3*(4), 292–318.

Williams, P. J. L. B., & Laurens, L. M. (2010). Microalgae as biodiesel & biomass feedstocks: Review & analysis of the biochemistry, energetics & economics. *Energy & Environmental Science*, *3*(5), 554–590.

Wollmann, F., Dietze, S., Ackermann, J. U., Bley, T., Walther, T., Steingroewer, J., & Krujatz, F. (2019). Microalgae wastewater treatment: Biological and technological approaches. *Engineering in Life Sciences*, *19*(12), 860–871.

Melih Onay

Chapter 2
Scope of the microalgae market: a demand and supply perspective

Abstract: Microalgae studies have recently become one of the most remarkable topics for humanity due to their wide range of uses. As the variety of products produced from microalgae increases, their usage areas are expanding and the demand is increasing. The most important microalgae cultivated to be used as a product today are *Spirulina*, *Chlorella*, *Dunaliella*, and *Haematococcus*. These microalgae can be used for some goals, such as the production of food and beverages, animal feeds, cosmetics, nutraceuticals, dietary supplements, bioremediation, agriculture, and biofuel. Biodiesel, bioethanol, biogas, and biojet fuel can be produced for the energy industry, while anti-inflammatory, anticancer, and antioxidant molecules can be used by the pharmaceutical industry. In addition, supplements, antioxidants, and vitamins contribute to the microalgae industry. Animal feeds, vegan foods, and colorants can be produced for the food industry. On the contrary, some molecules such as fucoxanthin, fucosterol, dolastatin, cannabinoids, and glucans will make significant contributions to the microalgae market in the future, and the microalgae market will expand with high production of these products. Before these products are obtained, microalgae are extracted and they go through many processes in order to meet the demand of the market. With the process integration and process intensification approaches, these can be produced as high-value-added products at a higher efficiency. Today, the value of microalgae-based products is approximately 1.25 billion dollars per year. This market will increase with the variety of algae products, supply chain features, and lower costs in algae production processes. From the point of view of the supply and demand chain, this chapter emphasis about how important the microalgae market is.

Keywords: algae products, microalgae market, microalgae demand supply, novel products

2.1 Introduction

Microalgae have recently come to the fore with the increase in their use in industry. The use of plants as food, global problems in food supply, and areal problems in growing plants have made microalgae attractive for industrial use (Kumar et al., 2022). We

Melih Onay, Department of Environmental Engineering, Van Yuzuncu Yil University, 65080 Van, Turkey, e-mail: melihonay@yyu.edu.tr

https://doi.org/10.1515/9783110781267-002

can divide them into three parts as high value, medium value, and low value when microalgae are considered from the perspective of supply and demand in the industry. While microalgae used for food and nutrition purposes are considered high value, those used for feed, agricultural purposes, and chemical raw materials are considered medium value. Microalgae used for biofuels are called low-value products because their use requires high amounts (Cheirsilp and Maneechote, 2022).

The fact that they are of high or low value does not change the difficulties in obtaining them. In all of them, it is necessary to provide biomass. The fractions until biomass is formed are similar in each microalgae species. Depending on whether microalgae are autotrophic, heterotrophic, or mixotrophic, different costs may occur (Castillo et al., 2021). In this type of farming, you can't use wastewater to make some high- and medium-value products. Because they are used as food and supplements, safety is important. There is no such problem with low-value products. Extraction and access to the product is another problem that determines the cost. There must be a positive advantage between the production cost and the selling price when determining the supply and demand situation. It is very difficult to use a product with a high production cost in the industry. Therefore, costs need to be reduced. While obtaining a product from microalgae, the higher the production steps, the higher the cost and the less its use in industry. In these processes, harvesting costs are another important issue. The cost of electricity usage while harvesting should be considered as a separate item. But if chemical harvesting is done, the low market price of this chemical is the reason for the preference to reduce the cost (Lopez-Sanchez et al., 2022). In these steps, purification of the product is another important factor. The purer the product, the higher the quality and the higher its value.

Operating costs, labor costs, and capital costs are of great importance in the formation of supply and demand in the industry. It is important in adjusting the balance between supply and demand in the transportation of the product (Mehariya et al., 2021). If the product is designed to be carried to nearby places, it will be easier for people to reach that product. When all of these factors are set up in a way that makes it easy to bring a product made from microalgae to market, it will have a positive effect on the balance of supply and demand.

When the market value of microalgae is considered globally, it is $1.25 billion per year. Microalgae can be used for food or nutraceutical purposes. They must be produced with great care in closed reactors. So, they are called "high-value products." Microalgae biomass is around £100,000 per t. Since their value-added is high, obtaining them in areas such as 1 ha is sufficient for production. There is no such thing as growing them in wastewater for safety reasons. When optimum conditions are provided, their production can increase by up to about 100 tons per year (Llamas et al., 2021; Torres-Tiji et al., 2020). The products extracted from microalgae according to supply and demand were given in Figure 2.1.

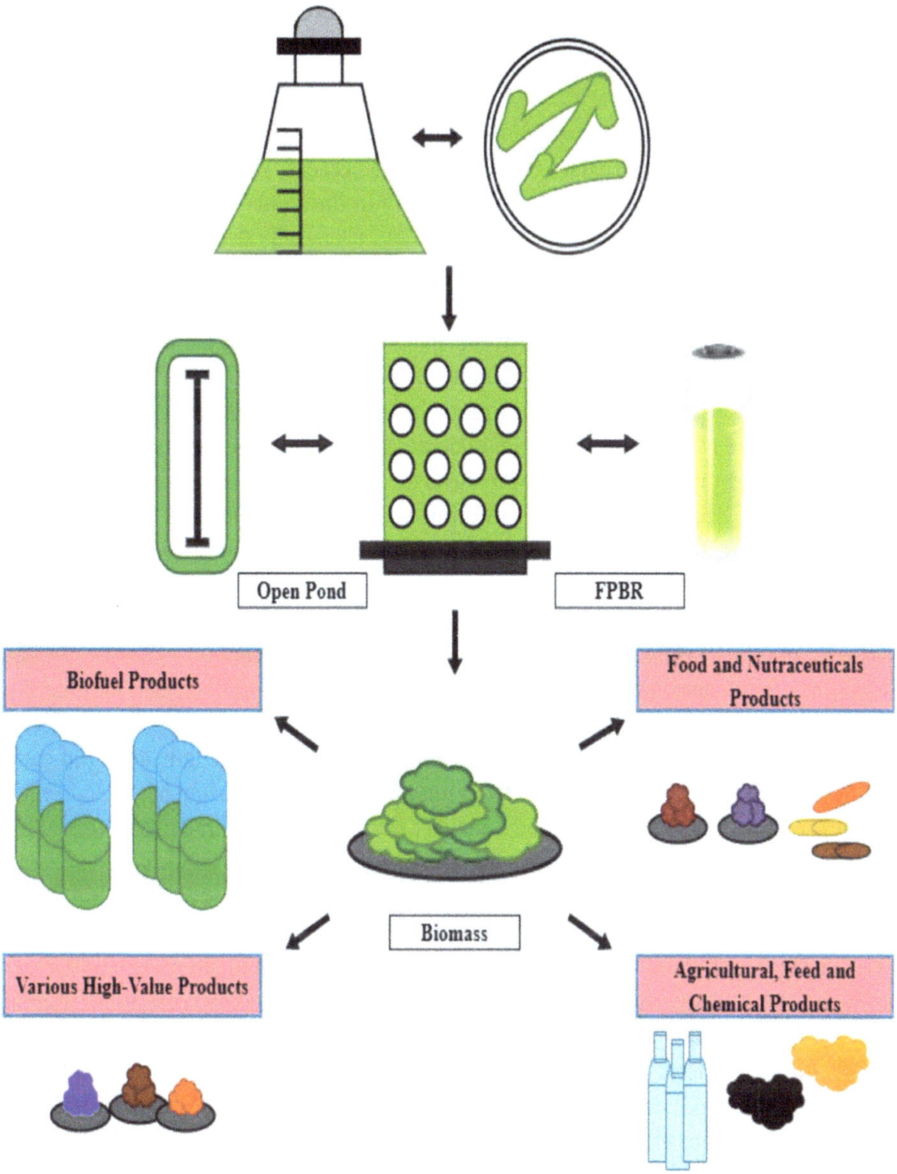

Figure 2.1: The products extracted from microalgae according to supply and demand.

2.2 Food and nutraceuticals products from microalgae

The first use of microalgae as food coincides with the Tenochtitlan period in 1520. *Spirulina maxima* is the first microalgae consumed in the records. The number of microalgae used as food has increased over time. Food safety in the United States is controlled by the Food and Drug Administration (FDA) and GRAS (Generally Recognized as Safe) permission is granted to safe foods. *Arthrospira platensis (Spirulina), Chlorella vulgaris, Euglena gracilis, Chlamydomonas reinhardtii, Dunaliella bardawil,* and *Auxenochlorella protothecoides* have this permission. In Europe, this situation is regulated by the European Food Safety Authority (EFSA) (Gohara-Beirigo et al., 2022). With the increasing human population, there is a shortage of food or transportation. Approximately 1 out of 9 people is faced with malnutrition. So, microalgae can often be used as a food source, and people need protein in their diet to survive. *Arthrospira platensis* has nearly 70% protein content. *Arthrospira platensis* (Spirulina) and *Chlorella vulgaris* can be used as protein sources. Also, *Spirulina* sp. and *Chlorella* sp. include all the amino acids needed by people. Their annual production is nearly 5,000 tons of dry weight (Torres-Tiji et al., 2020). Also, microalgae can be used as a thickener or gelling agent in food. *Arthrospira platensis* and *Tetraselmis suecica* have these properties (Bernaerts et al., 2019). Similar to proteins, lipids and fatty acids can also be synthesized by microalgae. The omega-3 fatty acids, docosahexaenoic acid (DHA), and eicosapentaenoic acid (EPA) are the most commonly used fatty acids. *Auxenochlorella protothecoides* may contain 70% lipid compared to its biomass. *Phaeodactylum tricornutum* and *Schizochytrium* sp. include high amounts of DHA and EPA. They have a market share of 550 million dollars (Torres-Tiji et al., 2020). Microalgae also contain vitamins and pigments. *Botryococcus braunii* can produce high amounts of vitamin E. *Dunaliella tertiolecta* is a source of vitamin B2, vitamin B7, and vitamin B7 (Mehariya et al., 2021). Also, *Porphyra* leaves are used to make sushi in Japan and these algae leaves have a market share of $1 billion per year (Pulz and Gros, 2004). Macroalgae, including alginates and carrageenans, have a $6.7 million per year market share (Wells et al., 2017). In addition, *Dunaliella salina* has β-carotene. It can be used as a food colorant. It has a market share of $200 million a year. Although it is difficult to produce, it can be produced in large quantities. Its market value is £1,000 per kg (Llamas et al., 2021). Also, astaxanthin is one of the most important carotenoids. It acts as a secondary metabolite and has antioxidant and feed properties in the food industry. It is a color additive agent (Onay, 2021). It has a market of 200 million a year. *Haematococcus pluvialis* is the microalgae species with the largest production of astaxanthin. It has an annual production of 300,000 tons per year and its market value is €10,000 per kg (Torres-Tiji et al., 2020; Llamas et al., 2021). Lutein is a yellow color pigment. It occupies an important place in the cosmetic industry (Onay, 2021). *Chlorella* and *Scenedesmus* species include high amounts of lutein. Another pigment, phycocyanin, is called phycobiliprotein, and it has antioxidant, anticarcinogenic, and anti-inflammatory effects. *Spirulina platensis* maintains high amounts of phycocyanin. It produces up to 20% phycocyanin content compared to the dry biomass of microalgae (Dey and Rathod, 2013).

Fucoxanthin, another pigment, prevents skin aging with its antioxidant properties and is used in the cosmetic industry (Onay, 2021). It can be extracted from *Undaria pinnatifida* and *Eisenia bicyclis*. Also, fuxoxanthine functions in the treatment of heart disease, hypertension, cancer, and osteoporosis. Its market price ranges from $40,000 to $80,000 per kg. Its trading volume in 2000 was $600 million (Abu-Ghosh et al., 2021). Some vitamins and elements can be extracted from microalgae. For example; vitamin D3 from *Nannochloropsis oceanica*, phycocyanin from *Spirulina* spp. and *Oscillatoria okeni*, chlorophyll from *Spirulina* spp., *Dunaliella salina*, and *Haematococcus pluvialis*, tocopherol from *Dunaliella* spp. and *Euglena glacilis*, and magnesium, calcium, iron, potassium, phosphorous, and zinc from *Spirulina* can be produced (Gohara-Beirigo et al., 2022). In summary, microalgae can be added to the structure of many foods for various purposes. *Chlorella vulgaris* is added to gelled desserts, bread, cookies, yogurt, broccoli soup, burgers, and sauces. *Nannochloropsis* sp. can be presented in bread, pasta, cookies, and tomato puree. In addition, *Arthrospira* sp. (Spirulina), *Scenedesmus almeriensis*, and *Tetraselmis suecica* can be found in bread mixtures (Bernaerts et al., 2019). Production and sales prices of microalgae used in the food industry differ according to cost and difficulties in production. The production costs of *Chlorella*, *Spirulina*, and *Schizochyrium* for astaxanthin, phycocyanin, β-carotene, EPA, and DHA are $5, $2, $2, $552, $46, $105, $39, and $39 per kg, respectively. In addition, the selling prices of *Chlorella*, *Spirulina*, and *Schizochyrium* for astaxanthin, phycocyanin, β-carotene, EPA, and DHA are $19, $8, $5.2, $2, $500, $548, $790, $100, and $120 per kg, respectively (Jacob-Lopes et al., 2019). In order for microalgae to take a more active role in the food industry, there are still many aspects that need to be improved. These include high biomass production, cheaper extraction methods, and more efficient use of harvesting. In addition, when we look at the situation in terms of food, the color and odor of microalgae still need to be improved. Another important issue is finding ways to get more of the products that can be taken from microalgae.

2.3 Agricultural, feed, and chemical products from microalgae

Microalgae used as agricultural, feed, and chemical products are called "medium-value microalgae." Their market value is around £10,000 per t. Open and closed reactors can be used for their growth and the safety level is medium (Llamas et al., 2021). Microalgae can be used as animal feed in industry. The market price of these may vary according to the amount of protein they contain. When the protein amount is 40%, a ton can be sold for €360, while when the protein amount is 70%, it can increase to €700 (Quinn and

Davis, 2015). Also, microalgae can be used as fish food. Its market value is £1,500 per t (Muller-Feuga, 2013). Animal feeds contain 47% cereals, 27% oil crops, 13% food by-products, and 3% mineral vitamins (Fawcett et al., 2022). *Spirulina* contains proteins, carbohydrates, and lipids necessary for an animal's nutrition. Also, *Spirulina* has some elements and vitamins such as calcium, iron, phosphorous, iodine, magnesium, zinc, selenium, copper, manganese, vitamin A, vitamin K, thiamine, riboflavin, niacin, vitamin B6, and vitamin B12 (Bature et al., 2022). Microalgae are often added to feed as a supplement in the form of biomass. Poultry (10%), swine (5.51%), lambs (20%), and cattle (1.18%) have different rates (Madeira et al., 2017). There are some studies about microalgal feed in the literature. Islam et al. (2022) studied three microalgae for their potential as feedstuffs and animal feed. *Nannochloropsis* sp. had potential feedstuff properties with biomass productivity (0.45 mg/L/d), protein content (35%) and carbohydrate content (22%). Also, *Tetraselmis* sp. included 40% protein and 17% carbohydrate. In another study, Ju et al. (2012) showed that *Haematococcus pluvialis* included 32% protein and 8.9% lipid and it can be used as animal feed for white shrimp. Also, *Aurantiochytrium limacinum* was carried out to feed pigs. It had 17.6 DHA/100 g of biomass and it can be cultivated to feed pigs needing DHA (Moran et al., 2018). Research by Li et al. (2021) carried out production of selenium-enriched microalgae and mixed microalgae were grown in domestic wastewater, and the highest protein content of microalgae was 48%, and selenium enhanced omega-3 and omega-6 fatty acid content. In conclusion, it can be used as a protein source instead of soybeans (Li et al., 2021). Onay (2020) examined *Nannochloropsis gaditana* for its metabolic content. The highest carbohydrate concentration was 21.3% in the wastewater (Onay, 2020). Also, *Picochlorum maculatum* and *Nannochloris atomus* were studied to evaluate techno-functional feed ingredients. *Picochlorum maculatum* and *Nannochloris atomus* had the highest biomass productivities of 32.9 g/m^2/d and 17.1 g/m^2/d, respectively. *Picochlorum maculatum* had 54.5% protein content. Both of them maintained toxicity assay tests and were found to be safe (Rasheed et al., 2022). *Phaeodactylum tricornutum* and *Tisochrysis lutea* were grown in 1 ha. Their biomass productivity was 13 t/ha/y, with a cost of €105 per kg/DW (Vázquez-Romero et al., 2022). Some studies on feed production from microalgae are given in Table 2.1.

Table 2.1: Some studies on feed production from microalgae.

Microalgae	Content	References
Nannochloropsis sp.	35% (Protein)	Islam et al., 2022
Tetraselmis sp.	40% (Protein)	Islam et al., 2022
Haematococcus pluvialis	32% (Protein)	Ju et al., 2012
Aurantiochytrium limacinum	17.6 DHA/100 g of biomass	Moran et al., 2018
Mixed microalgae	40% (Protein)	Li et al., 2021
Picochlorum maculatum	54.5% (Protein)	Rasheed et al., 2022

In addition, microalgae can be used as biopesticides in the agriculture industry due to the chemicals they contain. Their values are £30,000 per t. Microalgae have phytohormones and stimulant chemicals. They act as biofertilizers in the agricultural industry. They increase the yield of plants and provide them with resistance. *Anabaena variabilis* can enhance rice yield. The market price of biofertilizers varies between \$300 and \$1,200 per t (Abu-Ghosh et al., 2021). Microalgae are used to make biocomposites, biopolymers, and bioplastics. Their approximate value is £2,000 per t (Llamas et al., 2021). The annual bioplastic production is 750,000 tons. Polylactic acid (PLA) accounts for 140,000 tons, while starch-based bioplastics account for 35,000 tons (Chong et al., 2022). PLA and polyhydroxyalkanoate (PHA) can be produced from microalgae. PHAs are added to the preparation of packaging films in shopping bags and containers (Chen et al., 2011). There are many studies about bioplastics in the literature. *Scenedesmus acutus* (UTEX B72) was used for bioplastic production, and the area used for growing microalgae is 300 ha. The selling price of bioplastic feedstock (BPFS) is between \$0.97 and \$3.96 per kg (Beckstrom et al., 2020). In another study, poly (3-hydroxyalkanoate) (PHB) was produced under phosphate limited conditions from mixed cyanobacteria. The maximum PHB content was 0.10 g/L. On the other hand, the highest carbohydrate concentration was 48% under carbon and phosphorus limitation (Arias et al., 2018). Das et al. (2018) studied bioplastic production from *Chlorella pyrenoidosa* in 1 L. Polyhydroxybutyrate (PHB) content was 27% compared to the biomass of microalgae. A strain of *Chlamydomonas reinhardtii* 11-32A was carried out and microalgae were grown in 30 L of tubular PBR. Starch-based bioplastics were produced with 49% w/w of starch (Mathiot et al., 2019). Algal-based bioplastics can be produced via protein biomass from microalgae. *Chlorella vulgaris* and *Spirulina platensis* were used for thermomechanical polymerization and produced polyethylene (PE). Both of them have 58% protein content (Zeller et al., 2013). *Synechococcus elongatus* produces PHA under mixotrophic nitrogen and phosphate conditions. The highest content of PHA was 17.15% (Mendhulkar and Shetye, 2017). Some studies on bioplastics production from microalgae are given in Table 2.2.

2.4 Biofuels from microalgae

Studies on the commercial use of microalgae in biofuel production are ongoing. They can be called low-value because of their production with minimal safety requirements. Biodiesel, bioethanol, biomethane, and jet fuel can be used mostly in the use of microalgae as biofuel. The market prices of biodiesel (£630 per t), bioethanol (£560 per t), and jet fuel (£740 per t) make their production attractive. While the biomass value for biodiesel is £60 per t, one for biomethane is £340 per t (Llamas et al., 2021).

Table 2.2: Some studies on bioplastics production from microalgae.

Microalgae	Content	Product	References
Scenedesmus acutus	40% (Protein) 35% (Carbohydrate) 15% (Lipid)	BPFS	Beckstrom et al., 2020
Mixed cyanobacteria	0.10 g/L PHB 48% (Carbohydrate)	PHB	Arias et al., 2018
Chlorella pyrenoidosa	27% PHB	PHB	Das et al., 2018
Chlamydomonas reinhardtii	49% w/w of starch	Starch-based bioplastics	Mathiot et al., 2019
Chlorella vulgaris and *Spirulina platensis*	58% (Protein)	PE	Zeller et al., 2013
Synechococcus elongatus	17.15% PHA	PHA	Mendhulkar and Shetye, 2017

2.4.1 Biodiesel from microalgae

Biodiesel production is one of the most used areas of microalgae. Diesel fuels lead to air pollution due to the pollutants they contain, but biodiesel is environmentally friendly because it does not contain these gases that cause greenhouse gases. Biodiesel can be defined as a biofuel produced from mono alkyl esters. For the production of biodiesel, esterification of triglyceride-derived oil with alcohol is required. This reaction is usually mediated by a catalyst such as an acid base or an enzyme such as a lipase (Chhandama et al., 2021). Many steps have to be taken until the production of biodiesel from microalgae is carried out. First, microalgae must be cultured and suitable conditions should be maintained for the growth of microalgae. Then, the harvesting procedure of microalgae can be applied and microalgal biomass is obtained. After this procedure, extraction of microalgae is provided for lipid production. Lipid extraction can be achieved by various methods, such as organic solvents, ionic liquids, supercritical fluids, and different press methods. These processes are called downstream processes (Khoo et al., 2020). Next, lipid molecules can be combined with suitable alcohol via a catalyst such as lipase, and biodiesel is formed. So, biodiesel can be affected by many factors. One of which is the engine performance of diesel. The engine performance of biodiesel produced by microalgae is still being studied by researchers. In one of them, engine emissions decreased but NO_x emissions were enhanced with biodiesel. Also, it maintained less torque and higher heat generation compared to a conventional diesel engine (Islam et al., 2015). In another study, B10, B20, and B50 in algae biodiesel engines decreased cylinder pressure and torque emissions by 4.5%. Low sulfur content can be reduced by 4% using algae biodiesel engines (Ferreira Mota et al., 2022). On the other hand, in diesel engines, the use of biodiesel (100%) causes a reduction in the emission of carbon monoxide, dioxide, and nitrogen compounds while reducing the effectiveness of some parameters, such as brake power and torque (Wei et al., 2017). The economic use of biodiesel was also influenced by its negative and positive effects. The majority of biofuel production and consumption takes place in the United States and European Union countries. They generally provide the use of first and second-generation raw materials. In microalgae, which is a third-generation biofuel raw material, production costs need to be reduced. *Botryococcus braunii*, *Nannochloropsis* sp., *Dunaliella primolecta*, *Chlorella* sp., and *Crypthecodinium cohnii* are the microalgae species that are most often used to make biodiesel (Ianda et al., 2022). Bello et al. (2012) carried out two different transesterification reactions and interpreted them according to economics. They used alkali-catalyzed transesterification (NaOH) and a noncatalytic supercritical transesterification method with a pressure of 128 bars and 280 °C. The biodiesel costs were £6.39 per L and £6.29 per L, respectively (Bello et al., 2012). According to Chia et al. (2018), the cost of producing biodiesel from microalgae containing 30% lipid is $2.8 per L, while the cost of producing conventional diesel is $1.1 per L in the United States (Chia et al., 2018). An economic model was developed for biodiesel production from marine microalgae in sub-Saharan countries. The Aspen Plus® V12 simulator provided information related to the production of 5.47 m^3/d of biodiesel. This amount was added to the petroleum-derived diesel as 3%. Calcium oxide was used as a catalyst. As a result, the cost of biodiesel per 1.5102 ha was $0.9 per kg (Ianda et al., 2022). Branco-Vieira et al. (2020) carried out *Phaeodactylum tricornutum* for biodiesel production on economic analysis, and they modeled the system in a bubble column photobioreactor. An alkali-catalyzed transesterification reaction was used in their study. Biodiesel cost was €0.33 per L in 15.247 ha with production of 171,705 L of biodiesel (Branco-Vieira et al., 2020). In another study, the cost of biodiesel production was estimated in China. The cost of biodiesel per 100 ha was $2.29 per kg (Sun et al., 2019). This result shows us that biodiesel production from microalgae is still a problematic mode of production in China. Santander et al. (2014) studied the cost of biodiesel production in the Atacama Desert of Chile. The base catalyst, NaOH, was used in their study for transesterification. The cost of the catalyst was very low in overall cost. A total of 75,000 tons of microalgae oil were esterified per year and the net cost was 159 million dollars with a discount rate of 12% in 15 years. This project was suitable for biodiesel production when some parameters such as fed oil price, glycerin price, biodiesel selling price, methanol price, and discount rate were calculated (Santander et al., 2014). Biodiesel from *Scenedesmus obliquus* was simulated in Northern Italy. As in other studies, CAPEX (capital costs) and OPEX (operating costs) costs were calculated, and Aspen Plus was used for the simulation of biodiesel costs. The biodiesel cost was $21.11 per gal of biodiesel with 1 km^2 of surface area (Tercero et al., 2014). These results maintain a very high cost when compared to conventional biodiesel production. The authors' suggestion was to increase the amount of oil (40%↑)

in the microalgae biomass. Delrue et al. (2012) studied the methodology of biodiesel production using some new criteria such as water footprint, net energy ratio, greenhouse gas emission rate, and biodiesel production cost. In 333.3 ha, the cost of biodiesel ranged between €1.94 and €3.35 per L and H_3PO_4 were used as catalysts. This cost was still high when compared to conventional diesel production (Delrue et al., 2012). Some studies on reducing the cost of biodiesel production from microalgae are given in Table 2.3.

Table 2.3: Some studies on biodiesel production costs from microalgae.

Microalgae	Lipid content (%)	Transesterification	Project size	Cost	References
Chlorella protothecoides	55.2	Alkali catalyzed	1 ton microalgae biomass/h	£6.39 per L	Bello et al., 2012
Chlorella protothecoides	55.2	Noncatalytic supercritical	1 ton microalgae biomass/h	£6.29 per L	Bello et al., 2012
Any microalgae including 30% lipid	30	Conventional methods	10,000 tonnes microalgae	$2.8 per L	Chia et al., 2018
Nannochloropsis salina sp.	40	Calcium oxide	1.5×10^2 ha	$0.9 per kg	Ianda et al., 2022
Phaeodactylum tricornutum	9.08	Alkali catalyzed	1,811 tons microalgae	€0.33 per L	Branco-Vieira et al., 2020
Botryococcus braunii	20–70	Alkali catalyzed	75,000 tons	$1.21 per L	Santander et al., 2014
Scenedesmus obliquus	40	N.A.	39.2 tons	$21.11 per gal	Tercero et al., 2014
Any microalgae including 20–50% lipid	20–50	Alkali catalyzed and Acid catalyzed	1000,000 m^3	€1.94 per L–€3.35 per L	Delrue et al., 2012

As can be seen from these studies, the production of biodiesel from microalgae varies greatly according to the conditions used. Operating costs (OPEX) such as cultivation, electricity consumption, maintenance, water cost, cost of raw materials, and labor affect biodiesel costs. CAPEX, such as equipment, cost of land, and various expenses play an important role in the production of biodiesel (Ianda et al., 2022). In addition, economic parameters such as financial debt, company credit, market risk, and demolition costs have to be arranged for biodiesel production (Sun et al., 2019). Another problem with the production of biodiesel from microalgae is the planning of the supply chains for the products. Ahn et al. (2015) studied deterministic mathematical programming for its supply chain network. This model sought answers to questions such as how much biomass and biodiesel would be produced, how it would be transported, and where and how many refineries would be set up. In their study, the cost of biodiesel was $1.56 per L (Ahn et al., 2015). Three factors are important in the biofuel supply chain. These are the type, capacity, and location of refineries and biomass allocation. As a result, reducing costs, increasing profits, and reducing investment cost risk are the main targets. Mathematical programming and heuristic programming can be used for the optimization of the biofuel supply chain. While mathematical models examine the minimum and maximum values of the variables, other models can be used to solve more complex problems (Ahn et al., 2015). In another study, Ahn et al. developed a two-stage stochastic model for the optimization of the biodiesel supply chain and reduced costs. The investigation of unknown variables for biodiesel affects diesel cost and demand. In their study, it was found that it was important to reduce the cost of carbon dioxide needed for biodiesel production and the establishment of the refinery that will provide it. In addition, like in other studies, the storage areas determined for the produced biodiesel to go to the end users were considered as another uncertain variable in the study and optimization was maintained (Ahn and Kim, 2021).

2.4.2 Bioethanol from microalgae

Bioethanol production from microalgae has been preferred in recent years due to its harmless nature, degradability, and ability to minimize greenhouse gas formation. Bioethanol can be produced from microalgae via saccharification and fermentation. In this process, carbohydrates are hydrolyzed with acid and base enzymatically. Then, sugars such as starch or cellulose can be fermented using bacteria and yeast (Onay, 2019). While bioethanol production was around 17 billion liters in the 2000s, it was 160 billion liters in 20 years. Bioethanol production from microalgae has gained importance in reducing costs such as fermentation, purification, and transportation of bioethanol, as well as increasing the applicability of downstream processes (selection of species, cultivation, pretreatment methods, carbohydrate content, saccharification) to increase the yield (Phwan et al., 2018). Supply and demand studies on bioethanol gained momentum by using biowaste resources. A two-stage stochastic programming model was developed for rice straw, wheat straw, corn stover, forest residues, municipal solid waste wood and paper, MSW yard, and cotton residuals were used. The cost of bioethanol dropped to 1.20 per gallon (Chen and Fan, 2012). A wide variety of microalgae species can be used for bioethanol production. These are *Chlorella minutissima*, *Chlorella vulgaris*, *Chlorella vulgaris* FSP-E, *Hindakia tetrachotoma*, *Nannochloropsis gaditana*, and *Scenedesmus* sp. Their carbohydrate contents range from 14 to 93% (Maia et al., 2020). Szulczyk and Tan (2022) applied a partial equilibrium model for bioethanol production from *Chlorella vulgaris*. Some variables such as farm expenses, labor, and greenhouse gas emissions were used for this study. Capital costs (bioethanol equipment, extractor, fermenter, land costs, electrical supply, and pond), operating costs (power for paddle wheels and water supply), nutrients, labor, and waste disposal were calculated. According to their estimation, 63.67 million liters of bioethanol will be produced in 2044 and its cost will be $0.65 per L (Szulczyk and Tan, 2022). In addition, Chong et al. (2020) studied *Eucheuma cottonii* (macroalgae). They showed that *Eucheuma cottonii* could be valuable for bioethanol production in terms of cost. The cost of bioethanol was $0.54 per kg (Chong et al., 2020). Onay (2018) studied *Nannochloropsis gaditana* for bioethanol production in different concentrations of wastewater. The highest bioethanol content was approximately 94 mg per g of biomass (Onay, 2018). Brown algae can be used for bioethanol production. In another study, *Laminaria* spp. produced bioethanol via *Escherichia coli.* These algae include mannitol, Iaminarin, alginates, cellulose, and fucoidan. Carbohydrate (60%) obtained from 400,000 tons/year of dry brown algae was converted to bioethanol with a conversion rate of approximately 90%. Operating costs were dry seaweed, carbon dioxide, steam, sulfuric acid, cellulose, ammonia, water, and wastewater. Capital costs were pretreatment, saccharification, fermentation, purification, and capital investments. The total cost changed between $2.08 and $2.85 per gal according to the selected pretreatment methods (Fasahati et al., 2015). *Chlorella vulgaris* was examined. Its carbohydrate concentration was 50%. Bioethanol was produced via microwave-assisted extraction and fermentation through enzymatic reactions and separation. The total production price was US $2.22 million per annum. Furthermore, the total annual selling cost of bioethanol was US $2.87 million (Hossain et al., 2019). Artificial intelligence supported studies were carried out to increase the amount of bioethanol. Onay (2022) studied optimum conditions of *Chlorella saccharophila* to provide the highest bioethanol concentration (11.17 g/L) via an artificial neural network (Onay, 2022). Also, algae biofuels were commercially produced by PetroSun in 2006, and Algenol produced about 8,000 gallons per year of bioethanol per year in the USA, and Sapphire Energy maintained 100,000 gallons per year of bioethanol. According to these results, biofuel based on algae will dominate 75% of the market by 2030 (Phwan et al., 2018). Some studies on bioethanol production costs from microalgae are given in Table 2.4.

2.4.3 Biogas from microalgae

Biogas production is one of the most popular areas where microalgae are used. Biogas generally contains approximately 65% methane and 35% carbon dioxide with an anaerobic method (Haider et al., 2022). The factors affecting the biomethane economy are the same as those affecting other biofuels. Downstream and upstream processes are effective in this process. In addition, biomethane purification, dewatering cost, and biomethane price affect this process. While the global market price of biomethane is $0.76, it is $0.59 in developing countries such as India (Kannah et al., 2021). *Sargassum* spp. was studied for biomethane and bioethanol production. Operating

Table 2.4: Some studies on bioethanol production costs from microalgae.

Microalgae	Carbohydrate content (%)	Fermentation	Project size	Cost	References
Chlorella vulgaris	50	Traditional method	63.67 million liters of bioethanol	\$0.65 per L	Szulczyk and Tan, 2022
Eucheuma cottonii	51 (carrageenan)	*Saccharomyces cerevisiae*	7,626 kg/h of anhydrous bioethanol	\$0.54 per kg	Chong et al., 2020
Laminaria spp.	60	Simple pretreatment (Hot water wash) + E. coli	400,000 ton/year of dry brown algae	\$2.39 and \$2.85 per gal	Fasahati et al., 2015
Laminaria spp.	60	Combined pretreatment (Acid thermal hydrolysis) + E. coli	400,000 ton/year of dry brown algae	\$2.08 and \$2.33 per gal	Fasahati et al., 2015
Chlorella vulgaris	55	Microwave assisted extraction and fermentation through enzymatic reactions and separation	220 tons microalgae biomass	US\$ 2.87 million per annum	Hossain et al., 2019

costs, capital costs, and biomethane profit were calculated. The highest biomethane yield and productivity were 149.6 L/kg VS and 13.6 L/kg VS/d, respectively. While the estimated biomethane production was 301930.2 m^3 $year^{-1}$, the estimated biomethane production cost was US\$0.039 per m^3 (Abomohra et al., 2021). Brigagão et al. (2019) carried out biogas production with pretreatment methods such as thermal and thermomechanical methods. The biomethane value was US\$7.58 per GJ (Brigagão et al., 2019). Kavitha et al. (2019) sought to increase the profits of biomethane by inducing nanoparticles. They used *Chlorella vulgaris* microalgae in this study and obtained less costly biomethane from 7.2 gL^{-1} microalgae biomass (Kavitha et al., 2019). Li et al. (2022) produced biogas using microalgae in wastewater. In this study, a source-separated nutrient delivery approach and nonseparated point nutrient delivery policies were used. The cost was 0.38 US dollars per m^{-3} (Li et al., 2022). Meyer and Weiss studied life cycle costs for biogas production. They calculated all the parameters for biogas production and the biogas cost was €13 per kg/DW algal biomass (Meyer and Weiss, 2014). Xiao et al. (2020) developed life cycle and economic assessments to reduce biogas costs, and they used hydrothermal pretreatment methods. The cost was \$0.17 per m^3 for biogas production from *Chlorella* sp. (Xiao et al., 2020). Some studies on biogas production costs from microalgae are given in Table 2.5.

Table 2.5: Some studies on biogas production costs from microalgae.

Microalgae	Project Size	Cost	References
Sargassum spp.	301,930.2 m^3 $year^{-1}$	US$0.039 m^3	Abomohra et al., 2021
Nannochloropsis salina	11.27 kg/s biogas	US$7.58 GJ	Brigagão et al., 2019
Chlorella vulgaris	7.2 g/L biomass	US$320	Kavitha et al., 2019
Mixed microalgae	7,665 kgd^{-1} biogas	US$0.38/$m^3$	Li et al., 2022
Chlamydomonas reinhardtii	1 ha 6 g DW/l PBR	€13 per kg/DW algal biomass	Meyer and Weiss, 2014
Chlorella sp.	25,000 kg/d 100 ha.	$0.17 per m^3	Xiao et al., 2020

2.4.4 Jet fuel

Biojet fuels can be produced from biomass with various lignocellulosic properties such as alcohol, oil, and syngas to jet conversion. Kerosene can be converted to jet fuel and its market price is €740 per kg (Mousavi-Avval and Shah, 2021; Llamas et al., 2021). Jet fuel can be produced with various methods such as Fischer-Tropsch, alcohol-to-jet, and hydroprocessed esters and fatty acids. All of them need catalysts such as Co, Fe, Pt, Pd, Ni, Mo-based catalysts, and zeolites (Goh et al., 2022). Renewable jet fuels are produced from plant biomass due to their ease of production. Mousavi-Avval and Shah (2021) studied pennycress to produce jet fuel. It includes 25–35% oil content. They used the area with 18.9 million Lyr^{-1} and they found that the most important parameter was biomass yield for the production of jet fuel (Mousavi-Avval and Shah, 2021). Huang et al. (2019) produced jet fuel from corn stover using three methods (FT, ATJ, and HTL). At $2.83 per gallon, Fischer-Tropsch was the most cost-effective method. This value was still much higher than conventional jet fuel (Huang et al., 2019). In the United States 2.8 billion gallons of petroleum-derived jet fuel are consumed annually. However, with the decrease in petroleum-derived resources, new resources began to be used. It became attractive to obtain jet fuel from local biomass. The cost was reduced to $0.78 per liter (Beal et al., 2021).

Xin et al. (2018) carried out bio-oil production from microalgae grown in wastewater. They used a microwave-assisted pyrolysis method. Microalgae productivity was 5,420 tons per day and its selling price was $1.85 per gallon (Xin et al., 2018). Wang (2019) examined the techno-economic analysis of some feedstocks in Taiwan. Hydro-processed renewable jets were obtained and their prices were between $0.91 and $2.74 per liter. They calculated the price of microalgae at $1.66 per L and the fuel yield at 11,697 L/ha/year (Wang, 2019). Fortier et al. (2014) developed a model with hydrothermal liquefaction to produce biojet fuel from microalgae. GHG emissions decreased by up to 76% (Fortier et al., 2014). Zhang et al. (2022) analyzed the production of jet fuel from microalgae by ecosystem simulation. *Dunaliella salina* was cultivated for simulation and the HTL method was used for jet fuel production. Bio oil yield increased by up to 49% (Zhang et al., 2022). Li et al. (2018) studied the Fischer-Tropsch method for the production of biojet fuel. They used mesoporous Y-type zeolites as catalysts and provided high amounts of jet fuel (Li et al., 2018). Some studies related to bio-jet fuel production from microalgae are given in Table 2.6.

Table 2.6: Some studies related with bio-jet fuel production from microalgae.

Sources	Methods	Price	References
Pennycress	Hydroprocessed Renewable Jet Fuel (HRJ)	NA	Mousavi-Avval and Shah, 2021
Corn stover	Fischer-Tropsch (FT)	$2.83 per gal	Huang et al., 2019
Microalgae	Microwave-assisted pyrolysis (MWP)	$1.85 per gal	Xin et al., 2018
Microalgae	Hydro-processed renewable jet (HPRJ)	$1.66 per L	Wang, 2019
Microalgae	Hydrothermal liquefaction (HL)	NA	Fortier et al., 2014
Microalgae	Hydrothermal liquefaction (HTL)	NA	Zhang et al., 2022
Microalgae	Fischer-Tropsch (FT)	NA	Li et al., 2018

In order for biofuels to be converted into products commercially, the production value must be below €1,000 per t. Biofuel production requires at least 100 ha of land. The cost of biomass must be below €1 per kg. If a product is made using wastewater, this value is expected to decrease to €0.6 per kg (Llamas et al., 2021). In conclusion, if you want to lower the cost of all types of biofuel, you need to choose the right microalgae species, lower the costs of extraction and harvesting, and use pretreatment methods to boost the metabolic content of microalgae.

2.5 Untraditional high-value microalgae products

The products extracted from the microalgae in this part are outside of traditional production as they are produced in low quantities. They are often referred to as secondary metabolites, and they are often used for diagnosis and treatment. Cannabidiol can be extracted from *Cannabis sativa*. It has anti-inflammatory, antioxidant, and hepatoprotective properties (Erukainure et al., 2022). Also, it can be used to reduce the effect of the cytokine storm that occurs during lung injuries (Kocherlakota et al., 2022). Its cost ranges from $60 to $200 per g. It is worth approximately 344 billion dollars in the market. Since cannabidiol is associated with fatty acid metabolism, it can be produced by microalgae with high fatty acid content (Abu-Ghosh et al., 2021). Glucans are another product that can be extracted from microalgae. They have more advantages over extraction from bacteria and fungi as they are safely derived from microalgae. Glucans are polysaccharides and consist of D-glucose units. They are found in the cell wall and act as a building polysaccharide (Torres-Tiji et al., 2020). Glucans have anticancer properties, and they create an immune response. Also, glucans show antioxidant activity by capturing free radicals. *Euglena gracilis* includes high amounts of glucans, and 160 tons of *Euglena gracilis* were produced in 2019 (Gohara-Beirigo et al., 2022; Abu-Ghosh et al., 2021). Dolastatin is another product. It is a small peptide and can be used for medicinal purposes. It has anticancer activity but can lead to neuropathy. Dolastatin can be extracted from *Symploca* and *Lyngbya* (Cyanobacteria). It is still difficult to produce because it is extracted from microalgae at very low rates. On the other hand, this feature makes it valuable and it is sold between $30,000 and $60,000 per g (Mondal et al., 2020; Abu-Ghosh et al., 2021).

2.6 Recommendations

Today, microalgae are frequently used for commercial purposes. The main reasons for this are the rapid growth of microalgae, their safety, and their valuable content. In addition, there are still limitations. The most important disadvantages are the difficulties in production and cost. This situation determines the supply and demand for microalgae. Downstream processes in microalgae have the same challenges for all industries. Reducing the chemical inputs in the growth and extraction of microalgae, the high prices of devices such as reactors, and the reduction of expenses such as electricity and water is a necessity for the entire industry. On the other hand, the demand for microalgae will increase with the reduction of prices for upstream inputs that vary according to each product and the developments in science related to them. Last, the demand for microalgae will go up if the problems above are fixed and if improvements are made to products that use a small amount of microalgae but have a high value.

References

Abomohra, A. E., El-hefnawy, M. E., Wang, Q., & Huang, J. (2021). Sequential bioethanol and biogas production coupled with heavy metal removal using dry seaweeds: Towards enhanced economic feasibility. *Journal of Cleaner Production*, *316*. https://doi.org/10.1016/j.jclepro.2021.128341

Abu-Ghosh, S., Dubinsky, Z., Verdelho, V., & Iluz, D. (2021). Unconventional high-value products from microalgae: A review. *Bioresource Technology*, *329*, 124895. https://doi.org/10.1016/j.biortech.2021.124895

Ahn, Y., Lee, I., Lee, K., & Han, J. (2015). Strategic planning design of microalgae biomass-to-biodiesel supply chain network: Multi-period deterministic model. *Applied Energy*, *154*, 528–542. https://doi.org/10.1016/j.apenergy.2015.05.047

Ahn, Y., & Kim, J. (2021). Economic design framework of microalga-based biodiesel supply chains under uncertainties in CO_2 emission and diesel demand. *Computers and Chemical Engineering*, *155*, 107538. https://doi.org/10.1016/j.compchemeng.2021.107538

Arias, D. M., Fradinho, J. C., Uggetti, E., García, J., Oehmen, A., & Reis, M. A. M. (2018). Polymer accumulation in mixed cyanobacterial cultures selected under the feast and famine strategy. *Algal Research*, *33*, 99–108. https://doi.org/10.1016/j.algal.2018.04.027

Bature, A., Melville, L., Rahman, K. M., & Aulak, P. (2022). Microalgae as feed ingredients and a potential source of competitive advantage in livestock production: A review. *Livestock Science*, *259*, 104907. https://doi.org/10.1016/j.livsci.2022.104907

Beal, C. M., Cuellar, A. D., & Wagner, T. J. (2021). Sustainability assessment of alternative jet fuel for the U.S. Department of Defense. *Biomass and Bioenergy*, *144*, 105881. https://doi.org/10.1016/j.biombioe.2020.105881

Beckstrom, B. D., Wilson, M. H., Crocker, M., & Quinn, J. C. (2020). Bioplastic feedstock production from microalgae with fuel co-products : A techno-economic and life cycle impact assessment. *Algal Research*, *46*, 101769. https://doi.org/10.1016/j.algal.2019.101769

Bello, B. Z., Nwokoagbara, E., & Wang, M. (2012). Comparative techno-economic analysis of biodiesel production from microalgae via transesterification methods. *Computer Aided Chemical Engineering*. https://doi.org/10.1016/B978-0-444-59519-5.50027-7

Bernaerts, T. M. M., Gheysen, L., Foubert, I., Hendrickx, M. E., & Van Loey, A. M. (2019). The potential of microalgae and their biopolymers as structuring ingredients in food: A review. *Biotechnology Advances*, *37*, 107419. https://doi.org/10.1016/j.biotechadv.2019.107419

Branco-Vieira, M., Mata, T. M., Martins, A. A., Freitas, M. A. V., & Caetano, N. S. (2020). Economic analysis of microalgae biodiesel production in a small-scale facility. *Energy Reports*, *6*, 325–332. https://doi.org/10.1016/j.egyr.2020.11.156

Brigagão, G. V., Wiesberg, I. L., Pinto, J. L., De Queiroz, O., Araújo, F., & De Medeiros, J. L. (2019). Upstream and downstream processing of microalgal biogas: Emissions, energy and economic performances under carbon taxation. *Renewable and Sustainable Energy Reviews*, *112*, 508–520. https://doi.org/10.1016/j.rser.2019.06.009

Castillo, T., Ramos, D., García-beltr´an, T., Brito-Bazan, M., & Galindo, E. (2021). Mixotrophic cultivation of microalgae: An alternative to produce high-value metabolites. *Biochemical Engineering Journal*, *176*, 108183. https://doi.org/10.1016/j.bej.2021.108183

Cheirsilp, B., & Maneechote, W. (2022). Insight on zero waste approach for sustainable microalgae biorefinery: Sequential fractionation, conversion and applications for high-to-low value-added products. *Bioresource Technology Reports*, *18*, 101003. https://doi.org/10.1016/j.biteb.2022.101003

Chen, G. Q., Wu, Q., Jung, Y. K., & Lee, S. Y. (2011). PHA/PHB. In Moo-Young, M. (Ed.), Comprehensive Biotechnology (Second Edition). Elsevier Inc, Burlington. pp. 217–227. https://doi.org/10.1016/B978-0-08-088504-9.00179-3

Chen, C., & Fan, Y. (2012). Bioethanol supply chain system planning under supply and demand uncertainties. *Transportation Research Part E*, *48*, 150–164. https://doi.org/10.1016/j.tre.2011.08.004

Chhandama, M. V. L., Satyan, K. B., Changmai, B., Vanlalveni, C., & Rokhum, S. L. (2021). Microalgae as a feedstock for the production of biodiesel: A review. *Bioresource Technology Reports*, *15*, 100771. https://doi.org/10.1016/j.biteb.2021.100771

Chia, S. R., Chew, K. W., Show, P. L., Yap, Y. J., Ong, H. C., Ling, T. C., & Chang, J. S. (2018). Analysis of economic and environmental aspects of microalgae biorefinery for biofuels production: A review. *Biotechnology Journal*, *13*(6), e1700618. doi:: 10.1002/biot.201700618

Chong, T. Y., Cheah, S. A., Ong, C. T., Wong, L. Y., Goh, C. R., Tan, I. S., & Lim, S. (2020). Techno-economic evaluation of third-generation bioethanol production utilizing the Macroalgae waste: A case study in Malaysia. *Energy*, *210*, 118491. https://doi.org/10.1016/j.energy.2020.118491

Chong, J. W. R., Tan, X., Khoo, K. S., Ng, H. S., Jonglertjunya, W., Yew, G. Y., & Show, P. L. (2022). Microalgae-based bioplastics: Future solution towards mitigation of plastic wastes. *Environmental Research*, *206*, 112620. https://doi.org/10.1016/j.envres.2021.112620

Das, S. K., Sathish, A., & Stanley, J. (2018). Production of biofuel and bioplastic from *Chlorella Pyrenoidosa*. *Materials Today: Proceedings*, *5*, 16774–16781. https://doi.org/10.1016/j.matpr.2018.06.020

Delrue, F., Setier, P., Sahut, C., Cournac, L., Roubaud, A., Peltier, G., & Froment, A. (2012). An economic, sustainability, and energetic model of biodiesel production from microalgae. *Bioresource Technology*, *111*, 191–200. https://doi.org/10.1016/j.biortech.2012.02.020

Dey, S., & Rathod, V. K. (2013). Ultrasound assisted extraction of β-carotene from *Spirulina platensis*. *Ultrasonics Sonochemistry*, *20*(1), 271–276. https://doi.org/10.1016/j.ultsonch.2012.05.010

Erukainure, O. L., Matsabisa, M. G., Salau, V. F., Olofinsan, K. A., Oyedemi, S. O., Chukwuma, C. I., Lum, A., & Islam, S. (2022). Cannabidiol improves glucose utilization and modulates glucose-induced dysmetabolic activities in isolated rats' peripheral adipose tissues. *Biomedicine and Pharmacotherapy*, *149*, 112863. https://doi.org/10.1016/j.biopha.2022.112863

Fasahati, P., Woo, H. C., & Liu, J. J. (2015). Industrial-scale bioethanol production from brown algae: Effects of pretreatment processes on plant economics. *Applied Energy*, *139*, 175–187. https://doi.org/10.1016/j.apenergy.2014.11.032

Fawcett, C. A., Senhorinho, G. N. A., Laamanen, C. A., & Scott, J. A. (2022). Microalgae as an alternative to oil crops for edible oils and animal feed. *Algal Research*, *64*, 102663. https://doi.org/10.1016/j.algal.2022.102663

Ferreira Mota, G., Germano de Sousa, I., Luiz Barros de Oliveira, A., Luthierre Gama Cavalcante, A., Da Silva Moreira, K., Thálysson Tavares Cavalcante, F., Erick da Silva Souza, J., Rafael de Aguiar Falcão, Í., Guimarães Rocha, T., Bussons Rodrigues Valério, R., Cristina Freitas de Carvalho, S., Simaõ Neto, F., De França Serpa, J., Karolinny Chaves de Lima, R., Cristiane Martins de Souza, M., & Dos Santos, J. C. S. (2022). Biodiesel production from microalgae using lipase-based catalysts: Current challenges and prospects. *Algal Research*, *62*. https://doi.org/10.1016/j.algal.2021.102616

Fortier, M. P., Roberts, G. W., Stagg-williams, S. M., & Sturm, B. S. M. (2014). Life cycle assessment of bio-jet fuel from hydrothermal liquefaction of microalgae. *Applied Energy*, *122*, 73–82. https://doi.org/10.1016/j.apenergy.2014.01.077

Goh, B. H. H., Chong, C. T., Ong, H. C., Seljak, T., Katrasnik, T., Jozsa, V., et al. (2022). Recent advancements in catalytic conversion pathways for synthetic jet fuel produced from bioresources. *Energy Conversion and Management*, *251*, 14974. https://doi.org/10.1016/j.enconman.2021.114974

Gohara-Beirigo, A. K., Chuei, M., Cezare-gomes, E. A., & De Car, M. (2022). Microalgae trends toward functional staple food incorporation: Sustainable alternative for human health improvement. *Trends in Food Science & Technology*, *125*, 185–199. https://doi.org/10.1016/j.tifs.2022.04.030

Haider, J., Abdul, M., Riaz, A., Naquash, A., Kazmi, B., Yasin, M., Nizami, A., Byun, M., Lee, M., & Lim, H. (2022). State-of-the-art process simulations and techno-economic assessments of ionic liquid-based biogas upgrading techniques: Challenges and prospects. *Fuel*, *314*, 123064. https://doi.org/10.1016/j.fuel.2021.123064

Hossain, N., Zaini, J., Meurah, T., & Mahlia, I. (2019). Life cycle assessment, energy balance and sensitivity analysis of bioethanol production from microalgae in a tropical country. *Renewable and Sustainable Energy Reviews*, *115*, 109371. https://doi.org/10.1016/j.rser.2019.109371

Huang, E., Zhang, X., Rodriguez, L., Khanna, M., De Jong, S., Ting, K. C., Ying, Y., & Lin, T. (2019). Multi-objective optimization for sustainable renewable jet fuel production: A case study of corn stover based supply chain system in Midwestern U.S. *Renewable and Sustainable Energy Reviews*, *115*, 109403. https://doi.org/10.1016/j.rser.2019.109403

Ianda, T. F., Kalid, R. A., Rocha, L. B., Padula, A. D., & Zimmerman, W. B. (2022). Techno-economic modeling to produce biodiesel from marine microalgae in sub-Saharan countries: An exploratory study in Guinea-Bissau. *Biomass Bioenergy*, *158*, 106369. https://doi.org/10.1016/j.biombioe.2022.106369

Islam, M. A., Rahman, M. M., Heimann, K., Nabi, M. N., Ristovski, Z. D., Dowell, A., Thomas, G., Feng, B., Von Alvensleben, N., & Brown, R. J. (2015). Combustion analysis of microalgae methyl ester in a common rail direct injection diesel engine. *Fuel*, *143*, 351–360. https://doi.org/10.1016/j.fuel.2014.11.063

Islam, Z., Khatoon, H., Redwanur, M., Kumar, S., Hossain, S., Zaman, S., Jahedul, S., & Hasan, J. (2022). Growth, productivity and proximate profiling of indigenous marine microalgae from southern coast of Bangladesh as potential feed stuff for animal feed. *Bioresource Technology Reports*, *18*, 101025. https://doi.org/10.1016/j.biteb.2022.101025

Jacob-Lopes, E., Maroneze, M. M., Depra, M. C., Sartori, R. B., Dias, R. R., & Zepka, L. Q. (2019). Bioactive food compounds from microalgae: An innovative framework on industrial biorefineries. *Current Opinion in Food Science*, *25*, 1–7. https://doi.org/10.1016/j.cofs.2018.12.003

Ju, Z. Y., Deng, D.-F., & Dominy, W. (2012). A defatted microalgae (*Haematococcus pluvialis*) meal as a protein ingredient to partially replace fishmeal in diets of Pacific white shrimp (*Litopenaeus vannamei*, Boone, 1931). *Aquaculture*, 354, 50–55. https://doi.org/10.1016/j.aquaculture.2012.04.028

Kannah, R. Y., Kavitha, S., Parthiba, O., Rene, E. R., Kumar, G., & Banu, J. R. (2021). A review on anaerobic digestion of energy and cost effective microalgae pretreatment for biogas production. *Bioresource Technology*, *332*, 125055. https://doi.org/10.1016/j.biortech.2021.125055

Kavitha, S., Schikaran, M., Kannah, R. Y., Gunasekaran, M., & Kumar, G. (2019). Nanoparticle induced biological disintegration: A new phase separated pretreatment strategy on microalgal biomass for profitable biomethane recovery. *Bioresource Technology*, *289*, 121624. https://doi.org/10.1016/j.biortech.2019.121624

Khoo, K. S., Chew, K. W., Yew, G. Y., Leong, W. H., Chai, Y. H., Show, P. L., & Chen, W. H. (2020). Recent advances in downstream processing of microalgae lipid recovery for biofuel production. *Bioresource Technology*, *304*, 122996. https://doi.org/10.1016/j.biortech.2020.122996

Kocherlakota, C., Nagaraju, B., Arjun, N., Srinath, A., Kumar, S., Kothapalli, D., & Brenna, J. T. (2022). Inhalation of nebulized omega-3 fatty acids mitigate LPS-induced acute lung inflammation in rats: Implications for treatment of COPD and COVID-19. *Prostaglandins, Leukotrienes and Essential Fatty Acids*, *179*, 102426. https://doi.org/10.1016/j.plefa.2022.102426

Kumar, R., Hegde, A. S., Sharma, K., Parmar, P., & Srivatsan, V. (2022). Microalgae as a sustainable source of edible proteins and bioactive peptides-current trends and future prospects. *Food Research International*, *157*, 111338. https://doi.org/10.1016/j.foodres.2022.111338

Li, J., He, Y., Tan, L., Zhang, P., Peng, X., Oruganti, A., Yang, G., Abe, H., Wang, Y., & Tsubaki, N. (2018). Integrated tuneable synthesis of liquid fuels via Fischer-Tropsch technology. *Nature Catalysis*, *1*, 787–793. https://doi.org/10.1038/s41929-018-0144-z

Li, J., Otero-gonzalez, L., Michiels, J., Lens, P. N. L., Du, G., & Ferrer, I. (2021). Production of selenium-enriched microalgae as potential feed supplement in high-rate algae ponds treating domestic wastewater. *Bioresource Technology*, *333*, 125239. https://doi.org/10.1016/j.biortech.2021.125239

Li, P., Luo, Y., & Yuan, X. (2022). Life cycle and techno-economic assessment of source-separated wastewater-integrated microalgae biofuel production plant: A nutrient organization approach. *Bioresource Technology*, *344*, 126230. https://doi.org/10.1016/j.biortech.2021.126230

Llamas, B., Suárez-Rodríguez, M. C., González-Lopez, C. V., Mora, P., & Acién, F. G. (2021). Techno-economic analysis of microalgae related processes for CO_2 bio-fixation. *Algal Research*, 57. https://doi.org/10.1016/j.algal.2021.102339

López-Sánchez, A., Silva-Gálvez, A. L., Aguilar-Juárez, Ó., Senés-Guerrero, C., Orozco-Nunnelly, D. A., Carrillo-Nieves, D., & Gradilla-Hernández, M. S. (2022). Microalgae-based livestock wastewater treatment (MbWT) as a circular bioeconomy approach: Enhancement of biomass productivity, pollutant removal and high-value compound production. *Journal of Environmental Management*, *308*, 114612. https://doi.org/10.1016/j.jenvman.2022.114612

Madeira, M. S., Cardoso, C., Lopes, P. A., Coelho, D., Afonso, C., Bandarra, N. M., & Prates, J. A. M. (2017). Microalgae as feed ingredients for livestock production and meat quality: A review. *Livestock Science*, *205*, 111–121.

Maia, J., Cardoso, J. S., Mastrantonio, D., Bierhals, C. K., Moreira, J. B., Costa, J. A., & Morais, M. G. (2020). Microalgae starch: A promising raw material for the bioethanol production. *International Journal of Biological Macromolecules*, *165*, 2739–2749. https://doi.org/10.1016/j.ijbiomac.2020.10.159

Mathiot, C., Ponge, P., Gallard, B., Sassi, J., Delrue, F., Le, N., Cadarache, C. E. A., & Biomasse, G. (2019). Microalgae starch-based bioplastics: Screening of ten strains and plasticization of unfractionated microalgae by extrusion. *Carbohydrate Polymers*, *208*, 142–151. https://doi.org/10.1016/j.carbpol.2018.12.057

Mehariya, S., Kumar, R., Parthiba, O., & Verma, P. (2021). Chemosphere Microalgae for high-value products: A way towards green nutraceutical and pharmaceutical compounds. *Chemosphere*, *280*, 130553. https://doi.org/10.1016/j.chemosphere.2021.130553

Mendhulkar, V. D., & Shetye, L. A. (2017). Synthesis of biodegradable polymer polyhydroxyalkanoate (PHA) in cyanobacteria Synechococcus elongates under mixotrophic nitrogen and phosphate-Mediated stress conditions. *Industrial Biotechnology*, 85–93. https://doi.org/10.1089/ind.2016.0021

Meyer, M. A., & Weiss, A. (2014). Life cycle costs for the optimized production of hydrogen and biogas from microalgae. *Energy*, *78*, 84–93. https://doi.org/10.1016/j.energy.2014.08.069

Mondal, A., Bose, S., Banerjee, S., Patra, J. K., Malik, J., Mandal, S. K., Kilpatrick, K. L., Das, G., Kerry, R. G., Fimognari, C., & Bishayee, A. (2020). Marine cyanobacteria and microalgae metabolites-A rich source of potential anticancer drugs. *Marine Drugs*, *18*(9), 476. https://doi.org/10.3390/md18090476

Moran, C. A., Morlacchini, M., Keegan, J. D., Delles, R., & Fusconi, G. (2018). Effects of a DHA-rich unextracted microalgae as a dietary supplement on performance, carcass traits and meat fatty acid profile in growing-finishing pigs. *Journal of Animal Physiology and Animal Nutrition*, *102*, 1026–1038. https://doi.org/10.1111/jpn.12911

Mousavi-Avval, S. H., & Shah, A. (2021). Life cycle energy and environmental impacts of hydroprocessed renewable jet fuel production from pennycress. *Applied Energy*, *297*, 117098. https://doi.org/10.1016/j.apenergy.2021.117098

Muller-Feuga, A. (2013). Microalgae for Aquaculture: The Current Global Situation and Future Trends. In Richmond, A. & Hu, Q. (Eds.), Handbook of Microalgal Culture: Applied Phycology and Biotechnology. Wiley Online Library, Oxford, UK, pp. 613–627. https://doi.org/10.1002/9781118567166.ch33

Onay, M. (2018). Bioethanol production from *Nannochloropsis gaditana* in municipal wastewater. *Energy Procedia*, *153*, 253–257. https://doi.org/10.1016/j.egypro.2018.10.032

Onay, M. (2019). Bioethanol production via different saccharification strategies from *H. tetrachotoma* ME03 grown at various concentrations of municipal wastewater in a flat-photobioreactor. *Fuel*, *239*, 1315–1323. https://doi.org/10.1016/j.fuel.2018.11.126

Onay, M. (2020). Enhancing carbohydrate productivity from *Nannochloropsis gaditana* for bio-butanol production. *Energy Reports*, *6*, 63–67. https://doi.org/10.1016/j.egyr.2019.08.019

Onay, M. (2021). Photo-bioreactors for Value-added Products from Microalgae. In: Hasdemir, B., Turhan, M., & Karaman, C. (Eds.), Research & Reviews in Engineering Volume I. Gece Publishing, Ankara, Turkey, pp: 91–120.

Onay, M. (2022). Sequential modelling for carbohydrate and bioethanol production from *Chlorella saccharophila* CCALA 258 : A complementary experimental and theoretical approach for microalgal bioethanol production. *Environmental Science and Pollution Research*, 14316–14332. https://doi.org/10.1007/s11356-021-16831-w

Phwan, C. K., Ong, H. C., Chen, W. H., Ling, T. C., Ng, E. P., & Show, P. L. (2018). Overview: Comparison of pretreatment technologies and fermentation processes of bioethanol from microalgae. *Energy Conversion and Management*, *173*, 81–94. https://doi.org/10.1016/j.enconman.2018.07.054

Pulz, O., & Gross, W. (2004). Valuable products from biotechnology of microalgae. *Applied Microbiology and Biotechnology*, *65*, 635–648. https://doi.org/10.1007/s00253-004-1647-x

Rasheed, R., Thaher, M., Younes, N., Bounnit, T., Schipper, K., Nasrallah, G. K., Al, H., Gifuni, I., Goncalves, O., & Pruvost, J. (2022). Solar cultivation of microalgae in a desert environment for the development of techno-functional feed ingredients for aquaculture in Qatar. *Science of the Total Environment, 835*, 155538. https://doi.org/10.1016/j.scitotenv.2022.155538

Santander, C., Robles, P. A., Cisternas, L. A., & Rivas, M. (2014). Technical–economic feasibility study of the installation of biodiesel from microalgae crops in the Atacama Desert of Chile. *Fuel Processing Technology, 125*, 267–276. https://doi.org/10.1016/j.fuproc.2014.03.038

Sun, J., Xiong, X., Wang, M., Du, H., Li, J., Zhou, D., & Zuo, J. (2019). Microalgae biodiesel production in China: A preliminary economic analysis. *Renewable and Sustainable Energy Reviews, 104*, 296–306. https://doi.org/10.1016/j.rser.2019.01.021

Szulczyk, K. R., & Tan, Y. (2022). Economic feasibility and sustainability of commercial bioethanol from microalgal biomass: The case of Malaysia. *Energy, 253*, 124151. https://doi.org/10.1016/j.energy.2022.124151

Quinn, J. C., & Davis, R. (2015). The potentials and challenges of algae based biofuels: A review of the techno-economic, life cycle, and resource assessment modeling. *Bioresource Technology, 184*, 444–452. https://doi.org/10.1016/j.biortech.2014.10.075

Tercero, E. A. R., Domenicali, G., & Bertucco, A. (2014). Autotrophic production of biodiesel from microalgae: An updated process and economic analysis. *Energy, 76*, 807e815. https://doi.org/10.1016/j.energy.2014.08.077

Torres-Tiji, Y., Fields, F. J., & May, S. P. (2020). Microalgae as a future food source. *Biotechnology Advances*, 41. https://doi.org/10.1016/j.biotechadv.2020.107536

Vázquez-Romero, B., Antonio, J., Pereira, H., Barbosa, M., & Ruiz, J. (2022). Techno-economic assessment of microalgae production, harvesting and drying for food, feed, cosmetics, and agriculture. *Science of the Total Environment, 837*. https://doi.org/10.1016/j.scitotenv.2022.155742

Wang, W. (2019). Techno-economic analysis for evaluating the potential feedstocks for producing hydro-processed renewable jet fuel in Taiwan. *Energy, 179*, 771–783. https://doi.org/10.1016/j.energy.2019.04.181

Wei, L., Cheung, C. S., & Ning, Z. (2017). Influence of waste cooking oil biodiesel on combustion, unregulated gaseous emissions and particulate emissions of a direct-injection diesel engine. *Energy, 127*, 175–185.

Wells, M. L., Potin, P., Craigie, J. S., Raven, J. A., Merchant, S. S., Helliwell, K. E., Smith, A. G., Camire, M. E., & Brawley, S. H. (2017). Algae as nutritional and functional food sources: Revisiting our understanding. *Journal of Applied Phycology, 29*, 949–982. https://doi.org/10.1007/s10811-016-0974-5

Xiao, C., Fu, Q., Liao, Q., Huang, Y., Xia, A., & Chen, H. (2020). Life cycle and economic assessments of biogas production from microalgae biomass with hydrothermal pretreatment via anaerobic digestion. *Renewable Energy, 151*, 70–78. https://doi.org/10.1016/j.renene.2019.10.145

Xin, C., Addy, M. M., Zhao, J., Cheng, Y., Ma, Y., Liu, S., Mu, D., Liu, Y., Chen, P., & Ruan, R. (2018). Waste-to-biofuel integrated system and its comprehensive techno-economic assessment in wastewater treatment plants. *Bioresource Technology, 250*, 523–531. https://doi.org/10.1016/j.biortech.2017.11.040

Zeller, M. A., Hunt, R., Jones, A., & Sharma, S. (2013). Bioplastics and their thermoplastic blends from Spirulina and Chlorella microalgae. *Journal of Applied Polymer Science*, 1–13. https://doi.org/10.1002/app.39559

Zhang, X., Shi, Y., Chen, Y., Hu, H., Cheng, F., & Li, R. (2022). Ecosystem simulation and environmental impact analysis of transforming microalgae to produce jet fuel. *Journal of Cleaner Production, 333*, 130100. https://doi.org/10.1016/j.jclepro.2021.130100

Rosangela Rodrigues Dias*, Adriane Terezinha Schneider,
Mariane Bittencourt Fagundes

Chapter 3
Challenges and opportunities for microalgae biotechnology development

Abstract: Science shows us that, to ensure a livable future, immediate and profound measures must be taken, and climate promises must stop being a mere litany. Indeed, never before has the need to build a sustainable society been greater than in this century. Sustainability has become imperative and microalgae biotechnology can make an important contribution to the transition to a climate-neutral global economy. This chapter presents microalgae as a next-generation resource capable of meeting the most urgent needs of today's world. We also provide an overview of the current status of microalgae biotechnology. The core of the chapter is a critical discussion of the hottest spots in microalgae biotechnology. In the end, recent developments capable of unlocking the full potential of these microorganisms are discussed.

Keywords: algae, applications, market, unresolved bottlenecks, recent developments

3.1 Introduction

Microalgae are extremely diverse and robust ubiquitous microorganisms. They tolerate a wide spectrum of many factors such as light, temperature, pH, and salinity and can be cultivated in wastewater (Kholssi et al., 2021). These microorganisms have attracted considerable interest in recent years due to their potential to serve as a source of a wide range of products that may be able to meet the growing demands for food and feed, energy, chemicals, and materials as well as pharmaceuticals and cosmetics (Rizwan et al., 2018).

Besides that, it is worth mentioning that the current attention given to microalgae is also due, in part, to concerns related to the depletion of natural resources, climate change, and population growth that is emerging as the main challenges of this century

***Corresponding author: Rosangela Rodrigues Dias**, Bioprocess Intensification Group, Federal University of Santa Maria, UFSM, Roraima Avenue 1,000, 97105-900 Santa Maria, RS, Brazil, e-mail: ro.rosangelard@gmail.com
Adriane Terezinha Schneider, Bioprocess Intensification Group, Federal University of Santa Maria, UFSM, Roraima Avenue 1000, 97105-900 Santa Maria, RS, Brazil
Mariane Bittencourt Fagundes, Interdisciplinary Centre of Marine and Environmental Research, CIIMAR, Portugal

https://doi.org/10.1515/9783110781267-003

(de Mendonça et al., 2021; Calijuri et al., 2022). However, globally, despite the great promise shown by microalgae, commercial achievements so far have been modest. The biggest focus of microalgae biotechnology is the production of unicellular protein and chemical specialties such as pigments and polyunsaturated fatty acids (Leu and Boussiba, 2014; Dias et al., 2020).

Notoriously, for most microalgae applications, the market is still developing. But it is expected that in the near future, through sophisticated upstream and downstream processing technologies, they can overcome the valley of death between basic research and successful innovation (Rumin et al., 2021). Furthermore, it is noted that, from a biotechnological point of view, microalgae are little-studied organisms. This is due to the myriad of species held in collections around the world, where only a few dozen have been investigated and of these, only a handful are in use. Considering, therefore, that microalgae are largely unexplored, they represent a valuable opportunity for the discovery of new compounds (Mobin and Alam, 2017).

In light of this, this chapter addresses the potential of microalgae to satisfy many of the global demands, as well as the state of the art of microalgae biotechnology. It also discusses the challenges that prevent upgrading technologies from the demonstration and pilot phase to the industrial phase and the technological advances that can help unlock the full biotechnological potential of microalgae.

3.2 Microalgae: the green gold of nature?

In the last years, the world population presented significant growth, according to the United States Department of Population Division, it was observed that in the last 60 years the population increased by approximately 68% reaching 7.9 billion people in 2021. The UN also reported that by 2050 there will be a population increase to 9.6 billion (United Nations Statistics Division, 2022). This significant increase simultaneously influences the intensification of agricultural processes, being necessary alternatives to meet the basic needs of humanity.

Microalgae presents solutions to reverse numerous sectors of the environment that have been and will be impacted due to increasing population density, highlighting the agriculture, and industrial sectors, demonstrating the potential to be applied in energy, food and feed, and pharmaceutical industries, and even in the metallurgical sector (Khan et al., 2018).

Conventional industries can make use of the microalgae metabolic mechanism, in all processes, from obtaining energy to treating effluents (Chew et al., 2017). A major challenge to the acceptance of microalgae use in industrial processes is associated with economic issues. With this regard, to overcome these problems, in economic feasibility studies, it is possible to perceive that the best alternative for the future of microalgae biotechnology is to integrate algae biorefinery in all the operating units, to take the production of endless products sustainably and economically, and also to explore this technology together with sustainable energy alternatives, for example, photovoltaic and onshore wind (Dias et al., 2022a).

In Figure. 3.1 presents the major areas of study and biotechnological applications of microalgae in recent years and highlights areas associated with energy, followed by environmental science.

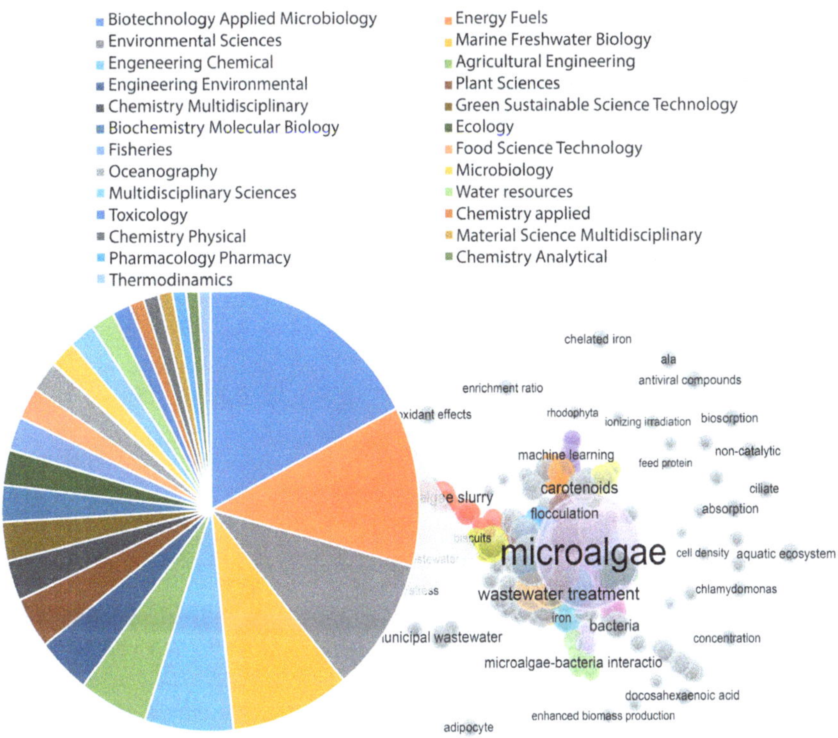

Figure 3.1: Microalgae biotechnology field scientometric analysis.

The bibliometric study allows the understanding of the latest research that has been mostly addressed in the literature. Thus, it is possible to create a networking system and verify the correlations between them. Therefore, as we can see in Figure 3.2, evaluating the last five years, the highest density of studies was focused on energy fuels, especially in 2018.

Studies on microalgae biodiesel were driven by problems related to climate change and due to the use of conventional energy and fossil fuels, which have high CO_2 emissions and are mainly responsible for global warming. According to IPCC, in October 2018, discussions on reducing CO_2 emissions continued, even after the 2015 International Paris Agreement to reduce greenhouse gas (GHG) emissions. The main discussion goals were a 50%

reduction in carbon emissions by 2030, 100% clean energy by 2035, and as one of the biggest global goals, net-zero carbon emission by 2050.

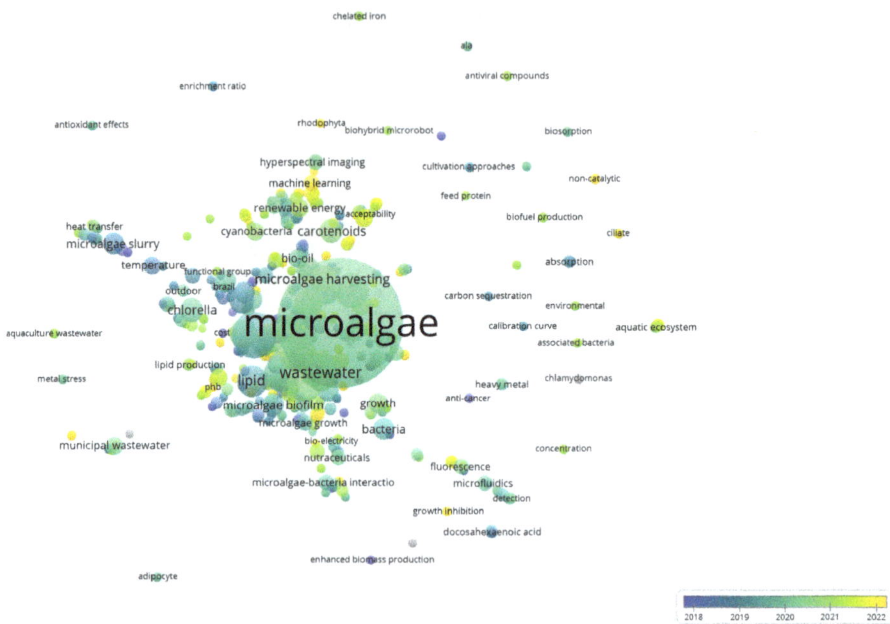

Figure 3.2: Scientometric analysis based on the last five years of study.

To mitigate the problems presented by the use of nonrenewable energy, there was a need to invest in studies on carbon capture, reduction of hydrogen and methane emissions, and the replacement of the use of fossil energy. This fact corroborates the bibliometric analysis which shows a significant increase in biodiesel, economy, and renewable energy studies in 2018, reflecting the current global concerns, and the microalgae's potential to act in these current aspects. From a biodiesel perspective, microalgae can replace petro-diesel. For this purpose, the lipid fractions are transesterified in two steps, being the most favorable profile is the one with the highest cetane number acquired, in addition to viscosity, density, cold filter plugging point, oxidative stability, and others (Chisti, 2007; Venkata Subhash et al., 2022).

Such parameters are acquired through specific microalgae cultures, and each microalga needs to be studied individually for this. Microalgae, compared to other sources of biodiesel, stand out for not depending on amounts of land, like plants, as well as not suffer fluctuations from the impact of the food market. Such characteristics make microalgae biomass a potential source (Slegers et al., 2020). However, the scenario for this replacement to occur is unfortunately far from being self-sufficient, as the microalgae industries for this purpose are economically unfeasible. In this sense, techno-economic feasibility studies have been carried out as a way to understand how to overcome this problem (Jacob et al., 2021). One of the most plausible and discussed alternatives used for this purpose is the use of photovoltaic energy to reduce such economic costs (Rodríguez-Roque et al., 2021; Dias et al., 2022a).

Another area explored is associated with applied microalgae biotechnology. As an example according to the cluster formed with the keywords lipids, in association with value-added compounds and astaxanthin, it is possible to see that the application of the microalgal lipid fraction is not only for studies on biodiesel but also the recovery of value-added metabolites that can be applied in the food industry, such as docosahexaenoic (DHA) and astaxanthin which are microalgae metabolites already widely produced (Figure 3.3).

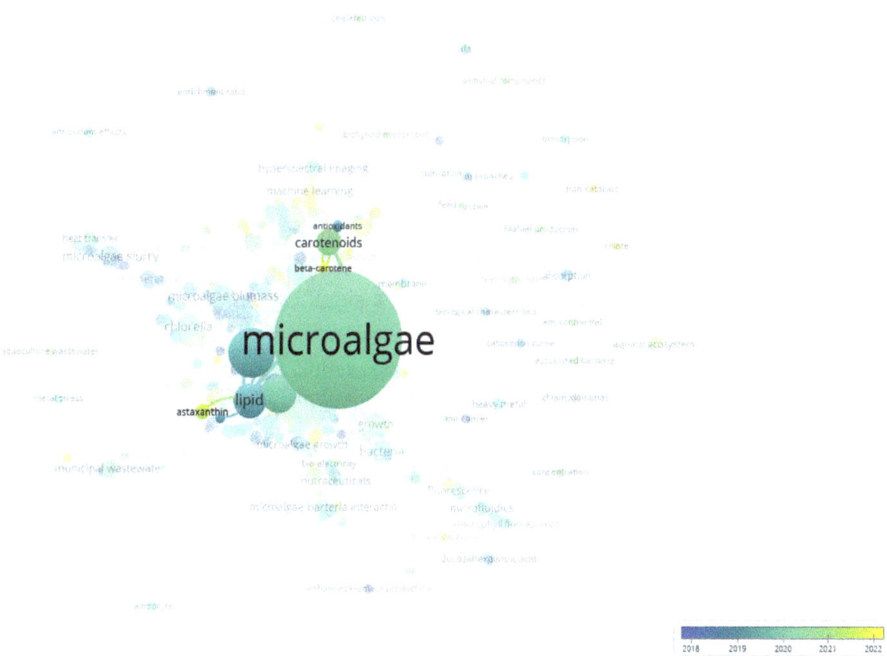

Figure 3.3: Clusters associated with recovery of high-value bioactive molecules formed in the last 5 years.

The same behavior can be seen about carotenoids, and also proteins, years ago, only studies associated with these compounds were observed. Today through the number of bioactive compounds being discovered in cyanobacteria and microalgae, among the pigments, phycobiliproteins, phycoferrins, and also chlorophylls stand out. Similarly, compounds such as terpenoids, sesquiterpenes, diterpenes, and also another class that has been studied, sterols, have already demonstrated the ability to be used for pharmacological development purposes.

From the lipidomic perspective, studies with sphingolipids and microalgae as a source of vitamins also stand out. This area is on the rise due to high demand from the pharmaceutical industry, especially after the outbreak of COVID-19 in 2019. This sector is one of the most promising being the search for metabolites the main step toward new drug discovery.

The importance of using natural substances from sustainable sources in the market is also related to their bioaccessibility, which includes bioavailability and bioactivity studies of natural extracts. In a comparative study between synthetic astaxanthin and the natural one isolated from microalgae, by evaluating the potential for scavenging free radicals, it was possible to see that microalgae extract is 20 times greater than synthetic (Capelli et al., 2013). This result corroborates the studies carried out in human cells using the *Haematococcus pluvialis* strain, being responsible for demonstrating a potential for protecting human vascular cells against oxidative stress (Régnier et al., 2015).

Recently, Fernandes et al. (2021) verified microalgae *Scenedesmus bijuga* and *Chlorella sorokiniana* bioaccessibility. The researchers observed correlations between the lipid compositions of each microalgae with their bioavailability (Fernandes et al., 2021). Among the numerous carotenoids mentioned in the study, it was found that the cis configuration was responsible for greater bioaccessibility.

Agro-industrial residues have been explored as another major line of research in microalgal biotechnology, as we can see when verifying the results from the scientometric analysis. According to statistical surveys, a significant increase in food loss of approximately 2.5 billion tons occurred in recent years, and this factor is directly correlated with GHG emissions (Srivastava and Bhaskar, 2022). To overcome these issues microalgae has been studied as an economical alternative compared to conventional treatments. Traditionally, heterotrophic culture was associated with tertiary treatments in industries (Mohsenpour et al., 2021). Currently, metabolic variations observed by using heterotrophic metabolism led to studies associated with different metabolites elucidations and the discovery of different metabolic paths for the formation of intermediates that can further be purified and verified according to their bioavailability and bioactivity (Rizwan et al., 2018).

For both cultivation modes microalgal, photoautotrophic, and heterotrophic (and their mixotrophic and photoheterotrophic variations), there is little elucidation of the metabolites produced, depending on the different cultivation conditions. With these variations, different classes of biomolecules can be produced, which have different applicability, among them associated with the protection of the skin against damage caused by the sun, as found in *Chlorella* extract (Dhandayuthapani et al., 2020). On the other hand, other advanced studies demonstrate the use of microalgal biomolecules for the production of vaccines against several viruses, such as HIV, Hepatitis B, Zika virus, and H1N1, among others, as observed in *Dunaliella salina, Chlamydomonas reinhardtii,* and *Schizochytrium* sp. (Ramos-Vega et al., 2021). Food science is another area that has been exploiting the use of microalgae; in the food industry the use of microalgae also stands out, as some strains have already been explored in food supplements, as well as used in conventional food formulations with the application of the 3D technology (Donn et al., 2022). However, harvesting and the scale-up for many strains remain difficult since each one needs to be evaluated in terms of downstream and upstream processes.

3.3 Current status of microalgae biotechnology

Currently, there are about 72,500 species of microalgae consistently cataloged (Jacob-Lopes et al., 2019). However, only a handful of them are actually used for commercial purposes. Furthermore, it is worth mentioning that few microalgae-based products have managed to overcome the valley of death and reach commercial exploitation (Mobin and Alam, 2017). Today, the most valuable commercial microalgae products are only the unicellular protein, the pigments β-carotene, astaxanthin, and phycocyanin, and the polyunsaturated fatty acids eicosapentaenoic, and DHA, with applications targeted for use as a supplement, dye, and food additive (Jacob-Lopes et al., 2019).

These products are of high-added value and as such outweigh the high cost of upstream and downstream processing. They are also, to this day, the main focus of the global microalgae market which, while modest, is solidly anchored due to its leaning slope toward health and wellness trends (Chen et al., 2022). In 2020, the global market for microalgae products was valued at US$977.3 million and is expected to reach approximately US$1,485.1 million by 2028, with a compound annual growth rate of 5.4%. Unfortunately, low awareness of the benefits of microalgae-based products and their complex production process can make it difficult, to a certain extent, for the growth of this market (Allied Market Research, 2021).

However, regardless of this, microalgae are metabolically versatile microorganisms that, in addition to containing a higher proportion of valuable chemical compounds when compared to conventional sources, can be the mainstream for solutions capable of meeting the most urgent needs of the agricultural and industrial sectors (Show, 2022; Castro and Cobos, 2022). The use of microalgae as biofactories for industrial use has, in fact, increased in recent years and promising results in this area depend on precise choices regarding the microalgae species and the growth conditions used (Fernandes and Cordeiro, 2021).

It is worth highlighting that the most significant challenges to making any process and product based on microalgae technically and economically viable are related to three main aspects, which are (i) improving the microalgae productivity, (ii) reducing the energy footprint of the upstream and downstream processes, especially for harvesting, drying, extraction, and purification, and (iii) making the most advantage of microalgae biomass in the context of a multiproduct biorefinery. The first and second challenges include a robust selection of microalgae strains with higher yields, low-cost cultivation media, and optimization throughout the entire process, especially in the main stages (Dias et al., 2021). About the third challenge, it includes the full recovery of all biomass fractions, and the improvement of downstream processes to enable the reduction of quality and yield losses and the improvement in the recovery of bioproducts (Severo et al., 2021).

Finally, it is worth mentioning that there are numerous microalgae-based products in different stages of development. Recombinant proteins, phycoerythrin, fucoxanthin, lutein, beta-glucans, and exopolysaccharides are some of the products in advanced development with great chances of consolidating their participation in the microalgae market shortly. In addition, there are also a large number of products in the early stages of development, including biofuels such as biodiesel, bioethanol, and biomethane, as well as polyhydroxyalkanoates, violaxanthin, zeaxanthin, prebiotics, and sterols (Dias et al., 2020; Rahpeyma and Raheb, 2019). However, although this is true, many technological nodes need to be untied before they can reach high levels of technological readiness. The unknowns about microalgae-based processes and products that need to be overcome also increase the investments in this field.

3.4 New or old challenges in microalgae biotechnology?

Microalgae can be an important part of the solution to the current problems that affect the various ecosystems on the planet, that is, the overcontamination of soil, water, and air. Besides that, they are a promising feedstock for products and solutions for nutrition (human and animal), aquaculture, and cosmetics. The immense application potential of these microorganisms for the pharmaceutical, biomaterial, and biofuel sectors has also been studied (Van der Voort et al., 2015). In relation to biofuels, different types of fuels can be produced from microalgae biomass, such as biodiesel, bioethanol, biobutanol, biohydrogen, biomethane, synthesis gas, and bio-oil. These microalgae-based biofuels could play a central role in the race for renewable energy (Deprá et al., 2018).

However, while all this is true, taking a look at the scientific literature, it is possible to assure that the focus of microalgae applications is on the production of high-added-value products (Levasseur et al., 2020). And the truth is that the entire microalgae biotechnology industry seems to be surviving on these products. This is because they counterbalance their high upstream and downstream processing costs (Severo et al., 2021).

Today, the biggest challenge for successful microalgae applications is related to the production cost, which is still economically and energetically expensive, especially for low-value products, such as biofuels (Fu et al., 2021). In the literature, although commercial microalgae plants do not disclose their production costs, which prevents a more comprehensive view of the situation, the consensus is that there is a long way to be paved for the establishment of a robust global microalgae bioeconomy. In fact, there are a series of challenges to be faced in order to deal with the high expectations created in the microalgae biotechnological field (de Mendonça et al., 2021).

The low productivity of cultivation and the high operating cost of facilities associated in part with their high energy demand are unanimously the most significant challenges (Dias et al., 2021). Recognizing the existence of these challenges is the first step toward solving them. However, the possibilities of improving the economic and environmental sustainability of microalgae-based processes and products have been discussed in detail by many researchers. Among the possibilities are the robust screening and selection of microalgae strains, modulation of growth conditions, low-cost culture media such as industrial waste, and integrated techniques for the main steps of the process. Other suggestions include biochemical engineering, genetic engineering, and transcription factor engineering strategies (Dias et al., 2022b).

Still about it, many of the bottlenecks present in microalgae-based processes, such as low culture productivity, have been known since the beginning of large-scale commercial production. In this sense, some decades later, it is controversial that even today they remain unresolved, showing that the proposed solutions are nothing more than a mere litany that in practice does not apply. However, it is clear that science over the years can alleviate the burden of already commercially consolidated processes. But, in theory, they remain plastered and far from an ideal standard process that keeps them competitive with other biological and synthetic feedstock.

In this narrative, since the main bottlenecks that affect the use of microalgae in different large-scale applications are already known, what is expected is that new solutions to old problems can unlock the full potential of microalgae biotechnology (Fernández et al., 2021). Today, a whole new range of technologies – including internet of things (IoT) automation technologies – is being applied to leverage microalgae-based technology and expresses hope for huge implications for its future competitiveness.

3.5 Advanced technologies: unlocking the biotechnological potential of microalgae

Advances in the area of microalgae cultivation are on the rise, as cultivation modifications imply increases in the levels of high-value compounds and intensification in the gene expression of enzymes associated with these metabolites, being the main aspect associated with the discovery of new biomolecules (Chandrasekhar et al., 2022).

The photosynthetic capacity of microalgae is modified according to the applied cultivation technology. As an example, in a study carried out by Tredici et al. (2015), it was possible to acquire up to 36 tons of dry biomass per hectare through the use of a Panel-II photobioreactor, developed by the researchers. The cultivation modes used in microalgal systems traditionally are subdivided into two: open systems and closed systems. Open systems are known as round ponds, and racetrack-type ponds, as a disadvantage, these systems have a high difficulty in controlling external variables, which makes them offer low productivity of microalgae biomass. However, many companies still today use these systems (Tan et al., 2020).

Closed systems, in turn, can overcome some of the bottlenecks presented by open systems. They have the advantage of a great mixing system, better energy transfer, minor contamination, and greater productivity. However, among its restrictions, problems of limiting light exposure in the system stand out (Sirohi et al., 2022). Among the closed systems, the plate photobioreactors provide a larger surface area, being a more stable system with better performance to be used on an industrial scale. Closed photobioreactor models can be presented in plates and tubes. The horizontal tubes have great productivity and stand out for being relatively cheap, but can present serious problems with the dissolved oxygen in the medium. However, when used with a gas sprinkler, it acts as a great gas-liquid transfer system, due to the higher transfer rate created by using it (Behera et al., 2022).

Microalgae harvesting is one point key to obtaining high productivity, being the most used methods: chemical methods (flocculation through inorganic and organic agents) and also physical (centrifugation, gravity sedimentation, filtration) and biological (bio flocculation with changes caused by pH, bacteria, and fungus).

Another technology applied to microalgae system currently to enhance the comprehension of pathway discovery, productivity, and elucidation of secondary metabolites is related with synthetic application, which has interpreted new regulatory systems from computational models. This is one of the aspects of metabolic engineering used as a way to study the metabolic pathways and gene relationships of these strains. Within the context of synthetic biology, techniques such as synthetic scaffold, CRISPR, CRISPRI, TALEN, ZFN, RNA synthetic, interference, and antisense are widely used for the genetic design (Grama et al., 2022). Consequently, in terms of genetic engineering, inside the field of biological systems, the latest techniques that have been applied are studies of metabolomics, transcriptomics, cytomics, and proteomics, which serve to assess the biological functions, predict, and optimize the discovery of new metabolites in these strains (Muthukrishnan, 2022).

In addition, today has been a huge discussion about the use of artificial intelligence in biorefineries process optimization, which includes the use of sensors applied in the process known as automation of the IoT (Lim et al., 2022). The IoT system acts through the use of hardware, aiming to explore the use of artificial intelligence as one of the factors to optimize culture systems and keep it stable to increase microalgal productivity throughout the entire microalgae biorefinery. The IoT automation system was associated with the term machine learning as expressed in Figure 3.4.

So far, the major advantages of these new cultivations systems are real-time monitoring of microalgae biorefinery process parameters, accompanying a low cost-biorefinery, with great systems predictions, and high efficiency. Besides the observed group, the term was mostly used only in 2022, which represents that today the research in machine learning on biorefineries started, but their actual implementation it is on its infancy yet.

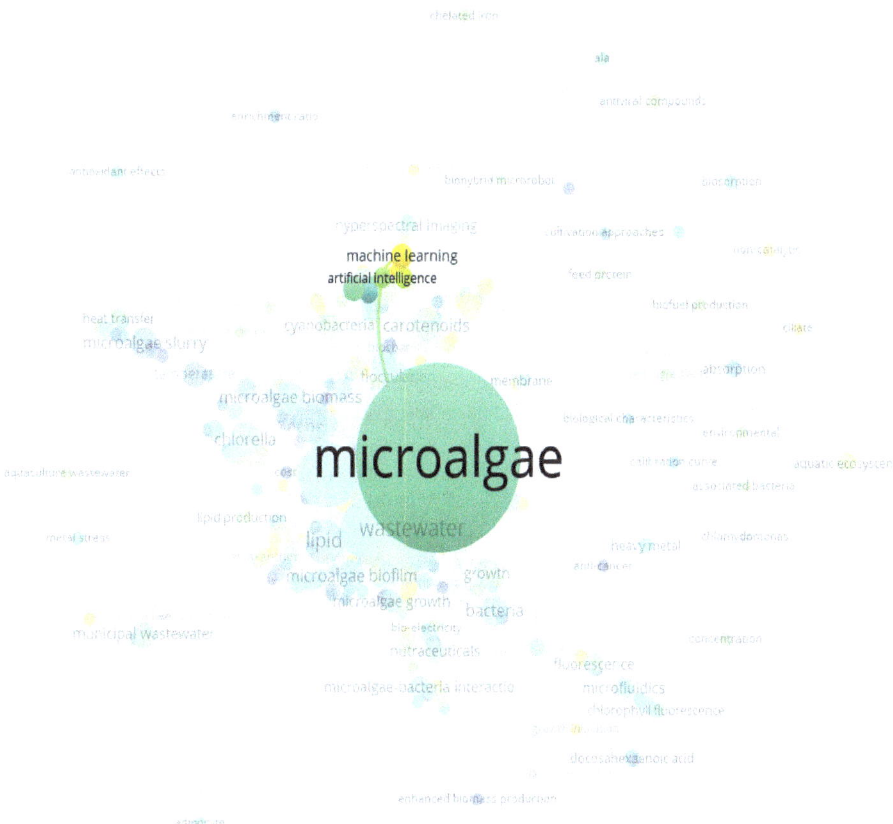

Figure 3.4: Cluster associated with artificial intelligence, machine learning, and internet of things.

3.6 Conclusion

The broad and near-universal view of the scientific community is that there are many technological knots that need to be untied before the use of microalgae in emerging applications becomes viable. Now, in the short term, to overcome the challenges it is necessary to truly understand how new technologies can, in practice, solve immediate problems without ignoring the risks to environmental and economic sustainability.

References

Allied Market Research. (2021). Microalgae Market by Type (*Spirulina*, *Chlorella*, *Dunaliella Salina*, and *Aphanizomenon Flos-Aquae)*, Application (Dietary Supplements, Food/Feed, Pharmaceutical, Cosmetic, and Others): Global Opportunity Analysis and Industry Forecast 2021–2028. Available in <https://www.alliedmarketresearch.com/microalgae-market-A13419>. Access in: 25/08/2022.

Behera, B. C., Dey, H., Jalan, R., Mishra, R. R., & Mohapatra, S. (2022). Chapter 1 – Photobioreactors for production of biodiesel from algae: A short review. (Thatoi, H., Mohapatra, S., S. K. B. T.-I. in F., & Das, P. T. Eds.).

Calijuri, M. L., Silva, T. A., Magalhães, I. B., De Paula Pereira, A. S. A., Marangon, B. B., De Assis, L. R., & Lorentz, J. F. (2022). Bioproducts from microalgae biomass: Technology, sustainability, challenges and opportunities. *Chemosphere*, 135508.

Capelli, B., Bagchi, D., & Cysewski, G. R. (2013). Synthetic astaxanthin is significantly inferior to algal-based astaxanthin as an antioxidant and may not be suitable as a human nutraceutical supplement. *Nutrafoods*, *12*(4), 145–152.

Castro, J. C., & Cobos, M. (2022). Biochemical Profiling, Transcriptomic Analysis, and Biotechnological Potential of Native Microalgae from the Peruvian Amazon. In Algal Biotechnology. Elsevier, pp. 305–321.

Chandrasekhar, K., Raj, T., Ramanaiah, S. V., Kumar, G., Banu, J. R., Varjani, S., . . . Kim, S.-H. (2022). Algae biorefinery: A promising approach to promote microalgae industry and waste utilization. *Journal of Biotechnology*, *345*, 1–16.

Chen, C., Tang, T., Shi, Q., Zhou, Z., & Fan, J. (2022). The potential and challenge of microalgae as promising fuical knots needture food sources. *Trends in Food Science & Technology*.

Chew, K. W., Yap, J. Y., Show, P. L., Suan, N. H., Juan, J. C., Ling, T. C., . . . Chang, J.-S. (2017). Microalgae biorefinery: High-value products perspectives. *Bioresource Technology*, *229*, 53–62.

Chisti, Y. (2007). Biodiesel from microalgae. *Biotechnology Advances*, *25*(3), 294–306.

de Mendonça, H. V., Assemany, P., Abreu, M., Couto, E., Maciel, A. M., Duarte, R. L., . . . Reis, A. (2021). Microalgae in a global world: New solutions for old problems? *Renewable Energy*, *165*, 842–862.

Deprá, M. C., Dos Santos, A. M., Severo, I. A., Santos, A. B., Zepka, L. Q., & Jacob-Lopes, E. (2018). Microalgal biorefineries for bioenergy production: Can we move from concept to industrial reality? *BioEnergy Research*, *11*(4), 727–747.

Dhandayuthapani, K., Malathy, S., Mulla, S., & Gupta, S. (2020). An Insight into the potential application of microalgae in pharmaceutical and nutraceutical production.

Dias, R. R., Deprá, M. C., Zepka, L. Q., & Jacob-Lopes, E. (2022a). *In silico* study of hybrid renewable energy in microalgae facilities: A path towards net-zero emissions. *Algal Research*, *63*, 102661.

Dias, R. R., Maroneze, M. M., Oliveira, Á. S. D., Caetano, P. A., Zepka, L. Q., & Jacob-Lopes, E. (2021). Bioconversion of Industrial Wastes into Biodiesel Feedstocks. In Sustainable Bioconversion of Waste to Value Added Products. Springer, Cham, pp. 109–120.

Dias, R. R., Sartori, R. B., Deprá, M. C., Lasta, P., & Maroneze, M. M. (2022b). Biochemical Engineering Approaches to Enhance the Production of Microalgae-based Fuels. In 3rd Generation Biofuels. Woodhead Publishing., pp. 65–90.

Dias, R. R., Severo, I. A., Deprá, M. C., Maroneze, M. M., Zepka, L. Q., & Jacob-Lopes, E. (2020). The Next-generation of Microalgae-based Products. In Microbial Enzymes and Biotechniques. Springer, Singapore, pp. 15–42.

Donn, P., Prieto, M. A., Mejuto, J. C., Cao, H., & Simal-Gandara, J. (2022). Functional foods based on the recovery of bioactive ingredients from food and algae by-products by emerging extraction technologies and 3D printing. *Food Bioscience*, *49*, 101853.

Fernandes, A. S., Nascimento, T. C., Pinheiro, P. N., Vendruscolo, R. G., Wagner, R., De Rosso, V. V., . . . Zepka, L. Q. (2021). Bioaccessibility of microalgae-based carotenoids and their association with the lipid matrix. *Food Research International, 148*, 110596.

Fernandes, T., & Cordeiro, N. (2021). Microalgae as sustainable biofactories to produce high-value lipids: Biodiversity, exploitation, and biotechnological applications. *Marine Drugs, 19*(10), 573.

Fernández, F. G. A., Reis, A., Wijffels, R. H., Barbosa, M., Verdelho, V., & Llamas, B. (2021). The role of microalgae in the bioeconomy. *New Biotechnology, 61*, 99–107.

Fu, Y., Chen, T., Chen, S. H. Y., Liu, B., Sun, P., Sun, H., & Chen, F. (2021). The potentials and challenges of using microalgae as an ingredient to produce meat analogues. *Trends in Food Science & Technology, 112*, 188–200.

Grama, S. B., Liu, Z., & Li, J. (2022). Emerging trends in genetic engineering of microalgae for commercial applications. *Marine Drugs*, Vol *20*.

Jacob, A., Ashok, B., Alagumalai, A., Chyuan, O. H., & Le, P. T. K. (2021). Critical review on third generation microalgae biodiesel production and its feasibility as future bioenergy for IC engine applications. *Energy Conversion and Management, 228*, 113655.

Jacob-Lopes, E., Maroneze, M. M., Deprá, M. C., Sartori, R. B., Dias, R. R., & Zepka, L. Q. (2019). Bioactive food compounds from microalgae: An innovative framework on industrial biorefineries. *Current Opinion in Food Science, 25*, 1–7.

Khan, M. I., Shin, J. H., & Kim, J. D. (2018). The promising future of microalgae: Current status, challenges, and optimization of a sustainable and renewable industry for biofuels, feed, and other products. *Microbial Cell Factories, 17*(1), 36.

Kholssi, R., Ramos, P. V., Marks, E. A., Montero, O., & Rad, C. (2021). Biotechnological uses of microalgae: A review on the state of the art and challenges for the circular economy. *Biocatalysis and Agricultural Biotechnology, 36*, 102114.

Leu, S., & Boussiba, S. (2014). Advances in the production of high-value products by microalgae. *Industrial Biotechnology, 10*(3), 169–183.

Levasseur, W., Perré, P., & Pozzobon, V. (2020). A review of high value-added molecules production by microalgae in light of the classification. *Biotechnology Advances, 41*, 107545.

Lim, H. R., Khoo, K. S., Chia, W. Y., Chew, K. W., Ho, S.-H., & Show, P. L. (2022). Smart microalgae farming with internet-of-things for sustainable agriculture. *Biotechnology Advances, 57*, 107931.

Mobin, S., & Alam, F. (2017). Some promising microalgal species for commercial applications: A review. *Energy Procedia, 110*, 510–517.

Mohsenpour, S. F., Hennige, S., Willoughby, N., Adeloye, A., & Gutierrez, T. (2021). Integrating micro-algae into wastewater treatment: A review. *Science of the Total Environment, 752*, 142168.

Muthukrishnan, L. (2022). Bio-engineering of microalgae: Challenges and future prospects toward industrial and environmental applications. *Journal of Basic Microbiology, 62*(3–4), 310–329.

Rahpeyma, S. S., & Raheb, J. (2019). Microalgae biodiesel as a valuable alternative to fossil fuels. *BioEnergy Research, 12*(4), 958–965.

Ramos-Vega, A., Angulo, C., Bañuelos-Hernández, B., & Monreal-Escalante, E. (2021). Microalgae-made vaccines against infectious diseases. *Algal Research, 58*, 102408.

Régnier, P., Bastias, J., Rodriguez-Ruiz, V., Caballero-Casero, N., Caballo, C., Sicilia, D., . . . Pavon-Djavid, G. (2015). Astaxanthin from *Haematococcus pluvialis* prevents oxidative stress on human endothelial cells without toxicity. *Marine Drugs, 13*(5), 2857–2874.

Rizwan, M., Mujtaba, G., Memon, S. A., Lee, K., & Rashid, N. (2018). Exploring the potential of microalgae for new biotechnology applications and beyond: A review. *Renewable and Sustainable, Energy Reviews, 92*, 394–404.

Rodríguez-Roque, M. J., Sánchez-Vega, R., Aguiló-Aguayo, I., Medina-Antillón, A. E., Soto-Caballero, M. C., Salas-Salazar, N. A., & Valdivia-Nájar, C. G. (2021). Chapter 12 – Bioaccessibility and Bioavailability of Bioactive Compounds Delivered from Microalgae. (Lafarga, T., & G. B. T.-C. M. for the Acién, F. I. Eds.).

Rumin, J., Gonçalves de Oliveira Junior, R., Bérard, J. B., & Picot, L. (2021). Improving microalgae research and marketing in the European Atlantic area: analysis of major gaps and barriers limiting sector development. *Marine Drugs*, *19*(6), 319.

Severo, I. A., Dias, R. R., Sartori, R. B., Maroneze, M. M., Zepka, L. Q., & Jacob-Lopes, E. (2021). Technological Bottlenecks in Establishing Microalgal Biorefineries. In Microalgal Biotechnology. pp. 118–134.

Show, P. L. (2022). Global market and economic analysis of microalgae technology: Status and perspectives. *Bioresource Technology*, 127329.

Sirohi, R., Kumar Pandey, A., Ranganathan, P., Singh, S., Udayan, A., Kumar Awasthi, M., . . . Sim, S. J. (2022). Design and applications of photobioreactors- a review. *Bioresource Technology*, *349*, 126858.

Slegers, P. M., Olivieri, G., Breitmayer, E., Sijtsma, L., Eppink, M. H. M., Wijffels, R. H., & Reith, J. H. (2020). Design of value chains for microalgal biorefinery at industrial scale: Process integration and techno-economic analysis. *Frontiers in Bioengineering and Biotechnology*, Vol *8*.

Srivastava, B., & Bhaskar, R. (2022). Transformation of India towards net zero targets: Challenges and opportunities: Transformation of India towards net zero. *Nimitmai Review Journal*, *5*(1), 1–8.

Tan, J. S., Lee, S. Y., Chew, K. W., Lam, M. K., Lim, J. W., Ho, S. H., & Show, P. L. (2020). A review on microalgae cultivation and harvesting, and their biomass extraction processing using ionic liquids. *BioEngineered*, *11*(1), 116–129.

Tredici, M. R., Bassi, N., Prussi, M., Biondi, N., Rodolfi, L., Chini Zittelli, G., & Sampietro, G. (2015). Energy balance of algal biomass production in a 1-ha “Green Wall Panel” plant: How to produce algal biomass in a closed reactor achieving a high Net Energy Ratio. *Applied Energy*, *154*, 1103–1111.

United Nations Statistics Division. (2022). Demographic and social Statistics. Available in <https://unstats.un.org/unsd/demographic-social/census/#methods>. Accessed in: 15/07/2022.

Van der Voort, M. P. J., Vulsteke, E., & De Visser, C. L. M. (2015). Macro-economics of algae products: Output WP2A7. 02. EnAlgae Swansea University.

Venkata Subhash, G., Rajvanshi, M., Raja Krishna Kumar, G., Shankar Sagaram, U., Prasad, V., Govindachary, S., & Dasgupta, S. (2022). Challenges in microalgal biofuel production: A perspective on techno economic feasibility under biorefinery stratagem. *BioResource Technology*, *343*, 126155.

Emmanuel Manirafasha*, Theoneste Ndikubwimana, Hanqing Fu, Mao Lin, Liangliang Zhang, Keju Jing*

Chapter 4 Major bottlenecks in industrial microalgae-based facilities

Abstract: Microalgae are natural, renewable, and sustainable micro-factories that produce various products with environmentally friendly applications in multiple industries, including food and nutraceuticals, cosmetics, energy, pharmaceuticals, biotechnology, agriculture, and medicines. Exploiting algal resources is a promising approach to solving different pressing world challenges, such as climate change, water scarcity, and food crises. Even though microalgae represent different potentials, associated bottlenecks hinder their efficiency and applicability. The major bottlenecks are classified into four categories: (i) inappropriate upstream and downstream technologies, (ii) financial funding and investments, (iii) cost-effectiveness and production life cycle assessment, and (iv) cultural misunderstandings in business models. The pivot purpose of this chapter is to foreground the main bottlenecks that impede the sustainability of industrial microalgae. This chapter also summarizes the industrial mapping approach that has overcome and/or reduced major bottlenecks in industrial microalgae for its salient and prominent achievements and future evolution. The collaboration among various stakeholders (in particular, researchers, investors, industrialists, and policymakers) is the key to the sustainability and competitiveness of the microalgae industry.

Financial support: The authors thank the National Natural Science Foundation of China (No. 21776232 and No. 21978244).

***Corresponding author: Emmanuel Manirafasha,** Department of Chemical and Biochemical Engineering, College of Chemistry and Chemical Engineering, and The Key Lab for Synthetic Biotechnology of Xiamen City, Xiamen University, Xiamen 361005, China; Alpha Natural Resources Company Limited (ANARECO Ltd.), Kigali, Rwanda; Xiamen Canco Biotech Co., Ltd., Xiamen 361000, China, e-mail: meonb2003@gmail.com
***Corresponding author: Keju Jing,** Department of Chemical and Biochemical Engineering, College of Chemistry and Chemical Engineering, and The Key Lab for Synthetic Biotechnology of Xiamen City, Xiamen University, Xiamen 361005, China, e-mail: jkj@xmu.edu.cn
Theoneste Ndikubwimana, Head of Department: General Higher Education Quality Standards Department, Higher Education Council (HEC), PO BOX 6311, Kigali, Rwanda, e-mails: ntheo05@gmail.com, tndikubwimana@hec.gov.rw
Hanqing Fu, Xiamen Canco Biotech Co., Ltd., Xiamen 361000, China, e-mail: 1512734847@qq.com
Mao Lin, Xiamen Canco Biotech Co., Ltd., Xiamen 361000, China; Fisheries College, Jimei University, Xiamen 361021, China, e-mail: linmao@jmu.edu.cn
Liangliang Zhang, Academy of Advanced Carbon Conversion Technology, Huaqiao University, Xiamen 361021, China, e-mail: zhangll@hqu.edu.cn

https://doi.org/10.1515/9783110781267-004

Keywords: bottlenecks, industrial mapping systems, microalgae industry, microfactories, natural resources

4.1 Introduction

As natural and renewable resources for a broad range of substances and bioactive compounds, microalgae possess the input potential to achieve a green economy and sustainable development (Manirafasha et al., 2019). Microalgae have most of the macro- and micronutrients required for human health and most of Earth's natural, ubiquitous, and renewable resources. The world is facing two major pressing challenges (i.e., climate change, food security, and nutrition), as well as their derived problems, such as extreme poverty and chronic diseases. Microalgae are promising and potential resources to mitigate those challenges due to their high-quality nutrients and other bioactive compounds. Furthermore, they can be produced without competing with conventional food production, as they do not require arable lands. They have other advantages, such as surviving harsh conditions, thus leading to the production of various metabolites in large amounts over short periods and all year round (Kaushik et al., 2022; Manirafasha et al., 2016). Moreover, microalgae biomass production also possesses advantages over plant biomass production in terms of higher yield, faster growth, and recovery of nutrients from wastewater (Benedetti et al., 2018).

Large-scale microalgae production consists of three central unit operations: upstream, midstream, and downstream. Each unit has various elements to be considered for optimized large-scale microalgae yield as depictured in Figure 4.1.

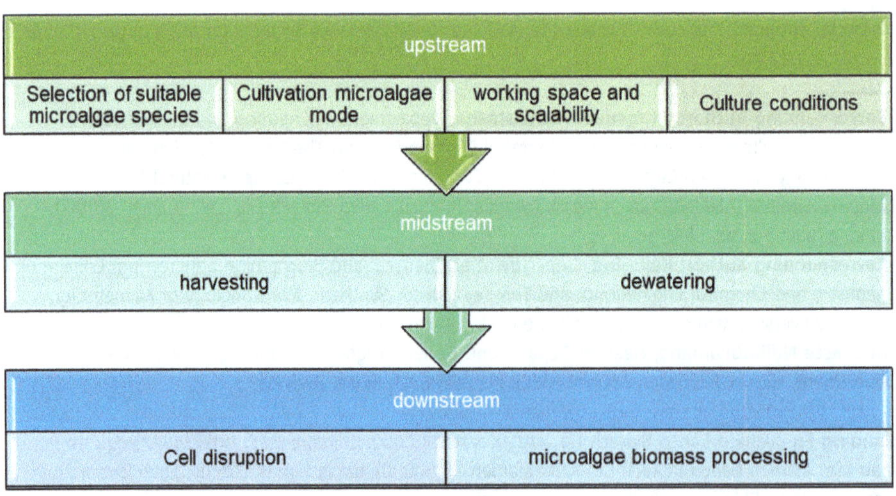

Figure 4.1: Industrial microalgae processing central unit operations.

Generally, those elements, which are behind the success of industrial microalgae production, can be classified into four main categories: microalgae seeds, working environment, working conditions, and advanced processing technologies. Many of those elements are found in the upstream unit: different species of microalgae; water source: freshwater, seawater, hypersaline water; cultivation modes: autotrophic, heterotrophic, and mixotrophic; working environment: closed systems (mostly known as photobioreactors (PBRs), including tubular, flat-panel PBRs, fermenters), open systems (open ponds, cascade raceways (CRW)). The Midstream unit combines two main elements: harvesting and dewatering. Some scholars consider midstream and downstream units as a single central unit operation, but separating them into two main unit operations is of much benefits. That slit is based on facts that biomass processing is industrially separated from harvesting and dewatering, where some companies deal with biomass production and sell harvested biomass to other companies for biomass processing. The downstream unit contains two critical elements: cell disruption and biomass processing into valued products. Consortia processing technologies transform simple aquatic microorganisms into several valued products and services to find solutions to a wide array of challenges.

Even though microalgae are natural, renewable, and ubiquitous resources with all those advantages, some drawbacks (after this mentioned as bottlenecks) still impede their full exploitation and integration into innovative industrial feedstock and other applications. This chapter aims to highlight the bottlenecks in industrial microalgae. This chapter categorizes those bottlenecks into four classes: inappropriate upstream and downstream technologies (e.g., contamination and dewatering), financial funding and investments, cost-effectiveness and production life cycle assessment, and cultural misunderstandings in business models. This chapter also summarizes the industrial mapping approach that has overcome and/or reduced major bottlenecks in industrial microalgae for its salient and prominent achievements and future evolution. The mapping approach strategically brings selected methods/elements from each major unit operation in the whole processing system (from upstream to application) on the synchronizing line of Life Cycle Life Assessments (LCA). The collaboration among various stakeholders (in particular, researchers, investors, industrialists, and policymakers) is the key to the sustainability and competitiveness of the microalgae industry.

4.2 The major bottlenecks in industrial microalgae and industrial mapping systems for tackling those bottlenecks

Science and Engineering are there for searching solutions to real problems that appear in daily life. Under the natural resources and technology deployment umbrella, collaboration among stakeholders, including academicians, researchers, engineers, investors, policy makers, and regulatory agencies, can contribute to sustainable solutions to current global pressing challenges. Microalgae are one of the natural resources that exhibit different potentials (Benedetti et al., 2018). Their production is environmentally sustainable by utilizing the sun's energy to capture carbon dioxide from the atmosphere and release oxygen. That production is all year round on nonarable land, which makes microalgae, the promising renewable resources with no competition on food production and disruption of natural habitats (Khan et al., 2018; Manirafasha et al., 2016). There is no mistake if someone spotlights those microalgae as driving innovation, referring to the microalgae being advantageous.

However, perfection is not for this world, and everything has advantages and disadvantages; in that line, this section describes the bottlenecks associated with industrial microalgae-based facilities and proposed solutions (Table 4.1). There are several bottlenecks in industrial microalgae, but this section mainly focuses on upstream (specifically, contamination, low productivity, working space, and scalability), midstream-harvesting and dewatering (microalgae sizes and low biomass concentration), and downstream and application (cell disruption and biomass processing). The overall strategy to overcome those bottlenecks is to consider industrial microalgae-based processing and application under one LCA (Figure 4.2); even when upstream and downstream processes and applications could occur in different companies or different workplaces. Paying attention to LCA is imperative because the unit operations in industrial microalgae processing are somehow concomitant. For example, the application of microalgal biomass and derived bioactive molecules have much influence on the selection of microalgae species to be applied to industrial microalgae production, as different microalgae species produce different macro- and micromolecules that refer to the bioactive molecules. Microalgae species that have to be involved in biofuels and bioenergy are preferred to be rich in lipids and fat.

In contrast, microalgae species that must be applied in food and nutrition should be rich in macro- and micronutrients, such as protein, vitamins, minerals, and other essential nutrients. Bringing choice of all elements of the industrial microalgae processing on one synchronizing line is part of the circular economy for transforming so-called waste and by-products from one application to another into valued products. At the same time, the synchronizing line of LCA coupled with techno-economic analysis can be considered industrial mapping systems that contribute to the depreciation of bottlenecks in case each unit operation is taken as a separate unit during industrial microalgae processing. For example, microalgae biomass to be applied to the food and feed industry should not be processed as biomass for material manufacturing. Microalgae biomass from wastes and wastewater treatment through phytoremediation that may contain some heavy metals and antibiotics waste may be applied in material manufacturing, such as shoes. Still, it cannot be applied in the food and feed industry.

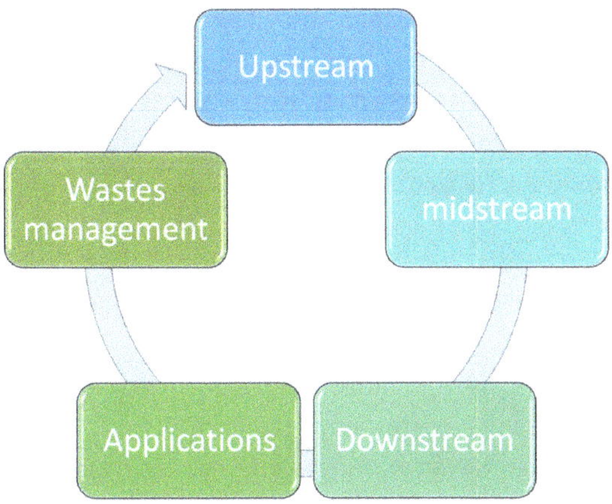

Figure 4.2: Synchronizing line of Life Cycle Assessment in industrial microalgae processing.

4.2.1 Bottlenecks in upstream processing of microalgae

Upstream processing is a baseline with many factors to be considered in microalgae upstream processing for optimum yield production (Daneshvar et al., 2021). The microalgae upstream processing consists of several operation subunits (elements) with trick aspects influencing microalgae cultivation for cell growth and biomass production. Researchers and engineers present bottlenecks as challenges and technology gaps in many research studies (Gifuni et al., 2019; Sivaramakrishnan et al., 2022). This section highlights the three biggest bottlenecks that hamper microalgae upstream processing: contamination, low productivity, working space, and scalability.

4.2.1.1 Biological contamination

Microalgae are cultivated for cell growth and biomass production. Harvested biomass can be used as feedstock for several industrial fields, such as nutraceuticals, pharmaceuticals, cosmeceuticals, human food, and animal feed supplements. Final applications influence the cultivation systems selection. Open cultivation systems dominate microalgae cultivation systems due to their low costs compared to closed cultivation systems, known as PBRs. Unfortunately, open microalgae cultivation systems are negatively affected by the presence of other biological contaminants. A major bottleneck in the microalgal upstream processing is preventing and controlling the culture system of biological contamination, which negatively affects cell growth and biomass production (Zhu et al., 2020). Eradication or minimization of biological contamination can be achieved by choosing a suitable cultivation mode. For example, the photoautotrophic cultivation mode can minimize the risk of biological contamination due to the absence of organic carbon. A photoautotrophic cultivation mode is a defensive approach against the heterotrophic bacteria as biological contaminants for the microalgae cultures. Therefore, the photoautotrophic mode is an adequate microalgae cultivation mode in open cultivation systems, commonly known as outdoor microalgae cultivation. In terms of processing cost, open microalgae cultivation systems are cheaper than closed microalgae cultivation systems.

4.2.1.2 Low cell growth rate and biomass production in industrial microalgae

The gap between theoretical (lab and small scale) and industrial production is another bottleneck in the industrialization of microalgae technology (Benedetti et al., 2018). Many factors affect microalgae production in terms of cell growth and biomass accumulation. Those factors can be termed as optimal growth conditions, such as culture medium nutrients concentration and environmental conditions. It is imperative to select the optimal culture conditions to overcome theoretical and industrial microalgae productivity gaps. The open (also known as outdoor) industrial (i.e., large scale or commercial) microalgae cultivation requires the photoautotrophic cultivation mode. It reduces risks of biological contamination, but its application is limited by light dependency.

The low cell growth rate and lower biomass accumulation of microalgae cultivated under the photoautotrophic mode, compared to the heterotrophic and mixotrophic cultivation modes, is a major bottleneck in upstream microalgae processing. Furthermore, some locations have limited sunlight irradiation depending on climatic conditions, season, geographical region, and emplacement. Sunlight irradiation as a sole energy source has a limiting factor of cell growth and biomass accumulation under the photoautotrophic microalgae cultivation mode. Photoautotrophic microalgae cultivation mode leads to lower biomass productivity in photoautotrophic cultivation due to the self-shading effect on the microalgal vertical distribution that prevents light availability for denser cultivation. Therefore, some commercial outdoor photoautotrophic microalgae cultivation is carried out with artificial light, but it is somehow energy consuming and expensive, which is another bottleneck in industrial microalgae processing. Lower cell growth rate and microalgae biomass accumulation can also result from the inhibition effect of biological contaminants at different growth stages. It is recommended to switch from open microalgae cultivation system with a photoautotrophic mode to a heterotrophic cultivation mode as a promising strategy to overcome those bottlenecks in the upstream processing. That swift can be done by adding organic and inorganic carbon sources in the culture medium mixed with some substrate that prevents negative and positive gam bacteria, such as potassium tellurite, as cultivation process optimization. The second alternative suitable technology is to apply renewable energy, such as wind or solar power, for artificial lights. It is also recommended to adopt genetic engineering strategies to overcome the lower cell growth and biomass accumulation in the upstream microalgae processing.

4.2.1.3 Working space and scalability

Commercial-scale microalgae cultivation still faces another major challenge: its economic feasibility, high cost, and energy consumption. Khor et al. (2022) highlighted a floating PBR for microalgae cultivation as a novel technology to reduce the cost effects of onshore land utilization with additional advantages, such as improved culture conditions and integrated ocean renewable energy (Khor et al., 2022).

Algae cultivation on rooftops is also considered a sustainable approach to prevent arable land competition and a good system that could help feed millions and create jobs. Unused space on roofs and backyards can be targeted to cultivate edible algae, such as *Spirulina* and *Chlorella*, that are high in nutritional value and easy to produce in homemade bioreactors. *Spirulina* is algae rich in a superior form of plant protein and micronutrients, including vitamins and essential minerals. Algae can convert greenhouse gases from the atmosphere into some metabolite and oxygen, where algae are considered as "micro-factories"; algae biomass can produce nutritious products. Algae are a sustainable source of protein accumulated through absorbing carbon dioxide from the atmosphere. It is estimated that a single bioreactor can grow about 1 kg of algae in a month, so every family can set up algae cultivating system that can supply them with plenty enough protein all year round. Therefore, in a creative and technology-driven solution, algae can be a sustainable, environmentally friendly business initiative where roofs of residential houses and hotels can be utilized to produce nutritious and enriched-micronutrients food and feed additives.

4.2.1.4 Harvesting and dewatering (midstream processing)

Harvesting and dewatering is a major unit operation in microalgae industrial-scale processing; it is also a critical bottleneck that needs much attention in the development of large-scale production of microalgae (Uduman et al., 2010). Midstream processing is a major bottleneck in industrial microalgae processing due to the small microalgal cell sizes, negative surface charge, and low biomass concentration (Muylaert et al., 2017). There is no single and straightforward method for harvesting and dewatering for all microalgae species, but selecting appropriate harvesting and dewatering technology depends on the microalgae characteristics, cultivation modes, and final biomass applications. The choice of proper harvesting and dewatering technique is one of the crucial successes for the microalgae industry. Most common harvesting and dewatering techniques require high energy consumption. Cocultivation of various microalgae species with different sizes or coculture between bacteria and microalgae species is one approach that does not require energy for biomass harvesting. Cocultivation approach has exhibited a promising solution where it facilitates the harvesting process through the formation of flocs followed by auto-settling (auto-flocculation) after stopping the air bubbling, in the process called auto-flocculation. Various microalgae species sizes initiate the formation of flocs. In the case of bio-flocculation through bacteria and microalgae coculture, the secretion of extracellular substances provokes the appearance of flocs. In other words, bio-flocculation is a harvesting method based on biologically excreted organic compounds (known as extracellular polymeric substances or extracellular polysaccharides (EPS)). Unfortunately, that auto-flocculation through EPS excretion by microalgae is not suitable for continuous cultures grown under optimal conditions for maximum productivity. That limitation in the continuous cultivation process is due to the EPS excretion that requires nonideal growth conditions (such as extreme temperature, pH, and nutrient stress conditions), which differ from optimal growth conditions (Lee et al., 2009). Microbial flocculation is a potentially low-cost harvesting technique for marine microalgae, even though it is somehow microalgae species-selective.

In addition to auto- and bio-flocculation (also known as microbial flocculation), there is another well-known harvesting method: electro-flocculation. Electro-flocculation is a physicochemical process that can be applied to any microalgae species. It is simpler to operate and offers more accurate and predictable results. Moreover, contrary to chemical flocculation, electro-flocculation does not introduce unnecessary anions, which can lead to a low pH value of the culture to be harvested and may destroy cells' vitality.

4.2.2 Downstream and application

The downstream in the industrial microalgae processing generally accounts for about 40% of the total cost. Several reasons are behind that expensive cost, including the dilute nature of the microalgae biomass to be processed. The low biomass content is also behind most industrial microalgae biorefinery unprofitability. The sustainable resolution of that downstream bottleneck should be regarded from each unit operation and taken on the line of synchronizing of Life Cycle Assessment because bottlenecks in industrial microalgae processing are counted in all unit operations. The design of downstream processes for a single main product is the second bottleneck downstream. Microalgae accumulate a broad range of active compounds and metabolites; single downstream processes for the single main product led to the remainder of metabolites as waste or by-products, implicating additional disposal costs. Coexploitable products are regarded as a promising approach to resolving that bottleneck. Selecting suitable

Table 4.1: Bottlenecks in industrial microalgae systems and promising solutions.

Unit	Bottleneck	Solutions
Upstream	Biological contamination	– selection of strain from the local environment – mixed-microalgae species cultivation – coculture bacteria and microalgae – selection of suitable cultivation mode
	Low cell growth rate and biomass accumulation	– selection of strain from the local environment – mixed-microalgae species cultivation – coculture bacteria and microalgae – selection of suitable cultivation mode
	Working space and scalability	– top roofs, lakes and sea shores, deserts
Midstream (harvesting and dewatering)	Low biomass concentration, high energy consumption, and the introduction of unnecessary anions can lead to a low pH value and damage to cells' vitality.	– mixed species with different morphological features for auto-flocculation/auto-settling – attached cultivation – electro-flocculation
Downstream and application (cell disruption and biomass processing)	High-cost, downstream processes that are designed for a single product.	– Selection for microalgae strains with hyperaccumulation capacity of the targeted primary product, cascade extraction approach for multiple products

microalgae strains that can optimize the target product accumulation and its recovery from total biomass, followed by the valorization of remaining biomass after primary extraction of the targeted product, is one of the promising approaches.

4.3 Conclusion, recommendation, and future perspectives

There is a productivity gap between the maximal theoretical estimations and industrial microalgae production outcomes. Therefore, identifying factors limiting biomass yield and tackling bottlenecks are critical development of sustainable strategies to make microalgae resources and associated technology profitable on the industrial scale. The technical and scientific feasibility of microalgae's upstream and downstream processes are essential by analyzing and focusing on process parameters that play an important role in the growth and accumulation of value-added chemicals in microalgae cells. The collaboration among various stakeholders (in particular, researchers, investors, industrialists, and policy makers) is another key to the sustainability and competitiveness of the microalgae industry. Apart from financial support in terms of capital and investment bottleneck that needs collaboration among stakeholders, other bottlenecks can be tackled by choosing appropriate elements and technology in each unit operation. This chapter suggests that an international fund can be established for industrial microalgae development. Furthermore, universities, research centers, industries, and various institutions should work together to fully exploit microalgae potentials. It will be imperative if that fund makes all microalgae-related intellectual properties available for free. Future microalgae resource exploitation perspectives will rely on the advantages and potentials of algal resources to make them prominent sustainable resources for various applications. The ample exploitation of microalgae resources can contribute to the potential change in producing a broad range of bioactive compounds. Phycoremediation is a novel technology that can contribute to environmental sustainability and green energy production at a low cost.

References

Benedetti, M., Vecchi, V., Barera, S., & Dall'Osto, L. (2018). Biomass from microalgae: The potential of domestication towards sustainable biofactories. *Microbial Cell Factories*, *17*(1), 173.

Daneshvar, E., Sik Ok, Y., Tavakoli, S., Sarkar, B., Shaheen, S. M., Hong, H., Luo, Y., Rinklebe, J., Song, H., & Bhatnagar, A. (2021). Insights into upstream processing of microalgae: A review. *Bioresource Technology*, *329*, 124870.

Gifuni, I., Pollio, A., Safi, C., Marzocchella, A., & Olivieri, G. (2019). Current bottlenecks and challenges of the microalgal biorefinery. *Trends in Biotechnology*, *37*(3), 242–252.

Kaushik, A., Sangtani, R., Parmar, H. S., & Bala, K. (2022). Algal metabolites: Paving the way towards new generation antidiabetic therapeutics. *Aquatic Life and Algal Research*, *69*, 102904.

Khan, M. I., Shin, J. H., & Kim, J. D. (2018). The promising future of microalgae: Current status, challenges, and optimization of a sustainable and renewable industry for biofuels, feed, and other products. *Microbial Cell Factories*, *17*(1), 36.

Khor, W. H., Kang, H.-S., Lim, J.-W., Iwamoto, K., Tang, C.-H.-H., Goh, P. S., Quen, L. K., Shaharuddin, N. M. R. B., & Lai, N. Y. G. (2022). Microalgae cultivation in offshore floating photobioreactor: State-of-the-art, opportunities and challenges. *Aquacultural Engineering*, *98*, 102269.

Lee, A. K., Lewis, D. M., & Ashman, P. J. (2009). Microbial flocculation, a potentially low-cost harvesting technique for marine microalgae for the production of biodiesel. *Journal of Applied Phycology*, *21*(5), 559–567.

Manirafasha, E., Ndikubwimana, T., Zeng, X., Lu, Y., & Jing, K. (2016). Phycobiliprotein: Potential microalgae derived pharmaceutical and biological reagent. *Biochemical Engineering Journal*, *109*, 282–296.

Manirafasha, E., Vangh, A., Murwanashyaka, T., Rugabirwa, B., & Ndikubwimana, T. (2019). Algal resources exploitation for green economy and sustainable development: A review. *Advances in Biochemical Engi neering/Biotechnology*, *7*, 1089.

Muylaert, K., Bastiaens, L., Vandamme, D., & Gouveia, L. (2017). 5 – Harvesting of Microalgae: Overview of Process Options and Their Strengths and Drawbacks. In: Gonzalez-Fernandez, C., & Muñoz, R. (Eds.),

Microalgae-based Biofuels and Bioproducts. Woodhead Publishing, Sawston, United Kingdom, pp. 113–132.

Sivaramakrishnan, R., Suresh, S., Kanwal, S., Ramadoss, G., Ramprakash, B., & Incharoensakdi, A. (2022). Microalgal biorefinery concepts' developments for biofuel and bioproducts: Current perspective and bottlenecks. *International Journal of Molecular Sciences*, *23*(5), 2623–2648 (1–25).

Uduman, N., Qi, Y., Danquah, M. K., Forde, G. M., & Hoadley, A. (2010). Dewatering of microalgal cultures: A major bottleneck to algae-based fuels. *Journal of Renewable and Sustainable Energy (JRSE)*, *2*(1), 012701.

Zhu, Z., Jiang, J., & Fa, Y. (2020). Overcoming the biological contamination in microalgae and cyanobacteria mass cultivations for photosynthetic biofuel production. *Molecules*, *25*(22), 5220–5233 (1–13).

Calvin Lo, Rene H. Wijffels, Iulian Boboescu, A. Kazbar,
Michel H. M. Eppink*

Chapter 5
Multimethod and multiproduct microalgae biorefineries: industrial scale feasibility: eutectic solvents as a novel extraction system for microalgae biorefinery

Abstract: Eutectic solvents (ES), including "deep eutectic solvents," hold great potential as an extraction system for microalgae. Besides being virtually inflammable, ES can be readily prepared from bioderived and biodegradable compounds. Several hydrophilic/hydrophobic ES can form pores or cracks in the cell wall of various microalgae, which enhanced the lipid extraction yield even without the cell disruption step. The tailorable properties of ES may also open possibilities for process integration. For instance, hydrophobic ES can be used to pretreat the biomass and extract the lipids simultaneously. However, the low volatility of ES complicates the process due to the challenging separation of lipids from the solvent. Therefore, this we aimed to develop suitable ES for lipid extraction from microalgae and further evaluate the feasibility of microalgae biorefinery. In this chapter, the major breakthroughs and challenges are summarized. Moreover, the future outlook on the ES application for microalgae biorefinery is discussed.

Keywords: microalgae, deep eutectic solvents (DES), lipids, extraction, biodegradable

5.1 Introduction

With the threatening issues of climate change and biodiversity loss, there is an urgent need to produce lipids in an environmental-harmless way. Compared to oleaginous terrestrial plants (like oil palm), several microalgae exhibit higher lipid productivity with-

Acknowledgment: This research is part of the MAGNIFICENT project, funded by the Bio-Based Industries Joint Undertaking under the European Union's Horizon 2020 research and innovation program (grant agreement no. 745754).

***Corresponding author: Michel H. M. Eppink,** Bioprocess Engineering, AlgaePARC, Wageningen University, PO Box 16, 6,700 AA Wageningen, The Netherlands, e-mail: michel.eppink@wur.nl
Calvin Lo, Iulian Boboescu, A. Kazbar, Bioprocess Engineering, AlgaePARC, Wageningen University, 6700 AA Wageningen, The Netherlands
Rene H. Wijffels, Bioprocess Engineering, AlgaePARC, Wageningen University, 6700 AA Wageningen, The Netherlands; Nord University, Faculty of Biosciences and Aquaculture, N-8049 Bodø, Norway

https://doi.org/10.1515/9783110781267-005

out the need for arable land (Wijffels et al., 2010; Ruiz Gonzalez et al., 2016). Therefore, there is growing attention on microalgae as lipid feedstock. However, the current lipid extraction methods from microalgae still involve series of energy-intensive pretreatments and nonrenewable organic solvents (Halim et al., 2012; Kumar et al., 2017; Lee et al., 2017; Halim et al., 2019; Günerken et al., 2015). These concerns make this process not only cost ineffective but unsustainable as well. Furthermore, to avoid biomass underutilization, the biorefinery concept (i.e., valorization of the entire valuable compounds from biomass) is necessary to be implemented.

A new class of extraction solvents for microalgae are eutectic solvents (ES), including "deep eutectic solvents". Besides being virtually inflammable, ES can be readily prepared from bioderived and biodegradable compounds which is a great benefit. Several hydrophilic ES can form pores or cracks in the cell wall of various microalgae, which enhanced the extraction yield of different components even without the cell disruption step (Lu et al., 2016; Pan et al., 2017). In this chapter it is aimed to give an overview of suitable ES for lipid extraction from microalgae and further evaluate the feasibility of microalgae biorefinery.

5.2 Begin with the end in mind: altering ES hydrophobicity

To develop a functional solvent process, the focus should address the extraction step and include solvent regeneration. Murphy's law of solvent states that the best solvent in any process would be bad for the subsequent step. This principle implies that while the strong affinity of hydrophobic ES toward the lipid solutes would benefit the extraction process, it would also cause the separation of lipids from the ES practically impossible. Therefore, we propose to use semihydrophobic ES, which were prepared by pairing hydrophobic and hydrophilic compounds (Lo, 2021a). For instance, the combination of imidazole and hexanoic acid was dissolved in model lipids and water. The solvent hydrophobicity was found to decrease with increasing imidazole concentration. Thus, by adding imidazole, model lipids could be recovered with relatively high purity (>85%). However, solvent regeneration via imidazole removal is not straightforward with this approach due to the strong association between imidazole and hexanoic acid.

Another approach shifting the solvent hydrophobicity is to use polar antisolvents, for example, water, methanol, and ethanol. The presence of antisolvents accentuated the ES hydrophilicity and thus reduced the solubility of model lipids. This approach offered a significantly simpler method to regenerate the ES, that is, by evaporating the antisolvents. Since a large amount of antisolvent can be loaded into the system, this approach reached higher recovery (>90%) and purity of the obtained lipids compared to the previous method (Lo, 2021a). Considering the recovery yield and the ease of regeneration, methanol was selected to be the best antisolvent.

5.3 ES on microalgae: lipids and beyond

Imidazole/hexanoic acid ES was found to extract lipid from undisrupted *Nannochloropsis oceanica*, which confirmed the hypothesis of using ES as "pretreatment" and extraction solvent (Lo et al., 2021b). At low imidazole content (≤15 mol%), the extraction yield using ES at 50 °C was comparable to the benchmark chloroform/methanol method. Interestingly, extraction on wet algae gave higher yields than the dried biomass. This finding suggests that both cell disruption and complete dehydration are unnecessary for the ES extraction, simplifying the microalgae processing. Besides that, it also implies that water enhanced the ES performance, contradicting the finding from (Lo et al., 2021b). Hence, it is hypothesized that water facilitates the ES penetration to the cell matrix. However, the interaction between the solvent with the biomolecules and how ES penetrated the cell wall is still unknown. Moreover, since cell wall structure is species-dependent, the ES penetration might be different between microalgal species. Additionally, the extracted lipid might undergo undesired reactions, such as hydrolysis, transesterification, or oxidation, which were undetected with the used analytical method.

Initial feasibility study of microalgae biorefinery based on the developed ES was performed (Lo et al., 2021b) and the proposed biorefinery process is shown in Figure 5.1. Solvent reusability and process scalability were evaluated for lipid extraction. Unlike the model lipids, the recovery of algal lipids with methanol reached a lower yield (~60%) even at low temperature (−20 °C). The lower recovery was due to the lower starting concentration of lipids; the concentration of model lipids was 10-fold higher than the extracted algal lipids. Besides that, the different fatty acid distribution between the algal extract and the model lipids might contribute to the lower recovery as well as the ES had a higher affinity toward certain fatty acids. Moreover, unlike the model lipids, which were mainly refined triacylglycerols, the extract from the algae could also contain polar lipids (PL), sterols, waxes, and pigments. These compounds might interact differently with the ES and interfere with the lipid recovery. Furthermore, despite the incomplete recovery, the ES could be reused for the three extraction cycles with consistent performance. Some losses of the solvents were observed during solid-liquid separations (i.e., after the extraction and the lipid recovery). In scaling up, the heterogeneity is a recurring issue for the extraction, while agitation and cooling rate influenced the recovery.

Besides lipids, biomass contains proteins, and carbohydrates, which remain inside the defatted biomass after the ES extraction (Lo et al., 2021b). Based on the protein analysis, the ES extraction was not mild since the proteins lost the native conformations and aggregated. The denaturation might be caused by the acidic and amphiphilic nature of the solvents, combined with the elevated extraction temperature. Moreover, the proteins and carbohydrates were isolated through aqueous extraction, which was enhanced by pH manipulation. However, despite giving the highest yield, the alkaline condition (i.e., pH 13) hydrolyzed the remaining proteins.

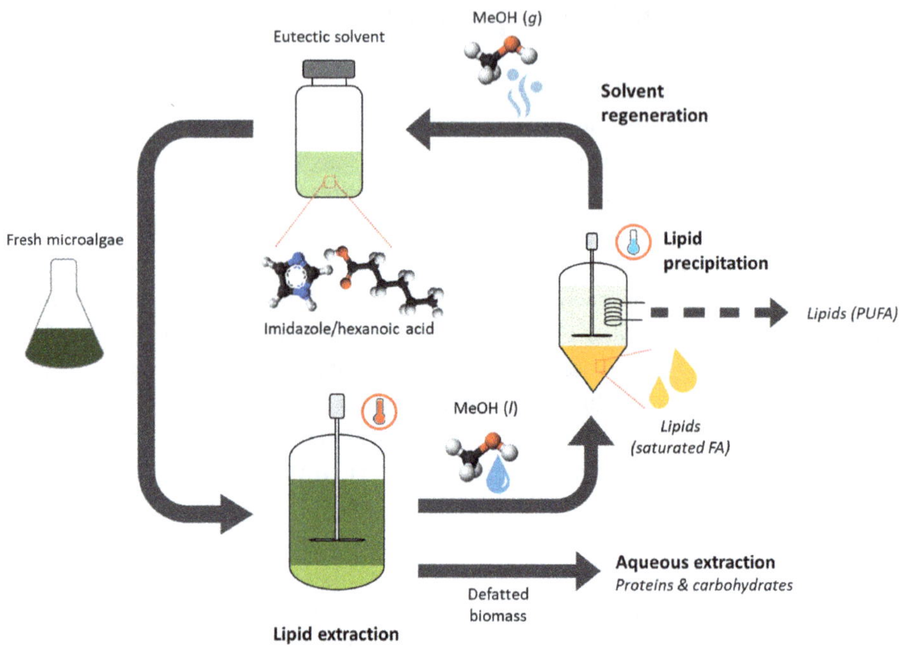

Figure 5.1: Schematic overview of the proposed biorefinery of *Nannochloropsis oceanica* based on the imidazole/hexanoic acid eutectic solvent.

5.4 Challenges and future perspectives

As mentioned above, applications of eutectic solvents on microalgae processes are still relatively new. In this work, several relevant challenges for biorefinery processing were discovered. Hence, we outline below the challenges to be addressed and future perspectives on microalgae biorefinery based on ES.

In this work, we mainly focused on the semihydrophobic ES as the basis of the microalgae biorefinery. This process, unfortunately, rendered the proteins denatured, which is associated with compromised functionality (Lo et al., 2021b). Arguably, the denatured proteins could also possess new functionalities, such as a gelling agent or as amino acids precursor. Moreover, from the Lowry analysis, some proteins were detected in the hexanoic acid and ES phase, which might indicate the isolation of hydrophobic proteins. If hydrophobic proteins were indeed extracted, then this study would be the first to extract hydrophobic proteins using ES. To date, little attention is given to this protein fraction, although the majority of microalgal proteins are insoluble (membrane-bound) (Dai et al., 2019). Recently, the insoluble proteins from microalga *Chlorella protothecoides* were used as an emulsifier (Dai et al., 2019). However, it is clear that this approach is not suitable to extract water-soluble proteins. Such as phycobiliproteins, the main pigment-protein complex as light absorber in cyanobacteria, or Rubisco (ribulose-11,5-biphosphate carboxylase oxygenase), the responsible enzyme for carbon assimilation in all photosynthetic organisms.

The protein denaturation might be caused by several factors, such as the solvent chemical property, low water activity, high extraction temperature, or their combinations. Depending on the triggering factors, strategies to improve the proposed process to be milder should be developed. Therefore, it is necessary to determine the actual cause of the denaturation.

Imidazole/hexanoic acid ES exhibited amphiphilicity (consisting of both hydrophilic and hydrophobic moieties), which is like detergent, could promote protein unfolding. Typically, the water-soluble proteins have a hydrophobic core and hydrophilic surface. Thus, the presence of amphiphiles could destabilize the protein structure and unfold the proteins (Otzen, 2002). Besides that, the high concentration of hexanoic acid, as a pure solvent or in ES, implies that the system is highly acidic. At high acidity (or low pH values in aqueous solutions), proteins are mostly positively charged and may change their conformation as similar charges repel each other (the electrostatic interaction). Furthermore, certain ions and compounds could also influence the surface charge of protein by either promoting (kosmotrope) or breaking (chaotrope) the hydrogen bonding network in water, the Hofmeister effect (Mazzini and Craig, 2017; Hyde et al., 2017). Typically, the protein stability is promoted by kosmotropic compounds or pairs of chaotropic cation and kosmotropic anion or kosmotropic compounds (Zhao, 2015). However, the categorization of ES based on the Hoffmeister effect has not yet been widely researched despite the extensive studies done on ionic liquids (Umapathi et al., 2018). In ionic liquids, imidazolium cations with shorter side chains tend to be chaotropic (Zhao, 2015; Umapathi et al., 2018), while carboxylate anion with longer alkyl chains became less kosmotropic (Sultana and Ismail, 2016). Therefore, there is a chance that the used ES destabilized the proteins.

That said, in the situation where the native structure of water-soluble proteins is desired, the use of hydrophilic ES could be beneficial. Not only they can weaken the microalgae cell wall (Lu et al., 2016), but they can stabilize proteins as well (Gertrudes et al., 2017; Sanchez-Fernandez et al., 2017; Sanchez-Fernandez et al., 2021). For instance, lipase could remain stable at choline chloride/urea ES despite the denaturing effect of urea (Monhemi et al., 2014). Moreover, ES choline chloride/urea and choline chloride/glycerol were reported to facilitate thermal refolding of lysozyme (Esquembre et al., 2013). Hydrophilic ES have also been applied for protein extraction in an aqueous two-phase system (ATPS). ATPS based on phosphate buffer and ES that made of organic salts (choline chloride or betaine) and hydrogen bond donors (polyols, sugars, or urea) were used to extract >98% of bovine serum albumin (BSA) (Xu et al., 2015; Li et al., 2016). The ATPS was further improved to enhance back-extraction of the protein (reaching 72% of efficiency) using a ternary ES tetramethylammonium chloride/glycerol/urea (Zhang et al., 2016).

Another approach would be implementing a biphasic system made of hydrophobic and hydrophilic phases. Such a system would simultaneously extract both hydrophobic and hydrophilic biomolecules without compromising the functionality. For instance, the combination of the semihydrophobic ES with salt or polymeric solution would form such a biphasic system. Thus, while the ES could directly permeabilize the cell wall and extract the lipids, the water-soluble components would migrate to the aqueous solution. Such a system was implemented for the separation of both hydrophobic and hydrophilic bioactive compounds from *Ginkgo biloba* leaves (Cao et al., 2018). The leaves contained flavonoids, terpene trilactone, procyanidine – which are hydrophilic – and polyprenyl acetate (hydrophobic). Using a biphasic system that was formed using three different ES, choline chloride/lactic acid (hydrophilic), choline chloride/malonic acid (hydrophilic), and methyltrioctylammonium chloride/capryl alcohol/octylic acid (hydrophobic) were used to separate those metabolites and reached ~80–95% of extraction efficiencies (Cao et al., 2018). However, to the furthest of our knowledge, the ES-based biphasic system has not yet been implemented for protein extraction.

As mentioned before, solvent acidity is an important parameter that can affect the protein charge and conformation. The use of a high concentration of unbuffered acid should be avoided. However, ES are a mixture of pure compounds where water is undesired. Thus, ES made of less acidic compound could greatly enhance the protein stability. For instance, instead of using hexanoic acid as the Brønsted acid, perhaps neutral hydrogen bond donors such as menthol could be used. However, since the concept of acidity in an aqueous solution would be different in the ES-rich environment, the charge dynamic of the protein surface needs to be studied.

Besides that, the Hofmeister effect – ion-specific interaction – should be considered (Mehringer et al., 2021; Zhao, 2015). For this purpose, further studies of how the solvent components interact with each other, water, and proteins should be understood. However, since the system is multicomponent, even the starting ES are already a mixture, instead of pure salts like ionic liquids, studying this system would be incredibly complex. Thus, it is important to implement a step-by-step approach with a model system. Starting with possibly formed kosmotropes and chaotropes in the ES, including their synergized effect, is recommended. Then, continue with the concentration of water.

Low water activity could also denature proteins. The polar groups of ES compete with proteins for water, while the latter requires hydration to stabilize their hydrophilic surface (Zhao et al., 2015). Moreover, without sufficient hydration of the ES polar groups (e.g., hexanoate anion) would interact strongly with the protein surface. Thus, at low water content, the hexanoate anion would be a chaotropic anion – destabilizing proteins – despite the kosmotropic effect in dilute aqueous solutions (Zhao et al., 2015). Furthermore, BSA and lysozyme were observed to be partially folded in pure ES of choline chloride/glycerol. When the ES was hydrated, the proteins retained their folded structure as in a phosphate buffer saline (Sanchez-Fernandez et al., 2017).

Thus, the strategy to tackle would be to remove less water during the harvesting or use one or more aqueous phases in the biphasic or ATPS system. The former would be economically attractive since less water removal would require less energy during this step. The latter could also be achieved by implementing ATPS or the biphasic system with a water-rich phase.

Protein denaturation may also be a thermal effect since hydrogen bond is weakened at higher temperatures. The ES extraction was performed optimally at 50 °C, and the lipid yield decreased at lower temperatures (Lo et al., 2021b). Thus, it is a trade-off between the lipid yield and the protein native state since lower temperature (≤35 °C) is necessary to ensure the latter. However, the actual upper limit for temperature might not be 35 °C and should be the denaturation temperature of the algal proteins with the presence of the ES. Thus, this upper limit needs to be determined. For instance, differential scanning calorimetry could be used to study the thermal stability (T_m) of proteins (Schön et al., 2017).

Simultaneously, the effect of heat for the ES extraction needs to be investigated so that the extraction temperature could be lowered without compromising the lipid yield. The lipid solubility in the ES with low imidazole content was already high, indicating the high solvent carrying capacity, even at room temperature. Thus, the higher temperature might improve the solvent penetration and the segregation of lipids from other biomolecules. Besides decreasing the ES viscosity (Hayyan et al., 2012; Abbot et al., 2004; Abbott et al., 2011), the high temperature could also weaken the cell wall, which served as the main barrier for solvent penetration. Hence, if the cell wall could be removed, disintegrated, or significantly weakened by physical or mechanical energy input, the compromised lipid yield could be compensated (Halim et al., 2019; Günerken et al., 2015; Yap et al., 2014). Previously, microwave treatments were used to enhance the cell wall-weakening effect of hydrophilic ES (Tommasi et al., 2017) and induce cell disruption before the extraction with a switchable hydrophilicity ES (Sed et al., 2018). However, the temperature could reach up to ≥100 °C during the microwave treatment, which would render the proteins denatured (Tommasi et al., 2017).

Other milder external forces, such as acoustic or electric fields, may be applied to reduce the cell wall integrity or to disrupt the cells. During ultrasonication, high-frequency acoustic waves decompress the liquid and induce cavitation, eventually collapsing and rupturing the cell wall (Günerken et al., 2015; Kurokawa et al., 2016; Zhang et al., 2020). The cell wall-weakening effect of the ES might also reduce the energy required to damage the cell wall. However, it is not yet clear how the cavitation in the ES would work since cavitation requires a pressure lower than the vapor pressure of ES, which are relatively nonvolatile. Besides that, the propagation of the sound wave, the optimal frequency and intensity, and the transmission of shear forces in the ES media should be investigated. On the other hand, electroporation by pulsed electric field (PEF) offers an alternative option to accelerate the solvent penetration. PEF require media with high conductivity to ensure the propagation of the electric field (Günerken et al., 2015; 't Lam et al., 2017; Parniakov et al., 2015). Thus, the

ES, when mixed with water, may give a beneficial effect due to the presence of ionic solutes with low viscosity (Dai et al., 2015). Alternatively, a short mechanical cell disruption by the conventional high-pressure homogenizer or bead milling technique could also be used (Suarez-Garcia et al., 2018). However, it is important to note that this additional treatment would increase the energy demand and the production cost. It is worth noting that with this extra treatment, the extraction time could be shortened. Currently, the lipid extraction from the intact biomass took place >8 h. With the ruptured cell wall, however, the lipids would be liberated and readily accessible for the solvent (Yap et al., 2014).

In addition, several improvement points on lipid extraction are also discussed. The operational parameters used in Lo et al. (2021b) were rather chosen arbitrarily due to the lack of information and it is highly likely to cause a suboptimal overall process. For instance, the solvent-to-biomass ratio of 10 mL g_{DW}^{-1} was used, resulting in a low lipid concentration and the low efficiency of lipid recovery. To design an optimal biorefinery process, process modeling is a powerful tool to predict the process outcome. In that regard, a combination of experimental data and robust mathematical models could be a good starting point. Our preliminary result using nonrandom two liquids thermodynamic model showed that the model could describe well the equilibria in the lipid extraction and precipitation.

Furthermore, in Lo et al. (2021b), the PUFA-rich PL fraction remained dissolved in the ES-rich fraction, with accumulation went on with the extraction cycles. It is economically and technically important to obtain this PUFA-rich fraction since the fraction would have a high added value and eventually decrease the yield of the next extraction cycle. Therefore, a strategy for this lipid recovery is required. High performance liquid chromatography techniques, both normal (polar stationary phase) and reverse phase (nonpolar stationary phase), may be useful to separate the lipid fractions from the solvent phase (Olsson et al., 2014).

Besides that, the main advantage of designer solvents, including ES, is their tailorable properties. In this thesis, we demonstrated that ES's physicochemical properties, particularly hydrophobicity, were influenced by the nature of their constituents and the composition (Lo et al., 2021a). Furthermore, additions of other compounds, such as lipids or water, would definitely affect the system property (Lo et al., 2021b). These insights were obtained through the empirical trial and error method. With this approach, although the molecular interactions could be deduced from the observable property, the exact interaction at the molecular level remains unknown. For instance, theoretically, imidazole and hexanoic acid could interact via several ways: (1) proton transfer, producing a protic ionic liquid (Anouti et al., 2009; Yoshizawa et al., 2003); (2) hydrogen bonding, which is typical for eutectic solvents (Abbott et al., 2003; Ashworth et al., 2016); (3) formation of other complexes, such as homo association of hexanoate anion and hexanoic acid (Yoshizawa et al., 2003; Martins et al., 2021; Johansson et al., 2008); and (4) combinations of above. Each of the mentioned interactions would implicate different lipid solubilization mechanisms and even recovery strategies.

In contrast to the empirical approach, the mechanistic approach could predict the observable macroscopic property based on the intermolecular forces. Thus, besides having a higher chance of designing the task-specific solvent, insight into the solvation mechanism could be acquired. Typically, solvatochromism is used to experimentally determine the molecular property of the ES, such as hydrogen bond donating and accepting capacity and polarity (Teles et al., 2017; Florindo et al., 2017; Pandey and Pandey, 2014; Martins et al., 2018). On the other hand, the molecular interactions in the ES system can also be simulated via computational chemistry modelings such as molecular dynamics (Zahn et al., 2016; Mohan et al., 2017) and quantum chemical calculation (Ashworth et al., 2016; Wagle et al., 2016; Stefanovic et al., 2017). The latter, particularly COSMO-RS (conductor-like screening model for realistic solvents), has been widely used in the field of ES (Martins et al., 2018; Silva et al., 2018; Fernandez et al., 2017; Kundu et al., 2020).

Computational chemistry like COSMO-RS is a powerful tool to study the interaction of ES components with biomolecules, for example, proteins (Mehringer et al., 2021) and plant secondary metabolites (Wojeicchowski et al., 2020; Jeliński et al., 2018) and cytotoxicity (Hayyan et al., 2016). Moreover, COSMO-RS has been used to develop biphasic eutectic solvents (hydrophilic: choline chloride/hexafluoroispropanol; and hydrophobic: trioctylmethylammonium chloride/menthol) for the extraction and separation of both polar and nonpolar natural compounds from *Artemisia annua* leaves (Tang and Row, 2020). Eventually, the obtained knowledge might also be used to predict the interaction of the solvents with more complex biomolecules, such as proteins and cell wall components. This prediction would enable designing the task-specific ES which suits the need of microalgae biorefinery.

Finally, solvent sustainability should not be taken for granted (Chen and Mu, 2021). Proper toxicity studies and life cycle assessments still need to be performed. Currently, few studies are available in the literature about the actual environmental impact and toxicity of ES. One study reported that ES made of choline chloride/acetic acid is more cytotoxic than the ionic liquid cholinium acetate (de Morais et al., 2015). Ironically, ionic liquids are more commonly associated with potential toxicity, whereas ES are perceived as environmentally benign. Furthermore, not all ES components in this thesis are categorized as renewables. While hexanoic acid is biodegradable and can be produced via fermentation, imidazole, despite being biodegradable, is currently fossil-derived and considered toxic for humans. Hence, it is necessary to find more sustainable and safe alternatives. Therefore, besides understanding the role of each component via COSMO-RS, having a database of sustainable and naturally available compounds would be advantageous for designing the task-specific green ES for microalgae biorefinery.

5.5 Recommendations

This chapter demonstrated the use of a new type of ES, semihydrophobic ES made of imidazole and hexanoic acid, for lipid extraction from microalgae without biomass pretreatments. Furthermore, the dissolved lipids can be recovered from the ES by the addition of methanol, which was later evaporated to regenerate the ES. We also performed a preliminary investigation of the use of the ES for microalgae biorefining. The solvent recyclability and scalability were feasible. However, despite the successful lipid extraction, the process was not sufficiently mild to maintain the native structure of proteins. Possible combinations of the ES chemical properties, the high extraction temperature, and the low water content might cause denaturation, which is undesired in the biorefinery context. Therefore, this issue is extensively discussed and several perspectives for the process improvement are suggested. The use of a biphasic system (hydrophobic and hydrophilic ES) with less acidic constituents, lower extraction temperatures, and the application of external physical fields might alleviate the problem and even accelerate the extraction process. The lipid extraction can be improved by recovering the PUFA-rich PL fraction from the ES phase. Moreover, the process parameters need to be optimized based on the process modeling (i.e., equilibrium-based liquid-liquid extraction). Besides that, the need to study the molecular interaction between the ES components, antisolvents, and biomolecules are emphasized. Computational chemistry modeling, like COSMO-RS, is an effective tool to understand the system's chemistry and design and tailor the suitable ES for microalgae biorefinery. Last, green and safe ES are prerequisite to have a sustainable microalgae biorefinery.

References

Abbott, A. P., Capper, G., Davies, D. L., Rasheed, R. K., & Tambyrajah, V. (2003). Novel solvent properties of choline chloride/ureamixtures. *Chemical Communication*, 70–71.

Abbott, A. P., Boothby, D., Capper, G., Davies, D. L., & Rasheed, R. K. (2004). Deep eutectic solvents formed between choline chloride and carboxylic acids: Versatile alternatives to ionic liquids. *Journal of the American Chemical Society*, *126*, 9142–9147.

Abbott, A. P., Harris, R. C., Ryder, K. S., Agostino, C. D., Gladden, L. F., & Mantle, M. D. (2011). Glycerol eutectics as sustainable solvent systems. *Green Chemistry*, *13*, 82–90.

Anouti, M., Jones, J., Boisset, A., Jacquemin, J., Caillon-Caravanier, M., & Lemordant, D. (2009). Aggregation behaviour in water of new imidazolium and pyrrolidinium alkycarboxylates protic ionic liquids. *Journal of Colloid and Interface Science*, *340*, 104–111.

Ashworth, C. R., Matthews, R. P., Welton, T., & Hunt, P. A. (2016). Double ionic hydrogen bond interactions within the choline chloride-urea deep eutectic solvent. *Physical Chemistry Chemical Physics*, *18*, 18145–18160.

Cao, J., Chen, L., Li, M., Cao, F., Zhao, L., & Su, E. (2018). Two-phase systems developed with hydrophilic and hydrophobic deep eutectic solvents for simultaneously extracting various bioactive compounds with different polarities. *Green Chemistry*, *20*, 1879–1886.

Chen, Y., & Mu, T. (2021). Revisiting greenness of ionic liquids and deep eutectic solvents. *Green Chemistry Engineering*, *2*, 174–186.

Dai, Y., Witkamp, G.-J., Verpoorte, R., & Choi, Y. H. (2015). Tailoring properties of natural deep eutectic solvents with water to facilitate their applications. *Food Chemistry*, *187*, 14–19.

Dai, L., Reichert, C. L., Hinrichs, J., & Weiss, J. (2019). Acid hydrolysis behavior of insoluble protein-rich fraction extracted from chlorella protothecoides. *Colloids and Surfaces A: Physicochemical and Engineering Aspects*, *569*, 129–136.

Esquembre, R., Sanz, J. M., Gerard Wall, J., Del Monte, F., Reyes Mateo, C., & Luisa Ferrer, M. (2013). Thermal unfolding and refolding of lysozyme in deep eutectic solvents and their aqueous dilutions. *Physical Chemistry Chemical Physics*, *15*, 11248–11256.

Fernandez, L., Silva, L. P., Martins, M. A., Ferreira, O., Ortega, J., Pinho, S. P., & Coutinho, J. A. P. (2017). Indirect assessment of the fusion properties of choline chloride from solid-liquid equilibria data. *Fluid Phase Equilibria*, *448*, 9–14.

Florindo, C., McIntosh, A. J. S., Welton, T., Branco, L. C., & Marrucho, I. M. (2017). A closer look into deep eutectic solvents: Exploring intermolecular interactions using solvatochromic probes. *Physical Chemistry Chemical Physics*, *20*, 206–213.

Gertrudes, A., Craveiro, R., Eltayari, Z., Reis, R. L., Paiva, A., Duarte, A. R. C., Rita, A., & Duarte, C. (2017). How do animals survive extreme temperature amplitudes? The role of natural deep eutectic solvents. *ACS Sustainable Chemistry and Engineering*, *5*, 9542–9553.

Günerken, E., D'Hondt, E., Eppink, M. H. M., Garcia–Gonzalez, L., Elst, K., & Wijffels, R. H. (2015). Cell disruption for microalgae biorefineries. *Biotechnology Advances*, *33*, 243–260.

Halim, R., Danquah, M. K., & Webley, P. A. (2012). Extraction of oil from microalgae for biodiesel production: A review. *Biotechnology Advances*, *30*, 709–732.

Halim, R., Hill, D. R. A., Hanssen, E., Webley, P. A., Blackburn, S., Grossman, A. R., Posten, C., & Martin, G. J. O. (2019). Towards sustainable microalgal biomass processing: Anaerobic induction of autolytic cell-wall self-ingestion in lipid-rich *nannochloropsis* slurries. *Green Chemistry*, *21*, 2967–2982.

Hayyan, A., Mjalli, F. S., Alnashef, I. M., Al-Wahaibi, Y. M., Al-Wahaibi, T., & Hashim, A. (2012). Glucose–based deep eutectic solvents: Physical properties. *Journal of Molecular Liquids*, *178*, 137–141.

Hayyan, M., Mbous, Y. P., Looi, C. Y., Wong, W. F., Hayyan, A., Salleh, Z., & Mohd-Ali, O. (2016). Natural deep eutectic solvents: Cytotoxic profile. *SpringerPlus*, *5*, 1–12.

Hyde, A. M., Zultanski, S. L., Waldman, J. H., Zhong, Y.-L., Shevlin, M., & Peng, M. (2017). General principles and strategies for salting-out informed by the hofmeister series. *Organic Process Research & Development*, *21*, 1355–1370.

Jeevan Kumar, S. P., Garlapati, V. K., Dash, A., Scholz, P., & Banerjee, R. (2017). Sustainable green solvents and techniques for lipid extraction from microalgae: A review. *Algal Research*, *21*, 138–147.

Jeliński, T., & Cysewski, P. (2018). Application of a computational model of natural deep eutectic solvents utilizing the COSMO-RS approach for screening of solvents with high solubility of rutin. *Journal of Molecular Modeling*, *24*, 180.

Johansson, K. M., Izgorodina, E. I., Forsyth, M., MacFarlane, D. R., & Seddon, K. R. (2008). Protic ionic liquids based on the dimeric and oligomeric anions: $[(aco)_x h_{x-1}]^-$. *Physical Chemistry Chemical Physics*, *10*, 2972–2978.

Kundu, D., Rao, P. S., & Banerjee, T. (2020). First principle prediction of kamlet-taft solvatochromic parameters of deep eutectic solvents using COSMO-RS model. *Industrial and Engineering Chemistry Research*, *59*, 11329–11339.

Kurokawa, M., King, P. M., Wu, X., Joyce, E. M., Mason, T. J., & Yamamoto, K. (2016). Effect of sonication frequency on the disruption of algae. *Ultrasonics Sonochemistry*, *31*, 157–162.

't Lam, G. P., Postma, P. R., Fernandes, D. A., Timmermans, R. A. H., Vermuë, M. H., Barbosa, M. J., Eppink, M. H. M., Wijffels, R. H., & Olivieri, G. (2017). Pulsed electric field for protein release of the microalgae *Chlorella vulgaris* and *Neochloris oleoabundans*. *Algal Research*, *24*, 181–187.

Lee, S. Y., Cho, J. M., Chang, Y. K., & Oh, Y.-K. (2017). Cell disruption and lipid extraction for microalgal biorefineries: A review. *Bioresource Technology*, *244*, 1317–1328.

Li, N., Wang, Y., Xu, K., Huang, Y., Wen, Q., & Ding, X. (2016). Development of green betaine-based deep eutectic solvent aqueous two-phase system for the extraction of protein. *Talanta*, *152*, 23–32.

Lo, C. (2021a). PhD Thesis "Eutectic Solvents as a Novel Extraction System for Microalgae Biorefinery". Wageningen University. ISBN: 978-94-6395-965-0. DOI: 10.18174/553274.

Lo, C., Wijffels, R. H., & Eppink, M. H. M. (2021b). Eutectic solvents with tunable hydrophobicity: Lipid dissolution and recovery. *RCS Advances*, *11*, 8142–8149.

Lu, W., Alam, M. A., Pan, Y., Wu, J., Wang, Z., & Yuan, Z. (2016). A new approach of microalgal biomass pretreatment using deep eutectic solvents for enhanced lipid recovery for biodiesel production. *Bioresource Technology*, *218*, 123–128.

Martins, M. A. R., Crespo, E. A., Pontes, P. V. A., Silva, L. P., Bülow, M., Maximo, G. J., Batista, E. A. C., Held, C., Pinho, S. P., & Coutinho, J. A. P. (2018). Tunable hydrophobic eutectic solvents based on terpenes and monocarboxylic acids. *ACS Sustainable Chemistry and Engineering*, *6*, 8836–8846.

Martins, M. A. R., Carvalho, P. J., Santos, L. M. N. B. F., Pinho, S. P., & Coutinho, J. A. P. (2021). The impact of oligomeric anions on the speciation of protic ionic liquids. *Fluid-Phase Equilibrium*, *531*, 112919.

Mazzini, V., & Craig, V. S. J. (2017). What is the fundamental ioin-specific series for anions and cations? Ion specificity in standard partical molar volumes of electrolytes and electrostriction in water and non-aqueous solvents. *Chemical Science*, *8*, 7052–7065.

Mehringer, J., Hofmann, E., Touraud, D., Koltzenburg, S., Kellermeier, M., & Kunz, W. (2021). Salting-in and salting-out effects of short amphiphilic molecules: A balance between specific ion effects and hydrophobicity. *Physical Chemistry Chemical Physics*, *23*, 1381–1391.

Mohan, M., Naik, P. K., Banerjee, T., Goud, V. V., & Paul, S. (2017). Solubility of glucose in tetrabutylammonium bromide based deep eutectic solvents: Experimental and molecular dynamics simulations. *Fluid-Phase Equilibrium*, *448*, 168–177.

Monhemi, H., Housaindokht, M. R., Moosavi-Movahedi, A. A., & Bozorgmehr, M. R. (2014). How a protein can remain stable in a solvent with high content of urea: Insights from molecular dynamics simulation of *Candida antarctica* lipase B in urea: Choline chloride deep eutectic solvents. *Physical Chemistry Chemical Physics*, *16*, 14882–14893.

de Morais, P., Gonçalves, F., Coutinho, J. A. P., & Ventura, S. P. M. (2015). Ecotoxicity of cholinium-based deep eutectic solvents. *ACS Sustainable Chemistry and Engineering*, *3*, 3398–3404.

Olsson, P., Holmbäck, J., & Herslöf, B. (2014). A single step reversed-phase high performance liquid chromatography separation of polar and non-polar lipids. *Journal of Chromatography A*, *1369*, 105–115.

Otzen, D. E. (2002). Protein unfolding in detergents: Effect of micelle structure, ionic strength, ph, and temperature. *Biophysical Journal*, *83*, 2219–2230.

Pan, Y., Alam, A., Wang, Z., Huang, D., Hu, K., Chen, H., & Yuan, Z. (2017). One-step production of biodiesel from wet and unbroken microalgae biomass using deep eutectic solvents. *Bioresource Technology*, *238*, 157–163.

Pandey, A., & Pandey, S. (2014). Solvatochromic probe behavior within choline chloride-based deep eutectic solvents: Effect of temperature and water. *Journal of Physical Chemistry B*, *118*, 14652–14661.

Parniakov, O., Barba, F. J., Grimi, N., Marchal, L., Jubeau, S., Lebovka, N., & Vorobiev, E. (2015). Pulsed electric field and pH assisted selective extraction of intracellular components from microalgae *Nannochloropsis*. *Algal Research*, *8*, 128–134.

Ruiz Gonzalez, J., Olivieri, G., De Vree, J., Bosma, R., Willems, P., Reith, H., Eppink, M. H. M., Kleinegris, D. M. M., Wijffels, R. H., & Barbosa, M. J. (2016). Towards industrial products from microalgae. *Energy and Environmental Sciences*, *9*, 3036–3043.

Sanchez-Fernandez, A., Edler, K. J., Arnold, T., Venero, D. A., & Jackson, A. J. (2017). Protein conformation in pure and hydrated deep eutectic solvents. *Physical Chemistry Chemical Physics*, *19*, 8667–8670.

Sanchez-Fernandez, A., Jackson, A. J., Prevost, S. F., Doutch, J. J., & Edler, K. J. (2021). Long–range electrostatic colloidal interactions and specific ion effects in deep eutectic solvents. *Journal of the American Chemical Society*, *143*, 14158–14168.

Schön, A., Clarkson, B. R., Jaime, M., & Freire, E. (2017). Temperature stability of proteins: Analysis of irreversible denaturation using isothermal calorimetry. *Proteins*, *85*, 2009–2016.

Sed, G., Cicci, A., Jessop, P. G., & Bravi, M. (2018). A novel switchable-hydrophilicity, natural deep eutectic solvent (NaDES)-based system for bio-safe biorefinery. *RSC Advances*, *8*, 37092–37097.

Silva, L. P., Fernández, L., Conceição, J. H. F. F., Martins, M. A. R., Sosa, A., Ortega, J., Pinho, S. P., & Coutinho, J. A. P. (2018). Design and characterization of sugar-based deep eutectic solvents using conductor-like screening model for real solvents. *ACS Sustainable Chemistry and Engineering*, *6*, 10724–10734.

Stefanovic, R., Ludwig, M., Webber, G. B., Atkin, R., & Page, A. J. (2017). Nanostructure, hydrogen bonding and rheology in choline chloride deep eutectic solvents as a function of the hydrogen bond donor. *Physical Chemistry Chemical Physics*, *19*, 3297–3306.

Suarez Garcia, E., V. Leeuwen, J., Safi, C., Sijtsma, L., Eppink, M. H. M., Wijffels, R. H., & Van den Berg, C. (2018). Selective and energy efficient extraction of functional proteins from microalgae for food applications. *Bioresource Technology*, *268*, 197–203.

Sultana, N., & Ismail, K. (2016). Specific ion effects of chloride vis-à-vis acetate, propionate and butyrate counterions on the cetylpyridinium headgroup at the micelle-solution and air-solution interfaces. *Journal of Molecular Liquids*, *213*, 145–152.

Tang, W., & Row, K. H. (2020). Design and evaluation of polarity controlled and recyclable deep eutectic solvent based biphasic system for the polarity drive extraction and separation of compounds. *Journal of Cleaner Production*, *268*, 122306.

Teles, A. R. R., Capela, E. V., Carmo, R. S., Coutinho, J. A. P., Silvestre, A. J. D., & Freire, M. G. (2017). Solvatochromic parameters of deep eutectic solvents formed by ammonium-based salts and carboxylic acids. *Fluid Phase Equilibria*, *448*, 15–21.

Tommasi, E., Cravotto, G., Galletti, P., Grillo, G., Mazzotti, M., Sacchetti, G., Samorì, C., Tabasso, S., Tacchini, M., & Tagliavini, E. (2017). Enhanced and selective lipid extraction from the microalgae *P. tricornutum* by dimethyl carbonate and supercritical CO_2 using deep eutectic solvents and microwaves as pretreatment. *ACS Sustainable Chemistry and Engineering*, *5*, 8316–8322.

Umapathi, R., Reddy, P. M., Rani, A., & Venkatesu, P. (2018). Influence of additives on thermoresponsive polymers in aqueous media: A case study of poly(*n*-isopropylacrylamide). *Physical Chemistry Chemical Physics*, *20*, 9717–9744.

Wagle, D. V., Deakyne, C. A., & Baker, G. A. (2016). Quantum chemical insight into the interactions and thermodynamics present in choline chloride based deep eutectic solvents. *Journal of Physical Chemistry B*, *120*, 6739–6746.

Wijffels, R. H., Barbosa, M. J., & Eppink, M. H. M. (2010). Microalgae for the production of bulk chemicals and biofuels. *Biofuels, Bioproducts and Biorefining*, *4*, 287–295.

Wojeicchowski, J. P., Ferreira, A. M., Abranches, D. O., Mafra, M. R., & Coutinho, J. A. P. (2020). Using COSMO-RS in the design of deep eutectic solvents for the extraction of antioxidants from rosemary. *ACS Sustainable Chemistry and Engineering*, *8*, 12132–12141.

Xu, K., Wang, Y., Huang, Y., Li, N., & Wen, Q. (2015). A green deep eutectic solvent-based aqueous two-phase system for protein extracting. *Analytica Chimica Acta*, *864*, 9–20.

Yap, B. H. J., Crawford, S. A., Dumsday, G. J., Scales, P. J., & Martin, G. J. O. (2014). A mechanistic study of algal cell disruption and its effect on lipid recovery by solvent extraction. *Algal Research*, *5*, 112–120.

Yoshizawa, M., W., X., & Angell, C. A. (2003). Ionic liquids by proton transfer: Vapor pressure, conductivity, and the relevance of pK_a from aqueous solutions. *Journal of the American Chemical Society*, *125*, 15411–15419.

Zahn, S., Kirchner, B., & Mollenhauer, D. (2016). Charge spreading in deep eutectic solvents. *Chemistry & Physics of Lipids, Colloids and Interfaces*, *17*, 3354–3358.
Zhang, H., Wang, Y., Xu, K., Li, N., Wen, Q., Yang, Q., & Zhou, Y. (2016). Ternary and binary deep eutectic solvents as a novel extraction medium for protein partitioning. *Analytical Methods*, *8*, 8196–8207.
Zhang, R., Lebovka, N., Marchal, L., Vorobiev, E., & Grimi, N. (2020). Pulsed electric energy and ultrasonication assisted green solvent extraction of bio-molecules from different microalgal species. *Innovative Food Science and Emerging Technologies*, *62*, 102358.
Zhao, H. (2015). Protein stabilization and enzyme activation in ionic liquids: Specific ion effects. *Journal of Chemical Technology and Biotechnology*, *91*, 25–50.

Samara C. Silva, Madalena M. Dias, M. Filomena Barreiro*

Chapter 6
What is next in microalgae research

Abstract: Due to the increasing competition with finite natural supplies and to environmental concerns, all segments of society seek for sustainable and efficient resources. In this framework, microalgae have attracted significant interest worldwide as they are fast-growing microorganisms and do not compete with traditional crops for arable land. They present a unique nutritional composition with high content of proteins, lipids, polysaccharides, enzymes, pigments, and bioactive compounds. Moreover, these microorganisms play an essential role in Earth's sustainability since they convert CO_2 into O_2 and can be cultivated in harsh conditions. They are applied in several fields, such as food, feed, health & well-being, and cosmetics, contributing to the global economic growth, with several promising applications continuously emerging. Examples include agricultural-based products and biomaterials, their exploitation for wastewater treatment and CO_2 removal from industrial flue gases, the use of their proteins as natural emulsifiers, and their pigments as natural colorants, where genetic engineering is being applied to potentiate the improvement of cultivation and specific characteristics. Due to all these scenarios, microalgae can be easily integrated within a biorefinery approach, being used as a feedstock for biofuels and bioenergy

Acknowledgments: The authors are grateful to the Foundation for Science and Technology (FCT, Portugal) for financial support through national funds FCT/MCTES (PIDDAC) to CIMO (UIDB/00690/2020 and UIDP/00690/2020), SusTEC (LA/P/0007/2021), LSRE-LCM (UIDB/50020/2020 and UIDP/00690/2020), and ALiCE (LA/P/0045/2020). FCT for the PhD research grant of Samara Cristina da Silva (SFRH/BD/148281/2019).

***Corresponding author: M. Filomena Barreiro,** Centro de Investigação de Montanha (CIMO), Instituto Politécnico de Bragança, Campus Santa Apolónia, 5300-253 Bragança, Portugal; Laboratório Associado para a Sustentabilidade e Tecnologia em Regiões de Montanha (LA SusTEC), Instituto Politécnico de Bragança, Campus de Santa Apolónia, 5300-253 Bragança, Portugal, e-mail: barreiro@ipb.pt

Samara C. Silva, Centro de Investigação de Montanha (CIMO), Instituto Politécnico de Bragança, Campus Santa Apolónia, 5300-253 Bragança, Portugal; Laboratório Associado para a Sustentabilidade e Tecnologia em Regiões de Montanha (LA SusTEC), Instituto Politécnico de Bragança, Campus de Santa Apolónia, 5300-253 Bragança, Portugal; Laboratory of Separation and Reaction Engineering – Laboratory of Catalysis and Materials (LSRE-LCM), Faculdade de Engenharia, Universidade do Porto, R. Dr. Roberto Frias, 4200-465 Porto, Portugal; ALiCE – Associate Laboratory in Chemical Engineering, Faculty of Engineering, University of Porto, Rua Dr. Roberto Frias, 4200-465 Porto, Portugal e-mail: samaras@ipb.pt

Madalena M. Dias, ALiCE – Associate Laboratory in Chemical Engineering, Faculty of Engineering, University of Porto, Rua Dr. Roberto Frias, 4200-465 Porto, Portugal; Laboratory of Separation and Reaction Engineering – Laboratory of Catalysis and Materials (LSRE-LCM), Faculdade de Engenharia da Universidade do Porto, R. Dr. Roberto Frias, 4200-465 Porto, Portugal, e-mail: dias@fe.up.pt

https://doi.org/10.1515/9783110781267-006

production, generating multiple products along the productive value chain, perfectly aligned with the bioeconomy framework. In this context, this chapter presents the microalgal research's past and current status by evaluating the evolution of the leading research fields over the years, supported by a bibliographic and bibliometric analysis. Particular emphasis is given to recent developments and perspectives driving the near future of the microalgae research.

Keywords: microalgae biotechnology, bioeconomy, microalgae biorefinery, wastewater treatment, microalgal proteins, microalgal pigments, bioenergy, biomaterials, biofertilizers, CO_2 removal

6.1 Introduction

Microalgae are a diverse group of prokaryotic and eukaryotic microorganisms conducting oxygen-evolving photosynthesis. The prokaryotic group refers to cyanobacteria (Cyanophyta), and the eukaryotic group includes green algae (Chlorophyta), red algae (Rhodophyta), diatoms (Bacillariophyta), golden-brown algae (Chrysophyta), yellow-green algae (Xanthophyta), brown algae (Phaeophyta), and some other divisions (Smith et al., 2021).

Regarding microalgae nutrition, they can be produced by the autotrophic, heterotrophic, or mixotrophic processes. Autotrophic microalgae production uses inorganic compounds as carbon sources (CO_2) and can use light as energy source (photoautotrophic) or oxidize inorganic compounds to obtain energy (chemoautotrophic). It is estimated that each ton of microalgae requires up to 2 ton of CO_2, 0.1 ton of N, 0.010 ton of P, and 0.015 ton of K, releasing 2 ton of O_2 during the process. On the other hand, heterotrophic microalgae production uses organic compounds as carbon sources, being photoheterotrophs or chemoheterotrophs. Generally, 2 kg of glucose are required to produce 1 kg of microalgal biomass. Some microalgae are mixotrophic and obtain energy from light using organic compounds as carbon sources (Fernández et al., 2021; Lee, 2008).

Microalgae can be cultivated in open or closed systems. Open systems are commonly used for species like *Arthrospira*, *Chlorella*, *Dunaliella*, and *Haematococcus* since these systems can lead to adequate quality biomass for human-related products with lower production costs. They consist of outdoor reactors directly exposed to sunlight and atmosphere contaminants, impeding a severe culture control. Even so, the raceway ponds are the most widely used open reactors, contributing to more than 90% of the autotrophic microalgal production worldwide. On the contrary, closed systems, also known as photobioreactors, can be made from different materials, for example, plastic or glass, being isolated from the atmosphere and thus avoiding environmental contaminations (Fernández et al., 2021; Silva et al., 2020).

Microalgae have attracted interest worldwide due to their nutritional rich composition, flexibility to adapt to different production conditions, and the fact of no arable land is needed for cultivation. Moreover, they are fast-growing microorganisms and use mainly sunlight as the energy source. Microalgae are rich in proteins containing all essential amino acids, lipids with omega-3 fatty acids, minerals, polysaccharides, enzymes, and pigments (chlorophylls, carotenoids, and phycobiliproteins). For these reasons, they are key candidates to generate high-value compounds through sustainable processes, contributing to the global economy (Fernández et al., 2021; Khan et al., 2018).

Some microalgal species have been already produced on a large scale, with the major ones being *Arthrospira* (Spirulina) *platensis* and *Chlorella* spp., used for nutritional purposes, and *Dunaliella salina* and *Haematococcus pluvialis*, used as a source of β-carotene and astaxanthin, respectively. Although large-scale production of microalgae is based mainly in open-air culture systems (especially raceway ponds), some companies have been applying closed photobioreactors (Borowitzka, 2016).

Several microalgae-derived products have been produced in different areas such as food and feed, pharmaceuticals, nutraceuticals, and cosmeceuticals. Microalgae have also been applied for energy production (e.g., bioethanol and biodiesel (Rempel et al., 2019; Sumprasit et al., 2017), wastewater treatment (Chavan and Mutnuri, 2019), obtainment of biomaterials (López Rocha et al., 2020), biostimulants, and biofertilisers (Suchithra et al., 2022; Varia et al., 2022). The increased interest in microalgae as a sustainable feedstock for biofuels production, combined with their high-value compounds, has led to a new focus on microalgae-based biorefinery approaches aiming at valorizing the whole biomass and obtaining multiproduct chains.

In recent years, many studies and global market reports have shown microalgae's high potential for diverse applications. The global microalgae market was valued at $977.3 million in 2020 and is projected to reach $1,485.1 million by 2028, at a CAGR of 5.4%. The Allied Market Research report also estimates that the microalgae market will have a stable growth in the coming years, which is mainly attributed to the increased application of algal proteins for food and nutrition-related applications (Kumar and Deshmukh, 2021).

This chapter covers the past, current, and future scenario of microalgae research. The work comprises a bibliographic and bibliometric review focused on the microalgae-related publications available on online scientific databases (Scopus and Web of Science) from 1960 to 2021. This allowed identifying three-time stages of publication growth and the most contributing countries. Different thematic areas were disclosed during the three-time stages revealing the microalgae research evolution over the years. Leading research domains were also identified by displaying their annual growth rate (AGR) and the number of publications across time. Challenges and bottlenecks limiting the microalgae market expansion were also tackled, along with recent developments related to improvements in microalgal strains and their cultivation conditions. Particular emphasis was given to the most relevant areas and trends driving the future of microalgae research.

6.2 Past and current status of microalgae research

6.2.1 Bibliographic and bibliometric analysis

Microalgae is a fast-growing field, as can be perceived by the rising number of publications. Two scientific reliable indexing databases, namely Scopus (Elsevier) and Web of Science Core Collection (WOS) (Clarivate), were used to perform a survey using the term "microalga*" in the title, abstract, and keywords (TITLE-ABS-KEY) for Scopus and Topic (title, abstract, author keywords, and Keyword Plus) for WOS as the search query. The asterisk "*" implies any group of characters (including no character), embracing singular "microalga" and plural "microalgae" terms. The survey was performed on May 13, 2022. Within a publication timespan from 1960 to 2021, 38,461 results were found from WOS, while 38,176 results were identified for Scopus. Whether different parameters are used, different results can be obtained. Besides, this method may have some gaps since the introduction of the keywords (author or database) may not suit the subject of the articles. No manual inspection was possible due to the high number of publications. The retrieved data were analyzed concerning the number of publications (Scopus and WOS) and subject categories defined by WOS.

Figure 6.1 depicts the evolution of the number of publications from 1960 to 2021 in the microalgae field for both scientific databases. The first recorded publication dates from 1962 for WOS and 1960 for Scopus, showing that this field has been investigated for at least 60 years. As it can be perceived from Figure 6.1, three-time stages (1, 2, and 3) can be identified during this period, which are characterized by linear fittings with different slopes, that is, different publication rates. The first stage corresponds to the period starting in 1960 (Scopus)/1962 (WOS) to 2005 and presents a slope of 10.46 ($R^2 = 0.781$) and 8.99 ($R^2 = 0.827$) for WOS and Scopus, respectively. For the second one (2005–2017), slopes 20 times higher were obtained (226.5 ($R^2 = 0.953$) (WOS) and 227.4 ($R^2 = 0.951$) (Scopus)). This growing publication rate trend is emphasized in the third stage (2017–2021, the last five years of research), where the slopes doubled compared with the previous stage, achieving values of 429.0 ($R^2 = 0.997$) and 420.2 ($R^2 = 0.995$), highlighting the boosted interest in the field in the last years. Garrido-Cardenas et al. (2018), based on Scopus database, identified two clear trends in the evolution of the number of publications from 1970 to 2017 regarding the microalgae research field, recognizing a significant rise from 2005 to 2017, translated by a slope value more than 15 times higher in comparison with the antecedent period of 1970–2005. This considerable increase may be associated with the intensified consumer's awareness regarding healthy habits and environmental issues. Consumers and industries have been trying to find eco-friendly and sustainable alternatives to a diversity of products and processes. Moreover, Rumin et al. (2020a) suggested that the fast acceleration from 2005 could be related to the outbreak of microalgae application as the third-generation raw material for biodiesel production, which started at the beginning of the twenty-first century. These findings are important indicators reinforcing the relevance of the microalgae field in the past, current, and future research.

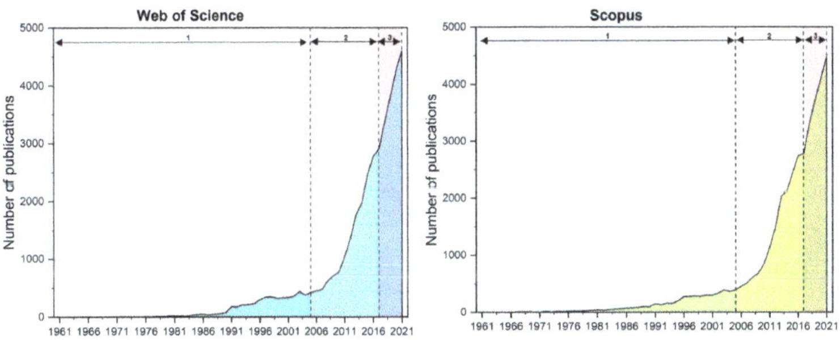

Figure 6.1: Evolution of the number of publications regarding microalgae research from 1960 to 2021 obtained from the Web of Science and Scopus database and reflecting three stages (1, 2, and 3) characterized by an increased rate of publication.

6.2.1.1 Most productive countries

Since WOS database presented a higher number of publications along the studied period, in comparison with Scopus, the subsequent analyses were performed using only the data retrieved from the Web of Science Core Collection.

Figure 6.2 displays the world map with the number of publications per country, from 1960 to 2021, translated in a graded colored scale. China and the United States of America (USA) are the leaders in the microalgae research field, corroborating the results of some published works (Garrido-Cardenas et al., 2018; Rumin et al., 2020a). However, according to these works USA was the leading country in number of publications, followed by China considering the analyzed timespan, namely up to 2017 (Garrido-Cardenas et al., 2018) and up to 2019 (Rumin et al., 2020a). The current work (up to 2021) shows that USA (5,595 publications) was surpassed by China, which is now the country with the highest publication productivity (6,699 publications), result for which the last 5 years have highly contributed. The third country in number of publications is Spain, showing 2,602 publications in the studied period.

6.2.1.2 Most published WOS subject areas

Concerning WOS subject areas, Figure 6.3 displays the 10 most published areas during the three-time stages defined in Figure 6.1. In the first period (1960–2005), the microalgae research was more centered on biological sciences, including phycology and physiology, marine ecology, and oceanography. WOS subject areas such as "Fisheries," "Ecology," "Plant Sciences," "Microbiology," and "Toxicology" also appeared during this period.

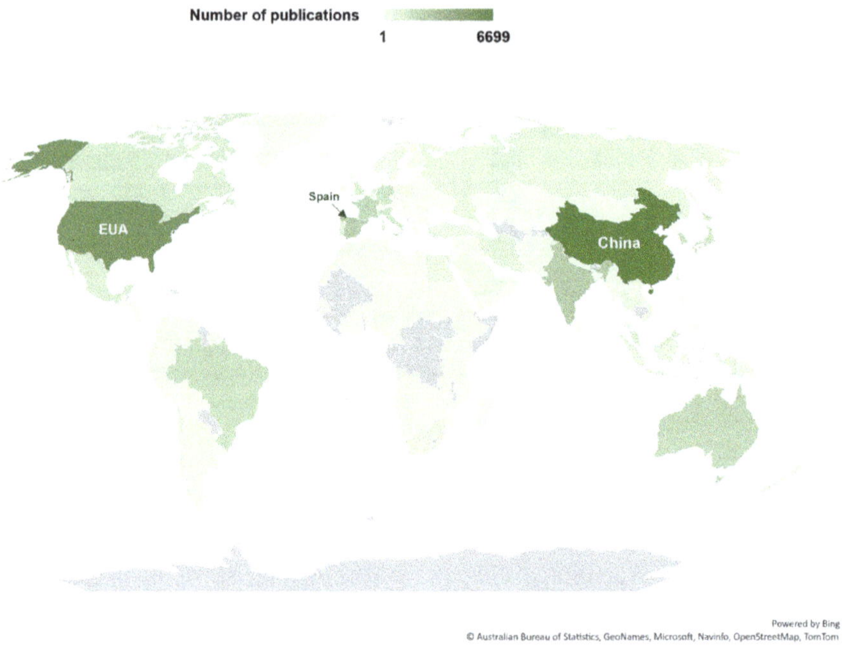

Figure 6.2: World map evidencing the leading countries in number of publications according to WOS database in the timeframe 1960–2021.

As the microalgae research evolved from the 1960–2005 to the 2005–2017 period, some WOS categories disappeared, giving rise to new ones, such as "Energy fuel," "Agricultural," "Environmental," and "Chemical Engineering." In the second period (2005–2017), applications of microalgae and their compounds started to be studied in different fields. Moreover, it is worth highlighting that the "Energy Fuel" WOS category appears as the second most important published subject area (Figure 6.3), corroborating the high interest in this topic during this period of time.

In the last five-year period (2017–2021), the "Marine Freshwater Biology" subject area, which was the third published area during 2005–2017, appears in the fifth position, while "Environmental Sciences" and "Chemical Engineering" emerged in the third and fourth position, respectively. This trend demonstrates that studies regarding the application of microalgae in different fields continued to raise interest, while more conceptual thematics started to decline as they become well-developed fields. Reinforcing, in the last five years, WOS subject areas like "Ecology" and "Plant Sciences" disappeared while "Green Sustainable Science Technology" and "Multidisciplinary Sciences" emerged as new subject categories.

"Biotechnology & Applied Microbiology" was the most published WOS subject area during the two studied periods (2005–2017 and 2017–2021). This WOS category covers several fields related to the use of living organisms or their compounds in dif-

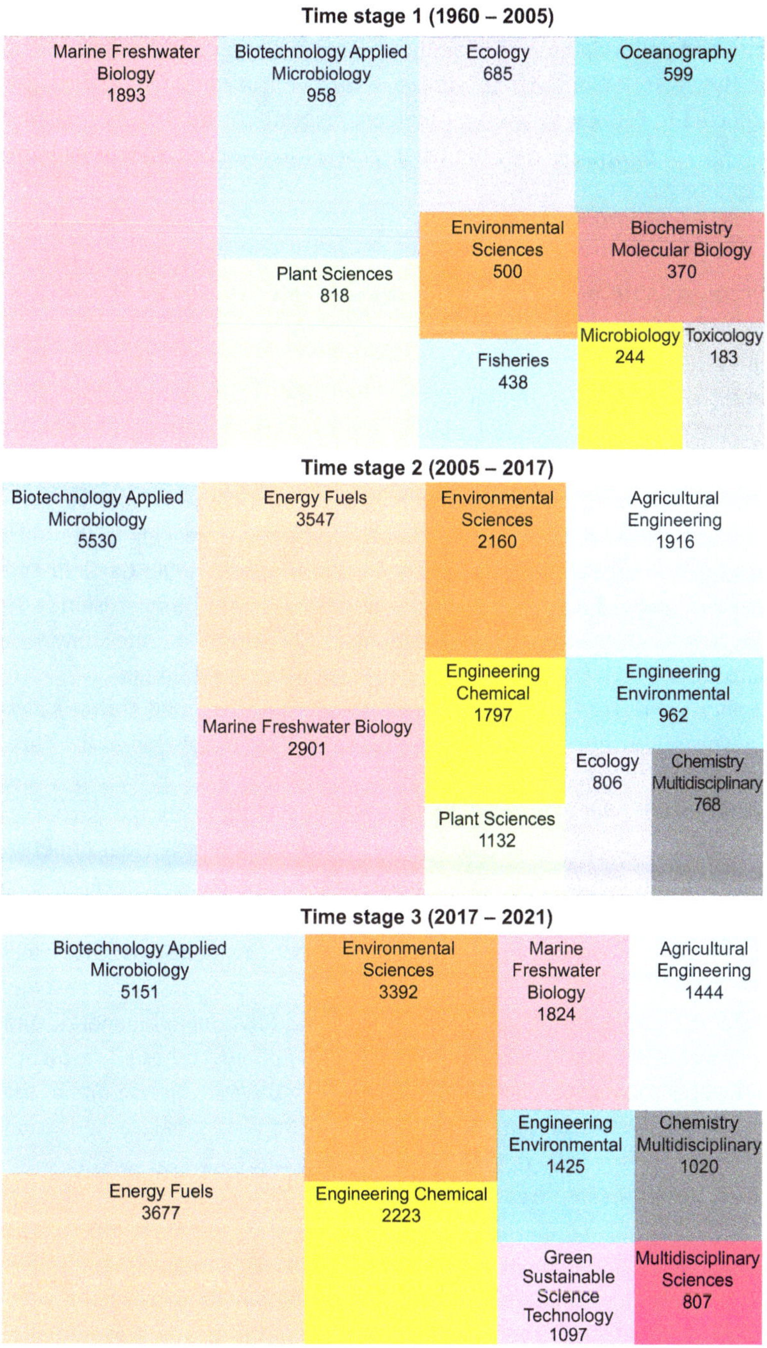

Figure 6.3: The 10 most published WOS categories during the three-time stages defined in Figure 6.1: 1 (1960–2005), 2 (2005–2017), and 3 (2017–2021) putting in evidence the evolution of their importance with time.

ferent applications, like industrial chemicals, food, flavors, fragrances, pesticides, waste treatment, and pollution bioremediation, to name just a few.

Looking at the evolution of the WOS subject areas over the years, the microalgae research is expected to remain focused on biotechnological approaches by applying integrated processes to minimize environmental issues and foster industrial-scale production and exploitation.

6.2.1.3 Most frequent author's keywords and *Keyword Plus*

To better identify the research trends in the microalgae field and their evolution over the studied timespan (1960–2021), the most frequent author's keywords and the *Keyword Plus* from WOS were explored to build the word clouds. The author's keywords correspond to terms provided by the authors as the ones better representing the manuscript content. The *Keyword Plus* resulted from an algorithm from the Clarivate Analytics database that is based on words or sentences frequently appearing in the titles of cited articles but not necessarily in the article's title (Garfield, 1990; Garfield and Sher, 1993). The retrieved data from WOS was analyzed using the open-source R-Studio software (www.rstudio.com), with a bibliometrix R-package and the shiny web interface (biblioshiny). The word clouds were generated using Prezi Design.

Figure 6.4 shows the word cloud graphs with the 50 most frequent author's keywords and *Keywords Plus* during the searched period. "Biodiesel," "biomass," "biofuel," "bio-oil," "biogas," "lipids," and "fatty acids" are words mainly related to the most significant research field of microalgae research, namely biofuels and bioenergy production. Concerning bioenergy technologies, words like "hydrothermal liquefaction," "pyrolysis," and "anaerobic digestion" appear as the most relevant ones.

Regarding microalgal species, *Chlorella vulgaris, Chlamydomonas reinhardtii, Haematococcus pluvialis, Scenedesmus obliquus,* and *Spirulina platensis* (which likely refers to *Arthrospira platensis*) appear as the most relevant ones, being extensively studied, and related to several applications areas, such as food, feed, bioenergy, and wastewater treatment. Keywords like "photobioreactor," "photosynthesis," "growth," "light," "temperature," "nitrogen," "accumulation," and "cultivation" arise due to the highest number of publications regarding the optimization of microalgal cultivation and growth parameters, especially in what concerns photobioreactors.

"Wastewater," "wastewater treatment," "bioremediation," "nutrient removal," and "carbon dioxide" are also important keywords centered on the use of microalgae for bioremediation and contaminants removal from wastewater treatment as well as CO_2 capture from the atmosphere. Moreover, the exploitation of microalgal carotenoids is also a relevant topic since keywords like "astaxanthin" and "carotenoids" were among the 50 most frequent ones. Biorefinery also emerged as one of the relevant keywords in the microalgae research field associated to the current investigation aligned with the circular bioeconomy.

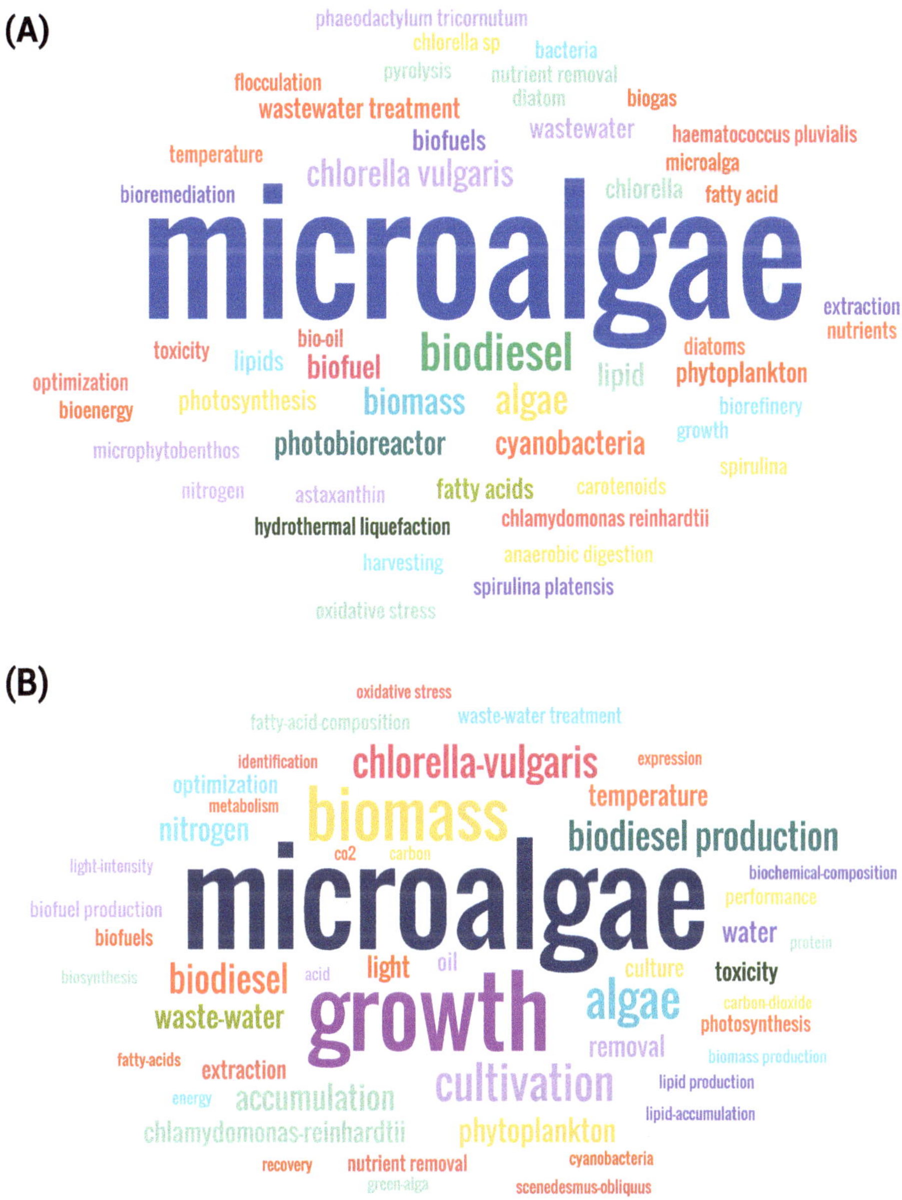

Figure 6.4: Word clouds of the most 50 frequent words from (A) author's keywords and (B) *Keyword Plus* from WOS, during the period 1960–2021.

6.2.1.4 Evolution of the microalgae-based research fields

Microalgae applications are extensively diverse, covering the following fields: (i) human nutrition and health (foods, pharmaceuticals, nutraceuticals, and cosmetics); (ii) animal feed (aquaculture, premix feed); (iii) energy (biofuels); (iv) biomaterials (chemical commodities, bioplastics); (v) agricultural (biofertilizers and biostimulants); and (vi) bioremediation (wastewater treatment, nutrient removal, and CO_2 capture from flue gases) (Vieira et al., 2022).

Thus, to better analyze the evolution of the number of documents published over the years regarding the different application areas, the word "microalga*" was combined with the respective research field, namely "biofuel or bioenergy," "food," "feed," "aquaculture," "nutraceutical*," "pharmaceutical," "pigment*," "colorant*," "emulsifier," "biomaterial," "biofertilizer," "biostimulant," "wastewater treatment," "CO_2 capture or carbon dioxide capture or CO_2 fixation," "genetic engineering," "bioremediation," "biorefinery," "bioeconomy," and "circular economy" and applied in topic (title, abstract, author's keywords, and *Keyword Plus*) as a search query in the WOS Core Collection from 1960 to 2021. The evolution of the number of publications concerning the different research areas is shown in Figures 6.5A.1 and 6.5A.2. The cumulative number of publications and the AGR (%) are displayed in Figure 6.5B.

To date, biofuel & bioenergy is the field with the highest number of published manuscripts, followed by food and feed areas (Figure 6.5A.1). As discussed in Section 6.2.1, the number of publications on microalgae research has significantly increased due to the interest in exploiting microalgae for biofuels production around 2005. Several companies like Origin Oil, Solix Biofuel, Sapphire, Energy, Solazyme, Petroalgae, and Aurora Biofuel have started investing in microalgal biofuels to make them competitive against their fossil counterparts. Nonetheless, due to the high costs and low feasibility, many companies have changed their productive efforts to microalgal biorefinery through their exploitation in the food and feed market as well as in value-added products (Behera et al., 2022).

Biofuel & bioenergy field shows a clear increase around 2006–2007, reaching a peak in 2020 and a slight decrease in 2021 (Figure 6.5A.1). On the other hand, food and feed areas seem to have continuously increased over the years. The fourth most published field is related to the wastewater treatment. This field also shows a continuous increase, especially from 2017 to 2021, which is likely to rise even more due to the excellent results deriving from the wastewater treatment integration in microalgal cultivation (Goswami et al., 2021; Hussain et al., 2021). Genetic engineering has been also another growing field in the past two years, owing to the positive impact of genetic-engineered microalgae strains on the production of specific compounds or cultivation conditions (Fayyaz et al., 2020; Spicer and Molnar, 2018). Similarly, Rumin et al. (2020b) have identified genetically modified microalgae as a fast-evolving technological domain within the European microalgae market. The field of the microalgae-derived pigments has also witnessed a significant increase over the years. This field has gained relevance due to food industry needs in replacing synthetic colorants, being microalgal pigments considered promising substitutes. Silva et al. (2020) concluded that the most relevant pigments in microalgae research were chlorophylls, phycocyanin, astaxanthin, and β-carotene. The most relevant sources were *Chlorella vulgaris, Arthrospira (Spirulina) platensis, Haematococcus pluvialis*, and *Dunaliella salina*, respectively. This statement corroborates with the results found in Section 6.2.1.3, where carotenoids and astaxanthin appear as the most 50 frequent words.

The AGR measures the number of publications over the years. The higher the AGR is, the higher is the field increase. In this context, the highest AGR was obtained for the fields of circular economy, bioeconomy, biostimulants, and biorefinery (Figure 6.5B). It is worth to highlight that the circular economy area presented an AGR higher than 100%, testifying its relevance in the microalgae research. The first manuscripts dealing with circular economy and bioeconomy microalgae dated from 2013; however, a significant increase was noticed during the last three years (2019–2021). Concerning the biorefinery microalgae-related manuscripts, the first ones started to appear around 2008, with a clear increase in the number of publications observed in 2012, resulting in more than 1,000 manuscripts until 2021. The biorefinery approach began to be associated with microalgae to decrease the costs related to their cultivation and dedicated biofuel production. This is a growing area with an AGR of more than 40%, indicating that microalgae biorefinery is under development and significant advances are expected in the near future. Even the biofuel & bioenergy field decreased the number of publications in 2021, a significant AGR (31.8%) was found, indicating that these studies integrated within the biorefinery concept can still influence the future of microalgae research.

Although biofertilizers, biostimulants, and biomaterials are fields with less publications, they will significantly impact the future of microalgae research, presenting significant AGRs. In fact, there is an increasing demand for materials, agricultural, cosmetic and food products with healthier, sustainable, and eco-friendly ingredients, being microalgae the best productive chain fitting all these requirements. Rumin et al. (2020b) also found that biofertilizers and biostimulants are recent research domains in Europe in their survey performed in 2019.

The exploitation of microalgae as natural emulsifiers is also an emerging field. The first publication dates from 2005, then with no publications until 2012. The number of publications increased during the last five years and is expected to increase even more owing to the potential of microalgal proteins to be used as natural emulsifiers (Böcker et al., 2021; Silva et al., 2022).

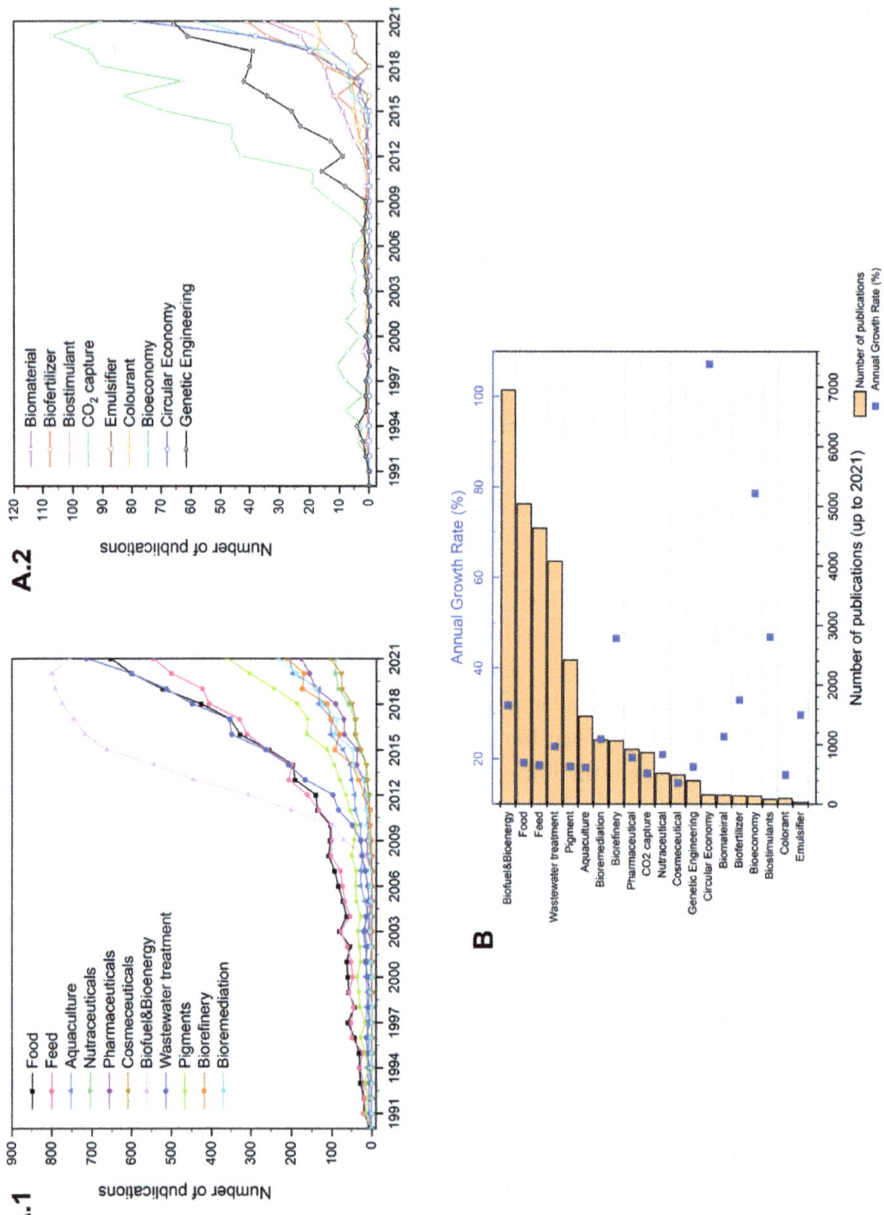

Figure 6.5: Evolution of the number of publications in different microalgae-related research fields from 1990 to 2021 (A.1 and A.2). Cumulative number of publications and the annual growth rate (%) of each microalgae-related field (B).

6.2.2 Challenges and bottlenecks

Although microalgae research has evolved throughout time, several fields have not yet been transposed to an industrial scale. Nowadays, the microalgae global market is segmented into dietary supplements, food (food ingredients, functional foods), feed (premix feeds), health (pharmaceutical and nutraceutical), and cosmetics. According to the Allied Market Research report, the largest market share corresponds to dietary supplements (powders, pills, and capsules) (Kumar and Deshmukh, 2021).

Microalgae are still produced at a small scale when compared to conventional crops like soy or fish-related products. Currently, most production facilities worldwide are small and medium-sized installations, with open systems ranging from 5 to 50 ha and closed photobioreactors from 10 to 500 m^3. To date, the microalgae production comprises 50,000 tons/year for five species: *Spirulina*, *Chlorella*, *Dunaliella*, *Haematococcus*, and *Nannochloropsis*. The majority of the production is directed to human consumption and nutritional products manufacturing since these sectors can cover the current biomass production costs (5–20 €/kg) (Fernández et al., 2021; Vieira et al., 2022).

Since microalgae can be produced by using open and closed systems, Banu et al. (2020) compared photobioreactors with open pond cultivation for biofuel production. They concluded that the first form accounts for 81.17% of the overall production costs while the latter one with only 45.73%. This is mainly due to the investment needed for the photobioreactors, which is four times higher than the one required for open pounds. On the other hand, closed systems can generate higher productivity compared to the open counterparts. The authors concluded that the market price of the target product plays an essential role in deciding the cultivation mode (Banu et al., 2020). When the final product is the microalgae biomass itself, not only cultivation technologies are required but also downstream processes such as harvesting, pretreatment (depending on the specie), and drying. However, once the purpose is to obtain a high-value product, processes like pretreatment, extraction, and purification are needed, increasing the production price (Silva et al., 2020).

Generally, the harvesting process, which refers to separating the biomass from the culture medium, is one of the major bottlenecks in microalgae production. According to the projections of Ruiz et al. (2016), the harvesting step contributes with circa 23% of the cultivation cost (1.2 €/kg) for raceways, while only 5–7% (0.2–0.3%/kg) of the total cost for closed systems, owing to the higher biomass concentration achieved in this latter production mode. Nonetheless, costs of raw materials (17–23%) and energy consumption (14–17%) turn out to be relevant in closed photobioreactors. Behera et al. (2022) also state that harvesting is the most challenging task sharing 20–30% of the total production costs (Barros et al., 2015).

Vieira et al. (2022) disclose the 10 most relevant bottlenecks for both macro- and microalgae production, as such (i) logistics (due to the complex infrastructures); (ii) contaminations; (iii) market demand (production costs restrict the potential market);

(iv) value chain; (v) labor force (lack of experienced professionals); (vi) investability (it takes a long time to return the investment); (vii) product technologies (high value and small scale); (viii) business model; (ix) bioprocessing (lengthy processing); and (x) political incentives. Moreover, the authors state that the listed bottlenecks are interrelated, needing to be addressed simultaneously (Vieira et al., 2022).

6.2.3 New developments in microalgal biotechnology

Advanced technologies are already in use in microalgae research with several benefits leading to new processes and marketable products. These include biotechnological approaches leading to microalgal strain improvements and application of digitalized approaches for the identification of microalgal species and optimization of their cultivation conditions.

Microalgae biomass is characterized by presenting a strong green color, odor, and taste. Due to these organoleptic properties, incorporating them into food products can be challenging since they can strongly impact consumer acceptance. In this framework, strategies aiming at isolating novel microalgal strains with superior organoleptic characteristics have been studied. Examples include the development of chlorophyll-deficient mutants of *Chlorella vulgaris* by applying chemically induced random mutagenesis, a nongenetically cell modification. As a result, they have obtained *C. vulgaris* mutants with yellow and white colors due to the decrease in their chlorophyll contents achieved under heterotrophic growth (Schüler et al., 2020). Recently, Allmicroalgae (www.allmicroalgae.com), a Portuguese company, started to sell these microalgal biomass powders as Clorela Honey® (yellow color) and Clorela White® (white color) for food applications.

Digitalization technologies have also attracted significant attention in the microalgae research field. Otálora et al. (2021) evaluated two models for microalgae identification using artificial neural networks. The main purpose for the first model was to characterize cultures composed of *Chlorella vulgaris* and *Scenedesmus almeriensis* based on their morphological data, whereas in the second model the focus was to characterize the microalgae using their cell images. The authors conclude that image analysis and deep learning techniques allow microalgae culture identification as the feature-based model presented high accuracy, widening the range of classification methods in the microalgae field (Otálora et al., 2021). del Rio-Chanona et al. (2019) suggested a deep learning model centered on a convolutional artificial neural network aiming at optimizing operation conditions and photobioreactor configuration in a pilot-scale microalgal biofuel production plant. Moreover, Teng et al. (2020) proposed that artificial intelligence (AI) algorithms can be used to extract crucial information and foresee molecular interactions concerning gene sequencing and editing. Furthermore, by reducing the number of experiments and optimizing the cultivation conditions, AI algorithms can enhance microalgae cultivation and conversion conditions (Teng et al., 2020).

6.3 Future perspectives

Even though microalgae production has not yet reached a very large scale (>100 ha, > 5,000 ton/year: open systems; >2,000 m^3, >50–70 ton/year: closed systems), researchers and entrepreneurs are working on solving bottlenecks and challenges aiming at decreasing the production costs and increasing market competitiveness. In this framework, the biorefinery and circular bio-economy concepts have been extensively studied and gained significance as great approaches to make profit microalgal commercialization (Behera et al., 2022; Fernández et al., 2021). 't Lam et al. (2018) state that a multi-product biorefinery is required to make microalgal production of bulk commodities economically viable meaning that all biomass fractions need to be valorized.

The results of the present work corroborate these statements by showing that biorefinery, and circular bio-economy presented the highest AGR regarding the number of publications. Besides, from a previous work of the group in which a bibliometric study on microalgal pigments was conducted, it was already perceived an emerging interest in the biorefinery concept applied to microalgae field (Silva et al., 2020). In fact, these research fields are expected to expand, bringing new developments in the coming years.

Microalgae production combined with wastewater treatment and CO_2 capture will also be explored as it is an integrated and sustainable system. Some political priorities also encourage the transition to a sustainable economy balancing the economic growth, environmental protection, and supporting the requirements of a growing global population (Araújo et al., 2021). For example, the European Bioeconomy Strategy targets to implement a sustainable and circular economy across Europe (European Comission, 2018). Moreover, the European Green Deal aims to make Europe climate-neutral by 2050 (European Commission, 2021) and contribute to the "farm to fork" strategy for fair, healthy, and sustainable food chains (European Commission, 2020).

Emerging fields like agricultural-related products (biostimulants and biofertilizers), biomaterials, emulsifiers, and genetic engineering strategies will also be targeted in the future of microalgae research. It is also forecasted that significant advances in the biofuels field toward a multiproduct biorefinery will be achieved by replacing the non-profit single-product facility. Moreover, the production of biofuels using wastewater as the cultivation medium will be intensified under the future circular economy context.

6.3.1 Circular economy

The circular economy is an economic model where resources are exploited to their full potential avoiding wastes, ideally achieving zero waste. The idea is basically to close loops in industrial processes and minimize waste disposal to the environment (Deutz, 2020; Fuentes-Grünewald et al., 2021). Microalgae can fit in the circular economy framework as they can bioremediate nutrient waste and be a source of biomass for several commercial products (Fuentes-Grünewald et al., 2021).

Some studies have shown that utilizing microalgae within a circular economy concept results in the generation of valuable products (Goswami et al., 2021; Kholssi et al., 2021). Although the products obtained by microalgal production using wastewater cannot be destined for human-related applications, products like biofuels (Li et al., 2021), biofertilizers (Mukherjee et al., 2016), and biomaterials (bioplastics) (López Rocha et al., 2020; Mastropetros et al., 2022) can be produced. Animal feed can also be considered only whether the safety and quality of the food, agricultural, and aquaculture waste are assured (Vieira et al., 2022).

Llamas et al. (2021) performed a techno-economic analysis to examine CO_2 biofixation and wastewater treatment in microalgae cultivation. The authors concluded that by integrating microalgae production with wastewater treatments, the average biomass production cost can be reduced from 1.6 to 0.50 €/kg as nutrients and water costs are saved. Moreover, they showed that by using flue gases to capture CO_2, the production cost can be reduced from 1.6 to 0.88 €/kg, in the case of open raceways. Sydney et al. (2019) proposed the integration of liquid and gaseous effluents from the bioethanol industry in the microalgae production. In this work, a high carbon transfer rate was obtained, reducing chemical and biological oxygen demand and turbidity, while a biomass production of 2.25 g/L was achieved during 15 days of cultivation. Since microalgae can remove nutrients from wastewater and thrive on greenhouse gases, these integrative scenarios enable to "close the loop" and thus generate a circular bioeconomy.

6.3.2 Biorefinery approach

Microalgal biorefinery is an important topic in microalgae research due to the richness of these microorganisms in several biological compounds (e.g., proteins, lipids, carbohydrates, and pigments). The biorefinery concept refers to biomass conversion into various products such as fuels, chemicals, materials, and food/feed goods. This approach minimizes waste since it valorizes the whole biomass contributing to reduce the high costs related to the up and downstream processes (Chew et al., 2017).

Several European projects aiming at producing microalgal biorefineries have been funded during the last years. D-Factory was an EU-funded project (2013–2017) that aimed to produce a sustainable biorefinery from *Dunaliella*. CYCLALG (2016–2019) was another EU project targeting an innovative microalgae biorefinery promoting a circular economy and zero-waste production. One of the main goals of ABACUS (2017–2020) project was to develop a new algal biorefinery to obtain several high-value products ranging from algal terpenes to long-chain terpenoids (carotenoids) for nutraceutical and cosmetic applications. SABANA (2016–2021), another relevant project, addressed a large-scale integrated microalgal biorefinery targeting the production of biostimulants, biopesticides, and feed additives by applying marine water and wastewater as cultivation media. More recent projects like SPIRALG (2018–2023) are focused on phycocyanin production from Spirulina with covalorization of the whole microalgae biomass, in line with an industrial biorefinery concept. MULTI-STR3AM (2020–2025) project intends to develop a sustainable multiproduct microalgal biorefinery by integrating industrial side streams to produce high-value products for the food, feed, and fragrances industry.

It is worth highlighting that some works aiming at extracting compounds using a cascade process within a biorefinery concept have been proposed. Sintra et al. (2021) performed a cascade recovery of C-phycocyanin and chlorophylls from *Anabaena cylindrica*. First, they extracted the C-phycocyanin from *Anabaena* fresh biomass and the chlorophylls from the biomass residues. Extraction recoveries of 90% and 55% of the total content in C-phycocyanin and chlorophylls were obtained, respectively. Monlau et al. (2021) also developed a cascade biorefinery process to obtain five bioproducts from *Chlorella protothecoides* biomass according to the following stages: (i) the lipid fraction of microalgal biomass was converted into fatty acids and then into biodiesel; (ii) the deoiled biomass was submitted to enzymatic hydrolysis and converted into a liquid hydrolysate composed by soluble amino acids and sugars; (iii) the leftover solid fraction was used as a substrate in an anaerobic digestion process to produce biogas, and the digestate was analyzed as a fertilizer. Therefore, upon this sequential processing of the microalgal biomass, a multiproduct chain was suggested as a possible way to improve microalgal biofuels profitability by integrating three conversion pathways.

6.3.3 Techno-economic analysis of microalgal biorefinery scenarios

During these years of microalgae research, different biorefinery scenarios were proposed to check the effect of different variables (e.g., cultivation system, location, and operational parameters). For this reason, techno-economic analysis and life cycle assessment have been employed to better understand the feasibility of microalgal biorefineries (Slegers et al., 2020). Table 6.1 depicts some microalgal biorefinery scenarios concerning their techno-economic analysis. Slegers et al. suggested four scenarios for a 10-kton microalgal dry weight per year biorefinery plant using photobioreactors in the south of Spain. The authors concluded that the biorefinery approach could notably increase the potential exploitation of the biomass into marketable products from 7–28 wt% to more than 97 wt%. Additionally, they stated that the cascade approach significantly increases biorefinery costs; however, it can be balanced by the overall revenue (Slegers et al., 2020).

Tejada Carbajal et al. (2020) developed a techno-economic analysis of five biorefinery scenarios for biodiesel production and glycerol valorization from *Scenedesmus dimorphus*. The authors applied raceway ponds suitable for the Mexican context. The best

Table 6.1: Techno-economic analysis of different microalgal biorefinery scenarios.

Scenario	Microalgae	Products/scenarios/routes	Complementary assumptions	Main results	Reference
1	*Nannochloropsis gaditana*	Soluble proteins, pigments, peptides, polysaccharides, monosaccharides, oil, and insoluble components	Microalgae production: 10 kton dry weight y^{-1}; photobioreactor; South of Spain; Benchmark level of 2 g/L biomass concentration.	Costs: 2.8 € $kg^{-1}{}_{biomass}$ Product revenue: 5–11.83 € $kg^{-1}{}_{biomass}$ Potential profit: 4.38 € $kg^{-1}{}_{biomass}$	Slegers et al., 2020
2	*Isochrysis galbana*	Pigments, proteins, peptides, oil, and insoluble components.		Costs: 2.3 € $kg^{-1}{}_{biomass}$ Product revenue: 3.59–8.61 € $kg^{-1}{}_{biomass}$ Potential profit: 1.61 € $kg^{-1}{}_{biomass}$	
3	*Nannochloropsis gaditana*	Oil, peptides, and insoluble components.		Costs: 1.03 € $kg^{-1}{}_{biomass}$ Product revenue: 6.19–16.44 € $kg^{-1}{}_{biomass}$ Potential profit: 7.76 € $kg^{-1}{}_{biomass}$	
4	*Nannochloropsis gaditana*	Oil, pigments, peptides, and insoluble components.		Costs: 1.5 € $kg^{-1}{}_{biomass}$ Product revenue: 6.12–16.48 € $kg^{-1}{}_{biomass}$ Potential profit: 7.26 € $kg^{-1}{}_{biomass}$	

5	*Scenedesmus dimorphus*	Biodiesel and glycerol production (heterogeneous catalytic transesterification in a reactive distillation column)	Raceway ponds; Adequate to the Mexican context (Mexico City)	CAPEX: USD 3.75 million OPEX: USD 2.4 million NPV: −17 USD million	Tejada Carbajal et al., 2020
6		Biodiesel and glycerol production (heterogeneous catalytic transesterification in a continuously stirred tank reactor)		CAPEX: USD 4.25 million OPEX: USD 2.6 Million NPV: −17.5 USD million	
7		Scenario 5 + biological oxidation of glycerol with *Gluconobacter oxydans* to produce DHA (dihydroxyacetone)		CAPEX: USD 5.5 million OPEX: USD 2.75 million NPV: 2 USD million IRR: 19.8%	
8		Scenario 6 + biological oxidation of glycerol with *Gluconobacter oxydans* to produce DHA (dihydroxyacetone)		CAPEX: USD 5.7 million OPEX: USD 2.8 million NPV: 1 USD million IRR: 14.9%	
9		Lipids extracted and commercialized as vegetable oil substituents and the deoiled microalgae cake destined to fishmeal after drying.		CAPEX: USD 3 million OPEX: USD 2.15 million NPV: −15 USD million	

(continued)

Table 6.1 (continued)

Scenario	Microalgae	Products/scenarios/routes	Complementary assumptions	Main results	Reference
10	*Chlorella vulgaris*	Proteins (hydrolysis process) and bio-oil (pyrolysis process from the extracted protein biomass)	Laboratory scale; energy consumption and chemicals used in the processes were considered; Biomass production steps not considered; estimated bio-oil market price: 0.34 USD/kg-oil; Estimated protein market price: 5.75 USD/kg (food)	Bio-oil extraction yield: 21.1% Protein recovery: 78.12% Product revenue (bio-oil): 0.02 USD/kg microalgae-raw material Product revenue (proteins): 2.18 USD/kg microalgae-raw material (food applications) Profit from protein extraction: 1.51 USD/kg microalgae-raw material (food applications)	Phusunti and Cheirsilp, 2020
11	*Haematococcus pluvialis*	Biodiesel, astaxanthin and poly-hydroxybutyrate (PHB)	Open ponds 43,800 t/year of biodiesel	NPV: $ 174.02 million Biodiesel production cost: $0.48/kg biodiesel.	García Prieto et al., 2017
12	*Dunaliella salina*	Basic: β-carotene and fertilizer	170 tons of dry weight/year Operation: 256 days/year Open pond Belgian conditions	Revenue (EUR/year): 16,746,464 NPV (EUR): 8,706,878	Thomassen et al., 2016
13	*Dunaliella salina*	Intermediate: β-carotene and fertilizer	170 tons of dry weight/year Operation: 256 days/year Membrane (medium recycle) Open pond Belgian conditions	Revenue (EUR/year): 16,746,464 NPV (EUR): 36,995,878	

14	*Dunaliella salina*	Advanced: β-carotene and fertilizer	170 tons of dry weight/year Operation: 256 days/year Photobioreactor (PBR) Belgian conditions	Revenue (EUR/year): 16,746,464 NPV (EUR): −13,771,402	
15	*Haematococcus pluvialis*	Alternative: Astaxanthin and fertilizer	170 tons of dry weight/year Operation: 256 days/year Photobioreactor (PBR) Belgian conditions	Revenue (EUR/year): 22,119,970 NPV (EUR): 8,347,261	
16	*Chlorella vulgaris*	Biodiesel + glycerol	1,000 ton dry algae/year Production rate: 30 g/m^2 d 340 days of early operation Raceway ponds	Net energy/ton algae (kWh/ton): 76.76 Net cost ($/year): −690,149.97	Ventura et al., 2013
17	*Chlorella vulgaris*	Scenario 16 + anaerobic digestion of the residuals after lipid extraction		Net energy/ton algae (kWh/ton): 842.92 Net cost ($/year): −648,413.89	
18	*Chlorella vulgaris*	Biogas (using the wet biomass)		Net energy/ton algae (kWh/ton): 540.95 Net cost ($/year): −744,936.13	
19	*Chlorella vulgaris*	Mixed gas (supercritical gasification) (using the wet biomass)		Net energy/ton algae (kWh/ton): 1,282.42 Net cost ($/year): −658,773.60	

CAPEX: capital investment; OPEX: operating expenditures; NPV: net present value; IRR: internal return rate.

scenario corresponded to biodiesel production by heterogeneous catalytic transesterification in a reactive distillation column and biological oxidation of glycerol with *Gluconobacter oxydans* to produce DHA (dihydroxyacetone), a compound commonly used in tanning creams. This scenario presented a positive and higher net present value (NPV) of 2 USD million, in comparison with a scenario where only biodiesel and glycerol were produced (NPV: −17 USD million), indicating that higher NPV can be observed when high-value products can be obtained within the same value chain (Table 6.1).

Thomassen et al. (2016) analyzed four scenarios, three for β-carotene and fertilizers production from *Dunaliella salina* and one for astaxanthin and fertilizers production from *Haematoccocus pluvialis* under a Belgian context. They evaluated the techno-economic analysis of β-carotene production in two scenarios, one using open raceways, with and without medium recycling (membrane process), and another using a photobioreactor. The astaxanthin production was evaluated only by applying photobioreactors. The most profitable scenario was the one using open raceways with medium recycling. The authors concluded that the use of photobioreactors can decrease the culture medium costs; however, the costs associated with the higher investments can lower the economic profit. Furthermore, they concluded that the pigments market volume and price were critical parameters for an economically feasible process (Thomassen et al., 2016).

In summary, all the results taken together demonstrate that the exploitation of microalgae within a multiproduct biorefinery enhances their value and profitability and balances the environmental impact. For this reason, microalgae research needs to stay focused on the complete valorization of the biomass and processes integration so that these scenarios can become an industrial reality in the near future.

6.4 Conclusions

Microalgae comprise a group of multiple microorganisms, presenting great potential to build sustainable processes by acting as feedstock for biofuels and bioenergy production as well as for high-value products. From the available online scientific databases (Scopus and Web of Science), microalgae-related publications, gathered from 1960 to 2021, evidenced that this research topic has a rising scientific field. More than 38,000 manuscripts were found, showing a growing trend with three clear time periods. During the first-time stage (1960–2005), the microalgae research was focused on biological sciences, then progressing to biotechnological applications, particularly as feedstock for biofuels and bioenergy production (2005–2017). Overall, the highly studied field in microalgae research is biofuels & bioenergy, emerging around 2005 and continuing to increase over the years. During the third-time stage (2017–2021, the last five years), engineering-related fields, for example, chemical, agricultural, and environmental engineering, started to gain importance reinforcing the focus on microalgal applications in different fields.

Although microalgae research has been widely addressed for various applications worldwide, for example, biofuels & bioenergy, food & feed, pharmaceutical, cosmeceutical, agriculture-based products and biomaterials, the food sector, mainly the area of dietary supplements, still leads the market. Many products and microalgal applications have not yet achieved the industrial scale owing to their high production costs and limited scale. In fact, there is still room for new developments and improvements, especially in the optimization of microalgal production, harvesting technologies and strain selection so that microalgae cultivation can be extended to a larger production scale guarantying competitiveness in different market fields.

The consolidation of the biorefinery concept to generate multiproducts in a single chain is also a step to go forward along with the integration of wastewater for microalgal cultivation and for biofuels/high-value products production. More comprehensive studies on economic and environmental conditions are needed to better establish the most profit and sustainable landscapes. The development of agricultural-related products (e.g., biostimulants and biofertilisers), biomaterials (e.g., bioplastics), emulsifiers, and pigments and the use of genetic engineering are considered as microalgae emerging fields on which progresses are anticipated. Summarizing, the microalgae research is expected to be driven by the use of biotechnological applications and integrative approaches so that the whole biomass can be valorized toward a more sustainable and zero-waste strategy, thus allowing their consolidation and expansion in the global economic market.

References

't Lam, G. P., Vermuë, M. H., Eppink, M. H. M., Wijffels, R. H., & van den Berg, C. (2018). Multi-product microalgae biorefineries: From concept towards reality. *Trends in Biotechnology*, *36*, 216–227.

Araújo, R., Vázquez Calderón, F., Sánchez López, J., Azevedo, I. C., Bruhn, A., Fluch, S., Garcia Tasende, M., Ghaderiardakani, F., Ilmjärv, T., Laurans, M., Mac Monagail, M., Mangini, S., Peteiro, C., Rebours, C., Stefansson, T., & Ullmann, J. (2021). Current status of the algae production industry in Europe: An emerging sector of the blue bioeconomy. *Frontiers in Marine Science*, *7*, 1–24.

Banu, J. R., Kavitha, S., Gunasekaran, M., & Kumar, G. (2020). Microalgae based biorefinery promoting circular bioeconomy-techno economic and life-cycle analysis. *Bioresource Technology*, *302*, 122822.

Barros, A. I., Gonçalves, A. L., Simões, M., & Pires, J. C. M. (2015). Harvesting techniques applied to microalgae: A review. *Renewable and Sustainable Energy Revolution*, *41*, 1489–1500.

Behera, B., Selvam, S. M., & Paramasivan, B. (2022). Research trends and market opportunities of microalgal biorefinery technologies from circular bioeconomy perspectives. *Bioresource Technology*, *351*, 127038.

Böcker, L., Bertsch, P., Wenner, D., Teixeira, S., Bergfreund, J., Eder, S., Fischer, P., & Mathys, A. (2021). Effect of Arthrospira platensis microalgae protein purification on emulsification mechanism and efficiency. *Journal of Colloid and Interface Science*, *584*, 344–353.

Borowitzka, M. A., 2016. Algal Physiology and Large-Scale Outdoor Cultures of Microalgae. In: Borowitzka, M. A., Beardall, J., & Raven, J. A. (Eds.), The Physiology of Microalgae, Developments in Applied Phycology 6. Springer, Switzerland, pp. 601–652.

Chavan, R., & Mutnuri, S. (2019). Tertiary treatment of domestic wastewater by Spirulina platensis integrated with microalgal biorefinery. *Biofuels*, *10*, 33–44.

Chew, K. W., Yap, J. Y., Show, P. L., Suan, N. H., Juan, J. C., Ling, T. C., Lee, D.-J., & Chang, J.-S. (2017). Microalgae biorefinery: High value products perspectives. *Bioresource Technology*, *229*, 53–62.

del Rio-Chanona, E. A., Wagner, J. L., Ali, H., Fiorelli, F., Zhang, D., & Hellgardt, K. (2019). Deep learning-based surrogate modeling and optimization for microalgal biofuel production and photobioreactor design. *AIChE Journal*, *65*, 915–923.

Deutz, P. (2020). Circular Economy. In: Kobayashi, A. (Ed.), International Encyclopedia of Human Geography (Second Edition). Elsevier, Amsterdam, The Netherlands, pp. 193–201.

European Comission. (2018). Comission Communication (2018/673). A Sustainable Bioeconomy for Europe: Strengthening the Connection between Economy, Society and the Environment, viewed 20 May 2022. https://eur-lex.europa.eu/legal-content/en/ALL/?uri=CELEX%3A52018DC0673

European Commission. (2020). Comission Communication (2020/381). A Farm to Fork Strategy: For a Fair, Healthy and Environmentally-Friendly Food System, viewed 20 May 2022. https://eur-lex.europa.eu/legal-content/EN/TXT/?uri=CELEX:52020DC0381

European Commission. (2021). Regulation (EU) 2021/1119 of the European Parliament and of the Council of 30 June 2021 establishing the framework for achieving climate neutrality and amending Regulations (EC) No 401/2009 and (EU) 2018/1999 ('European Climate Law'). Official Journal of the European Union, viewed 20 May 2022. https://eur-lex.europa.eu/legal-content/EN/TXT/?uri=CELEX:32021R1119

Fayyaz, M., Chew, K. W., Show, P. L., Ling, T. C., Ng, I. S., & Chang, J. S. (2020). Genetic engineering of microalgae for enhanced biorefinery capabilities. *Biotechnology Advances*, *43*, 107554.

Fernández, F. G. A., Reis, A., Wijffels, R. H., Barbosa, M., Verdelho, V., & Llamas, B. (2021). The role of microalgae in the bioeconomy. *New Biotechnology*, *61*, 99–107.

Fuentes-Grünewald, C., Ignacio Gayo-Peláez, J., Ndovela, V., Wood, E., Vijay Kapoore, R., & Anne Llewellyn, C. (2021). Towards a circular economy: A novel microalgal two-step growth approach to treat excess nutrients from digestate and to produce biomass for animal feed. *Bioresource Technology*, *320*, 124349.

García Prieto, C. V., Ramos, F. D., Estrada, V., Villar, M. A., & Diaz, M. S. (2017). Optimization of an integrated algae-based biorefinery for the production of biodiesel, astaxanthin and PHB. *Energy*, *139*, 1159–1172.

Garfield, E. (1990). KeyWords Plus: ISI's. *Current Comments*, *13*, 295.

Garfield, E., & Sher, I. H. (1993). KeyWords Plus [TM]-algorithmic derivative indexing. *Journal of the American Society for Information Science*, *44*, 298–299.

Garrido-Cardenas, J. A., Manzano-Agugliaro, F., Acien-Fernandez, F. G., & Molina-Grima, E. (2018). Microalgae research worldwide. *Algal Research*, *35*, 50–60.

Goswami, R. K., Mehariya, S., Verma, P., Lavecchia, R., & Zuorro, A. (2021). Microalgae-based biorefineries for sustainable resource recovery from wastewater. *Journal of Water Process Engineering*, *40*, 101747.

Hussain, F., Shah, S. Z., Ahmad, H., Abubshait, S. A., Abubshait, H. A., Laref, A., Manikandan, A., Kusuma, H. S., & Iqbal, M. (2021). Microalgae an ecofriendly and sustainable wastewater treatment option: Biomass application in biofuel and bio-fertilizer production. A review. *Renewable and Sustainable Energy Revolution*, *137*, 110603.

Khan, M. I., Shin, J. H., & Kim, J. D. (2018). The promising future of microalgae: Current status, challenges, and optimization of a sustainable and renewable industry for biofuels, feed, and other products. *Microbial Cell Factories*, *17*, 36.

Kholssi, R., Ramos, P. V., Marks, E. A. N., Montero, O., & Rad, C. (2021). Biotechnological uses of microalgae: A review on the state of the art and challenges for the circular economy. *Biocatalysis and Agricultural Biotechnology*, *36*, 102114.

Kumar, S., & Deshmukh, R. (2021). Microalgae Market by Type (Spirulina, Chlorella, Dunaliella Salina, and Aphanizomenon Flos-Aquae), Application (Dietary Supplements, Food/Feed, Pharmaceutical, Cosmetic, and Others): Global Opportunity Analysis and Industry Forecast 2021–2028, viewed 09 June 2022. https://www.alliedmarketresearch.com/microalgae-market-A13419

Lee, R. E. (2008). Phycology (Fourth Edition). Cambridge University Press, New York.

Li, G., Zhang, J., Li, H., Hu, R., Yao, X., Liu, Y., Zhou, Y., & Lyu, T. (2021). Towards high-quality biodiesel production from microalgae using original and anaerobically-digested livestock wastewater. *Chemosphere*, *273*, 128578.

Llamas, B., Suárez-Rodríguez, M. C., González-López, C. V., Mora, P., & Acién, F. G. (2021). Techno-economic analysis of microalgae related processes for CO2 bio-fixation. *Algal Research*, *57*, 102339.

López Rocha, C. J., Álvarez-Castillo, E., Estrada Yáñez, M. R., Bengoechea, C., Guerrero, A., & Orta Ledesma, M. T. (2020). Development of bioplastics from a microalgae consortium from wastewater. *Journal of Environmental Management*, *263*, 110353.

Mastropetros, S. G., Pispas, K., Zagklis, D., Ali, S. S., & Kornaros, M. (2022). Biopolymers production from microalgae and cyanobacteria cultivated in wastewater: Recent advances. *Biotechnology Advances*, *60*, 107999.

Monlau, F., Suarez-Alvarez, S., Lallement, A., Vaca-Medina, G., Giacinti, G., Munarriz, M., Urreta, I., Raynaud, C., Ferrer, C., & Castañón, S. (2021). A cascade biorefinery for the valorization of microalgal biomass: Biodiesel, biogas, fertilizers and high valuable compounds. *Algal Research*, *59*, 102433.

Mukherjee, C., Chowdhury, R., Sutradhar, T., Begam, M., Ghosh, S. M., Basak, S. K., & Ray, K. (2016). Parboiled rice effluent: A wastewater niche for microalgae and cyanobacteria with growth coupled to comprehensive remediation and phosphorus biofertilization. *Algal Research*, *19*, 225–236.

Otálora, P., Guzmán, J. L., Acién, F. G., Berenguel, M., & Reul, A. (2021). Microalgae classification based on machine learning techniques. *Algal Research*, *55*, 102256.

Phusunti, N., & Cheirsilp, B. (2020). Integrated protein extraction with bio-oil production for microalgal biorefinery. *Algal Research*, *48*, 101918.

Rempel, A., de Souza Sossella, F., Margarites, A. C., Astolfi, A. L., Steinmetz, R. L. R., Kunz, A., Treichel, H., & Colla, L. M. (2019). Bioethanol from *Spirulina platensis* biomass and the use of residuals to produce biomethane: An energy efficient approach. *Bioresource Technology*, *288*, 121588.

Ruiz, J., Olivieri, G., De Vree, J., Bosma, R., Willems, P., Reith, J. H., Eppink, M. H. M., Kleinegris, D. M. M., Wijffels, R. H., & Barbosa, M. J. (2016). Towards industrial products from microalgae. *Energy and Environmental Sciences*, *9*, 3036–3043.

Rumin, J., Nicolau, E., de Oliveira, R. G., Fuentes-Grünewald, C., Flynn, K. J., & Picot, L. (2020a). A bibliometric analysis of microalgae research in the world, Europe, and the European Atlantic area. *Marine Drugs*, *18*, 79.

Rumin, J., Nicolau, E., de Oliveira, R. G., Fuentes-Grünewald, C., & Picot, L. (2020b). Analysis of scientific research driving microalgae market opportunities in Europe. *Marine Drugs*, *18*, 264.

Schüler, L., Greque de Morais, E., Trovão, M., Machado, A., Carvalho, B., Carneiro, M., Maia, I., Soares, M., Duarte, P., Barros, A., Pereira, H., Silva, J., & Varela, J. (2020). Isolation and characterization of novel Chlorella vulgaris mutants with low chlorophyll and improved protein contents for food applications. *Frontiers in Bioengineering and Biotechnology*, *8*, 1–10.

Silva, S. C., Almeida, T., Colucci, G., Santamaria-echart, A., Manrique, Y. A., Dias, M. M., Barros, L., Colla, E., & Barreiro, M. F. (2022). Spirulina (*Arthrospira platensis*) protein-rich extract as a natural emulsifier for oil-in-water emulsions : Optimization through a sequential experimental design strategy. *Colloids and Surfaces A: Physicochemical and Engineering Aspects*, *648*, 129264.

Silva, S. C., Ferreira, I. C. F. R., Dias, M. M., & Barreiro, M. F. (2020). Microalgae-derived pigments: A 10-year bibliometric review and industry and market trend analysis. *Molecules*, *25*, 1–24.

Sintra, T. E., Bagagem, S. S., Ghazizadeh Ahsaie, F., Fernandes, A., Martins, M., Macário, I. P. E., Pereira, J. L., Gonçalves, F. J. M., Pazuki, G., Coutinho, J. A. P., & Ventura, S. P. M. (2021). Sequential recovery of C-phycocyanin and chlorophylls from *Anabaena cylindrica*. *Separation and Purification Technology*, *255*, 117538.

Slegers, P. M., Olivieri, G., Breitmayer, E., Sijtsma, L., Eppink, M. H. M., Wijffels, R. H., & Reith, J. H. (2020). Design of value chains for microalgal biorefinery at industrial scale: Process integration and techno-economic analysis. *Frontiers in Bioengineering and Biotechnology*, *8*, 1–17.

Smith, A. G., Tredici, M. R., Boussiba, S., Verdelho, V., Cadoret, J.-P., Davey, M. P., Huete-Ortega, M., Acien, F. G., Schmid-Staiger, U., Rodriguez, H., Benemann, J., Leu, S., Rodolfi, L., Biondi, N., & Meinerz, L. (2021). What are algae ? (Version 2.0). *European Algae Biomass Association (EABA)*, 1–11.

Spicer, A., & Molnar, A. (2018). Gene editing of microalgae: Scientific progress and regulatory challenges in Europe. *Biology*, *7*, 21.

Suchithra, M. R., Muniswami, D. M., Sri, M. S., Usha, R., Rasheeq, A. A., Preethi, B. A., & Dinesh kumar, R. (2022). Effectiveness of green microalgae as biostimulants and biofertilizer through foliar spray and soil drench method for tomato cultivation. *South African Journal of Botany*, *146*, 740–750.

Sumprasit, N., Wagle, N., Glanpracha, N., & Annachhatre, A. P. (2017). Biodiesel and biogas recovery from *Spirulina platensis*. *International Biodeterioration and Biodegradation*, *119*, 196–204.

Sydney, E. B., Neto, C. J. D., de Carvalho, J. C., Vandenberghe, L. P. de S., Sydney, A. C. N., Letti, L. A. J., Karp, S. G., Soccol, V. T., Woiciechowski, A. L., Medeiros, A. B. P., & Soccol, C. R. (2019). Microalgal biorefineries: Integrated use of liquid and gaseous effluents from bioethanol industry for efficient biomass production. *Bioresource Technology*, *292*, 121955.

Tejada Carbajal, E. M., Martínez Hernández, E., Fernández Linares, L., Novelo Maldonado, E., & Limas Ballesteros, R. (2020). Techno-economic analysis of *Scenedesmus dimorphus* microalgae biorefinery scenarios for biodiesel production and glycerol valorization. *Bioresource Technology Reports*, *12*, 100605.

Teng, S. Y., Yew, G. Y., Sukačová, K., Show, P. L., Máša, V., & Chang, J. S. (2020). Microalgae with artificial intelligence: A digitalized perspective on genetics, systems and products. *Biotechnology Advances*, *44*, 107631.

Thomassen, G., Egiguren Vila, U., Van Dael, M., Lemmens, B., & Van Passel, S. (2016). A techno-economic assessment of an algal-based biorefinery. *Clean Technologies and Environmental Policy*, *18*, 1849–1862.

Varia, J., Kamaleson, C., & Lerer, L. (2022). Biostimulation with phycocyanin-rich Spirulina extract in hydroponic vertical farming. *Science of Horticulture*, *299*, 111042.

Ventura, J. R. S., Yang, B., Lee, Y. W., Lee, K., & Jahng, D. (2013). Life cycle analyses of CO_2, energy, and cost for four different routes of microalgal bioenergy conversion. *Bioresource Technology*, *137*, 302–310.

Vieira, V. V., Cadoret, J. P., Acien, F. G., & Benemann, J. (2022). Clarification of most relevant concepts related to the microalgae production sector. *Processes*, *10*, 175.

Diva S. Andrade*, Tiago Pellini, Karla C. T. T. Rodrigues, Danilo A. Silvestre, Heder Asdrubal Montañez Valencia, Jerusa S. Andrade, Freddy Zambrano Gavilanes, Tiago S. Telles

Chapter 7 Microalgae supply chains

Abstract: Microalgae produce several bioactive metabolites in their cells which are considered as innovative sources for suitable food additives, human and animal nutrition products, and cosmetics and pharmaceutical industries feedstock, showing strong potential to boost the bioeconomy. In most South American (SA) countries, the microalga supply chain is still emerging because there is not sufficient robust knowledge and know-how in basic techniques in microalga biomass production to be transferred to entrepreneurs aiming to carry out production at a commercial scale. However, the microalgae production chain contributes to the development of new bioindustries focused on sustainability, with potential for regional and economic development in all SA countries. Thus, from a new market vision, the microalgae supply chain, related to the circular economy and bioeconomy, can represent and expand the possibility of new businesses. The implementation of public policies for environmental services arising from the cultivation of microalgae that promote significant carbon dioxide fixation and also by the bioremediation of wastewater may contribute to the consolidation of microalgae supply chain in SA.

Keywords: animal feed, biofertilizer, circular economy, human nutrition, microalgae biomass, phycoremediation, protein

7.1 Introduction

It is expected that the world population will be 8.6 billion in 2030, 9.8 billion in 2050, and 11.2 billion by 2100 (United Nations, 2017). The increasing population requires the expansion of food production to meet its consumption needs, and it is estimated that agricultural yield should raise by 60%, that is, produce 1.3 billion tons a year, in order to feed the 3,000 million more people (FAO, 2018). In this context, microalgal biomass

***Corresponding author: Diva S. Andrade** Instituto de Desenvolvimento Rural do Paraná – IAPAR-EMATER, 86047-902 Londrina, Paraná, Brazil, e-mail: diva@idr.pr.gov.br
Jerusa S. Andrade, Instituto Nacional de Pesquisas da Amazônia – INPA, Manaus, Brazil
Freddy Zambrano Gavilanes, Departamento de Agronomía, Facultad de Ingeniería Agronómica, Universidad Técnica de Manabí, Portoviejo, Manabí, Ecuador
Tiago Pellini, Karla C. T. T. Rodrigues, Danilo A. Silvestre, Heder Asdrubal Montañez Valencia, Tiago S. Telles, Instituto de Desenvolvimento Rural do Paraná – IAPAR-EMATER, 86047-902 Londrina, Paraná, Brazil

https://doi.org/10.1515/9783110781267-007

emerges as a potential option as a source of protein and carbohydrates for human and animal diets (Nethravathy et al., 2019) and other bioactive metabolites.

These characteristics result in opportunities for the development of a sustainable industry based on microalgae whose productivity is independent of soil fertility and less dependent on water purity, providing ecological services such as greenhouse gases mitigation because their potential for carbon dioxide (CO_2) sink (Pardo-Cárdenas et al., 2013).

Microalgae also may act as bioremediation agents for wastewater reducing environmental impacts and treatment costs in other economic sectors (Melo et al., 2022). It has been estimated that these microorganisms can fix approximately 1.83 kg CO_2 per kg of biomass produced (Gendy and El-Temtamy, 2013; Ray et al., 2022).

Besides, microalgae cultivation systems has been highlighted as an alternative raw material for the production of biofuels (Correa et al., 2020) and, biofertilizers, plant biostimulants, and livestock feeds (Siedenburg, 2022). However, in the case of biodiesel, it was learned that viability would require larger scale, regionally located, processing plants, a feature that favors the integrated and sequential production of several products and reduces logistics costs in biorefinery facilities.

In this chapter, we will focus in the following aspects of the microalgae supply chain, with a view to South American countries: research in identification of microalgae species with potential for commercial cultivation based on biochemical and genetic characterization; the main products and environmental services provided by companies in the microalgal business; and also highlight the challenges and trends to consolidate a microalgae supply chain. An overview of the microalgae supply chain in some South American countries where there was public information about microalgal research development and innovation available (Argentina, Bolivia, Brazil, Chile, Colombia, Ecuador, French Guyana, Guyana, Paraguay, Peru, Suriname, Venezuela, and Uruguay) are presented.

7.2 Microalgae in the circular economy

The microalgae supply chain represents a set of phases, upstream and downstream biomass production, in which the various inputs undergo some type of transformation until the constitution of a final product and/or services. The first step of the microalgal biomass production is segmented into two parts: first, cultivation to multiply cells and, second, harvest/dewatering cellular biomass which is often required prior to the production of commercial products (Ubando et al., 2022). Microalgae biotechnology is still evolving, but is increasing in global widespread, with purposes ranging from pharmaceuticals and cosmetics industries to climate change action via their agri-food applications (Siedenburg, 2022).

Since the early 1970s, the concept of a circular economy where resources are used as long as possible has gained momentum. It began and successfully applied on a small scale in the 1990s and has recently become practical in various industries. The concept seeks to close cycles in production activities and minimize waste, reduce emissions released into ecosystems, including discharge of nutrients that cause eutrophication in water bodies, in addition to reducing greenhouse gas emissions, reducing the discharge of nutrients that cause eutrophication in water bodies and helping to create sustainable jobs (Fuentes-Grünewald et al., 2021). An overall scheme of the sources of nutrients and CO_2, a variety of production and extraction processes, and various transformed products from microalgal biomass are represented in Figure 7.1.

Figure 7.1: Outline of a general scheme of the microalgae production main stages.

In Colombia, one of the early studies on wastewater treatment with microalgae highlighted the high potential of *Chlorella vulgaris* and *Scenedesmus dimorphus* for nutrient removal in effluents from dairy and swine industries (González et al., 1997) and using coimmobilization of *C. vulgaris* in alginate beads with *Azospirillum brasilense* (de-Bashan et al., 2002). In Brazil, *Arthrospira platensis* presented higher effectivity in removing nitrogen sources than phosphates from cassava processing wastewater (Araujo et al., 2021), and *C. sorokiniana* cultivated in wastewater from industrial processing of instant coffee, dairy products, and cassava flour/starch cassava, instant coffee produced higher biomass than in synthetic medium, which resulted in greater profit potential (Melo et al., 2022). Also, studies of the biodiesel production from microalgae biomass (Pardo-Cárdenas et al., 2013) suggest its potential to mitigate greenhouse gases emissions, although biofuel production from microalgae involves energy-intensive processes, mainly in harvesting and lipid extraction (Ubando et al., 2022).

Intensive cultivation of microalgae in closed systems for biomass production is an approach that generates innovative and high value bioactive products (e.g., pigments, enzymes, sugars, and lipids), but in open systems that use wastewater the cost of biomass product is lower due to reduction in the efficiency use of nutrients contained. Laboratory trials in laboratory (Melo et al., 2022) and small-scale pilot plants suggest that there is potential to explore the use of microalgae integrated with the phycoremediation of wastewater from the agroindustry processing within the eco-parks to solve environmental problems a vast field of applications (Mayhead et al., 2018; Melo et al., 2022), such as wastewater treatment (Baldev et al., 2021; de Carvalho et al., 2022; Padri et al., 2020), eutrophication, or treatment of agroindustry by-products (Andrade et al., 2020; Nagarajan et al., 2020), which represent one of the largest sources of waste produced in the world.

Most of the South American countries have great biodiversity of microalgae, availability of area, and high insolation that are highly favorable to the cultivation. These characteristics along with abounding availability of sources of raw materials to microalgal cultivation make South American countries potential leaders in the future scenario of establishing biofactories for microalgae production in a circular economy context.

7.3 Species and biochemical composition of microalgal

Microalgae are single-cell or colonial photosynthetic organisms living in different aquatic/wet environments, including rivers, lakes, oceans, and soils. They can be used as a source for the synthesis of various products, such as fuels, chemicals, materials, cosmetics, animal feed, and food supplements. Microalgae biomass has considerable advantages over traditional raw materials, such as (i) high productivity – generally 10–100 times higher than traditional agricultural crops; (ii) highly efficient carbon capture; (iii) high content of lipids or starch, which can be used to produce biodiesel or ethanol, respectively; (iv) cultivation in seawater, brackish water, or even in wastewater; and (v) production on nonarable land (Andrade, et al., 2021; Sarwer et al., 2022).

The selection of microalgae species is an essential step to carry out an efficient microalgal production process. There are gaps and challenges in the knowledge of microalgal biodiversity in SA countries, for example, in Paraguay (Rosset et al., 2020), a minuscule percentage within the almost million species of microalgae that could exist (Guiry, 2012).

In South American countries, the most microalgae cultivated are *Arthrospira* sp. (*Spirulina*), *Chlorella vulgaris*, *C. sorokiniana*, and *Haematococcus pluvialis* (Table 7.1).

Table 7.1: Cultivated microalgae species in SA and their high-value products/molecules for different uses.

Microalgae/ environment	Product/molecule	Use/application	References
Arthrospira spp./ freshwater	Biomass; C-phycocyanin; chitosan, phycolbiliproteins; indole-3-acetic acid (IAA), γ-linolenic acid (GLA)	Biofertilizer, cosmetics, human nutrition, supplements	Pulz and Gross (2004); Spolaore et al. (2006); Martínez and Ramírez (2017); Zapata et al. (2021); Sukhinov et al. (2021)
Chlamydomonas sp./marine	Lutein	Human health	Oliveira et al. (2022)
Chaetoceros sp./ marine	Fatty acid	Human nutrition	Martínez and Ramírez (2017); Quiroga et al. (2019)
Chlorella sp./ freshwater	Biomass, lutein, carbohydrate, and β-*galactosidase*	Feed surrogates, human health	Pulz and Gross (2004); de Morais et al. (2015); Sampathkumar and Gothandam (2019); Santos et al. (2021)
Chlorococcum sp./marine	Biofertilizers	Crop production	Renuka et al. (2016)
Chroococcus sp./marine	Metabolites	Antimicrobial activity	Gutiérrez et al. (2022b)
Chroomonas sp./ marine	Carbohydrate, exopolysaccharide, protein	Nutrition	Bermúdez et al.; (2004)
Cryptheconidium cohnii/marine	Docosahexaenoic acid (DHA)	Nutritional supplements/ aquaculture	Spolaore et al. (2006)
Dunaliella salina/marine	B-carotene	Aquaculture; cosmetic, human nutrition, and pharmaceuticals	Spolaore et al. (2006)
Desmodesmus sp./freshwater	Biomass	Aquaculture	Oliveira, et al. (2022)
Euglena sp./ freshwater and saltwater	Biomass, β-glucans	Bioremediation, food products, and cosmetics	Barsanti, et al. (2021); Gutiérrez et al. (2022b)

Table 7.1 (continued)

Microalgae/ environment	Product/molecule	Use/application	References
Golenkiniopsis sp./freshwater	Biomass	Bioremediation	Gutiérrez et al. (2022b)
Haematococcus pluvialis/ freshwater	Astaxanthin, lutein, zeaxanthin, canthaxanthin, lutein, β-carotene, and oleic acid	Aquaculture; feed additives, health food, and pharmaceuticals	Pulz and Gross (2004); Lemoine and Schoefs (2010); de Morais et al. (2015)
Isochrysis galbana		Biodiesel, biogas	Sánchez-Bayo, et al. (2020)
Nannochloropsis sp./marine	Eicosapentaenoic acid (EPA)	Aquaculture	Zittelli et al. (1999); Mitra et al. (2015);
Neochloris oleaobundans/ freshwater and salt water	Lipids, linolenic acid (GLA)	Aquaculture, biofuels; animal feed	Skjånes et al. (2013); Silva et al. (2016)
Nitzschia/marine	Eicosapentaenoic acid (EPA), lipids	Nutritional supplements Aquaculture	Kandasamy, et al. (2022)
Nostoc sp./ freshwater	Polysaccharides; cyanophycin, Stoyneva-Gärtner et al. fixing	Biostimulants and biofertilizers, plant nutrition	Stoyneva-Gärtner et al.; (2020); Trentin et al. (2023)
Phaeodactylum/ marine	Eicosapentaenoic acid (EPA), fucoxanthin	Nutritional supplements Aquaculture	Kim et al. (2012); Golubkina et al. (2022)
Porphyridium/ freshwater	Arachidonic acid (ARA)	Infant nutrition	Spolaore et al. (2006)
Scenedesmus sp./freshwater	Phytohormones	Crop production	Ronga et al. (2019)
Schizochytrium limacinum/ marine	Docosahexaenoic acid (DHA)	Aquaculture/ nutritional supplements	Oliveira et al. (2022); Oliveira and Bragotto (2022)
Tetraselmis sp./ *marine*	Biomass (P, C)	Fish farming, animal nutrition	Martínez and Ramírez (2017); Quiroga et al. (2019)
Thalassiosira sp./marine	Proteins, carbohydrates	Aquaculture	Oliveira et al. (2022)

Arthrospira is a genus of cyanobacteria in the order Oscillatoriales (commercially known as *Spirulina*) that encompass several species (e.g., *A. platensis* and *A. maxima*) that are widely cultivated in SA countries, aiming at the production of biomass which has several biotechnological purposes due to its high nutritional value in terms of protein, carbohydrates, vitamins, antioxidants, and essential fatty acids (Rosset et al., 2020; Rumin et al., 2021; Sukhinov et al., 2021).

The microalgae *Arthrospira* and *Chlorella* are source to produce a great variety of products, including categories such as vitamin and mineral supplements; bioactive substances and probiotics; new foods; foods with claims of functional properties; supplements for athletes; food supplements for pregnant and lactating women; and specific over-the-counter medications (Oliveira and Bragotto, 2022). *Spirulina* is used in human nutrition in Bolivia, Chile, and Colombia (Martínez and Ramírez, 2017; Revista-Aqua, 2016) and as biofertilizers for rice in Chile (Pereira et al., 2009). In Ecuador besides *Spirulina* sp., *Thalassiosira, Chaetoceros, and Tetraselmis* are used as human food and animal feed (Martínez and Ramírez, 2017) (Aquatropical, 2022). In French Guiana *Spirulina* sp. is used to produce biofuels (Pruvost, 2022).

In Colombia, *Spirulina* and *Chlorella have been cultivated* to increase fish farming, and production of biofertilizers, spirulina biomass, bio-inputs for livestock feed (pets, poultry, pigs, and cattle), or beauty products, such as dyes, ageing retardants, and biolubricants (Leal, 2001; Valero et al., 2018).

There are other species with applications in phycoremediation, and for biogas, biodiesel, and biomass production *such as Cryptheconidium cohnii* human nutrition (P, C) (Pereira et al., 2004). *Euglena* sp., *Gonium* sp., *Chroococcus* sp., *Chlorococcum* sp., *Scenedesmus* sp., *Golenkiniopsis* sp., *Chlamydomonas* sp., *Haematococcus* sp., and *Desmodesmus* sp. (Gutiérrez et al., 2022b).

In Paraguay, *Arthrospira platensis,* which is a native microalga from the Paraguayan Chaco, has been studied as an alternative for food purposes and (Alderete, 2018); however, there are gaps on the diversity of microalgae in this country with few published studies (Rosset et al., 2020). In Peru the mostly cultivated microalgal are *Spirulina platensis, Haematococcus pluvialis, Dunaliella salina, Chlorella vulgaris,* and *Nostoc* sp., which are used for animal and human nutrition (Martínez and Ramírez, 2017). In Uruguay only *Spirulina* sp. is *Spirulina* sp. For biofuels production, *Neochloris oleaobundans* has been cultivated in Suriname (Biobased, 2022) while in Venezuela a microalgae *Chroomonas* sp. is commercially cultivated (Bermúdez et al., 2004). In Argentina *Nannochloropsis* sp., *Isochrysis galbana* and *Chaetoceros* sp., and *Chlorella* are cultivated to use as feed in aquaculture (Prodiesel, 2022; Quiroga et al., 2019).

In order to choose the most efficient location for a microalgae production business it is required to identify the factors that affect all components of the supply chain from the acquisition of raw material to the distribution of the product to the consumer. The main factors are availability of supplies, basic services, and labor, market, means (modals) and cost of transport, environmental factor, trash deposit, geographical factors, and public legislation and regulations. The choice of microalgae species depends on several factors, starting with the definition of which products and uses are targeted in the market.

Microalgae are able to enhance the nutritional content of conventional food preparations and, hence, to positively affect the health of humans, animals, and to be a source of nutrient for plant nutrition that is due to their original chemical composition (Figure 7.2).

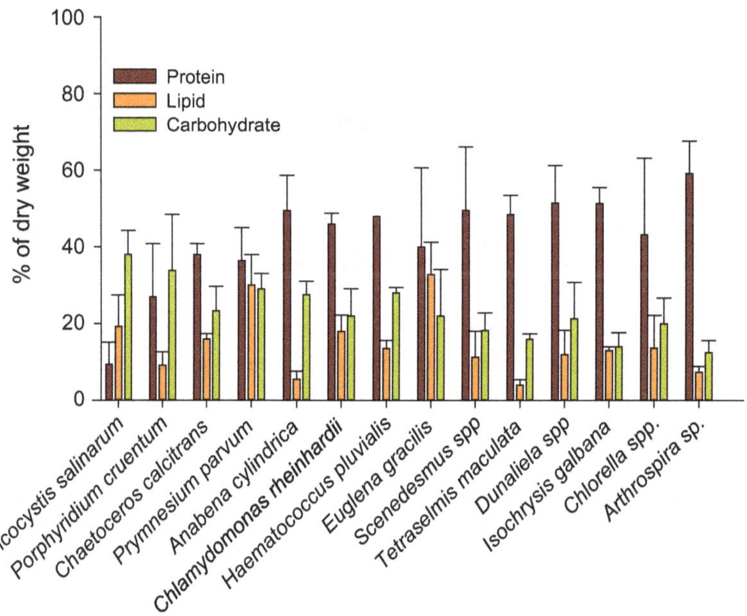

Figure 7.2: Biochemical composition of microalgae in % of protein, lipids, and carbohydrate dry weight (DW). Lines on bars are standard deviation [adapted from Delgado et al. (2021), Gouveia et al. (2008), Koyande et al. (2019), and Quiroga et al. (2019)].

Most of the microalgae species used as feedstock for the food industry have been chosen based on their main biochemical composition whose determination depends on previous research and development activities. Microalgal are generally used in the field of human and animal nutrition due to their high-value molecules (Mehariya et al., 2021).

Microalgae molecules of particular interest are used to prepare valuable products, like omega-3 fatty acids such as docosahexaenoic acid (DHA), and pigments such as carotenoids and phycobiliproteins (Spolaore et al., 2006). Microalga also produces exopolysaccharides that comprise a group of important high-molecular-weight biopolymers (Gaignard et al., 2019; Liu et al., 2016).

Regulatory procedures for registration and marketing of microalgae-based products depend on their purpose and on the country or region involved. For human and or animal consumption, there usually are stricter requirements involving their health safety (Jacob-Lopes, et al., 2019). The use as biostimulants in agriculture has specific rules to be followed for product registration (Backer et al., 2018), regarding their agronomic efficiency. Before commercialization of microalgae-based products for food or feed applications, it is necessary to identify their safety aspect as "no toxin known" (Jacob-Lopes et al., 2019).

7.4 Processes of microalgal cultivation

The great challenge in the microalgae biomass production is to gain commercial scale keeping production costs viable. In general, microalgae biomass has been produced either in closed photobioreactor (PHB) with air injection or CO_2 or in open culture systems the so-called "raceyway" which are open tanks that can be covered with or without system to stir the culture medium (Gupta et al., 2015; Pulz and Gross, 2004; Zittelli et al., 1999). Closed PHB provides more precise environmental control of temperature, light, CO_2, and pH and has lower risk of contamination than in an open system; however has higher costs when compared to the latter.

Nannochloropsis sp. biomass production had higher total cost in closed tube-type PHB, followed by the horizontal plate PHB than in the open raceway tanks (Jorquera et al., 2010). But, in the raceway there is a greater risk of contamination by either bacteria or other microalgae species, losing quality of the biomass. To overcome this bottleneck, the selection of species that have a rapid initial growth and are resistant to contamination is a strategy for microalgae production in open systems (raceway, tank). Nowadays, there are studies on genetic manipulation of microalgae to select resistance genes that result in reduced risks of contamination. At the same time, semi-closed cultivation systems, with partial control of environmental variables, can provide lower risk for contamination.

Regarding the culture medium, there are two main types: (i) photoautotrophic cultivation in clean culture using defined mineral nutrients such as nitrogen, phosphorus, potassium, calcium, magnesium, light, and CO_2 in freshwater; and (ii) wastewater systems by the use of effluents with high organic carbon contents (Soeder, 1980). Wastewaters, including those from agro-industry processing, are excellent sources to supply nutrients for microalgal growth in cultivation.

In addition, this technology of growing microalgae in wastewater may contribute to the solution of an environmental problem-related generation and treatment of residues from agro-industrial process. Wastewater from agricultural products industry processing, such as milk, coffee, sugar cane, and cassava, which is potentially a pollutant, has high potential to cultivate microalgae in mixotrophic or heterotrophic growth with reduced production costs in terms of providing nutrients and water as feedstock (Barajas-Solano et al., 2014; Melo et al., 2022; Quintero-Dallos et al., 2019; Zapata et al., 2021).

7.5 Industrialization of microalgae products

The industrial production of microalgae biomass in the different countries of South America, as in other parts of the world, has been based on models aimed at increasing biomass productivity with reduced costs, by means of the use of low cost sources of nutrients for cultivation and low energy input, seeking a circular economy with less impact on the environment. Other promising opportunities are expected from innovations such as the heterotrophic microalgae culture method in reactors. Also, the viability of biomass production may come from microalgae and microalgae packs selected to decontaminate effluents, which are synergistic mixtures of microorganisms and adapted to degrade organic compounds present in effluents, generating secondarily to the cleaning process, useful biomass. An example of this is the use of *Chlorella vulgaris*, in addition to cleaning the waterways rich in nitrogen (N), phosphorus (P), chemical oxygen demand (COD), and also generating biomass cake rich in nutrients, oils, and derivatives.

Currently, there are several companies in the SA countries whose activity is the cultivation, harvesting, drying, and commercialization of microalgae products with some countries having a greater number of microalgae companies than others, possibly, among other reasons, due to their more advanced research development in this area.

In Argentina there are companies that work with microalgae. It was found that *Spirulina* sp. is used for food by the company Fox Oil, *Tetraselmis* sp. and *Nannochloropsis* sp. are used for the aquaculture industry by the company "Microalgas," while the company "Enlasa" uses *Chlorella* sp. for extraction of oils, proteins and sugars, and biomass production; the microalga *Chlorella vulgaris* biomass is produced by the Prodiesel Company that has focus on microalgae species for bioremediation.

A study by Morales et al. (2009) shows that despite an enormous potential for the contribution of new microorganisms to science and industry in Bolivia, the lack of interest in microalgae production and biotechnology severely hinders the development of such activities in this country, which could help improving its economy.

The diversity and large extension of its coastline favored the different companies installed in Chile for the multiplication and industrial use of microalgae (Bravo-Fritz et al., 2015). Examples of Chilean microalgae companies are (i) Atacama Bio Natural Products S.A. that cultivated *Haematococcus pluvialis; Spirulina* sp. and *Dunaliella salina* in closed and open photobioreactors specially to extract healthy ingredients for human well-being (https://www.atacamabionatural.com/) (Lim, 2022); (ii) Pigmentos Naturales S.A.; (iii) Solabia-Algatech Nutrition (https://www.nutritioninsight.com/news/solabia-algatech-nutrition-debuts-astaxanthin-algae-gummies-with-vitamin-c-ahead-of-supplyside-west.html); (iv) Solarium Biotechnology S.A. (*Spirulina maxima*/Open Food supplements in capsules and powder (Martínez-Angulo and Ramírez-Mérida, 2017); (v) Algae Fuels S.A. (*Spirulina* sp./Open Enriched flour (Martínez-Angulo and Ramírez-Mérida, 2017).

Mexican investors moved to the north of Chile in 2004 and created a plant in the Pampa del Tamarugal (Atacama Desert), where it is produced from 15 to 25 tons of *Arthrospira* per year and sold in capsules and in powder, locally and internationally, under the brand "Spirulina Mater" (Revista-Aqua, 2016).

Also in Chile, the application of Cyanobacteria in rice crops has increased grain yields (Pereira et al., 2009). The results reveal an increase in grain yield between 15 and 20% in field experiments (Innok et al., 2009). Inoculation with four cyanobacteria (*Anabaena iyengariivar, Nostoc commune, Nostoc linckia*, and *Nostoc* sp.) in rice soils in Chile allowed a decrease of up to 50% in the use of nitrogenous fertilizers (equivalent to 50 kg N/ha) without reducing the 7.4 t/ha average yields of rice grain (Pereira et al., 2009).

In French Guyana, the energy supply company SARA (Société Anonyme de Raffinage des Antilles) is part of the PIAN project (Intensive production of natural algae) together with the Nantes University which is investing in the production of *Spirulina* microalgae to obtain biofuels (Pruvost, 2022).

In Brazil, likewise in other SA countries, the development in microalgae cultivation activities is increasing due to its representative cost and suitable climate conditions, along with the enormous availability of fresh and saltwater sources (Andrade et al., 2020; Brasil et al., 2017; Valenti et al., 2021). Brazil is among the top five exporters of microalgae products along with China, Ireland, South Korea, and France (CredenceResearch, 2023). Currently, there are around 10 companies working with microalgae: Algae Biotecnologia; Claeff Engenharia; AlgaSul; AllGA (Biomass and its application); Cia das Algas (Cosmetics and agriculture, https://www.ciadasalgas.com.br/); Bioalgas Análise e Consultoria Ambiental LTDA; Algasbras (Agriculture, food industry, pharmaceutical, and cosmetics, https://algasbras.com.br/); Séston Biotecnologia (Human and animal nutrition, food and chemistry industry, pharmaceutical, and cosmetics, http://www.seston.com.br/pt/); Syntalgae (Cosmetics, human and animal nutrition, agriculture/biofuels, cosmetics, and bioplastics, https://www.syntalgae.com.br/); Algabloom microalgas (Food based on *Artemia and Spirulina* microalgae biomass, https://algabloom.com.br/); Ocean Drop (Trade of several products from microalgae, https://www.oceandrop.com.br/ocean-box/p); Corbion (Human and animal nutrition, https://www.corbion.com/Products/Algae-ingredients-products).

Biotechnology companies are using microalgae that have heterotrophic or mixotrophic growth, consuming sucrose as a source of carbon to obtain biomass, in Brazil, in the adjacent sugarcane mill to produce high value-added fatty oils for nonenergy purposes. These fatty acids are sold to industrial companies for use in cosmetic and personal care formulations. There are also markets for *Spirulina* and *Chlorella*-based products being marketed for human food supplements. In this context, startup companies have emerged with various activities in the microalgae sector, such as the biofixation of carbon emissions and bioremediation of industrial effluents through the cultivation of microalgae, whose biomass is later used for biofertilizers.

In the category of dietary supplements, the microalgae *Arthrospira* and *Chlorella* are the most commonly found commercially in Brazil (Oliveira and Bragotto, 2022). The Instituto Fazenda Tamanduá, a company located in the Brazilian northeast region, is a pioneer in the commercial cultivation of *Spirulina platensis* in Brazil for biofertilizers such as SpiroFert® (Alves, et al., 2016; Ronga et al., 2019). Its main advantage is having the plant facilities situated very close to the Equator line, which guarantees heat, strong sunlight, and long days (in fact, average daylight hours do not vary in equatorial regions) throughout the year. It is important to highlight that Brazilian production of *Spirulina* is insufficient to meet its domestic demand, and most of this product is imported from China.

It is expected that the consolidation and growth of the microalgal bioindustry in Brazil, now based on high value-added products for food and cosmetic industries, will pave the way, in the near future, to boost microalgae biomass production to generate low value-added products, such as biofuels, and other high value-added ones, as biomaterials. Although technological and market barriers still need to be overcome this scenario to become a reality, the potential gains arising from the microalgal exploitation increasingly stimulate investments in this sector.

Notwithstanding the great diversity of microalgae species distributed along the coastal profile of Ecuador (Ballesteros et al., 2021), there are only two companies dedicated to the microalgae production that are the companies Andes Spirulina CA using *Spirulina* sp. to produce food supplements (Martínez-Angulo and Ramírez-Mérida, 2017) and the aquatropical using species of diatoms of *Thalassiosira, Chaetoceros* sp., and *Tetraselmis* (https://aquatropicalsa.com/).

7.6 Microalgae companies, marketing, and services

Currently, products made from microalgae mainly supply the cosmetics, personal care, and human and animal nutrition markets (Show, 2022; Oliveira and Bragotto, 2022). These are value-added products that are produced on a small and medium scale in most countries. The main cultivated species belong to the genera *Arthrospira* and *Chlorella*, being used as sources of protein and pigments for the cosmetics industry or as protein supplements for human food and aquaculture feed stuff.

The species *Dunaliella salina* and *Haematococcus pluvialis* are used in human health products and as a source of pigments and antioxidants, such as the carotenoids astaxanthin, canthaxanthin, and beta-carotene (Koyande et al., 2019). Omega-3 and omega-6 polyunsaturated fatty acids (PUFAs), such as EPA (eicosapentaenoic acid) and DHA, are also produced from microalgae and make up infant nutritional formulations, beverages, and dietary supplements. However, the economic viability of large-scale cultivation of microalgae for the production of low value-added products (such as chemical commodities, biomaterials, and energy) has not been reached yet.

Thus, although the technical feasibility of using microalgae to produce bioplastics, polymers, and biofuels, such as biodiesel, ethanol, and biokerosene, using microalgae has already been demonstrated, such processes still do not present competitive production costs with equivalent derivatives from the petrochemical industry. The current technological challenges to scale up microalgal production consist mainly in the genetic improvement of strains, development of efficient methods to control pests, in cultivation methods, and optimization of cell harvesting processes and extraction of compounds of commercial value. In order to overcome these bottlenecks, considerable and growing investments in research and demonstration (precommercial) pilot-plants are being carried out in several SA countries.

The expectation is that the production of microalgae in the world will continue to grow in the coming years, leading to an increase in the scale of this industry and overcoming the current bottlenecks. An example would be associating the already existent extraction of oil from microalgae for products with high value-added (e.g., PUFAs) and the use of residual biomass to generate products with low value-added, but which supply larger markets, such as animal nutrition.

Something similar already occurs in the current supply chains of soybean, sugarcane, and corn, in which products serve as raw material for the concomitant production of food, biofuels, and other by-products. In a medium/long-term horizon, a similar model could be taken place for the large-scale production of microalgae, for example, chlorophytes, aiming at the simultaneous production of products such as beta-carotene, animal feed, and biofuel.

Microalgae can be integrated into a scalable, zero-carbon circular process that exploits their capabilities for CO_2 fixation and wastewater remediation, which results in the production of usable biomass as feedstock for renewable energy and bioproducts such as biofertilizers (Serrà et al., 2020).

South America has advantageous climatic and natural resources conditions for microalgae cultivation and may become one of the main sources of its biomass for the industries. There is land, sunlight, and water availability and a great microalgae diversity that can be used for different applications.

That is the case of Colombia, which has aspects that favor microalgae crops such as water resources and a wealth of light, being a contribution to industrial production and the use of microalgae, a factor that according to the National Federation of Biofuels. Currently, several Colombian companies and universities are using this eco-friendly technology under the premise of "restoring by producing." In Colombia, bio-prospecting, development, and application of microalgae in water bodies would be an excellent strategy for the recovery of these ecosystems, improving the quality of life of their surrounding communities (Aranguren Díaz et al., 2022).

One of the production factors in Colombia is the use of microalgae for agribusiness promoted by universities and companies such as Ecotec and Phycore. Because of their useful and easy production and the fact that is a group of wide-spread microorganisms that are present in the soil, in bodies of water, and in most ecosystems, this type of production is increasing more and more (Gutiérrez et al., 2022b).

La Guajira region, based on a theoretical model (Kang et al., 2022), has potentially significant competitive advantages for the use and exploitation of microalgae in various biotechnological processes due to the great biodiversity of ecosystems in the region, high lighting rates throughout the year, the availability of seawater for the cultivation of microalgae and potential use of unproductive land. Due to the favorable conditions that the country has for the cultivation of microalgae, their use has been used to fish farming and some approaches to the exploitation of *Spirulina* and *Chlorella* in the Colombian Caribbean.

An interesting aspect of microalgae use in a pilot-scale high-rate algal pond is to remove pharmaceutical compounds from domestic wastewater in the city of Santiago de Cali as described by Jiménez-Bambague et al. (2020). Highly with high competitive and nonharmful microalgae species have been used in drinking water reservoirs as an ecological tool to neutralize cyanobacterial proliferation and mitigate the risks of cyanotoxins for animals and humans (Gutiérrez et al., 2022a).

Docosahexaenoic acid (DHA, 22:6) is one of the most important PUFAs since it constitutes a major component of the gray matter and the retina. It is key in the human neurological development as well as in cardiac tissue (Pereira et al., 2004). The Colombian company Market-DHASCO produces 40–50% dry weight DHA from *Cryptheconidium cohnii* cultures and its production in 2003 reached 240 tons (Vanegas and Hernández, 2018). Other polyunsaturated acids present in microalgae are the following: eicosapentanoic acid (EPA, C20:5), GAL (C18:3), linoleic acid (ALA, C18:3 n-3), and docosapentanoic acid (DPA, C22:5 n-3) (Jacob-Lopes et al., 2019; Santos Montes et al., 2014).

Microalgae can be used for the production of protein concentrates for the aquaculture industry, the production of biofertilizers, spirulina biomass, bioinputs for livestock feed, pets, poultry, pigs, cattle, or beauty products (dyes, aging retardants, biolubricants) by companies such as Nutre SAS and Naturela (https://connectamericas.com/company/nutre-sas) located in the municipality of Cumaral, Colombia.

Peru's geographic location in the tropics has favorable climate conditions to produce microalgae. Peru is a superpower in the production of fishmeal and has potential in the production of microalgae to supply the animal nutrition market. Microalgae are also considered sources of human nutrition (Cobos et al., 2017; Salazar-Torres et al., 2012; Soeder, 1980). However, the raw material for *Spirulina*-based products is almost entirely imported from Mexico, Chile, China, and the USA (Reynaga, 2019).

In Peru, *Arthrospira* is mainly used for human consumption as an additive in capsule or powder form. But there are other species being produced and commercialized, for example, *Nannochloropsis oculata* (Droop) Hibberd, which contains a large amount of PUFAs, such as EPA, arachidonic acid (ARA), and DHA of great importance for animal and human nutrition (Cobos et al., 2020).

There are several companies in Peru whose activity is the cultivation, harvesting, drying, and commercialization of microalgae, for example, the Andexs Biotechnology SRL and the BIONUTREC (Marketing of *Spirulina* imported from Mexico, Chile, China, and the USA. Human nutrition, https://www.bionutrec.com/bo/espirulina-en-potosi/). The predominant production system of microalgae biomass of these Peruvian companies is the employment of open concrete or plastic tanks, containing culture medium; cells are harvested and dried in an oven with the control of temperature, and there is greater investment in hermetic packaging, advertising, and distribution (Cárdenas et al., 2010).

The commercialization of *Spirulina* in Peru that is carried out by intermediary traders or directly by the producer has been oriented toward a market, predominantly, for human consumption (Reynaga, 2019). It is expected that the market of microalgae products will grow (Reynaga, 2019). An investment input of around US$ 3 million in microalgae research in Peru is focused on the *Spirulina* genus. In this sense, advances in high-level biotechnology, with very good productivity, have allowed companies to the development of microalgal activities in this country (https://www.redagricola.com/pe/las-microalgas-pueden-ser-el-gran-nuevo-producto-de-exportacion).

Currently, there is no facility in Suriname to produce biofuels from microalgae, but local industrial partners such as Suriname Staatsolie have participated in a consortium (Algae PARC) with the Wageningen University aiming to develop research knowledge and technology to make large-scale algae cultivation feasible. After several testing phases, CaribAlgae is ready to scale up to a fully productive facility on Curacao.

7.7 Challenges and trends in the microalgae supply chain

For consolidation of the microalgae supply chain, growth and stability of production (products and by-products) are required. Therefore, public policies to support its integration in industrial eco-parks are essential. The use of renewable raw materials from microalgae biomass and the integration of industrial processes in a biorefinery concept are seen as potential sustainable solutions to meet part of the economy´s demands for energy, food, chemicals, and materials. In a biorefinery, the processes convert cellular biomass into various marketable products and energy, optimizing the use of resources and minimizing the generation of waste. Thus, there is a need to expand the range of bio-based products in order to replace or/and reduce petroleum derivatives. Awareness of these challenges is driving investments into research and commercial production of alternative feedstock from microalgae.

Microalgal biomass has a wide application in several sectors. In biotechnology they are the sources of chemical inputs, including pigments, fatty acids, oils, polysaccharides, as well as nutraceuticals, cosmetics, and drugs. They can be used in the production of food, animal feed, biofertilizers, biomaterials, and chemicals. Microalgae are a potential raw material for the production of biofuels, including biodiesel, bioethanol, biomethane, and biohydrogen, which raises expectations of action in the renewable energy sector in the future.

Studies highlight some growing fields related to microalgae: (i) in the industry with *Spirulina* replacing collagen as raw material for capsule shells due to the polyhydroxybutyrate (PHB) in its composition (Dianursanti, et al., 2020); (ii) a new field in vaccine production based on the use of antigen selection for their protective effects through studies with in silico tools (Ramos-Vega et al., 2021); and (iii) in skin regeneration, demonstrating the positive effects of rapid healing resulting from their different biological activities (Miguel et al., 2021).

The global microalgae market was worth around 10.2 (USD billion) in 2021, and it is expected to grow approximately 18.3 (USD billion) by 2028, with Compound Annual Growth Rate of about 8.2% (Facts&FactorsResearch, 2022).

However, technologies involving either the selection of microalgae species or the cultivation, harvesting and processing systems of microalgae still need improvements to increase their capacity to compete with other raw materials such as soybean. The high cost of installing pilot plants and operating them, microalgal strains with protein-rich contents, difficulties in growing outdoors due to variations in weather conditions, and commercial-scale harvesting seem to be major challenges for microalgae supply chain to evolve.

Another obstacle is related to the extraction of functional bioactive compounds from microalgae biomass due to the need to apply emerging technologies and the high cost of energy required in production for these extractive processes. To solve this problem, there is a demand to develop low-cost processes, which can be solved with scientific studies in cooperation with research institutes, aiming at better economic viability (Savio et al., 2021). The legislation on microalgae products is an aspect that needs to be clearly established in the South American countries. For instance, in Brazil, according to the Resolution of the Collegiate Board (RDC 240/2018), foods containing microalgae are exempt from the obligation of sanitary registration, except those containing enzymes or probiotics must be registered with National Health Surveillance Agency (ANVISA) (Brasil, 2018). A broad and comprehensive regulation framework for the production and use of these innovative products and public policies supporting research and development and favoring entrepreneurship are paramount to the development of a microalgal supply chain in SA.

References

Alderete, C. I. V. (2018). Bioprospección de *Arthrospira platensis* nativa del chaco paraguayo como propuesta alternativa para fines alimentarios. Master Science, Universidad Nacional de Itapúa Facultad De Ciencias Y Tecnología, Encarnación-Paraguay.

Andrade, D. S., Amaral, H. F., Gavilanes, F. Z., Morioka, L. R. I., Nassar, J. M., & Melo, J. M. D., et al. (2021). Microalgae: Cultivation, Biotechnological, Environmental, and Agricultural Applications. In: Maddela, N., Cruzatty, L. G., & Chakraborty, S. (Eds.), Advances in the Domain of Environmental Biotechnology. Springer, Singapore, pp. 635–701

Andrade, D. S., SantosTelles, T., & Castro, G. H. L. (2020). The brazilian microalgae production chain and alternatives for its consolidation. *Journal of Cleaner Production*, *250*, 119526. doi:https://doi.org/10.1016/j.jclepro.2019.119526.

Aquatropical. (2022). Aquatropical S. A. Guayaqui.

Aranguren Díaz, Y., Monterroza Martínez, E., Carillo García, L., Serrano, M. C., & Machado Sierra, E. (2022). Phycoremediation as a strategy for the recovery of marsh and wetland with potential in colombia. *Resources*, *11*, 15.

Araujo, G. S., Santiago, C. S., Moreira, R. T., Dantas Neto, M. P., & Fernandes, F. A. N. (2021). Nutrient removal by *Arthrospira platensis* cyanobacteria in cassava processing wastewater. *Journal of Water Process Engineering*, *40*, 101826. doi:https://doi.org/10.1016/j.jwpe.2020.101826.

Backer, R., Rokem, J. S., Ilangumaran, G., Lamont, J., Praslickova, D., & Ricci, E., et al. (2018). Plant growth-promoting rhizobacteria: context, mechanisms of action, and roadmap to commercialization of biostimulants for sustainable agriculture. *Frontiers in Plant Science*, *9* doi:https://doi.org/10.3389/fpls.2018.01473.

Baldev, E., Mubarak Ali, D., Pugazhendhi, A., & Thajuddin, N. (2021). Wastewater as an economical and ecofriendly green medium for microalgal biofuel production. *Fuel*, *294*, 120484. doi:https://doi.org/10.1016/j.fuel.2021.120484.

Ballesteros, I., Terán, P., Guamán-Burneo, C., González, N., Cruz, A., & Castillejo, P. (2021). DNA barcoding approach to characterize microalgae isolated from freshwater systems in Ecuador. *Neotropical Biodiversity*, *7*, 170–183. doi:https://doi.org/10.1080/23766808.2021.1920296.

Barajas-Solano, A., Gonzalez-Delgado, A., & Kafarov, V. (2014). Effect of thermal pre-treatment on fermentable sugar production of *Chlorella vulgaris*. *Chemical Engineering Transaction*, *37*, 655–660. doi: https://doi.org/10.3303/CET1437110.

Barsanti, L., Ciurli, A., Birindelli, L., & Gualtieri, P. (2021). Remediation of dairy wastewater by *Euglena gracilis* WZSL mutant and β-glucan production. *Journal Applied Phycology*, *33*, 431–441. doi:https://doi.org/10.1007/s10811-020-02314-x.

Bermúdez, J., Rosales, N., Loreto, C., Briceño, B., & Morales, E. (2004). Exopolysaccharide, pigment and protein production by the marine microalga *Chroomonas* sp. in semicontinuous cultures. *World Journal of Microbiology & Biotechnology*, *20*, 179–183. doi:https://doi.org/10.1023/B:WIBI.0000021754.59894.4a.

Biobased. (2022). Biobased economy.

Brasil (2018). RESOLUÇÃO DA DIRETORIA COLEGIADA – RDC N° 240, DE 26 DE JULHO DE 2018. Altera a Resolução – RDC n° 27, de 6 de agosto de 2010, que dispõe sobre as categorias de alimentos e embalagens isentos e com obrigatoriedade de registro sanitário ANVISA, Brasilia.

Brasil, B. S. A. F., Silva, F. C. P., & Siqueira, F. G. (2017). Microalgae biorefineries: The Brazilian scenario in perspective. *New Biotechnology*, *39*, 90–98. doi:https://doi.org/10.1016/j.nbt.2016.04.007.

Bravo-Fritz, C. P., Sáez-Navarrete, C. A., Herrera Zeppelin, L. A., & Ginocchio Cea, R. (2015). Site selection for microalgae farming on an industrial scale in chile. *Algal Research*, *11*, 343–349. doi:https://doi.org/10.1016/j.algal.2015.07.012.

Cárdenas, J., Díaz, M., & Vizcaíno, M. (2010). Industrialización del alga Spirulina. *Revisiones de la Ciencia, Tecnología e Ingeniería de los Alimentos*, *10*, 1–41.

Cobos, M., Paredes, J. D., Maddox, J. D., Vargas-Arana, G., Flores, L., & Aguilar, C. P., et al. (2017). Isolation and characterization of native microalgae from the peruvian amazon with potential for biodiesel production. Energies, *10* doi:https://doi.org/10.3390/en10020224.

Cobos, M., Pérez, S., Braga, J., Vargas-Arana, G., Flores, L., & Paredes, J. D., et al. (2020). Nutritional evaluation and human health-promoting potential of compounds biosynthesized by native microalgae from the Peruvian Amazon. *World Journal of Microbiology and Biotechnology*, *36*, 121. doi: https://doi.org/10.1007/s11274-020-02896-1.

Correa, D. F., Beyer, H. L., Possingham, H. P., García-Ulloa, J., Ghazoul, J., & Schenk, P. M. (2020). Freeing land from biofuel production through microalgal cultivation in the neotropical region. *Environmental Research Letters*, *15*, 094094. doi:https://doi.org/10.1088/1748-9326/ab8d7f.

CredenceResearch. (2023). Algae products market by type (lipids, carrageenan, carotenoids, algae protein, alginates, other types) by application (food & beverage, nutraceutical & diet supplement, animal feed, personal care, pharmaceutical, other applications) by source (Brown algae, Blue-green algae, Green algae, Red algae, Other sources) by form (solid, liquid) – Growth, Share, Opportunities & Competitive Analysis, 2016 – 2028.

de-Bashan, L. E., Moreno, M., Hernandez, J.-P., & Bashan, Y. (2002). Removal of ammonium and phosphorus ions from synthetic wastewater by the microalgae chlorella vulgaris coimmobilized in alginate beads with the microalgae growth-promoting bacterium *Azospirillum brasilense*. *Water Research*, *36*, 2941–2948. doi:https://doi.org/10.1016/S0043-1354(01)00522-X.

de Carvalho, J. C., Molina-Aulestia, D. T., Martinez-Burgos, W. J., Karp, S. G., Manzoki, M. C., & Medeiros, A. B., et al. (2022). Agro-industrial wastewaters for algal biomass production, bio-based products, and biofuels in a circular bioeconomy. *Fermentation*, *8* doi:http://dx.doi.org/10.3390/fermentation8120728.

de Morais, M. G., Vaz, B. D. S., De moraisde morais, E. G., & Costa, J. A. V. (2015). Biologically active metabolites synthesized by microalgae. *BioMedicine Research International*, *2015*, 835761. doi: http://dx.doi.org/10.1155/2015/835761.

Delgado, R. T., Guarieiro, M. D. S., Antunes, P. W., Cassini, S. T., Terreros, H. M., & Fernandes, V. D. O. (2021). Effect of nitrogen limitation on growth, biochemical composition, and cell ultrastructure of the microalga *Picocystis salinarum*. *Journal Applied Phycology*, doi:http://dx.doi.org/10.1007/s10811-021-02462-8.

[2020] Utilization of microalgae spirulina platensis as a raw material for making capsule shell. The 4th International Tropical Renewable Energy Conference, 2020. AIP Publishing LLC.

Facts&FactorsResearch (2022). Global microalgae market – the forecast period of 2022 to 2028. *Facts and Factors Research*.

FAO, F.a.A.O.o.t.U.N. (2018). The Future of Food and Agriculture – Alternative Pathways to 2050 FAO, Rome, pp. 224.

Fuentes-Grünewald, C., Ignacio Gayo-Peláez, J., Ndovela, V., Wood, E., Vijay Kapoore, R., & Anne Llewellyn, C. (2021). Towards a circular economy: a novel microalgal two-step growth approach to treat excess nutrients from digestate and to produce biomass for animal feed. *Bioresource Technology*, *320*, 124349. doi:https://doi.org/10.1016/j.biortech.2020.124349.

Gaignard, C., Laroche, C., Pierre, G., Dubessay, P., Delattre, C., & Gardarin, C., et al. (2019). Screening of marine microalgae: investigation of new exopolysaccharide producers. *Algal Research*, *44*, 101711. doi: https://doi.org/10.1016/j.algal.2019.101711.

Gendy, T. S., & El-Temtamy, S. A. (2013). Commercialization potential aspects of microalgae for biofuel production: an overview. *Egyptian Journal of Petroleuml*, *22*, 43–51. doi:https://doi.org/10.1016/j.ejpe.2012.07.001.

George Alves, D., Railene Herica Carlos, R., Josinaldo Lopes, A., José Franciraldo, L., & Wellinghton Alves, G. (2016). Growth, yield, and postharvest quality in eggplant produced under different foliar fertilizer (*Spirulina platensis*) treatments. *Semina: Ciências Agrárias*, *37*, 3893–3902. doi:http://dx.doi.org/10.5433/1679-0359.2016v37n6p3893.

Golubkina, N., Sekara, A., Tallarita, A., Sellitto, V. M., Torino, V., & Stoleru, V., et al. (2022). Microalgae in agricultural crop production. *Italus Hortus*, *29*, 94–114. doi:http://dx.doi.org/10.26353/j.itahort/2022.1.94114.

González, L. E., Cañizares, R. O., & Baena, S. (1997). Efficiency of ammonia and phosphorus removal from a Colombian agroindustrial wastewater by the microalgae *Chlorella vulgaris* and *Scenedesmus dimorphus*. *Bioresource Technology*, *60*, 259–262. doi: https://doi.org/10.1016/S0960-8524(97)00029-1.

Gouveia, L., Batista, A. P., Sousa, I., Raymundo, A., & Bandarra, N. M. (2008). Microalgae in Novel Food Products. In: P. KN (Ed.), Food Chemistry Research Developments. Nova Science Publishers, Inc., NY, pp. 75–111.

Guiry, M. D. (2012). How many species of algae are there? *Journal of Phycology*, *48*, 1057–1063. doi: https://doi.org/10.1111/j.1529-8817.2012.01222.x.

Gupta, P. L., Lee, S.-M., & Choi, H.-J. (2015). A mini review: photobioreactors for large scale algal cultivation. *World Journal of Microbiology and Biotechnology*, *31*, 1409–1417. doi:https://doi.org/10.1007/s11274-015-1892-4.

Gutiérrez, J. E., Gutiérrez-Hoyos, N., Gutiérrez Benedetti, J. S., Vives, M. J., & Sivasubramanian, V. (2022a). Clarification of cyanotoxins in El Guajaro Reservoir, Colombia using a microalgae-based consortium MPMC. *Journal of Chemical Technology and Biotechnology*, *97*, 1468–1481. doi:https://doi.org/10.1002/jctb.7016.

Gutiérrez, J. E., Gutiérrez-Hoyos, N., Gutiérrez, J. S., Vives, M. J., & Sivasubramanian, V. (2022b). Bioremediation of a sewage-contaminated tropical swamp through bioaugmentation with a microalgae-predominant microbial consortium. *Indian Journal of Microbiology*, *62*, 307–311. doi: http://dx.doi.org/10.1007/s12088-021-00990-y.

Innok, S., Chunleuchanon, S., Boonkerd, N., & Teaumroong, N. (2009). Cyanobacterial akinete induction and its application as biofertilizer for rice cultivation. *Journal of Applied Phycology*, *21*, 737. doi: https://doi.org/10.1007/s10811-009-9409-x.

Jacob-Lopes, E., Maroneze, M. M., Deprá, M. C., Sartori, R. B., Dias, R. R., & Zepka, L. Q. (2019). Bioactive food compounds from microalgae: an innovative framework on industrial biorefineries. *Current Opinion in Food Science*, *25*, 1–7. doi:https://doi.org/10.1016/j.cofs.2018.12.003.

Jiménez-Bambague, E. M., Madera-Parra, C. A., Ortiz-Escobar, A. C., Morales-Acosta, P. A., Peña-Salamanca, E. J., & Machuca-Martínez, F. (2020). High-rate algal pond for removal of pharmaceutical compounds from urban domestic wastewater under tropical conditions. Case study: Santiago de Cali, Colombia. *Water Science and Technology*, *82*, 1031–1043. doi:https://doi.org/10.2166/wst.2020.362.

Jorquera, O., Kiperstok, A., Sales, E. A., Embiruçu, M., & Ghirardi, M. L. (2010). Comparative energy life-cycle analyses of microalgal biomass production in open ponds and photobioreactors. *Bioresource Technology*, *101*, 1406–1413. doi:10.1016/j.biortech.2009.09.038.

Kandasamy, S., Zhang, B., He, Z., Bhuvanendran, N., El-Seesy, A. I., & Wang, Q., et al. (2022). Microalgae as a multipotential role in commercial applications: current scenario and future perspectives. *Fuel*, *308*, 122053. doi:https://doi.org/10.1016/j.fuel.2021.122053.

Kang, S., Realff, M. J., Yuan, Y., Chance, R., & Lee, J. H. (2022). Global evaluation of economics of microalgae-based biofuel supply chain using gis-based framework. *Korean Journal of Chemical Engineering*, doi:https://doi.org/10.1007/s11814-021-1053-4.

Kim, S. M., Jung, Y.-J., Kwon, O.-N., Cha, K. H., Um, B.-H., & Chung, D., et al. (2012). A potential commercial source of fucoxanthin extracted from the microalga *Phaeodactylum tricornutum*. *Applied Biochemistry and Biotechnology*, *166*, 1843–1855. doi:http://dx.doi.org/10.1007/s12010-012-9602-2.

Koyande, A. K., Chew, K. W., Rambabu, K., Tao, Y., Chu, D.-T., & Show, P.-L. (2019). Microalgae: a potential alternative to health supplementation for humans. *Food Science and Human Wellness*, *8*, 16–24. doi: https://doi.org/10.1016/j.fshw.2019.03.001.

Leal, E. (2001). Biotecnología Microalgal en Colombia. Memorias del II Curso: Microalgas y Cianobacterias. Aislamiento, Cultivo y Fisioecología. Instituto de Investigaciones Marinas y Costeras, Santa Marta, p. 52.

Lemoine, Y., & Schoefs, B. (2010). Secondary ketocarotenoid astaxanthin biosynthesis in algae: a multifunctional response to stress. *Photosynthesis Research*, *106*, 155–177. doi:10.1007/s11120-010-9583-3.

Lim, G. Y. (2022). Chilean Astaxanthin Firm Eyes Potential in South East Asia. Chile-based Atacama Bio Natural Products S.A. Which Grows Algae for Astaxanthin, Is Looking to Expand in Asia, with a More Natural and Sustainable Process and Product. Chile https://www.nutraingredients-asia.com/Article/2019/2010/2014/Chilean-astaxanthin-firm-eyes-potential-in-South-East-Asia?utm_source=copyright&utm_medium=OnSite&utm_campaign=copyright.

Liu, L., Pohnert, G., & Wei, D. (2016). Extracellular metabolites from industrial microalgae and their biotechnological potential. *Marine Drugs*, *14*, 191. doi:http://dx.doi.org/10.3390/md14100191.

Loke Show, P. (2022). Global market and economic analysis of microalgae technology: status and perspectives. *Bioresource Technology*, *357*, 127329. doi:https://doi.org/10.1016/j.biortech.2022.127329.

Martínez-Angulo, L. D., & Ramírez-Mérida, L. G. (2017). Estado actual de las empresas productoras de microalgas destinadas a alimentos y suplementos alimenticios en América Latina. *Revista Venezolana de Ciencia y Tecnología de Alimentos*, *8*, 130–147.

Martínez, A. L., & Ramírez, M. L. (2017). Estado actual de las empresas productoras de microalgas destinadas a alimentos y suplementos alimenticios en América Latina. *Revista Venezolana de Ciencia y Tecnología de Alimentos*, *8*, 130–147.

Mayhead, E., Silkina, A., Llewellyn, C. A., & Fuentes-Grünewald, C. (2018). Comparing nutrient removal from membrane filtered and unfiltered domestic wastewater using *Chlorella vulgaris*. *Biology*, *7*, doi: http://dx.doi.org/10.3390/biology7010012.

Mehariya, S., Goswami, R. K., Karthikeysan, O. P., & Verma, P. (2021). Microalgae for high-value products: a way towards green nutraceutical and pharmaceutical compounds. *Chemosphere*, *280*, 130553. doi: https://doi.org/10.1016/j.chemosphere.2021.130553.

Melo, J. M., Telles, T. S., Ribeiro, M. R., De carvalho junior, O., & Andrade, D. S. (2022). *Chlorella sorokiniana* as bioremediator of wastewater: Nutrient removal, biomass production, and potential profit. *Bioresource Technology Reports*, *17*, 100933. doi:https://doi.org/10.1016/j.biteb.2021.100933.

Miguel, S. P., Ribeiro, M. P., Otero, A., & Coutinho, P. (2021). Application of microalgae and microalgal bioactive compounds in skin regeneration. *Algal Research*, *58*, 102395. doi:https://doi.org/10.1016/j.algal.2021.102395.

Mitra, M., Patidar, S. K., George, B., Shah, F., & Mishra, S. (2015). A euryhaline nannochloropsis gaditana with potential for nutraceutical (EPA) and biodiesel production. *Algal Research*, *8*, 161–167. doi: https://doi.org/10.1016/j.algal.2015.02.006.

Morales, E. A., Fernandez, E., & Kociolek, P. J. (2009). Epilithic diatoms (Bacillariophyta) from cloud forest and alpine streams in Bolivia, South America 3: diatoms from Sehuencas, Carrasco National Park, Department of Cochabamba. *Acta Botanica Croatica*, *68*, 263–283.

Nagarajan, D., Lee, D.-J., Chen, C.-Y., & Chang, J.-S. (2020). Resource recovery from wastewaters using microalgae-based approaches: a circular bioeconomy perspective. *Bioresource Technology*, *302*, 122817. doi:https://doi.org/10.1016/j.biortech.2020.122817.

Nethravathy, M. U., Mehar, J. G., Mudliar, S. N., & Shekh, A. Y. (2019). Recent advances in microalgal bioactives for food, feed, and healthcare products: commercial potential, market space, and sustainability. *Comprehensive Reviews in Food Science and Food Safety*, *18*, 1882–1897. doi:http://dx.doi.org/10.1111/1541-4337.12500.

Oliveira, A. P. F. D., & Bragotto, A. P. A. (2022). Microalgae-based products: food and public health. *Future Foods*, *6*, 100157. doi:https://doi.org/10.1016/j.fufo.2022.100157.
Oliveira, C. Y. B., Jacob, A., Nader, C., Oliveira, C. D. L., Matos, Â. P., & Araújo, E. S., et al. (2022). An overview on microalgae as renewable resources for meeting sustainable development goals. *Journal of Environmental Management*, *320*, 115897. doi:https://doi.org/10.1016/j.jenvman.2022.115897.
Padri, M., Boontian, N., Piasai, C., & Phorndon, T. (2020). Coupling wastewater treatment with microalgae biomass production: focusing on biomass generation and treatment efficiency. *Engineering Journal*, *6*, 11–29. doi:https://doi.org/10.4186/ej.2020.24.6.11.
Pardo-Cárdenas, Y., Herrera-Orozco, I., González-Delgado, Á.-D., & Kafarov, V. (2013). Environmental assessment of microalgae biodiesel production in colombia: comparison of three oil extraction systems. *Ciencia, Tecnología y Futuro*, *5*, 85–100. doi:http://www.scielo.org.co/pdf/ctyf/v5n2/v5n2a07.pdf.
Pereira, I., Ortega, R., Barrientos, L., Moya, M., Reyes, G., & Kramm, V. (2009). Development of a biofertilizer based on filamentous nitrogen-fixing cyanobacteria for rice crops in Chile. *Journal Applied Phycology*, *21*, 135–144. doi:http://dx.doi.org/10.1007/s10811-008-9342-4.
Pereira, S. L., Leonard, A. E., Huang, Y. S., Chuang, L. T., & Mukerji, P. (2004). Identification of two novel microalgal enzymes involved in the conversion of the omega3-fatty acid, eicosapentaenoic acid, into docosahexaenoic acid. *The Biochemical Journal*, *384*, 357–366. doi:https://doi.org/10.1042/BJ20040970.
Prodiesel (2022). Prodiesel Energia – Cultivamos Micro Alga Argentina.
Pruvost, J. (2022). Alimentation, énergie, dépollution. . . Les microalgues, une solution aux enjeux de demain? Rapport d'activités 2016–2018 – Project: Microalgal biomass refining and valorisation: 187. https://doi.org/10.13140/RG.2.2.31946.44483.
Pulz, O., & Gross, W. (2004). Valuable products from biotechnology of microalgae. *Applied Microbiology and Biotechnology*, *65*, 635–648. doi:http://dx.doi.org/10.1007/s00253-004-1647-x.
Quintero-Dallos, V., García-Martínez, J. B., Contreras-Ropero, J. E., Barajas-Solano, A. F., Barajas-Ferrerira, C., & Lavecchia, R., et al. (2019). Vinasse as a sustainable medium for the production of *chlorella vulgaris* UTEX 1803. *Water*, *11*, 1526. doi:https://doi.org/10.3390/w11081526.
Quiroga, M. L. R., Avaro, M. G., Gittardi, A., Vivar, M. E. A. D. D., Rodrigo, G. S., & Molen, S. V. D. (2019). Are temperature and culture scale determinant factors in biochemical composition and fatty acid profile in microalgae? (in Spanish). *Naturalia Patagónica*, *15*, 94–114.
Ramos-Vega, A., Angulo, C., Bañuelos-Hernández, B., & Monreal-Escalante, E. (2021). Microalgae-made vaccines against infectious diseases. *Algal Research*, *58*, 102408. doi:https://doi.org/10.1016/j.algal.2021.102408.
Ray, S., Abraham, J., Jordan, N., Lindsay, M., & Chauhan, N. (2022). Synthetic, photosynthetic, and chemical strategies to enhance carbon dioxide fixation. *Journal of Carbon Research*, *8*, 18.
Renuka, N., Prasanna, R., Sood, A., Ahluwalia, A. S., Bansal, R., & Babu, S., et al. (2016). Exploring the efficacy of wastewater-grown microalgal biomass as a biofertilizer for wheat. *Environmental Science and Pollution Research*, *23*, 6608–6620. doi:http://dx.doi.org/10.1007/s11356-015-5884-6.
Revista-Aqua. (2016). Cultivo de microalgas: Pequeñas de alta sofisticación.
Reynaga, E. J. D. (2019). Plan de negocio para cultivo de microalga *Arthrospira platensis* como aditivo alimenticio para pollos de engorde. Master, Universidad ESAN – Programa de Maestría en Administración Lima, Peru.
Ronga, D., Biazzi, E., Parati, K., Carminati, D., Carminati, E., & Tava, A. (2019). Microalgal biostimulants and biofertilisers in crop productions. *Agronomy*, *9*, 192. doi: http://dx.doi.org/10.3390/agronomy9040192.
Rosset, V. K., Bartozek, E. C. R., Lambrecht, R. W., Auricchio, M. R., Dos Santos, M., & Peres, C. K. (2020). Gaps and challenges in the knowledge of algal biodiversity in Paraguay. *Phycologia*, *59*, 571–577. doi: http://dx.doi.org/10.1080/00318884.2020.1830597.

Rumin, J., Gonçalves de oliveira junior, R., Bérard, J.-B., & Picot, L. (2021). Improving microalgae research and marketing in the European Atlantic area: Analysis of major gaps and barriers limiting sector development. *Marine Drugs*, *19*, 319. doi:https://doi.org/10.3390/md19060319.

Salazar-Torres, G., Huszar, & Moraes, V. L. D. (2012). Microalgae community of the Huaytire wetland, an Andean high-altitude wetland in peru. *Acta Limnologica Brasiliensia*, *24*, 285–292. doi:https://doi.org/10.1590/S2179-975X2012005000046.

Sampathkumar, S. J., & Gothandam, K. M. (2019). Sodium bicarbonate augmentation enhances lutein biosynthesis in green microalgae *chlorella pyrenoidosa*. *Biocatalysis and Agricultural Biotechnology*, *22*, 101406. doi:https://doi.org/10.1016/j.bcab.2019.101406.

Sánchez-Bayo, A., López-Chicharro, D., Morales, V., Espada, J. J., Puyol, D., & Martínez, F., et al. (2020). Biodiesel and biogas production from *Isochrysis galbana* using dry and wet lipid extraction: a biorefinery approach. *Renewable Energy*, *146*, 188–195. doi:http://dx.doi.org/10.1016/j.renene.2019.06.148.

Santos, M. J. B. D. A., Andrade, D. S., Bosso, A., Murata, M. M., Morioka, L. R. I., & Silva, J. B. D., et al. (2021). *Chlorella sorokiniana* cultivation in cheese whey for β-galactosidase production. *Research, Society and Development*, *12* doi:https://doi.org/10.33448/rsd-v10i12.20727.

Santos Montes, A. M., González Arechavala, Y., & Martín Sastre, C. (2014). Uso y aplicaciones potenciales de las microalgas. *Anales de mecánica y electricidad*, *91*(1), 20–28.

Sarwer, A., Hamed, S. M., Osman, A. I., Jamil, F., Al-Muhtaseb, A. A. H., & Alhajeri, N. S., et al. (2022). Algal biomass valorization for biofuel production and carbon sequestration: a review. *Environmental Chemistry Letters*, *20*, 2797–2851. doi:https://doi.org/10.1007/s10311-022-01458-1.

Savio, S., Congestri, R., & Rodolfo, C. (2021). Are we out of the infancy of microalgae-based drug discovery? *Algal Research*, *54*, 102173. doi:https://doi.org/10.1016/j.algal.2020.102173.

Serrà, A., Artal, R., García-Amorós, J., Gómez, E., & Philippe, L. (2020). Circular zero-residue process using microalgae for efficient water decontamination, biofuel production, and carbon dioxide fixation. *Chemical Engineering Journal*, *388*, 124278. doi:https://doi.org/10.1016/j.cej.2020.124278.

Siedenburg, J. (2022). Could microalgae offer promising options for climate action via their agri-food applications? *Frontiers in Sustainable Food Systems*, *6*. doi:https://doi.org/0.3389/fsufs.2022.976946.

Silva, H. R., Prete, C. E. C., Zambrano, F., De mello, V. H., Tischer, C. A., & Andrade, D. S. (2016). Combining glucose and sodium acetate improves the growth of *Neochloris oleoabundans* under mixotrophic conditions. *AMB Express*, *6*, 1–11. doi:https://doi.org/10.1186/s13568-016-0180-5.

Skjånes, K., Rebours, C., & Lindblad, P. (2013). Potential for green microalgae to produce hydrogen, pharmaceuticals and other high value products in a combined process. *Critical Reviews in Biotechnology*, *33*, 172–215. doi:https://doi.org/10.3109/07388551.2012.681625.

Soeder, C. J. (1980). Massive cultivation of microalgae: Results and prospects. *Hydrobiologia*, *72*, 197–209. doi:https://doi.org/10.1007/BF00016247.

Spolaore, P., Joannis-Cassan, C., Duran, E., & Isambert, A. (2006). Commercial applications of microalgae. *Journal of Bioscience and Bioengineering*, *101*, 87–96. doi:http://dx.doi.org/10.1263/jbb.101.87.

Stoyneva-Gärtner, M., Uzunov, B., & Gärtner, G. (2020). Enigmatic microalgae from aeroterrestrial and extreme habitats in cosmetics: the potential of the untapped natural sources. *Cosmetics*, *7*, 1–22. doi: https://doi.org/10.3390/cosmetics7020027.

Sukhinov, D. V., Gorin, K. V., Romanov, A. O., Gotovtsev, P. M., & Sergeeva, Y. E. (2021). Increased c-phycocyanin extract purity by flocculation of *Arthrospira platensis* with chitosan. *Algal Research*, *58*, 102393. doi:https://doi.org/10.1016/j.algal.2021.102393.

Trentin, G., Piazza, F., Carletti, M., Zorin, B., Khozin-Goldberg, I., & Bertucco, A., et al. (2023). Fixing N_2 into cyanophycin: Continuous cultivation of *Nostoc* sp. *PCC 7120*. *Applied Microbiology and Biotechnology*, *107*, 97–110. doi:http://dx.doi.org/10.1007/s00253-022-12292-4.

Ubando, A. T., Anderson, E., Ng, S., Chen, W.-H., Culaba, A. B., & Kwon, E. E. (2022). Life cycle assessment of microalgal biorefinery: a state-of-the-art review. *Bioresource Technology*, *360*, 127615. doi: https://doi.org/10.1016/j.biortech.2022.127615.

United Nations. (2017). World Population Projected to Reach 9.8 Billion in 2050, and 11.2 billion in 2100.

Valenti, W. C., Barros, H. P., Moraes-Valenti, P., Bueno, G. W., & Cavalli, R. O. (2021). Aquaculture in Brazil: past, present and future. *Aquaculture Reports*, *19*, 100611. doi:https://doi.org/10.1016/j.aqrep.2021.100611.

Valero, N. O., Gómez, L. C. G., Vanegas, J., & Hernández-Benítez, R. E. (2018). Actividad Promotora de Crecimiento Vegetal por Cianobacterias en Ambientes Semiáridos, Caso La Guajira. In: V. J. y., & B. RH. (Eds.), Potencial Biotecnológico de las Microalgas en Zonas Áridas. Bogotá, Colombia.

Vanegas, J., & Hernández, L. H. (2018). Potencial biotecnológico de las microalgas zonas áridasServicio Nacional de Aprendizaje (SENA) Bogota.

Zapata, D., Arroyave, C., Cardona, L., Aristizábal, A., Poschenrieder, C., & Llugany, M. (2021). Phytohormone production and morphology of *spirulina platensis* grown in dairy wastewaters. *Algal Research*, *59*, 102469. doi:https://doi.org/10.1016/j.algal.2021.102469.

Zittelli, G. C., Lavista, F., Bastianini, A., Rodolfi, L., Vincenzini, M., & Tredici, M. R. (1999). Production of eicosapentaenoic acid by *nannochloropsis* sp. cultures in outdoor tubular photobioreactors. *Journal of Biotechnology*, *70*, 299–312. doi:http://dx.doi.org/10.1016/S0168-1656(99)00082-6.

Part II: **Process integration applied to microalgae-based systems**

Ihana A. Severo*, Diego de O. Corrêa, Wellington Balmant, Juan C. Ordonez, André B. Mariano, José V. C. Vargas

Chapter 8 Energy and heat integration applied to microalgae-based systems

Abstract: Microalgae culture and downstream biomass processing are exceedingly energy-intensive stages. The bottleneck inherent in developing microalgae processes on a scale is limited because of their very high cost and negative energy balance, which inevitably jeopardizes the sustainability and profitability of bioproducts. A strategy to circumvent the limitation related to the energetic problem of these systems would be to minimize the use of expensive and impacting inputs of processes that use too much energy. Energy and heat integration can provide a series of surplus flows that can be promptly recovered, distributed, and used during microalgae production in the cultivation, drying, and extraction steps. In this sense, this book chapter addresses the possibilities of energy and heat integration in microalgae-based systems. The chapter covers topics about state of the art, including energy systems, a short introduction to aspects of energy and thermodynamics, culture systems, microalgal biomass production, energy demand in microalgae-based systems, and opportunities for energy and heat integration in microalgae-based systems.

Keywords: microalgae, process integration, energy integration, heat recovery, renewable energy

8.1 Introduction

The world needs energy to support population growth, economic progress, limitations in the amount from different sources, and fluctuations in prices. But its supply is threatened. The dilemma about responsible energy use is not something new. In this sense,

***Corresponding author: Ihana A. Severo,** Sustainable Energy Research and Development Center (NPDEAS), Federal University of Paraná (UFPR), Curitiba, PR 81531-980, Brazil; Department of Mechanical Engineering, Energy and Sustainability Center and Center for Advanced Power Systems (CAPS), Florida State University, Tallahassee, FL 32310-6046, USA, e-mail: ihana.aguiar@gmail.com
Diego de O. Corrêa, Wellington Balmant, André B. Mariano, Sustainable Energy Research and Development Center (NPDEAS), Federal University of Paraná (UFPR), Curitiba, PR 81531-980, Brazil
Juan C. Ordonez, Department of Mechanical Engineering, Energy and Sustainability Center and Center for Advanced Power Systems (CAPS), Florida State University, Tallahassee, FL 32310-6046, USA
José V. C. Vargas, Sustainable Energy Research and Development Center (NPDEAS), Federal University of Paraná (UFPR), Curitiba, PR 81531-980, Brazil

https://doi.org/10.1515/9783110781267-008

the process integration concept emerged in the 1950s in response to the oil crisis. The first contributions of this approach were naturally in the energy and heat integration field. Its main objectives are to achieve efficiency, economy, and sustainability related to the delivery of energy-based resource delivery. This perception was first and foremost used in the petrochemical industry. Afterward, it began to reach some maturity with several implementations in the food, beverage, power generation, pharmaceutical, and chemical sectors (Klemeš, 2013).

Energy integration is founded on the principles of thermodynamics and considers different types of energy (e.g., thermal, electrical, chemical, and mechanical). It is a set of systematic methods for combining the heating and cooling demands within each process or operation to reduce the consumption of heat and cold utilities systems by maximizing the recovery and use of heat and associated footprints (Friedler, 2010). Thus, whatever the energy system, it is equally applicable to small, medium, and large industrial plants. Additionally, the energy and heat integration approach can be widespread to integrate renewable energy sources, including biomass, biofuels, and bioenergy combined into heating and cooling cycles of various capacities, sizes, and designations. According to the International Energy Agency (IEA), renewable energy integration aims to incorporate heat, distributed generation, storage, thermally activated technologies, and demand in the heat distribution and transmission system. It is also an important step in achieving low-carbon production (IEA, 2023).

Therefore, bio-based processes, including microalgae-based ones, have been considered potential technological routes toward renewable energy integration (Dias et al., 2022a, 2022b). In recent years, microalgal biotechnology has received new investments and developments in the areas of engineering, biology, physics, and chemistry related to the design and operation of the bioreactor, biochemistry, and upstream and downstream biomass processing. Interest in this research field has increased because microalgae are microorganisms with a distinct capacity for cell growth per unit area, eliminating the requirements for extensive hectares of arable land. In addition, they grow with the input of nutrients from wastewater or gaseous emissions, generating biomass rich in compounds that can be exploited commercially (Dourou et al., 2020).

However, the scale-up of microalgae mass production is still challenging, making the sale of bulk products with lower values, such as biofuels, biomaterials, and animal feed, unfeasible. One of the most critical factors is due to the fact that microalgae processes are excessive in energy, in addition to issues of profitability, resource demand, and environmental impact of production. Therefore, microalgae commercialization consequently remains restricted to a few species and bioproducts, such as food supplements and pigments (Baala Harini and Rajkumar, 2022).

One of the potential methods of reducing costs associated with microalgae systems is through the generation of various by-products in the context of an integrated biorefinery, allowing the integration of flows and surplus of mass, water (Fresewinkel et al., 2014), and especially energy. Even so, these production platforms must operate close to the limits of lucrativeness and environmental gains, with elevated levels of uncertainty (Severo et al., 2019).

The identification of energy integration opportunities or combination with industrial facilities could supply numerous of the resources needed for staggered biomass output, such as nutrients, CO_2, and heat. In such systems, there is the possibility to recover, recycle, and reuse energy flows from processes and products. Different energy vectors are available, that is, the energy load can be distributed among all products, normally calculated by allocation procedures. The main objective is to develop or improve energy conversion technologies, thus reducing costs and environmental impact compared to fossil ones (Chowdhury et al., 2012; Severo et al., 2020, 2022).

Given this scenario, this chapter aims to provide an overview of energy and heat integration applied to microalgae-based systems. The chapter presents issues about state-of-the-art, including energy systems, a short introduction to aspects of energy and thermodynamics, culture systems, microalgal biomass production, energy demand in microalgae-based systems, and opportunities for energy and heat integration in microalgae-based systems.

8.2 State of the art

8.2.1 Energy systems: a brief overview

Energy can be defined as the capacity to produce work, that is, of two systems interacting with each other. According to the law of conservation, energy cannot be created or destroyed, although it can be changed from one form to another. There are several types of energy and energetic resources that derive from thermal energy, electrical energy, chemical energy, kinetic energy, and mechanical energy. Already, energy integration is one of the process integration methodologies, which is fundamentally based on thermodynamic principles, from energy and exergy balance insights (Sinnott and Towler, 2020). **Box 1** presents an overview of the fundamental aspects of energy and thermodynamics.

Box 1: A short introduction to aspects of energy and thermodynamics

Energy is an essential constant of the universe. In general, energy is the capacity of matter to produce work, such as manufacturing molecules or moving substances, through a transformation in different forms. Although there is no exact definition for energy, it can be said that it is a quantity, assigned mathematically, rather than being considered a substance or element. Besides, it derives from thermodynamics, applying to all areas of science and engineering.

Forms of energy and applications

In the universe, energy is available in many forms, whether internal or transient. It is a field of thermodynamics, which can be classified into two groups: macroscopic and microscopic energy. Macroscopic forms, related to the quantities of kinetic and potential energies, are attributed to the overall energy of the reference system. It is directly influenced by external effects in which the system exists, such as gravity, electricity, magnetism, and surface tension. On the other hand, microscopic forms, related to internal energy, are attributed to the molecular structure of a system and are independent of the external conditions in which it is found. The internal energy depends, however, on the qualities or properties of the materials in the system (physical form and composition, environmental parameters, such as pressure, temperature, and electric and magnetic fields). There are many forms of microscopic energy, such as thermal, chemical, electrical, and mechanical, among others. The energy seems almost tangible to us since it is present in all daily activities and involves the transfer and alteration of power. All industrial processes have energy (e.g., steam generation in a unit operation). Energy applications energy are broad, extending to the extremely complex world of biotechnology and bioprocesses.

Thermodynamics

Thermodynamics is a branch of the physical sciences, which is based on the interactions of temperature and heat with the other different forms of energy. All thermodynamic aspects – both energy, exergy, and entropy approaches – play a key role in the evaluation of processes, systems, and components in which the conversion, transfer, and use of energy occur. These issues, to some extent, make an energy balance, for example, more complicated than a mass balance. The development of a robust thermodynamic concept for industrial biotechnology has been one of the most significant challenges for the scientific community when designing an efficient, rapid, economical, and sustainable bioprocess. There are, therefore, four experimentally established laws of thermodynamics that determine the physical amounts of a system. Although all of them are important, the first and the second are fundamental for the analysis, design, development, evaluation, and improvement of microalgae-based thermal processes [adapted from Dincer and Bicer (2020)].

In microalgae facilities, energy integrations are used to define the heating, cooling, and power needs of given equipments, unit operations, thermal systems, or products. At an operational level, this will help show the energy usage pattern in the production chain and identify sites where there is a need to conserve and save energy forms. In this sense, the overall target is to achieve maximum integration levels for improved energetic efficiency and reduce costs (Aziz et al., 2014). Table 8.1 summarizes some of the different forms of energy in microalgae-based systems.

Energy applications in microalgae processes vary widely, including power generation plants (e.g., coproduction of biogas and biomass combustion), energetic products (solid, liquid, and gaseous biofuels; bioenergy), and heating and cooling systems in different equipment (e.g., heat waste valorization), and so on. These are just some of the many examples of applications. Energy is relevant to several processes, with applicability in different areas. A proper elucidation of this study topic is necessary to improve new or consolidated designs and the performance of microalgae-based energetic systems.

Table 8.1: Examples of energy in microalgae-based processes.

Forms	Description	Application
Light	Refers to all electromagnetic radiation of frequency and wavelength within the range of the visible spectrum.	Use of solar energy for the thermal maintenance of outdoor microalgal cultures.
Thermal	Internal heat that is directly associated with the absolute temperature of a system.	Direct combustion of biomass to generate heat.
Electrical	Energy based on the generation of differences in electrical potential between two points, which allow the establishment of an electric current between them.	Creation of electrical power from thermal energy through the heat recovered from the drying step.
Chemical	A type of potential energy stored in the chemical bonds between the atoms of matter, being released by breaking these bonds.	Photosynthetic mechanism of microalgae.
Mechanical	Capacity of an object to produce work.	Use a bead milling to apply a shear force during the microalgae cell disruption.

8.2.2 Microalgae culture systems

Commercial microalgae cultivation has been conducted in open systems, while closed systems (photobioreactors) are often used to produce high-value products due to better control of growing conditions. Additionally, there are hybrid systems, more robust designs developed to bring together the strengths of open and closed systems, aimed at increasing efficiency and reducing capital costs (Narala et al., 2016).

Open systems were initially developed for nitrogen and phosphorus conversion from wastewater. They consist of raceway ponds or circular tanks, in which the synthetic medium used for microalgal cultivation is replaced by wastewater with substantial concentrations of organic matter, nitrogen, and total phosphorus (Morillas-España et al., 2021). High-rate ponds are simple arrangements with low capital and operational costs. However, the construction of these outdoor systems requires large areas for implementation, since the depth of the ponds varies between 0.15 and 0.5 m, resulting in low volumes of work; in addition, performance is strongly associated with local weather conditions (Pires et al., 2012). Thus, the choice of the geographical position of the production plant must be taken into account, including solar irradiation, temperature, and precipitation. Besides, the availability of resources such as energy, water, nutrients, and CO_2 are key factors to having high yields in a given local climate (Boruff et al., 2015).

On the other hand, closed systems, including bubble column, tubular, airlift, flat plate, and big-bag photobioreactors, are preferable because each of those different configurations has efficient and robust parameters, providing conditions for better control and monitoring of the culture medium. Additionally, other notable advantages of photobioreactors are less propensity to contamination, less hydrodynamic stress, greater surface/volume (S/V) and height/diameter (A/D) ratio, and considerable productivity. The main factors that must be considered to avoid poor cell growth performance in photobioreactors are light, temperature, pH, nutrient supply, and CO_2/O_2 balance and mixing. Beisides, the ideal photobioreactor design for industrial applications should take into account the species of microalgae used, process yield, production costs, and the product obtained (Sathinathan et al., 2023).

8.2.2.1 Microalgal biomass production: theory × practice

Actual biomass productivity contradicts the theory in practice. It differs greatly from one cultivation system to another. The theoretical maximum photosynthetic efficiency of microalgae is measured as the ratio between the energy supplied to cells by light energy and the energy content of the biomass produced by photosynthesis. It has typically been recorded around 0.1–10%, and the maximum theoretical value is approximately 12%. However, growth experiments vary a lot from these values. For example, the theoretical biomass yield of microalgae was reported as 100–200 g/m^2/day (dry weight) and the practical productivity rate was 15–30 g/m^2/day (dry weight) (Sun et al., 2018).

Therefore, the real biochemical energy yield of light input to cultures is undoubtedly a hot spot that prevents high-density and energy-efficient cultivation of microalgae at scale. Determination of photosynthetic efficiency should be a priority during demonstration trial experiments to get a complete elucidation of the factors that affect this parameter as well as to assist in the development of effective models of area growth.

8.2.3 Energy demand in microalgae-based systems

Energy is required in all stages of microalgae biomass processing. Agitation is a necessary operation in photosynthetic microorganism cultivation, as it ensures the spatial uniformity of the reaction vessels, favoring cell exposure to light, heat transfer, and thermal stratification, in addition to improving gas exchange. Proper mixing further minimizes the formation of cell aggregates that increase the overall inefficiency of the bioreactor. Microalgae bioreactors are usually equipped with pneumatic aeration and mechanical agitation systems or even a combination of these systems. Paddle wheels are normally used to mix the bioreactors and the energy demand depends on several factors, including mainly the lagoon depth, in open systems, and liquid velocity, affecting biomass productivity. For example, a deeper raceway pond has the potential to increase volumetric productivity; however, paddle wheels require more energy to mix the broth. These parameters must be optimized to obtain a deficit in electricity consumption per unit of biomass generated. The energy expenditure to turn the paddle wheels ranges from 18 to 290 MJ ha/day, which represents 0.25–4 MJ/kg of dry biomass (Clarens et al., 2010; Collet et al., 2011; Deprá et al., 2020a).

In addition, the gas injection system for mixing/aerating in open systems also consumes a lot of energy (0.09–0.15 MJ/kg of dry biomass), making this demand very expensive. Studies on the electricity requirements in this type of cultivation system report values ranging from 1.5 to 8.4 W/m^3 (Mendoza et al., 2013; Barceló-Villalobos et al., 2018), being much lower than those of closed photobioreactors (Lehr and Posten, 2009).

In contrast, energy demands in closed photobioreactors depend exclusively on the model type. Tubular systems require energy to aerate and pump the culture medium, whose consumption varies from 2,000 to 3,000 W/m^3. This configuration is more complex and may impair cell growth due to the shear force caused by the pumps. Flat plate systems, on the other hand, require less energy input, around 55 W/m^3, as the electricity is only for the CO_2/air injection. Both models have a negative energy balance (Jorquera et al., 2010).

Outdoor cultivation also demands additional energy inputs for lighting. Aspects such as locality, seasonality, variations over the light photoperiod, and occurrence of light/dark photoperiods are the main issues of naturally lit systems. The light input must be sufficient, both qualitatively and quantitatively, to support adequate growth of microalgal cultures. For example, the source of artificial lighting and the energy cost are elements to be well-defined in the implementation of cultivation systems. In the study by Blanken et al. (2013), the total electricity costs resulting from artificial lighting represent 26.7 and 25.3 USD per kg of dry biomass for LEDs and high-intensity discharge lamps, respectively.

Energy is also needed to pump the inoculum into the primary culture. Other points of energy consumption include unit operations such as harvesting, where the centrifuge is the device that can consume up to 85% of the total energy required in the process due to the high moisture content (Aziz et al., 2014). Stages such as drying, extraction, evaporation and solvent recovery, and biomass thermochemical conversion also require additional heat.

Finally, the use of fertilizers in microalgae cultivation presents considerable energy demands. For example, if 88 g and 12 g of N and P, respectively, are required for the production of 1 kg of microalgae biomass, the predicted energy input would be almost 5 MJ and 0.70 MJ for N and P, respectively (considerations: ammonium nitrate (N) = 51 MJ/kg; phosphate (P) = 58.9 MJ/kg). The biomass contains on average 24 MJ of energy, indicating 19% and 3% of the energy contained in the biomass from N and P, respectively. Therefore, reducing or withdrawing the use of fertilizers in the microalgae mass culture is a hot spot for the sustainable output of bioenergy or chemical raw material, as long as the energy balance and related emissions are favorable (Mayers et al., 2016).

8.2.4 Opportunities for energy and heat integration in microalgae-based systems

Heat integration in microalgae-based systems can be done with the help of Pinch analysis (see **Box 2**), which is applied to minimize the primary energy needs of the process, being one of the most used methods in energy systems.

Box 2: What is Pinch analysis?

Pinch analysis is a set of methods for resource conservation supported by thermodynamic principles. It is considered a tool that calculates energy savings in processes and optimizes heat recovery systems, energy supply, and operational conditions. Pinch analysis was established when the concept of process integration emerged, and currently, it has been widely implemented in the planning of the energy sector, mainly in the development of renewable energy. The method is based on an evaluation of the heat exchanger flows and streams, determining the sites where such exchanges are restricted. Thus, it is possible to allocate the energy near the point "pinch." This is particularly important when dealing with batch processes of different operation times. The analysis can be done through the heat load, that is, the specific product of enthalpy, as a function of temperature. The combination of these data will provide a diagram containing all heat streams, both hot (heat produced) and cold (heat required), as shown in Figure 8.1.

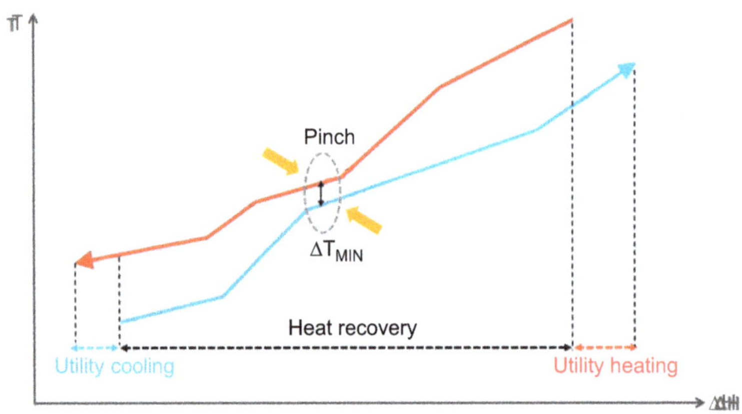

Figure 8.1: Diagram of temperature × enthalpy [known as the composite curves; adapted from Klemeš (2013)].

The point of approximation between the hot and cold flow curves is the location of the Pinch point. The temperature difference between the two curves is minimal. In the division of the composite curves, above the pinch, it is necessary heating (i.e., steam), while below the pinch, cooling (i.e., cooling water). This means that when designing a process, the energy location is identified, and heat recovery can be achieved. There is a tendency to design the best equipment, operations, technologies, systems, and processes of energy to obtain improved overall performance. The ideal design, appropriate resource use, efficiency, sustainability, environmental impact, economy, and security of energy in microalgae-based processes can be investigated through the fundamental aspects of thermodynamics.

As cultures are exposed to the surrounding environment, the need for temperature control in the growing medium (heating or cooling) may be required to keep the temperature in the ideal range. Adequate performance of photosynthesis is generally in the mesophilic range, that is, between 20 and 30 °C, where it generally constitutes the majority of microalgae with potential for commercial exploitation. Cell growth rates can decrease at temperatures below 20 °C and above 30–35 °C (Nwoba et al., 2019). Despite this, some strains are known for their ability to develop in extremophile conditions, that is, organisms that can survive in environments of extreme conditions, including the temperature factor. For example, species considered to be thermophilic, including *Mastigocladus laminosus* and *Synechococcus*, can grow at a high temperature, above 70 °C. On the other hand, psychrotrophic species such as *Chloromonas nivalis* and *Raphidonema nivale* develop on ice or snow (Varshney et al., 2015).

Depending on the geospatial location of outdoor cultivation, climatic factors, atmospheric scattering, altitude, and latitude influence temperature. Culture systems are prone to daily climatic fluctuations and seasonal variations that will have a drastic impact on microalgae productivity (Dias et al., 2022a). By way of exemplification, in middle-latitude countries (considered to be temperate climate locations), although these regions most often favor microalgae mass production, the temperature can reach up to 45 °C. This thermal range is well above that optimum, frequently claimed by most commercial strains. Based on the seasons, on a typical summer working day, photobioreactors are subject to high solar irradiation, and the temperature rises considerably. The overheating problem of these systems is inevitable, causing phenomena of photo-inhibition and photo-saturation. Conversely, during the winter, outdoor microalgae cultures can be photo-limited and generally need to be heated. In these cases, a temperature control unit is essential to keep it stable. Some solutions based on thermal regulation, especially related to heating systems, have been used for a long time, while others are under development to achieve this purpose. Heat exchangers and the adaptation of photovoltaic (PV) panels are examples of heating regulators (Nwoba et al., 2020a). As for cooling, passive evaporative systems are used with freshwater spray, direct submersion in thermoregulated water pools, dark sheets for shading, overlapping tubes, heat exchanger, greenhouses, and infrared filtering (Nwoba et al., 2020b). However, there are some pros and cons of heating/cooling microalgae systems in terms of economy and environmental impact, usually attributed to the electricity consumption. It should be noted that the adoption of any temperature control system is an expensive component on a large scale.

Thus, energy integration through waste heat recirculation from other hot sources would be a promising strategy. This would be possible with the installation of alternative devices, such as PV panels, to generate electricity (Fresewinkel et al., 2014). Tredici et al. (2015) demonstrated photovoltaic integration and the potential energy gain exceeded 600 GJ ha/year. Since this is a sustainable process, at first glance this strategy seems incoherent to microalgae, but reasonable to the extent that 15% of the light-to-electricity conversion efficiency of this auxiliary source would cover conventional electrical power demands. In practice, it will be more sustainable to employ PV cells, as sunlight cannot be depleted in contrast to coal-derived electricity. However, these scenarios reveal the critical implications of trying to control the temperature of outdoor cultures implications of trying to control the temperature of outdoor microalgae biomass production (Dias et al., 2022b).

Downstream processes also demand additional heat. Depending on the target product, microalgal biomass processing goes through many operations, including equipment and procedures thermal, which are often energy-intensive. This could be for drying microalgae biomass after harvesting, upgrading product extraction efficiency, evaporation, and solvent recovery after extraction or processing the biomass through thermochemical conversion processes (Carvalho et al., 2020).

Regardless, the quantity and quality (temperature) of heat demanded for each of these processes, whether in the upstream (cultivation) or downstream phase, will be quite different. For example, as mentioned earlier, (i) maintaining the temperature of a raceway ponds or pilot photobioreactor a few degrees above ambient temperature may require hot water of approximately 60 °C; (ii) biomass drying may require heating >80 °C; (iii) while thermochemical biomass conversion routes (e.g., direct combustion, pyrolysis, and gasification) usually operate above 300 °C.

The opportunity to supply heat from other industrial process sources in the form of hot gas integration, including exhaust gases (typically >120 °C) or process cooling water, has the potential to be recovered for use in one of these three approaches, which demand heat. According to Aziz et al. (2014), the integration of drying with gasification and energy generation based on a combined cycle for microalgae *Chlorella* sp. was proposed. The system under study is based on an approach that includes exergy recovery and process integration. In the first case, exergy recovery is obtained by efficiently coupling each type of heat. Process integration, on the other hand, is implemented to use the remaining energy of one process, which can be effectively reused in other processes. This research, using Pinch analysis, demonstrated that the energy required for drying can be significantly reduced – a maximum drying coefficient of performance of ~18.5 can be achieved. In this case, almost all the energy involved in drying can be recovered with minimal exergy destruction.

Figure 8.2 illustrates a schematic diagram that exemplifies that probably the heat integration of the downstream biomass processing steps is the most viable technological route for this purpose. For example, coupling in an industrial plant, recovering exhaust gases, can provide CO_2 for strain cultivation and heat for temperature maintenance. In this scenario, as the gases contain many impurities (CO, NO_x, SO_x, particulate matter), a cleaning step is necessary to avoid failures in growth performance and thus low productivity. In addition, gas temperatures will differ, likely requiring cooling. Alternatively, hot gas streams can be used first to dry the biomass and then to heat the culture and CO_2 input.

Integration with an industrial facility that provides surplus heat for biomass drying can decrease the primary energy requirements of microalgae systems, which can lead to economic gains compared to autonomous processes. Nevertheless, the type of dryer appropriate for use with residual heat must be considered since the biomass will be directly exposed to incident gases that may be toxic or corrosive. In addition, the environmental impact of heat integration levels, which influence GHG emissions and the economics of microalgae processes, must be considered (Mayers et al., 2016).

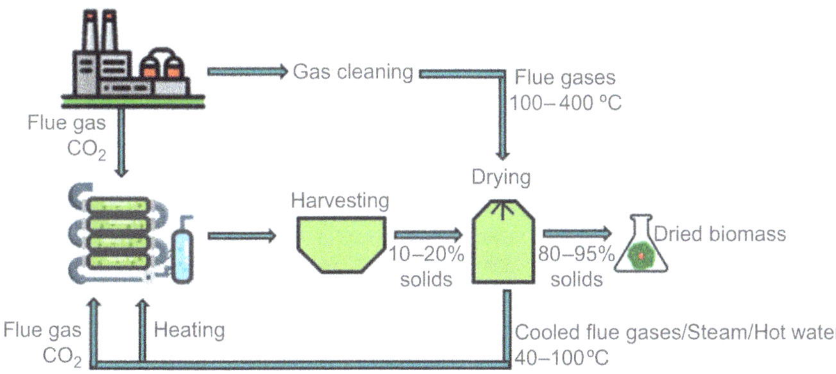

Figure 8.2: Heat integration in an industrially integrated microalgae system.

8.3 Conclusions and recommendations

Despite the definition of industrial energy/heat integration and similar concepts varying greatly, this approach is undoubtedly an opportunity that can be exploited to maximize the overall energy efficiency of an industrially integrated microalgae system.

The recovery of surplus heat from power plants, oil refineries, and chemical facilities in general, which are currently operating in daily life for thermal purposes, is being explored. However, not in the way that it could be to obtain maximum benefits. The idea is that the energy and heat integration from industrial processes is facing a comprehension bottleneck not only from the aspect of microalgae cultivation, biology, or bioreactor engineering but also from a myriad of industrial plants. The prediction of the potential recovery and energy and heat integration can be done through tools such as Pinch analysis and energy balances and other thermodynamic principles. But even so, there is uncertainty in relation to the actual provision of heat derived from a microalgae process because it is highly specific and depends on factors such as the geographic positioning of the installation, which includes qualitative and quantitative aspects of the region's climate.

Thus, understanding the energy/heat integration requires the implementation of processes at demonstration scales, for more in-depth research, including life cycle assessments and technical-economic feasibility studies, with input data as precise and comprehensive as possible.

References

Aziz, M., Oda, T., & Kashiwagi, T. (2014). Integration of energy-efficient drying in microalgae utilization based on enhanced process integration. *Energy, 70*, 307–316. https://doi.org/10.1016/j.energy.2014.03.126

Baala Harini, A., & Rajkumar, R. (2022). Development of sustainable bioproducts from microalgae biomass: Current status and future perspectives. *BioResources, 17*(4), 7285–7312. 10.15376/biores.17.4.Harini

Barceló-Villalobos, M., et al. (2018). Analysis of mass transfer capacity in raceway reactors. *Algal Research, 35*, 91–97. https://doi.org/10.1016/j.algal.2018.08.017

Blanken, W., Cuaresma, M., Wijffels, R. H., & Janssen, M. (2013). Cultivation of microalgae on artificial light comes at a cost. *Algal Research, 2*(4), 333–340. https://doi.org/10.1016/j.algal.2013.09.004

Boruff, B. J., Moheimani, N. R., & Borowitzka, M. A. (2015). Identifying locations for large-scale microalgae cultivation in Western Australia: A GIS approach. *Applied Energy, 149*, 379–391. https://doi.org/10.1016/j.apenergy.2015.03.089

Carvalho, J. C. de, et al (2020). Microalgal biomass pretreatment for integrated processing into biofuels, food, and feed. *Bioresource Technology, 300*, 122719. https://doi.org/10.1016/j.biortech.2019.122719

Chowdhury, R., Viamajala, S., & Gerlach, R. (2012). Reduction of environmental and energy footprint of microalgal biodiesel production through material and energy integration. *Bioresource Technology, 108*, 102–111. https://doi.org/10.1016/j.biortech.2011.12.099

Clarens, A. F., Resurreccion, E. P., White, M. A., & Colosi, L. M. (2010). Environmental life cycle comparison of algae to other bioenergy feedstocks. *Environmental Science and Technology, 44*, 1813–1819. https://doi.org/10.1021/es902838n

Collet, P., Hélias, A., Lardon, L., et al. (2011). Life-cycle assessment of microalgae culture coupled to biogas production. *Bioresource Technology, 102*, 207–214. https://doi.org/10.1016/j.biortech.2010.06.154

Deprá, M. C., Severo, I. A., dos Santos, A. M., Zepka, L. Q., & Jacob-Lopes, E. (2020). Environmental impacts on commercial microalgae-based products: Sustainability metrics and indicators. *Algal Research, 51*, 102056. https://doi.org/10.1016/j.algal.2020.102056

Dias, R. R., Deprá, M. C., Severo, I. A., Zepka, L. Q., & Jacob-Lopes, E. (2022a). Smart override of the energy matrix in commercial microalgae facilities: A transition path to a low-carbon bioeconomy. *Sustainable Energy Technologies and Assessments, 52*, 102073. https://doi.org/10.1016/j.seta.2022.102073

Dias, R. R., Deprá, M. C., Zepka, L. Q., & Jacob-Lopes, E. (2022b). *In silico* study of hybrid renewable energy in microalgae facilities: A path towards net-zero emissions. *Algal Research*, 102661. https://doi.org/10.1016/j.algal.2022.102661

Dincer, I., & Bicer, Y. (2020). Chapter 2 – Fundamentals of Energy Systems. In: Dincer, I., & Bicer, Y. (Eds.), Integrated Energy Systems for Multigeneration. Elsevier Science, pp. 33–83. https://doi.org/10.1016/B978-0-12-809943-8.00002-9

Dourou, M., et al. (2020). High-added value products from microalgae and prospects of aquaculture wastewaters as microalgae growth media. *FEMS Microbiology Letters, 367*, fnaa081. https://doi.org/10.1093/femsle/fnaa081

Fresewinkel, M., Rosello, R., Wilhelm, C., Kruse, O., Hankamer, B., & Posten, C. (2014). Integration in microalgal bioprocess development: Design of efficient, sustainable, and economic processes. *Engineering in Life Sciences*, *14*, 560–573. https://doi.org/10.1002/elsc.201300153

Friedler, F. (2010). Process integration, modelling and optimisation for energy saving and pollution reduction. *Applied Thermal Engineering*, *30*, 2270–2280. https://doi.org/10.1016/j.applthermaleng.2010.04.030

IEA. International Energy Agency. Available at: https://www.iea.org/areas-of-work/programmes-and-partnerships/clean-energy-transitions-programme. Access in: January 15, 2023.

Jorquera, O., Kiperstok, A., Sales, E. A., Embiruc¸u, M., & Ghirardi, M. L. (2010). Comparative energy life-cycle analyses of microalgal biomass production in open ponds and photobioreactors. *Bioresource Technology*, *101*, 1406–1413. https://doi.org/10.1016/j.biortech.2009.09.038

Klemeš, J. J. (2013). Handbook of Process Integration (PI). Minimisation of Energy and Water Use, Waste and Emissions. A volume in Woodhead Publishing Series in Energy, Woodhead Publishing, Sawston, Cambridge, England.

Lehr, F., & Posten, C. (2009). Closed photo-bioreactors as tools for biofuel production. *Current Opinion in Biotechnology*, *20*, 280–285. https://doi.org/10.1016/j.copbio.2009.04.004

Mayers, J. J., Nilsson, A. E., Svensson, E., & Albers, E. (2016). Integrating microalgal production with industrial outputs – Reducing process inputs and quantifying the benefits. *Industrial Biotechnology*, *12*, 4. https://doi.org/10.1089/ind.2016.0006

Mendoza, J. L., et al. (2013). Oxygen transfer and evolution in microalgal culture in open raceways. *Bioresource Technology*, *137*, 188–195. https://doi.org/10.1016/j.biortech.2013.03.127

Morillas-España, A., et al (2021). Year-long evaluation of microalgae production in wastewater using pilot-scale raceway photobioreactors: Assessment of biomass productivity and nutrient recovery capacity. *Algal Research*, *60*, 102500. https://doi.org/10.1016/j.algal.2021.102500

Narala, R. R., Garg, S., Sharma, K. K., Thomas-Hall, S. R., Deme, M., Li, Y., & Schenk, P. M. (2016). Comparison of microalgae cultivation in photobioreactor, open raceway pond, and a two-stage hybrid system. *Frontiers in Energy Research*, *4*, 29. https://doi.org/10.3389/fenrg.2016.00029

Nwoba, E. G., et al. (2019). Light management technologies for increasing algal photobioreactor efficiency. *Algal Research*, *39*, 101433. https://doi.org/10.1016/j.algal.2019.101433

Nwoba, E. G., et al. (2020a). Energy efficiency analysis of outdoor standalone photovoltaic-powered photobioreactors coproducing lipid-rich algal biomass and electricity. *Applied Energy*, *275*, 115403. https://doi.org/10.1016/j.apenergy.2020.115403

Nwoba, E. G., et al. (2020b). Pilot-scale self-cooling microalgal closed photobioreactor for biomass production and electricity generation. *Algal Research*, *45*, 101731. https://doi.org/10.1016/j.algal.2019.101731

Pires, J. C. M., Alvim-Ferraz, M. C. M., Martins, F. G., & Simões, M. (2012). Carbon dioxide capture from flue gases using microalgae: Engineering aspects and biorefinery concept. *Renewable and Sustainable, Energy Reviews*, *16*, 3043–3053. https://doi.org/10.1016/j.rser.2012.02.055

Sathinathan, R., et al. (2023). Photobioreactor design and parameters essential for algal cultivation using industrial wastewater: A review. *Renewable and Sustainable, Energy Reviews*, *173*, 113096. https://doi.org/10.1016/j.rser.2022.113096

Severo, I. A., Siqueira, S. F., Deprá, M. C., Maroneze, M. M., Zepka, L. Q., & Jacob-Lopes, E. (2019). Biodiesel facilities. What can we address to make biorefineries commercially competitive? *Renewable and Sustainable Energy Reviews*, *112*, 686–705. https://doi.org/10.1016/j.rser.2019.06.020

Severo, I. A., Deprá, M. C., Dias, R. R., & Jacob-Lopes, E. (2020). Chapter 26 – Process Integration Applied to Microalgae-based Systems. In: Handbook of Microalgae-Based Processes and Products. Fundamentals and Advances in Energy, Food, Feed, Fertilizer, and Bioactive Compounds, pp. 709–735. https://doi.org/10.1016/B978-0-12-818536-0.00026-9

Severo, I. A., Jacob-Furlan, B., Vargas, J. V. C., & Mariano, A. B. (2022). 3rd generation biofuels – Disruptive technologies to enable commercial production. *Woodhead Publishing Series in Energy*, 249–267. https://doi.org/10.1016/B978-0-323-90971-6.00029-2

Sun, H., Zhao, W., Mao, X., Li, Y., & Chen, F. (2018). High-value biomass from microalgae production platforms: Strategies and progress based on carbon metabolism and energy conversion. *Biotechnology for Biofuels*, *11*, 227. https://doi.org/10.1186/s13068-018-1225-6

Tredici, M. R., Bassi, N., Prussi, M., Biondi, N., Rodolfi, L., Chini Zittelli, G., & Sampietro, G. (2015). Energy balance of algal biomass production in a 1-ha "Green Wall panel" plant: How to produce algal biomass in a closed reactor achieving a high net energy ratio. *Applied Energy*, *154*, 1103–1111. https://doi.org/10.1016/j.apenergy.2015.01.086

Varshney, P., Mikulic, P., Vonshak, A., Beardall, J., & Wangikar, P. P. (2015). Extremophilic micro-algae and their potential contribution in biotechnology. *Bioresource Technology*, *184*, 363–372. http://dx.doi.org/10.1016/j.biortech.2014.11.040

Jalelys Liceth Leones-Cerpa, Eduardo Luis Sánchez-Tuirán,
Karina A. Ojeda-Delgado*

Chapter 9
Mass integration applied to microalgae-based systems

Abstract: In recent years, the imminent shortage of fossil fuels has increased global concern to explore renewable energy sources derived from sustainable chemical processes. Biofuels are emerging as a promising alternative, particularly third-generation fuels obtained from microalgae. The importance of these types of biofuel is that they are ecological, and microalgae have high reproducibility, performance, and adaptability. Additionally, these represent an opportunity to obtain value-added products through biorefinery processes. Microalgae-based systems need cultivation and extraction technologies, but also the application of biorefinery platforms and process integration methodologies to reduce production costs minimizing waste and emissions from the process. Currently, mass integration is an important tool used to optimize processes in microalgae biorefineries by taking advantage of water and other components streams, to improve the efficiency, and to reduce the impact on the environment. In this chapter, a review of microalgae cultivation and extraction technologies is displayed. Also, the main concepts related to the integration of processes oriented to mass integration, microalgae biorefineries, and the application of mass integration within biorefinery design through a case study are presented.

Keywords: microalgae, biorefineries, optimization, third-generation biofuels, mass integration

9.1 Introduction

Currently, with the global energy crisis, the search for sustainable renewable energy sources is a necessity. Third-generation biofuels are widely studied because they use

***Corresponding author: Karina A. Ojeda-Delgado**, Process Design and Biomass Utilization Research Group (IDAB), University of Cartagena, Chemical Engineering Program, Av. El Consulado Street 30 #48-150, Cartagena, Colombia, e-mail: kojedad@unicartagena.edu.co
Jalelys Liceth Leones-Cerpa, Process Design and Biomass Utilization Research Group (IDAB), University of Cartagena, Chemical Engineering Program, Av. El Consulado Street 30 #48-150, Cartagena, Colombia, e-mail: jleonesc@unicartagena.edu.co
Eduardo Luis Sánchez-Tuirán, Process Design and Biomass Utilization Research Group (IDAB), University of Cartagena, Chemical Engineering Program, Av. El Consulado Street 30 #48-150, Cartagena, Colombia, e-mail: esanchezt2@unicartagena.edu.co

https://doi.org/10.1515/9783110781267-009

microalgae (biomass) as raw materials. Microalgal biomass has many advantages over other biomasses such as its yield and use of carbon dioxide for its reproduction; however, it implies increasing water and energy consumption (Hernández-Pérez et al., 2019). Various technologies have been developed for the stages of cultivation, harvesting, and extraction of lipids from microalgae (Tan et al., 2020). From microalgae, it is possible to obtain a wide variety of products used as raw material for various industries, for which they can be integrated into a biorefinery (Chew et al., 2018; Thomassen et al., 2018). On the other hand, process optimization through process integration is used to solve problems such as reduction in the consumption of resources (raw materials) and services (El-Halwagi, 2016; Klemeš et al., 2013). Specifically, mass integration is used to achieve more efficient processes (El-Halwagi, 2016). This study is aimed to present a review of microalgae cultivation and extraction technologies, the main concepts related to the integration of processes oriented to mass integration, microalgae biorefineries, and the application of mass integration within the design of biorefineries through a case study.

9.1.1 Technologies for biofuels production

Traditional energy sources satisfy more than three-quarters of the world's energy requirements, a dependency that affects the environment and contributes to climate change (Vasistha et al., 2021). Environmental concern over climate change led to the search for cleaner, renewable, and more sustainable energy production alternatives (Debnath et al., 2021). Biofuels emerged to solve environmental problems due to their second-generation characteristics using undervalued raw materials and contributing to food security (Zhu et al., 2022). In general, technologies such as fermentation, transesterification, esterification, anaerobic digestion, hydrogenation, pyrolysis, and hydrogenation have been used to obtain biofuels (Ha et al., 2020; Halim et al., 2022).

Transesterification has been a favorable technique to obtain biofuels due to the effects on the viscosity (Silitonga et al., 2020), new technologies have emerged in the production of biofuels that are responsible for improving the transesterification process such is the case of the microwave-assisted technique and ultrasonic-assisted method in the transesterification reaction, and they emerge as alternatives to reduce energy costs and increase reaction speed and yield of biofuel (Amani et al., 2022; Yusup et al., 2019). Magnetic-assisted transesterification seeks to recover organic (such as enzymes) and inorganic catalysts by magnetic decantation (Ali et al., 2020). Plasma-assisted transesterification is a method developed for catalytic and noncatalytic processes (it does not depend on a catalyst), which takes advantage of the energy produced by the collision of high-energy electrons to carry out and reduce the reaction time in the transesterification process (Amani et al., 2022; Istadi et al., 2020).

Also, technologies for the production of raw materials as the application of transgenic technologies for the genetic improvement of plants and algae to obtain biofuels and high value-added products using synthetic biology have been studied (Chen et al., 2019; Shaikh et al., 2020). Membrane technology has recently been implemented to purify biofuels after the esterification reaction where other by-products are generated (water, soap, glycerides, among others), which could lower their quality (Pal, 2020).

Nanotechnology in biofuels production can be used in various nanomaterials such as nanofibers, nanocomposites, metallic nanoparticles, and carbon nanotubes (Manikandan et al., 2022). The interaction of nanomaterials in the transesterification process with nanocatalysts accelerates the reaction for the production of bioethanol and biodiesel, and this technology generates environmentally friendly fuels; in addition, it is likely to improve the characteristics, quality, thermal, and mechanical properties of biofuels (Ali et al., 2020; Khan et al., 2022).

9.1.2 Third-generation biofuels: the importance, advantages, and disadvantages

The sources of raw material for obtaining biofuels have been extensively studied, especially biomass with high production yields, sustainability, and economics. Microalgae are positioned as one of the most complete biomasses due to their high reproducibility, high carbon dioxide fixation capacity, and biomolecular composition (Kasani et al., 2022; Yang et al., 2022), specifically their concentrations of lipids and carbohydrates to produce biodiesel and bioethanol, respectively (Ojeda et al., 2020). Third-generation fuels are obtained by processing microalgal biomass (Debnath et al., 2021).

Comparing microalgae and lignocellulosic biomass, the first generally contains more lipids and less lignin (Vuppaladadiyam et al., 2018). For the cultivation of microalgae, it is possible to use land that is not available for planting (Sánchez et al., 2011).

Microalgae have advantages over other sources for obtaining biofuels, such as their ability to reproduce in various water sources, including fresh water, salad, and wastewater (Chew et al., 2018; Ojeda et al., 2020). Wastewater affects the characteristics of surface water in ecosystems mainly due to its high nutrient content, which is why it has become a culture medium for microalgae (Khan et al., 2022). Microalgae biodiesel is part of the third-generation biofuels widely studied for its technical-economic and environmental feasibility (Ianda et al., 2022; Li et al., 2022).

Various technologies have been developed for the cultivation, harvesting, and extraction of lipids from microalgae, which influence the characteristics of biodiesel (Ojeda et al., 2020). However, the stages of cultivation, extraction, and processing of microalgae face various difficulties such as estimating the appropriate conditions for their growth, harvesting, and the lipids present in their structure (Gao et al., 2022). On the other hand, the cultivation of microalgae generates large investments, and little is known about its impact on the environment (Thomassen et al., 2018). Specifically for the production of biodiesel from lipids, standard primary techniques such as pyrolysis, microemulsion, oil blending, and transesterification are commonly used, of which the most studied is transesterification (Amani et al., 2022).

9.2 Current state of cultivation and extraction technologies

Microalgae require several stages and conditions for their optimal development, including the establishment of a culture media (De Carvalho et al., 2019). But first, it is necessary to select cultivation techniques for microalgae according to strain type, source of nutrients, carbon dioxide capture, and investment (Chew et al., 2018; Klinthong et al., 2015).

There are four cultivation conditions for algae production, the first is photoautotrophic, which considers the necessary light and investment for nutrients (inorganic substrates); the second is heterotrophic, which is based on maximum production in less time without a light source and uses organic substrates, and the third is the mixotrophic, which uses the combination of the two previous systems where microalgae need light and organic and inorganic compounds, and the fourth is photoorganitrophy cultivation where light is essential for photo-metabolic growth (Chew et al., 2018; Vuppaladadiyam et al., 2018).

Microalgae are cultivated in open and closed systems (photobioreactors, PBR) (Mantzorou and Ververidis, 2019; Tan et al., 2020). The classification of open systems can be categorized into three distinct types, namely the unstirred pond, race track-type pond, and circular pond (Chew et al., 2018). The advantages of this particular system include its low investment and operational cost, easy handling, whereas its disadvantages consist of the presence of microorganisms contamination, a loss of carbon dioxide due to evaporation, and the high cost of collection due to low production rates (Veera-badhran et al., 2021).

Closed systems are divided into vertical tubes, horizontal tubes, stirred tanks, and flat-panel PBR (Chew et al., 2018). These systems have advantages such as controlled variables, little loss of water and carbon dioxide, high production (concentrations), and less pollution (Huang et al., 2017), and disadvantages such as high production costs (energy and operation), complexity in the design, and establishment of parameters in PBR (Sevda et al., 2017).

Studies have been carried out on microalgal biofilms as the future of microalgae cultivation techniques, it is expected that they replace microalgae suspended in liquid media and solid surfaces are used, solving the current problems associated with the cultivation of microalgae (Mantzorou and Ververidis, 2019). Also, studies on the application of nano-additives in the cultivation of microalgae to increase their productivity have been reported (Hossain et al., 2019).

After establishing the microalgae cultivation techniques, it is necessary to identify the cultivation parameters (light, carbon source, and nutrients), the optimum operating temperature, salinity, and pH, which will arise from the type of strain; subsequently, determine a Harvesting technique such as coagulation, filtration, centrifugation, flotation, electroflotation, among others must be determined, according to investment costs and required energy efficiency (Enamala et al., 2018).

Lipid extraction is one of the most important process steps in the production of microalgae products such as biodiesel (Mansour et al., 2019). Lipid extraction methodologies take into account factors such as efficiency, duration, investment, safety, and process waste (Goh et al., 2019; Islam et al., 2014). Microalgae are made up of cells with rigid and thick walls, and it is necessary to break (cell disruption) them to ensure the extraction of the greatest amount of lipids (Wu et al., 2021; Zhou et al., 2022).

Several techniques have been used for the extraction of lipids from microalgae, including chemical and solvents methods, and mechanical processes, some of which are shown in Table 9.1. In some of these methods they use organic solvents such as hexane and ethanol (Peralta-Ruiz et al., 2013), green solvents (De Jesus et al., 2018), and more recently eutectic solvents (Rui et al., 2022).

Table 9.1: Microalgae lipids: extraction techniques.

Chemical and solvents	Mechanical-biomechanical process
Soxhlet extraction	Expeller press
Folch extraction	Bead beating
Bligh-Dyer extraction	Ultrasound-assisted extraction (UAE)
Supercritical fluid extraction (SFE)	Microwave-assisted extraction (MAE)
Pressurized liquid extraction (PLE)	Electroporation – pulse electric field (PEF)
Ionic liquids extraction (ILs)	Osmotic technique
Lipid extraction by single step procedure	Enzyme-assisted extraction
Nano-additives	Photoelectrical system
Green solvents	

Source: Modified from Enamala et al. (2018) and Zhou et al. (2022).

9.3 Microalgae biorefineries

Microalgae are positioned above other biomasses as ideal for obtaining biofuels, within an integrated model of biorefineries, comparing them with those of oil whose difference lies in the raw material, the equipment used, and the products obtained (Kasani et al., 2022).

One of the most important characteristics of microalgae is that a wide variety of products requested by various industries can be obtained from them, a scenario that envisions the implementation of biorefineries for the production of quality raw materials, in a feasible technical-economic system, with the generation of higher income and the reduction of impacts on the environment (Chew et al., 2017; Thomassen et al., 2018).

There is great potential for microalgae for biorefinery processes because it is possible to extract various value-added products that are required by several industries such as pharmaceuticals, fertilizers, energy, and food, with the intention of not generating waste, reducing global warming, and creating alternatives for sustainable production (Cheirsilp and Maneechote, 2022). Figure 9.1 shows the most important applications and value-added products obtained from microalgae.

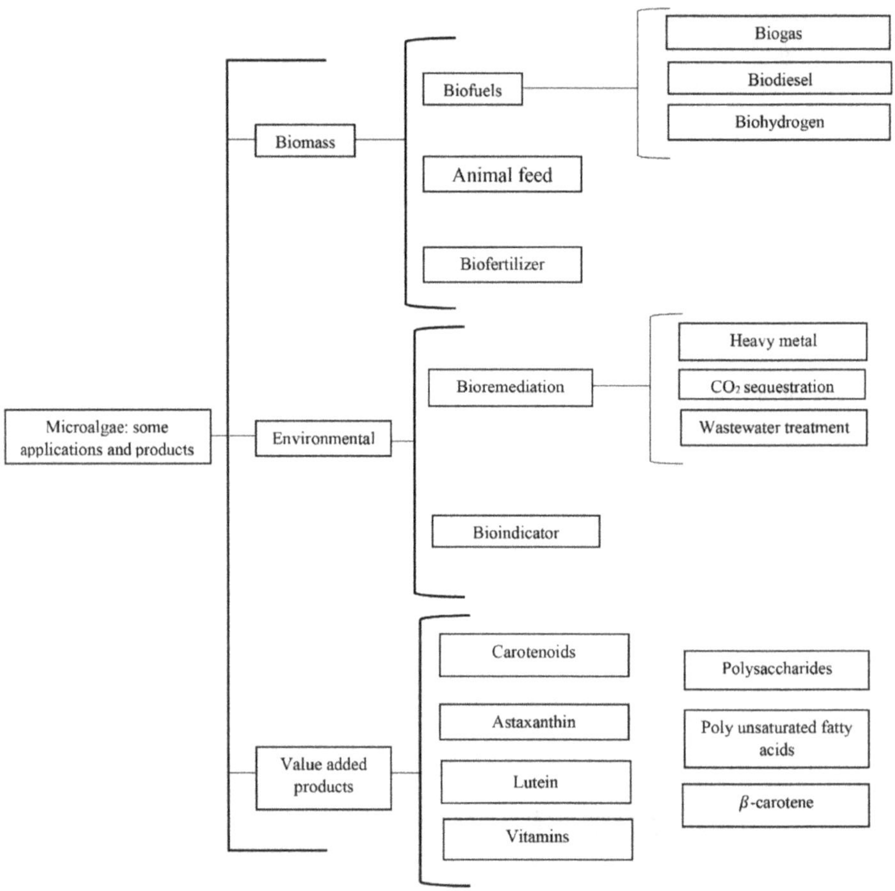

Figure 9.1: Most important applications and value-added products obtained from microalgae.
Source: Pandey et al. (2021), Oleszek and Krzemińska (2021), Kumari and Singh (2021), feed (Tomaluski et al., 2021; Suchithra et al., 2022; Pavithra et al., 2020; Delrue et al., 2016; Ighalo et al., 2022), O'Neill and Rowan (2022), De Jesus et al. (2013), D'Alessandro and Antoniosi (2016), Faraone et al. (2020), Chen et al. (2018), Rajasekar et al. (2019), Ljubic et al. (2020), and Sayegh et al. (2016).

Microalgae biorefineries at the industrial level depend on technologies and process design, which generally increase production costs and their economic viability (Draaisma et al., 2013; Tejada Carbajal et al., 2020).

Several studies related to microalgae biorefineries have been developed. Pinedo et al. (2016) designed and optimized a microalgae biorefinery to obtain biodiesel and estimation of associated risks using two alternatives for algae cultivation and lipid extraction. Hemalatha et al. (2019) worked in a microalgae biorefinery for the treatment of dairy wastewater (removal of organic carbon and nutrients) and the subsequent high production of bioethanol. Studies focused on zero waste have been carried out in the search for the sustainability of microalgae biorefineries and the obtaining of various value-added products (Cheirsilp and Maneechote, 2022). Haghpanah et al. (2022) carried out multiobjective optimization using a superstructure in a microalgae biorefinery to produce biofuels, coupling economic and environmental criteria. Bibi et al. (2022) evaluated the different methods for the cultivation of microalgae and the abiotic growth factors to apply them to microalgae biorefineries in biodiesel production.

9.4 Mass integration in microalgae biorefineries

9.4.1 Main concepts of process integration

Optimization is a concept that implies the integration of resources coming from or required by industrial processes. Process integration is responsible for finding alternatives for the use of currents and energy to reduce the number of resources required and greatly reduce the polluting effluents that alter the nature of ecosystems (Klemeš et al., 2013).

Process integration represents a generalized vision to study the dynamics of production processes and design optimization methods from the energy and mass approaches (Dunn and El-Halwagi, 2003). To determine the sustainability of industrial processes, integration is used to optimize and combine the systems of a process globally, highlighting the integration of properties, mass, and energy (El-Halwagi, 2016).

The application of process integration provides advantages such as improving productivity and reducing investment costs (Morar and Agachi, 2010). El-Halwagi et al. (2011) used process integration tools in Life Cycle Analysis to obtain various biofuels (with different technologies) and thus create scenarios where mass and energy consumption are reduced. Moncada et al. (2014) carried out the analysis of oil palm biorefinery where different scenarios to produce biodiesel, ethanol, poly-3-hydroxybutyrate, and multiproduct were modeled and evaluated using mass integration, for which the use of raw materials and raw sewage was reduced. Process integration has been applied to biorefineries design using commercial software (Aspen Plus) based on systems methodology (Kokossis et al., 2015), energy integration for first and second-generation ethanol, and bioelectricity production, achieving a decrease in the costs of public services (Oliveira et al., 2016).

Recently, process integration has been applied to biorefineries within the concept of "Integrated biorefinery" under mathematical models that involve process optimization. Qiao et al. (2022) used lignocellulosic biorefineries routes for biofuels production (biohydrogen, bioethanol) and chemical products based on cellulose, reducing costs, and improving the feasibility of biofuel production. Pedrozo et al. (2022) designed an integrated microalgae biorefinery to produce sorbitol, isosorbide, biofuels, and fertilizers with energy savings of more than 50%. Process integration has important applications in obtaining biofuels and biorefineries. Although its greatest application is toward energy integration, mass integration is a great opportunity to rethink process design and the inclusion of other methodologies.

9.4.2 Mass integration: formulation and applications

Mass integration is responsible for providing methodologies that seek to achieve more efficient processes through the management of currents and species characterized from a global vision of the process (El-Halwagi, 2016).

To carry out a mass integration of a process a flow diagram is needed to locate the species and tag them as sources and sinks (Bahy Noureldin and El-Halwagi, 1999; El-Halwagi et al., 1996). For this case, the flows are the "sources" and the process units the "sinks" which have specific conditions, it is also necessary to modify the currents or add new "sinks" (El-Halwagi, 2016).

It is possible to use different mass integration alternatives to achieve production objectives, among which are low, moderate, and high-cost strategies (design of new technologies) (El-Halwagi, 2006, 2012, 2016).

Considering the case of direct recycling where it is sought to evaluate the process streams without the addition of new units, it is necessary to use the material recycling Pinch diagram shown by El-Halwagi et al. (2003), for which the fluxes and composition of all sources and sinks must initially be determined, defined as:

$$\text{Load of impurities} = \text{Flow rate of the source} * \text{Composition of impurities in the source} \tag{9.1}$$

In the construction of the load versus flow diagram, the sources and sinks are represented, and the "material recycle pinch point" and the three important targets must be identified (El-Halwagi, 2016).

Another mass integration problem uses mass exchange networks (El-Halwagi and Manousiouthakis, 1989) using mass separation agents, following the example to transfer a specie of a rich stream to the jth lean stream, written as

$$y^* = m_j x_j^* + b_j \tag{9.2}$$

where y^* is the composition of the transferable species in the rich phase and x^* is the composition of the transferable species in the lean phase, m_j is the slope, and b_j is the intersection in the plot of the rich phase composition versus lean phase composition (El-Halwagi, 2016).

To calculate the maximum composition in the lean phase, the coefficient minimum allowable composition difference, ε_j, defined as

$$x_j^{\text{max,practical}} = x_j^* - \varepsilon_j \tag{9.3}$$

With the above, a graph of mass exchanged against composition is created to secure the minimum amount of external agents and the optimization of raw materials (El-Halwagi, 2016, 2017). Previously, mass integration was used in overall mass targets for benchmarking process (El-Halwagi, 2012).

9.4.3 Mass integration applied to microalgae biorefineries

Microalgae biorefineries have been studied through total chain integration systems to minimize energy and resource costs, improve product conditions, and reduce environmental damage by optimizing processes (Budzianowski and Postawa, 2016). Process integration has been studied in microalgae biorefineries for the total chain with the objective of recovering, recirculating, or recycling some currents and energy requirements, differentiating the mass integration to estimate the global flow of mass in the process and thus reduce wastewater and water consumption considering environmental and economic impacts (Deprá et al., 2018).

Mass integration was defined within the concept of integration by using wastewater and carbon dioxide from a lignocellulosic biorefinery in an algae biorefinery and in turn the waste streams of the latter are recycled for the former, and the integration was carried out through a mathematical model for the optimization of a superstructure developed in advanced interactive multidimensional modeling (AIMMS) software (Galanopoulos et al., 2019). The choice of an integrated system for process optimization in microalgae biorefineries was investigated for different stages of cultivation and harvesting describing the advantages and disadvantages of each mentioned software tool (Kasani et al., 2022).

9.5 Case studies

9.5.1 Biodiesel production from microalgae *Chlorella vulgaris* using $ZnCl_2$ pretreatment and basic transesterification with ethanol

In this case study, 280,000 kg/day of *C. vulgaris* microalgae oil (MAO) containing 10% wt. of free fatty acids (FFA) to obtain biodiesel was used. The use of ethanol instead of traditional methanol is one of its main features to increase the possibilities of process integration since microalgae residues can be used to obtain this alcohol. Figure 9.2 presents the process:

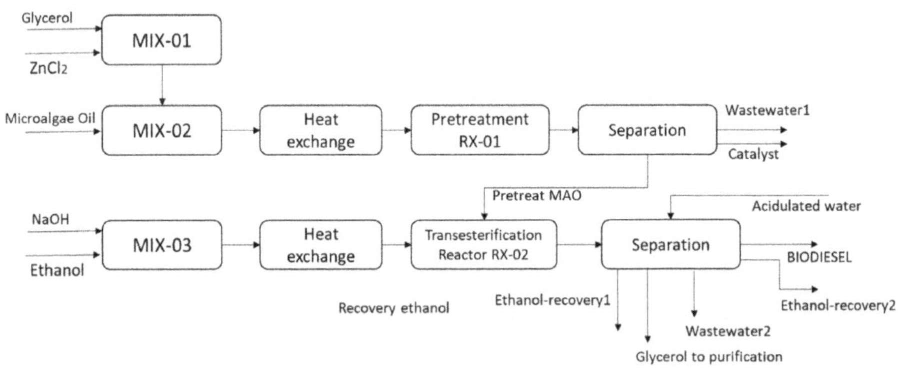

Figure 9.2: Process diagram of biodiesel production using $ZnCl_2$ pretreatment and transesterification with ethanol.
Source: Modified from Sanchez (2012).

To predict the behavior of the streams two thermodynamics fluid packages were selected following Carlson (1996) recommendations. First, the RK-Soave model was used in the liquid–liquid separation equipment to represent the phase equilibrium. Also, the NRTL model was used for all the remaining equipments such as heat exchangers, reactors, mixers, distillation towers, and flashing tanks, among others. For this process, a pretreatment with $ZnCl_2$ to reduce the acidity of the MAO stream to at least 2% wt. was included. This will avoid the undesired soap formation in the base-catalyzed transesterification with ethanol. This pretreatment consists of the reaction of 3,205 kg/day of glycerol and 28,000 kg/day of the FFA in the *C. vulgaris* MAO in the presence of $ZnCl_2$ (1% wt. of the FFA's) at 200 °C in the RX-01 reactor to obtain the triglycerides (TG) and water (Van Gerpen et al., 2004). As mentioned before, the water present in the products can cause several technical issues in the transesterification reaction. Therefore, the pretreated stream is sent to a flashing stage to remove 100% of the water. The bottom stream of tower is sent to decanter where 98.5% of $ZnCl_2$ is sent out of the process. Pretreated MAO has a mass flow of 281,337.2 kg/day of TG and is sent to reactor RX-02. In this reactor the transesterification takes place at 60 °C with ethanol: TG molar ratio of 6:1 using NaOH (1% wt. respect to the TG). After transesterification, the product stream is sent to decantation to segregate the heavy phase (rich in glycerol) from the light phase (rich in ethyl esters). Since the affinity between glycerol and ethanol is not as strong as that with methanol, only 56.7% of the ethanol in the stream is removed.

In a flashing stage, 24,439.1 kg/day of ethanol is separated (Ethanol-Recovery1) from glycerol. The glycerol obtained after the flashing is neutralized with a sulfuric acid solution. The stream has a mass flow of 38,218.6 kg/day and is mostly glycerol with a concentration of 77.5% wt. The stream rich in biodiesel obtained is sent to the tower where 98.6% (wt.) of ethanol is removed. The top stream of the tower contains mostly ethanol with a mass flow of 20,241.7 kg/day (Ethanol-Recovery2). After this, the biodiesel stream goes to washing with hot acidic water and then sent to a decanter. In this stage, most of the acidic water, catalyst, and other impurities are removed (Wastewater2) with a mass flow of 68,100 kg/day and a water concentration of 99.69% wt. After decantation, the stream undergoes a final distillation in tower T-05 (six stages, reflux ratio 2.0) where impurities are removed and a biodiesel-rich stream of 297,645.6 kg/day and 96.8% (wt.) is obtained.

In the following table, the properties of the biodiesel obtained and the ASTM and EN standards are reported.

Table 9.2: Biodiesel properties obtained in the process.

Properties	ASTM	EN	Case study
Kinematic viscosity @40 °C, (cSt)	1.9–6.0	3.5–5.0	11.7
Cetane number	47 min.	51 min.	45.1
Free glycerol, % wt.	0.020 max.	–	0
Total glycerol, % wt.	0.240 max.	–	0.3
Cloud point, °C	Report	–	−0.2
Density @15 °C, kg/m^3	–	860–900	958.1
Water content, mg/kg	–	500 max.	0
Esters content, % wt.	–	96.5	96.8
Methanol content, % wt.	–	0.2 max.	0
Triglycerides content, % wt.	–	0.2 max.	2.6
Alkaline metals (Na + K), mg/kg	–	5 max.	0

Source: Modified from Sanchez (2012).

From Table 9.2, we can confirm that the biodiesel meets the free glycerol demanded by the ASTM standards but kinematic viscosity, cetane number, and total glycerol. The biodiesel also meets the water content, esters content, methanol content, and alkaline metals content demanded by the EN standards; however, it does not comply with density and TG content. It is worth noting that the methyl esters on linolenic acid content do not apply to this biodiesel since its production comes from the transesterification of MAO with ethanol instead of methanol. The biodiesel obtained does meet the content of alkaline metals since the pretreatment with $ZnCl_2$ avoids the NaOH inclusion in the process used typically to neutralize the sulfuric acid used in the esterification. Nevertheless, the remaining specifications attain quite similar values.

9.5.2 Mass integration

In order to reduce the consumption of raw materials and to improve the efficiency of the process, a mass integration analysis was conducted using the streams' mass flows and compositions.

The outlet streams of the process with flow streams greater than 5× 103 kg/day were considered first as potential sources of components in a mass integration initiative. In this way, the streams Ethanol-Recovery1, Ethanol-Recovery2, and Wastewater2 met the mass flow specifications.

Then, the procedure for identifying minimum waste discharge (El-Halwagi, 2006) was applied and an adjustment in design conditions was made to minimize the fresh load. The recycle of the ethanol after decantation (Ethanol-Recovery1) and distillation (Ethanol-Recovery2) was maximized to guarantee that the number of impurities entering the equipment would be kept equal to or under 1% wt. The same procedure was applied to the Wastewater2 stream obtained after the neutralization of biodiesel in the tank T-04 and the separation of the heavy phase in the decanter DEC-03. Table 9.3 shows the information on the mass flows and components of the streams used in the mass integration:

Table 9.3: Mass flows and components of the streams used in the mass integration.

	Ethanol-Recovery1	Ethanol-Recovery2	Wastewater2
Massflow, $\times 10^3$ (kg/day)	24.4	20.2	68.1
Temperature, °C	112	78.3	50
Components		Mass fractions	
TG160	0	0.0029	0
TG161	0	0.0004	0
TG182	0	0.0009	0
TG183	0	0.0009	0
Glycerol	0.0012	0	0.0030
Water	0	0	0.9969
Ethanol	0.9987	0.9948	0.0001

These streams were used to replace fresh raw materials used in transesterification and biodiesel washing.

From the mass balances of the process the mass flow of raw materials with and without recycle were calculated (Table 9.4).

Table 9.4: Comparison of results without and with integration.

	Without mass integration	With mass integration
Ethanol	Massflow, $\times 10^3$ kg/day	
	136.6	92
	Impurities, % wt.	
	0	0.07
	Without mass integration	**With mass integration**
Water	Mass flow, $\times 10^3$ kg/day	
	70.5	2.4
	Impurities, % wt.	
	0	1

Also the percentage of reduction in the consumption using the following mathematical model:

$$\%\,\text{Reduction} = \left(1 - \frac{\text{Consumption with mass integration}}{\text{Consumption without mass integration}}\right) \times 100\%$$

Table 9.5: Total reduction of fresh streams for the process.

Stream	% Reduction
Ethanol	32.7
Water	96.6

Applying simple concepts of mass integration to the process, we were able to significantly reduce the consumption of fresh raw materials such as ethanol and water as shown in Table 9.5. Also, the impurities in the integrated scenario were kept under a reasonable limit of 1% wt. or less in order to avoid downstream inconvenience. The mass integration approach will also help in the economics of the process since the operational costs will be reduced even though some minor modifications in the pipeline, equipment, and controls would be required in order to implement it.

References

Ali, R., Elkatory, M. R., & Hamad, H. A. (2020). Highly active and stable magnetically recyclable cufe2o4 as a heterogenous catalyst for efficient conversion of waste frying oil to biodiesel. *Fuel*, *268*, 1–14.

Ali, S., Shafique, O., Mahmood, S., Mahmood, T., Khan, B. A., & Ahmad, I. (2020). Biofuels production from weed biomass using nanocatalyst technology. *Biomass and Bioenergy*, *139*, 1–7.

Amani, A., Rahmati, S., Fakhlaei, R., Barati, B., Wang, S., Doherty, W., & Ostrikov, K. (2022). Emerging technologies for biodiesel production : Processes, challenges, and opportunities. *Biomass and Bioenergy*, *163*, 1–18.

Bahy Noureldin, M., & El-Halwagi, M. M. (1999). Interval-based targeting for pollution prevention via mass integration. *Computers & Chemical Engineering*, *23*(10), 1527–1543.

Bibi, F., Jamal, A., Huang, Z., Urynowicz, M., & Ishtiaq Ali, M. (2022). Advancement and role of abiotic stresses in microalgae biorefinery with a focus on lipid production. *Fuel*, *316*, 1–13.

Budzianowski, W. M., & Postawa, K. (2016). Total chain integration of sustainable biorefinery systems. *Applied Energy*, *184*, 1432–1446.

Carlson, E. (October 1996). Don't gamble with physical properties for simulations. *Chemical Engineering Progress*, *92*(10), 35–46.

Cheirsilp, B., & Maneechote, W. (2022). Insight on zero waste approach for sustainable microalgae biorefinery: Sequential fractionation, conversion and applications for high-to-low value-added products. *Bioresource Technology Reports*, *18*, 101003.

Chen, C. H., Chang, J. H., Hsu, C. H., Chiu, N. C., Peng, C. C., Jim, W. T., Chang, H. Y., & Lee, K. S. (2018). A 12-year-experience with tracheostomy for neonates and infants in northern Taiwan: Indications, hospital courses, and long-term outcomes. *Pediatrics and Neonatology*, *59*(2), 141–146.

Chen, H., Li, T., & Wang, Q. (2019). Ten years of algal biofuel and bioproducts: Gains and pains. *Planta*, *249* (1), 195–219.

Chew, K. W., Chia, S. R., Show, P. L., Yap, Y. J., Ling, T. C., & Chang, J. S. (2018). Effects of water culture medium, cultivation systems and growth modes for microalgae cultivation: A review. *Journal of the Taiwan Institute of Chemical Engineers*, *91*, 332–344.

Chew, K. W., Yap, J. Y., Show, P. L., Suan, N. H., Juan, J. C., Ling, T. C., Lee, D. J., & Chang, J. S. (2017). Microalgae biorefinery: High value products perspectives. *Bioresource Technology*, *229*, 53–62.

D'Alessandro, E. B., & Antoniosi, N. R. (2016). Concepts and studies on lipid and pigments of microalgae: A review. *Renewable and Sustainable, Energy Reviews*, *58*, 832–841.

De Carvalho, J. C., Sydney, E. B., Assú Tessari, L. F., & Soccol, C. R. (2019). Culture media for mass production of microalgae, in: Pandey, A., Chang, J., Soccol, R., Lee, D., Chisti, Y. (Eds.), *Biofuels from Algae* (Second Edition). Elsevier: Amsterdam, The Netherlands, pp. 33–50.

De Jesus, M. F., De Morais, R. M. S. C., & De Morais, A. M. M. B. (2013). Health applications of bioactive compounds from marine microalgae. *Life Science*, *93*(15), 479–486.

De Jesus, S. S., Ferreira, G. F., Fregolente, L. V., & Maciel Filho, R. (2018). Laboratory extraction of microalgal lipids using sugarcane bagasse derived green solvents. *Algal Research*, *35*, 292–300.

Debnath, C., Bandyopadhyay, T. K., Bhunia, B., Mishra, U., Narayanasamy, S., & Muthuraj, M. (2021). Microalgae: Sustainable resource of carbohydrates in third-generation biofuel production. *Renewable and Sustainable Energy Reviews*, *150*, 1–21.

Delrue, F., Álvarez-Díaz, P. D., Fon-Sing, S., Fleury, G., & Sassi, J. F. (2016). The environmental biorefinery: Using microalgae to remediate wastewater, a win-win paradigm. *Energies*, *9*(3), 1–19.

Deprá, M. C., Dos Santos, A. M., Severo, I. A., Santos, A. B., Zepka, L. Q., & Jacob-Lopes, E. (2018). Microalgal biorefineries for bioenergy production: Can we move from concept to industrial reality? *Bioenergy Research*, *11*, 727–747.

Draaisma, R. B., Wijffels, R. H., Slegers, P. M., Brentner, L. B., Roy, A., & Barbosa, M. J. (2013). Food commodities from microalgae. *Current Opinion in Biotechnology*, *24*(2), 169–177.

Dunn, R. F., & El-Halwagi, M. M. (2003). Process integration technology review: Background and applications in the chemical process industry. *Journal of Chemical Technology and Biotechnology*, *78*(9), 1011–1021.

El-Halwagi, M. M. (2006). Process integration. *Process Sistems Engineering*, *7*(2), Elsevier, 21–37.

El-Halwagi, M. M. (2012). Combining Mass-Integration Strategies, in: El-Halwagi, M. M. S. (Eds.), *Sustainable Design Through Process Integration*, Butterworth-Heinemann, pp. 133–145.

El-Halwagi, M. M. (2017). Combining Mass-Integration Strategies, in: El-Halwagi, M. M. (Eds.), *Sustainable Design Through Process Integration (Second Edition)*, Butterworth-Heinemann, pp. 199–213.

El-Halwagi, M. M., & Manousiouthakis, V. (1989). Synthesis of mass exchange network. *Huagong Xiandai/ Modern Chemical Industry*, *21*(6), 1233–1244.

El-Halwagi, M. M., Chouinard-Dussault, P., Bradt, L., & Ponce-Ortega, J. M. (2011). Incorporation of process integration into life cycle analysis for the production of biofuels. *Clean Technologies and Environmental Policy*, *13*(5), 673–685.

El-Halwagi, M. M., Gabriel, F., & Harell, D. (2003). Rigorous graphical targeting for resource conservation via material recycle/reuse networks. *Industrial and Engineering Chemistry Research*, *42*(19), 4319–4328.

El-Halwagi, M. M., Hamad, A. A., & Garrison, G. W. (1996). Synthesis of waste interception and allocation networks. *AIChE Journal*, *42*(11), 3087–3101.

Enamala, M. K., Enamala, S., Chavali, M., Donepudi, J., Yadavalli, R., Kolapalli, B., Aradhyula, T. V., Velpuri, J., & Kuppam, C. (2018). Production of biofuels from microalgae – A review on cultivation, harvesting, lipid extraction, and numerous applications of microalgae. *Renewable and Sustainable Energy Reviews*, *94*, 49–68.

Faraone, I., Sinisgalli, C., Ostuni, A., Armentano, M. F., Carmosino, M., Milella, L., Russo, D., Labanca, F., & Khan, H. (2020). Astaxanthin anticancer effects are mediated through multiple molecular mechanisms: A systematic review. *Pharmacological Research*, *155*, 1–13.

Galanopoulos, C., Kenkel, P., & Zondervan, E. (2019). Superstructure optimization of an integrated algae biorefinery. *Computers and Chemical Engineering*, *130*, 106530.

Gao, F., Zhang, X. L., Zhu, C. J., Huang, K. H., & Liu, Q. (2022). High-efficiency biofuel production by mixing seawater and domestic sewage to culture freshwater microalgae. *Chemical Engineering Journal*, *443*, 136–361.

Goh, B. H. H., Ong, H. C., Cheah, M. Y., Chen, W. H., Yu, K. L., & Mahlia, T. M. I. (2019). Sustainability of direct biodiesel synthesis from microalgae biomass: A critical review. *Renewable and Sustainable Energy Reviews*, *107*, 59–74.

Ha, G. S., El-Dalatony, M. M., Kim, D. H., Salama, E. S., Kurade, M. B., Roh, H. S., El-Fatah Abomohra, A., & Jeon, B. H. (2020). Biocomponent-based microalgal transformations into biofuels during the pretreatment and fermentation process. *Bioresource Technology*, *302*, 1–8.

Haghpanah, T., Sobati, M. A., & Pishvaee, M. S. (2022). Multi-objective superstructure optimization of a microalgae biorefinery considering economic and environmental aspects. *Computers and Chemical Engineering*, *164*, 1–46.

Halim, S. A., Mohd, N. A., & Razali, N. A. (2022). A comparative assessment of biofuel products from rice husk and oil palm empty fruit bunch obtained from conventional and microwave pyrolysis. *Journal of the Taiwan Institute of Chemical Engineers*, *134*, 1–10.

Hemalatha, M., Sravan, J. S., Min, B., & Venkata Mohan, S. (2019). Microalgae-biorefinery with cascading resource recovery design associated to dairy wastewater treatment. *Bioresource Technology*, *284*, 424–429.

Hernández-Pérez, L. G., Sánchez-Tuirán, E., Ojeda, K. A., El-Halwagi, M. M., & Ponce-Ortega, J. M. (2019). Optimization of microalgae-to-biodiesel production process using a metaheuristic technique. *ACS Sustainable Chemistry and Engineering*, *7*, 8490–8498.

Hossain, N., Mahlia, T. M. I., & Saidur, R. (2019). Latest development in microalgae-biofuel production with nano-additives. *Biotechnology for Biofuels*, *12*, 1–16.

Huang, Q., Jiang, F., Wang, L., & Yang, C. (2017). Design of photobioreactors for mass cultivation of photosynthetic organisms. *Engineering, 3,* 318–329.

Ianda, T. F., Kalid, R. D. A., Rocha, L. B., Padula, A. D., & Zimmerman, W. B. (2022). Techno-economic modeling to produce biodiesel from marine microalgae in Sub-Saharan countries: An exploratory study in guinea-bissau. *Biomass and Bioenergy, 158,* 1–12.

Ighalo, J. O., Dulta, K., Kurniawan, S. B., Omoarukhe, F. O., Ewuzie, U., Eshiemogie, S. O., Ojo, A. U., Rozaimah, S., & Abdullah, S. (2022). Progress in microalgae application for CO_2 sequestration. *Cleaner Chemical Engineering, 3,* 1–15.

Islam, M. A., Brown, R. J., O'Hara, I., Kent, M., & Heimann, K. (2014). Effect of temperature and moisture on high pressure lipid/oil extraction from microalgae. *Energy Conversion and Management, 88,* 307–316.

Istadi, I., Riyanto, T., Buchori, L., Anggoro, D. D., Saputra, R. A., & Muhamad, T. G. (2020). Effect of temperature on plasma-assisted catalytic cracking of palm oil into biofuels. *International Journal of Renewable Energy Development, 9*(1), 107–112.

Kasani, A. A., Esmaeili, A., & Golzary, A. (2022). Software tools for microalgae biorefineries: Cultivation, separation, conversion process integration, modeling, and optimization. *Algal Research, 61,* 1–18.

Khan, S., Naushad, M., Iqbal, J., Bathula, C., & Al-muhtaseb, A. H. (2022). Challenges and perspectives on innovative technologies for biofuel production and sustainable environmental management. *Fuel, 325,* 1–16.

Klemeš, J. J., Varbanov, P. S., & Kravanja, Z. (2013). Recent developments in process integration. *Chemical Engineering Research and Design, 91*(10), 2037–2053.

Klinthong, W., Yang, Y. H., Huang, C. H., & Tan, C. S. (2015). A review: Microalgae and their applications in CO2 capture and renewable energy. *Aerosol and Air Quality Research, 15,* 712–742.

Kokossis, A. C., Tsakalova, M., & Pyrgakis, K. (2015). Design of integrated biorefineries. *Computers and Chemical Engineering, 81*(2015), 40–56.

Kumari, N., & Singh, R. K. (2021). Bio-diesel production from airborne algae. *Environmental Challenges, 5* (March), 1–10.

Li, P., Wang, X., Luo, Y., & Yuan, X. (2022). Sustainability evaluation of microalgae biodiesel production process integrated with nutrient close-loop pathway based on emergy analysis method. *Bioresource Technology, 346,* 1–8.

Ljubic, A., Jacobsen, C., Holdt, S. L., & Jakobsen, J. (2020). Microalgae nannochloropsis Oceanica as a future new natural source of vitamin D3. *Food Chemistry, 320,* 1–7.

Manikandan, S., Subbaiya, R., Biruntha, M., Krishnan, R. Y., Muthusamy, G., & Karmegam, N. (2022). Recent development patterns, utilization and prospective of biofuel production: Emerging nanotechnological intervention for environmental sustainability – A review. *Fuel, 314,* 1–12.

Mansour, E. A., Abo El-Enin, S. A., Hamouda, A. S., & Mahmoud, H. M. (2019). Efficacy of extraction techniques and solvent polarity on lipid recovery from domestic wastewater microalgae. *Environmental Nanotechnology, Monitoring and Management, 12,* 1–8.

Mantzorou, A., & Ververidis, F. (2019). Microalgal biofilms: A further step over current microalgal cultivation techniques. *Science of the Total Environment, 651,* 3187–3201.

Moncada, J., Tamayo, J., & Cardona, C. A. (2014). Evolution from biofuels to integrated biorefineries: Techno-economic and environmental assessment of oil palm in Colombia. *Journal of Cleaner Production, 81,* 51–59.

Morar, M., & Agachi, P. S. (2010). Review: Important contributions in development and improvement of the heat integration techniques. *Computers and Chemical Engineering, 34*(8), 1171–1179.

O'Neill, E. A., & Rowan, N. J. (2022). Microalgae as a natural ecological bioindicator for the simple real-time monitoring of aquaculture wastewater quality including provision for assessing impact of extremes in climate variance – A comparative case study from the republic of Ireland. *Science of the Total Environment, 802,* 1–10.

Ojeda, K. A., Sánchez-Tuirán, E., Gonzalez-Diaz, J., Gomez-Ochoa, M., & Kafarov, V. (2020). Exergy Analysis Applied to Microalgae-based Processes and Products. In: Handbook of Microalgae-Based Processes and Products: Fundamentals and Advances in Energy, Food, Feed, Fertilizer, and Bioactive Compounds, Jacob-Lopes, E., Manzoni, M., Queiroz, M., Queiroz, L. (Eds). Elsevier Inc. Academic Press, USA, pp. 841–859.

Oleszek, M., & Krzemińska, I. (2021). Biogas production from high-protein and rigid cell wall microalgal biomasses: Ultrasonication and FT-IR evaluation of pretreatment effects. *Fuel*, *296*, 1–7.

Oliveira, C. M., Cruz, A. J. G., & Costa, C. B. B. (2016). Improving second generation bioethanol production in sugarcane biorefineries through energy integration. *Applied Thermal Engineering*, *109*, 819–827.

Pal, P. (2020). Membrane-based Green Technology in Biofuel Production. In: Membrane-based Technologies for Environmental Pollution Control, Elsevier Inc. Butterworth-Heinemann, pp. 703–725.

Pandey, A., Sinha, P., & Pandey, A. (2021). Hydrogen production by sequential dark and photofermentation using wet biomass hydrolysate of *Spirulina platensis*: Response surface methodological approach. *International Journal of Hydrogen Energy*, *46*(10), 7137–7146.

Pavithra, K. G., Kumar, P. S., Jaikumar, V., Vardhan, K. H., & SundarRajan, P. S. (2020). Microalgae for biofuel production and removal of heavy metals: A review. *Environmental Chemistry Letters*, *18*(6), 1905–1923.

Pedrozo, H. A., Casoni, A. I., Ramos, F. D., Estrada, V., & Soledad, M. (2022). Simultaneous design of macroalgae-based integrated biorefineries and their heat exchanger network. *Computers and Chemical Engineering*, *164*, 107885. https://doi.org/10.1016/j.compchemeng.2022.107885

Peralta-Ruiz, Y., González-Delgado, A. D., & Kafarov, V. (2013). Evaluation of alternatives for microalgae oil extraction based on exergy analysis. *Applied Energy*, *101*, 226–236.

Pinedo, J., García Prieto, C. V., D'Alessandro, A. A., Ibáñez, R., Tonelli, S., Díaz, M. S., & Irabien, Á. (2016). Microalgae biorefinery alternatives and hazard evaluation. *Chemical Engineering Research and Design*, *107*, 117–125.

Qiao, J., Cui, H., Wang, M., Fu, X., Wang, X., Li, X., & Huang, H. (2022). Integrated biorefinery approaches for the industrialization of cellulosic ethanol fuel. *Bioresource Technology*, *360*, 1–13.

Rajasekar, P., Palanisamy, S., Anjali, R., Vinosha, M., Elakkiya, M., Marudhupandi, T., Tabarsa, M., You, S. G., & Prabhu, N. M. (2019). Isolation and structural characterization of sulfated polysaccharide from spirulina platensis and its bioactive potential: In vitro antioxidant, antibacterial activity and zebrafish growth and reproductive performance. *International Journal of Biological Macromolecules*, *141*, 809–821.

Rui, H., Yu, H., Xianrui, Y., Yujie, Y., Wenlu, S., Weijuan, Y., & Jun, C. (2022). Disintegration of wet microalgae biomass with deep-eutectic-solvent-assisted hydrothermal treatment for sustainable lipid extraction. *Green Chemistry*, *24*, 1615–1626.

Sanchez, E. (2012). Desarrollo de Un Proceso Para El Aprovechamiento Integral de Microalgas Para la Obtención de Biocombustibles. Tesis Doctoral. Universidad Industrial de Santander, Colombia.

Sánchez, E., Ojeda, K., El-Halwagi, M., & Kafarov, V. (2011). Biodiesel from microalgae oil production in two sequential esterification/transesterification reactors: Pinch analysis of heat integration. *Chemical Engineering Journal*, *176–177*, 211–216.

Sayegh, F., Elazzazy, A., Bellou, S., Moustogianni, A., Elkady, A. I., Baeshen, M. N., & Aggelis, G. (2016). Production of polyunsaturated single cell oils possessing antimicrobial and anticancer properties. *Annals of Microbiology*, *66*(3), 937–948.

Sevda, S., Bhattacharya, S., Reesh, I. M. A., Bhuvanesh, S., & Sreekrishnan, T. R. (2017). Challenges in the design and operation of an efficient photobioreactor for microalgae cultivation and hydrogen production, in: Singh, A., Rathore, D. (Eds), *Biohydrogen Production: Sustainability of Current Technology and Future Perspective*. Springer, New Delhi, pp. 147–162.

Shaikh, K. M., Mariam, I., Nesamma, A. A., Abdin, M. Z., & Jutur, P. P. (2020). Application of Transgenic technologies in biofuel production through photosynthetic chassis-new paradigms from gene mining to genome editing, in: Kiran, U., Abdin, M. (Eds), *Transgenic Technology Based Value Addition in Plant Biotechnology*. Academic Press, pp. 227–245.

Silitonga, A. S., Shamsuddin, A. H., Mahlia, T. M. I., Milano, J., Kusumo, F., Siswantoro, J., Dharma, S., Sebayang, A. H., Masjuki, H. H., & Ong, H. C. (2020). Biodiesel synthesis from *Ceiba pentandra* oil by microwave irradiation-assisted transesterification: ELM modeling and optimization. *Renewable Energy*, *146*, 1278–1291.

Suchithra, M. R., Muniswami, D. M., Sri, M. S., Usha, R., Rasheeq, A. A., Preethi, B. A., & Dinesh Kumar, R. (2022). Effectiveness of green microalgae as biostimulants and biofertilizer through foliar spray and soil drench method for tomato cultivation. *South African Journal of Botany*, *146*, 740–750.

Tan, J. S., Lee, S. Y., Chew, K. W., Lam, M. K., Lim, J. W., Ho, S. H., & Show, P. L. (2020). A review on microalgae cultivation and harvesting, and their biomass extraction processing using ionic liquids. *Bioengineered*, *11*, 116–129.

Tejada Carbajal, E. M., Martínez Hernández, E., Fernández Linares, L., Novelo Maldonado, E., & Limas Ballesteros, R. (2020). Techno-economic analysis of scenedesmus dimorphus microalgae biorefinery scenarios for biodiesel production and glycerol valorization. *Bioresource Technology Reports*, *12*, 1–10.

Thomassen, G., Van Dael, M., & Van Passel, S. (2018). The potential of microalgae biorefineries in Belgium and India: An environmental techno-economic assessment. *Bioresource Technology*, *267*, 271–280.

Tomaluski, C. R., Baggio, C., Campigotto, G., Baldissera, M. D., Souza, C. F., Da Silva, A. S., & Zotti, C. A. (2021). Use of *Schizochytrium* spp. microalgae in suckling Holstein calves at different periods after birth. *Livestock Science*, *245*, 1–6.

Van Gerpen, J., Shanks, B. and Pruszko, R. (2004) Biodiesel production technology. Iowa State University, Ames, pp. 30–78.

Vasistha, S., Khanra, A., Clifford, M., & Rai, M. P. (2021). Current advances in microalgae harvesting and lipid extraction processes for improved biodiesel production: A review. *Renewable and Sustainable Energy Reviews*, *137*, 1–17.

Veerabadhran, M., Natesan, S., MubarakAli, D., Xu, S., & Yang, F. (2021). Using different cultivation strategies and methods for the production of microalgal biomass as a raw material for the generation of bioproducts. *Chemosphere*, *285*, 1–12.

Vuppaladadiyam, A. K., Prinsen, P., Raheem, A., Luque, R., & Zhao, M. (2018). Microalgae cultivation and metabolites production: A comprehensive review. *Biofuels, Bioproducts and Biorefining*, *12*, 1–21.

Wu, Y., Xiang, W., Li, L., Liu, H., Zhong, N., Chang, H., & Rittmann, B. E. (2021). A novel photoelectrochemical system to disrupt microalgae for maximizing lipid-extraction efficiency. *Chemical Engineering Journal*, *420*, 2–8.

Yang, L., Su, Q., Si, B., Zhang, Y., Zhang, Y., Yang, H., & Zhou, X. (2022). Enhancing bioenergy production with carbon capture of microalgae by ultraviolet spectrum conversion via graphene oxide quantum dots. *Chemical Engineering Journal*, *429*, 1–9.

Yusup, S., Bokhari, A., Trinh, H., Shahbaz, M., Patrick, D. O., Cheah, K. W., Azizan, M. T., Ramli, A., Ameen, M., Osman, N., Shuhaili, A. F. A., & Singh, H. K. G. (2019). Emerging Technologies for Biofuels Production, in: Pandey, A., Larroche, C., Dussap, C., Gnansounou, E., Khanal, S., Ricke, S. (Eds), *Biofuels: Alternative Feedstocks and Conversion Processes for the Production of Liquid and Gaseous Biofuels* (Second Edition). Academic Press, pp. 45–76.

Zhou, J., Wang, M., Saraiva, J. A., Martins, A. P., Pinto, C. A., Prieto, M. A., Simal-Gandara, J., Cao, H., Xiao, J., & Barba, F. J. (2022). Extraction of lipids from microalgae using classical and innovative approaches. *Food Chemistry*, *384*, 1–15.

Zhu, H., Saddler, J., & Bi, X. (2022). An economic and environmental assessment of biofuel produced via microwave-assisted catalytic pyrolysis of forest residues. *Energy Conversion and Management*, *263*, 1–13.

Alberto Reis, Tiago F. Lopes*

Chapter 10 Water integration applied to microalgae-based systems

Abstract: Microalgae-based systems have emerged as a promising solution for sustainable production of food, feed, and biofuels. However, water stress and scarcity are major challenges that limit the viability and scalability of microalgae production. To address this challenge, water integration has been proposed as a means to optimize water use efficiency and reduce the environmental impact of microalgae-based systems. This chapter reviews the current state of knowledge on water integration in microalgae-based systems, with a focus on different types of microalgae cultivation systems, process integration for water optimization, and Life Cycle Assessment (LCA) of microalgae-based systems. The chapter concludes with research gaps and future directions in water integration and LCA of microalgae-based systems.

Keywords: microalgae, water stress, water scarcity, process integration, water footprint, sustainability, Life Cycle Assessment, acidification, eutrophication, ecotoxicity

10.1 Introduction

Microalgae-based systems have gained increasing attention in recent years as a promising solution for sustainable production of food, feed, nutraceuticals, and biofuels (Acién Fernández et al., 2021; Chua et al., 2022). Microalgae are photosynthetic microorganisms that can grow rapidly and efficiently using sunlight and carbon dioxide, while producing a range of valuable products such as fine proteins, such as phycocyanin as pigment, fine lipids such as omega-3 highly polyunsaturated fatty acids (EPA, DHA), and carbohydrates (Acién Fernández et al., 2021; Siddiki et al., 2022). However, the intensive water consumption and associated environmental impact of microalgae production pose a significant challenge to the sustainability and scalability of this technology (Acién et al., 2017). To address this challenge, water integration has been proposed as a mean of optimizing water use efficiency and reduce the environmental impact of microalgae-based systems.

***Corresponding author: Tiago F. Lopes,** LNEG – UBB – National Laboratory of Energy and Geology I.P., Bioenergy and Biorefineries Unit, Estrada do Paço do Lumiar 22, 1649-038 Lisbon, Portugal, e-mail: tiago.lopes@lneg.pt
Alberto Reis, LNEG – UBB – National Laboratory of Energy and Geology I.P., Bioenergy and Biorefineries Unit, Estrada do Paço do Lumiar 22, 1649-038 Lisbon, Portugal

https://doi.org/10.1515/9783110781267-010

Water integration is the process of integrating different water-using processes in a system to reduce water use and improve efficiency (Severo et al., 2020). In the context of microalgae-based systems, water integration can involve the integration of different cultivation systems and the reuse or recycling of water within the system. This chapter reviews the current state of knowledge on water integration in microalgae-based systems, with a focus on different types of microalgae cultivation systems, process integration for water optimization, and Life Cycle Assessment (LCA) of microalgae-based systems.

10.2 Water stress and scarcity in microalgae production

Although our planet's hydrosphere stores an impressive amount of water (roughly 1,386 million km^3), a significant part of this quantity (97.5%) are saline waters, thus remaining 2.5% as fresh water. From the latest fraction, 68.7% is ice and permanent snow cover, followed by 29.9% as fresh groundwaters and just 0.26% are most easily accessible for our economic activities and essential for aquatic ecosystems, being subjected to massive threat due to global warming and climate change (UNESCO, 1998).

Water scarcity has been a crucial factor in many regions of the world, requiring an efficient and urgent management of water resources (Vieira de Mendonça et al., 2021). Either the uneven geographical distribution of water resources or climate change is triggering more seasonal restrictions on places that did not have this concern (Vieira de Mendonça et al., 2021). Water stress is a growing concern worldwide as many regions face water scarcity due to increasing population, climate change, and unsustainable use of water resources. According to the United Nations, by the year 2030, it is projected that if progress continues at the current pace, approximately 1.6 billion individuals will not have access to safely managed drinking water, 2.8 billion people will lack access to safely managed sanitation, and 1.9 billion people will not have access to basic hand hygiene facilities (United Nations, 2022). The impact of water stress is far-reaching, including reduced crop yields, increased risk of waterborne diseases, and conflicts over access to water resources. Figure 10.1 depicts the projected ratio of water withdrawals to water supply by 2040. In the next few decades, it is likely that extreme water stress will primarily affect developed countries that have favorable climate and environmental conditions for microalgae cultivation and sufficient financial capacity to invest in these technologies. These countries are therefore more likely to benefit from adopting microalgae cultivation as a potential solution for addressing water scarcity. Addressing water stress requires a multifaceted approach, including improving water infrastructure, promoting water conservation, and implementing policies to ensure equitable access to water resources. Additionally, innovative technologies and practices, such as desalination and rainwater harvesting, can help alleviate water stress. However, there are significant challenges to overcome, including financing for water infrastructure, changing behaviors and attitudes toward water use, and addressing the root causes of water scarcity, such as climate change and unsustainable water use practices.

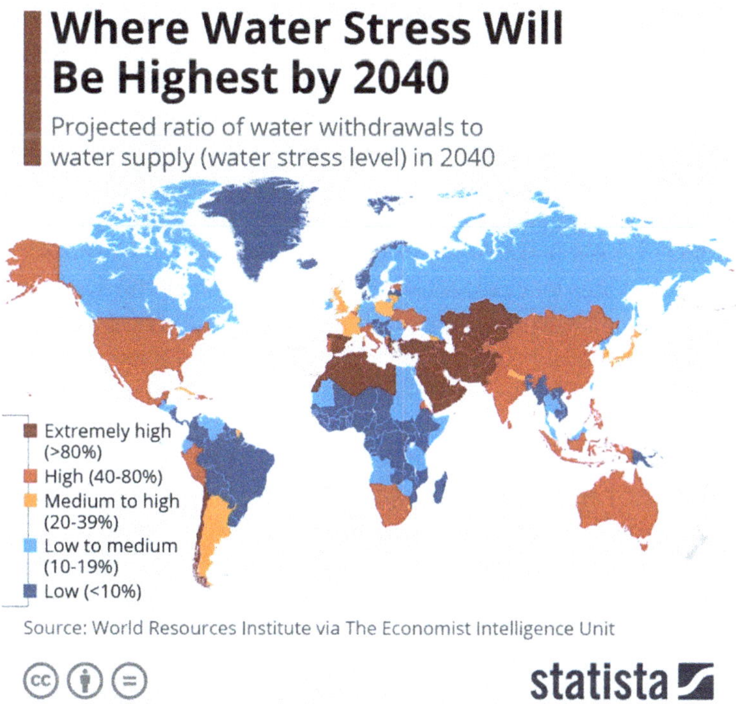

Figure 10.1: Water stress distribution by 2040 [adapted from Statista (2022)].

Water is a crucial resource for microalgae cultivation, as it provides the necessary medium for microalgae growth and nutrient uptake. Despite their potential, microalgae-based systems have faced a significant obstacle in their development due to the excessive water demand and the costs associated with it. Therefore, reducing water use and improving water use efficiency are important for the sustainable and economically viable production of microalgae-based products.

Nonetheless, microalgae cultivation can be used to complement traditional agriculture and help conserve freshwater resources. Microalgae have a high growth rate and can be harvested multiple times per year, resulting in a higher yield of biomass per unit of water used. Despite challenges regarding the water demand in microalgae cultivation systems, they still have the potential to be a sustainable and efficient solution for addressing water scarcity and food insecurity.

10.2.1 Water demand in microalgae cultivation

The water demand for microalgae cultivation varies depending on the cultivation system, the microalgae species, and the growth phase. The growth and nutrient uptake of microalgae usually require a significant volume of water. For example, open pond systems typically require large volumes of water, ranging from 1,000 to 10,000 L of water per kg of dry biomass produced (Mata et al., 2010). Similarly, closed photobioreactor (PBR) systems can also have substantial water requirement, with water consumption ranging from 1.5 to 100 L per kg of dry biomass produced (Norsker et al., 2011).

This intensive water consumption in microalgae cultivation is attributed to several factors, including the need for maintaining a suitable growth environment, such as light, temperature, and nutrient availability, and the requirement for water exchange to remove metabolic waste and maintain nutrient balance (Sialve et al., 2009). Therefore, reducing water use and improving water use efficiency are important for the sustainable and economically viable production of microalgae-based products.

10.2.2 Strategies for reducing water use in microalgae cultivation

There are several strategies for reducing water use in microalgae cultivation, including water reuse, wastewater-based cultivation, and integration with water treatment and resource recovery systems.

10.2.2.1 Water reuse

Water reuse is an effective strategy for reducing water use in microalgae cultivation and downstream processing. In closed PBR systems, the water can be recycled and reused by circulating the culture medium through the system (Yen et al., 2013). In open pond systems, the water can be reused by using a recirculation system to reduce the amount of water needed for makeup and minimize water loss due to evaporation (Costa et al., 2019). Water reuse can significantly reduce water use and improve water use efficiency in microalgae cultivation. A simplified scheme of water reuse in microalgae-based systems is represented in Figure 10.2.

10.2.2.2 Wastewater-based cultivation

Wastewater-based cultivation is another strategy for reducing water use in microalgae cultivation. Municipal wastewater, agricultural wastewater, and industrial wastewater can serve as a nutrient source for microalgae cultivation, reducing the need for freshwater and nutrient supplements (Plöhn et al., 2021). Moreover, wastewater-based

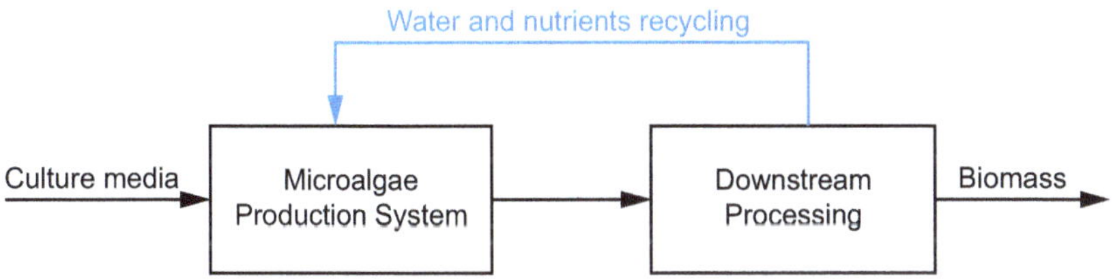

Figure 10.2: Simplified scheme for exemplifying water (and nutrients) reuse in microalgae production.

cultivation can provide a solution for wastewater treatment and resource recovery, contributing to the development of a circular economy. Although the use of wastewater for microalgae cultivation can help reduce the demand for freshwater and nutrient supplements, it must be managed carefully to ensure the water quality and nutrient balance are suitable. Additionally, there is a risk of pathogen transmission associated with this practice, which is especially important to consider if the microalgae or microalgae-based products will be used in human food production (Nur and Buma, 2019). A simplified scheme of microalgae cultivation using wastewater is represented in Figure 10.3.

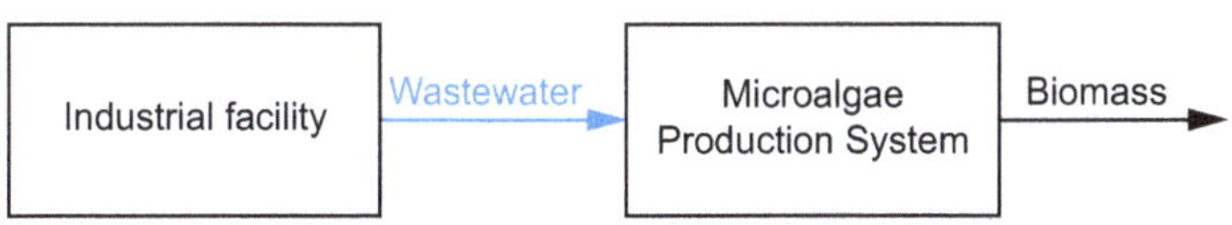

Figure 10.3: Simplified scheme for exemplifying microalgae cultivation with wastewater.

10.2.2.3 Integration with water treatment and resource recovery systems

Integration with water treatment and resource recovery systems is an innovative strategy for reducing water use and improving the sustainability of microalgae-based systems. Microalgae can be integrated with various water treatment processes, such as anaerobic digestion, biological nutrient removal, and reverse osmosis, to recover nutrients and clean water (Mohsenpour et al., 2021). For instance, Li et al. (2022b) present a review of the use of microalgae for swine wastewater treatment, highlighting its potential for nutrient recovery and discussing key microbial communities and current challenges. Figure 10.4 exemplifies how to integrate microalgae cultivation (as a biological treatment) in a wastewater treatment plant.

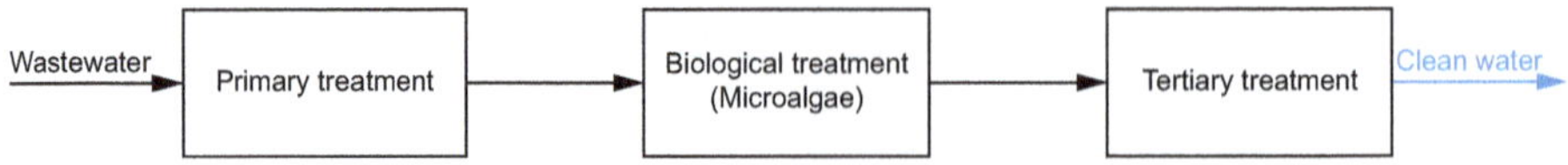

Figure 10.4: Simplified scheme for exemplifying the integration of microalgae cultivation within wastewater treatment systems.

10.3 Advances in water use efficiency in microalgae cultivation

In addition to reducing water use, improving water use efficiency is also essential for the sustainability and economic viability of microalgae-based systems. Advances in water use efficiency can be achieved through various strategies, such as optimizing cultivation conditions, improving biomass productivity, and reducing water loss.

10.3.1 Optimizing cultivation conditions

Optimizing cultivation conditions, such as light intensity, temperature, pH, and nutrient availability, can improve water use efficiency in microalgae cultivation. For example, adjusting light intensity can increase biomass productivity and reduce water use per unit of biomass produced (Dębowski et al., 2020). Similarly, optimizing nutrient availability can enhance microalgae growth and reduce the amount of water needed for nutrient supplementation (Figueroa-Torres et al., 2021). Therefore, understanding the optimal cultivation conditions for different microalgae species and growth phases is important for improving water use efficiency.

10.3.1.1 Improving biomass productivity

Improving biomass productivity is another important strategy for improving water use efficiency in microalgae cultivation. By increasing the biomass productivity, the water use intensity is reduced since more biomass can be produced per unit of water used. Various approaches, such as genetic engineering (Fayyaz et al., 2020; Ng et al., 2017), strain selection (Sydney et al., 2019), coculture of microalgae with other microorganisms (Santos and Reis, 2014), and cultivation optimization (Chu, 2017; Nagappan et al., 2019), have been applied to improve biomass productivity in microalgae cultivation and reviewed by Chu (2017). For instance, Dash and Banerjee (2017) showed a significant enhancement in biomass (2.6–3.9-fold) and lipid yields (3.4–5.1-fold) when *Chlorella minutissima* and *Aspergillus awamori* were cocultured in comparison with axenic monocultures. On the other hand, Santos et al. (2013) studied, for the first time, the culture of yeasts and microalgae in two separate reactors connected by their gas phases, taking advantage of their complementary nutritional metabolisms, that is, respiration (heterotrophy) and photosynthesis (autotrophy), respectively.

For such purpose, the yeast *Rhodosporidium toruloides* was chosen for lipid production, yielding a CO_2-enriched outlet gas stream which in turn was utilized to enhance the autotrophic growth of *Chlorella protothecoides* in a vertical-alveolar-panel (VAP) PBR.

The microalgal biomass and lipid productivities showed an increase of 94% and 87%, respectively, as compared to a control culture aerated with air. The CO_2 bio-fixed by the microalgae was 1.9-fold higher compared with the control VAP.

10.3.1.2 Reducing water loss

Algae cultivation can have a significant impact on freshwater resources as a result of the water demand associated with the process. This is particularly concerning when algae are cultivated in open ponds, where there is a high risk of water loss through free surface water evaporation. Additionally, if a substantial amount of process water is not recycled back into the pond after biomass harvesting, the water demand can further increase (Guieysse et al., 2013). Therefore, reducing water loss through various approaches, such as shading, covering, and reducing the surface area of the pond, can significantly improve water use efficiency in open pond systems (Ciardi et al., 2022). Moreover, improving water circulation and reducing the number of water exchanges can also reduce water loss in closed PBR systems (Daiek et al., 2022).

10.4 Water requirements and challenges in microalgae-based systems

Microalgae are photosynthetic microorganisms that require sunlight, carbon dioxide (CO_2), nutrients, and water to grow. They can be cultivated in various types of systems, including open ponds, closed PBRs, and hybrid systems. The choice of cultivation system can affect the water demand, efficiency, and environmental impact of microalgae production and thus is an important consideration in water integration. Each of these systems has different water requirements, which can range from less than 1 L/kg of dry biomass for closed PBRs to several thousand liters for open ponds. The water requirements of microalgae-based systems depend on several factors, such as the species of microalgae, the cultivation method, the geographical location, and the climatic conditions.

Open ponds are the simplest and most cost-effective system for microalgae cultivation, but they also have the highest water demand and environmental impact (Leite et al., 2013). Open ponds typically require large volumes of freshwater, which can be subject to evaporation, and contamination. In addition, open ponds can be vulnerable to extreme weather events and algae contamination, which can reduce the productivity and quality of the microalgae biomass (Costa and de Morais, 2014).

In contrast, closed PBRs offer a more controlled and efficient system for microalgae cultivation. PBRs can be designed to minimize the water demand and environmental impact of microalgae production by controlling factors such as light intensity, temperature, nutrient supply, and CO_2 concentration. PBRs can also enable the reuse or recycling of water within the system, which can improve water efficiency and reduce the environmental impact (Narala et al., 2016). However, PBRs can also have higher capital and operating costs than open ponds and may require specialized skills and equipment for construction and maintenance (Leite et al., 2013).

Hybrid systems that combine the advantages of open ponds and PBRs have also been proposed for microalgae cultivation. This approach can reduce the water demand and environmental impact of microalgae production while maintaining high productivity and quality of the biomass (Narala et al., 2016).

One of the main challenges in microalgae-based systems is the management of water resources. The large volumes of water required for microalgae cultivation can strain the available water resources, especially in arid and semiarid regions where water scarcity is a major concern (Seckler et al., 1999; Tzanakakis et al., 2020). In addition, the disposal of wastewater generated from microalgae-based systems can cause environmental problems, such as eutrophication and contamination of water bodies with nutrients and pathogens (Acién Fernández et al., 2018).

10.5 Water integration in microalgae-based systems

Water integration is a concept that involves the optimization of water use and the reduction of water waste by integrating the water flows of different processes within a system. Water integration can be achieved through several strategies, such as reuse, recycle, and recovery of water (Severo et al., 2020). The application of water integration in microalgae-based systems can result in significant benefits, such as:

- *Reduction of water use*: Water integration can reduce the overall water demand of microalgae-based systems by minimizing water losses and optimizing water use. For example, the reuse of wastewater generated from microalgae-based systems can reduce the freshwater demand by up to 90%.
- *Reduction of environmental impact*: Water integration can minimize the discharge of wastewater and the associated environmental impacts, such as eutrophication and contamination of water bodies. In addition, the reuse of nutrients and or-

ganic matter contained in wastewater can reduce the need for fertilizers in agriculture.
- *Economic benefits*: Water integration can result in economic benefits by reducing the operating costs of microalgae-based systems, such as the costs of water supply and wastewater treatment. In addition, the recovery of valuable products from wastewater, such as lipids and pigments, can generate additional revenue streams.

10.6 Process integration for water optimization

Process integration is a key strategy for optimizing water use efficiency and reducing the environmental impact of microalgae-based systems (Severo et al., 2020). Process integration involves the design and operation of different processes in a system to maximize the use of resources and minimize waste and emissions (Wan Alwi and Abd Manan, 2023). In the context of microalgae-based systems, process integration can involve the integration of different cultivation systems, the reuse or recycling of water within the system, and the integration of microalgae production with other processes such as wastewater treatment and CO_2 capture.

Some strategies for water integration in microalgae-based systems include:
- *Using alternative water sources*: Microalgae can grow in different water sources, including freshwater, seawater, and wastewater. Using alternative water sources can reduce the pressure on freshwater resources and provide nutrients and minerals that can enhance the growth and quality of microalgae. However, using wastewater and other low-quality water sources may also pose risks of contamination and toxicity and require proper treatment and management.
- *Recycling and reusing water*: Microalgae cultivation generates wastewater that contains nutrients and biomass residues that can be recycled and reused for cultivation. Recycling and reusing water can reduce the water use and the environmental impacts of microalgae-based systems, but may also require additional treatment and management to avoid contamination and nutrient imbalance.
- *Optimizing the cultivation system*: The choice of the cultivation system can affect the water use and the environmental impacts of microalgae-based systems. Optimizing the design and operation of the cultivation system can enhance the water use efficiency, reduce the risk of contamination and pollution, and increase the productivity and profitability of the system. For example, using closed PBRs with high-efficiency lighting and temperature control can reduce the water use and improve the quality and yield of microalgae, but may also require higher capital and operating costs.
- *Recovering and valorizing nutrients and biomass*: Microalgae-based systems can generate biomass and nutrients that can be recovered and valorized for other ap-

plications, such as food, feed, fertilizer, cosmetics, and bioenergy. Recovering and valorizing these by-products can reduce the environmental impacts of microalgae-based systems and create additional revenue streams but may also require additional processing and transportation costs.

Water reuse and recycling are important aspects of process integration for water optimization in microalgae-based systems. The reuse and recycling of water can reduce the water demand and environmental impact of microalgae production, while also providing opportunities for resource recovery and waste reduction. For example, treated wastewater can be used as a source of nutrients and water for microalgae cultivation, and the biomass can be harvested and processed for use as biofuels or other products (Li et al., 2022a; Nur and Buma, 2019).

Another approach for process integration in microalgae-based systems is the integration of microalgae production with other processes such as wastewater treatment (Mohsenpour et al., 2021) and CO_2 capture (Molazadeh et al., 2019). By using this approach, it becomes possible to utilize waste streams and greenhouse gas emissions as inputs for microalgae cultivation. Additionally, it offers a solution for waste treatment and reduction of emissions.

10.7 Life Cycle Assessment of microalgae-based systems

10.7.1 Overview

Life Cycle Assessment (LCA) is a useful tool for evaluating the environmental impact of a product or process throughout its entire life cycle from raw material extraction to end-of-life disposal (Hauschild et al., 2018). LCA can be applied to microalgae-based systems to assess the environmental impact of different cultivation systems, process integration strategies, and end products.

LCA can provide insights into the environmental impact of microalgae-based systems, including water use and scarcity, acidification, eutrophication, and ecotoxicity. LCA can also help to identify areas for improvement and optimization in microalgae production systems (Ali, 2022; Gnansounou et al., 2017; Reijnders, 2020).

The water footprint is a key indicator in LCA for assessing the water use and scarcity of microalgae-based systems. The water footprint of a product or process is the volume of water consumed, evaporated, or polluted during its production (Martins et al., 2018; Wang et al., 2021). In microalgae production, water footprint varies depending on the cultivation system, the source and quality of the water, and the degree of water reuse and recycling. LCA can be used to compare the water footprint of different microalgae production systems and identify opportunities for water optimization (Batan et al., 2013; Gerbens-Leenes et al., 2013; Martins et al., 2018).

In addition to water use and scarcity, LCA can also assess the impact of microalgae-based systems on other environmental indicators such as acidification, eutrophication, and ecotoxicity. Acidification is the process by which acids are deposited in the atmosphere and soils, leading to soil and water acidification and damage to ecosystems. Eutrophication is the process by which excess nutrients such as nitrogen and phosphorus lead to the overgrowth of algae and other aquatic plants, resulting in oxygen depletion and fish kills. Ecotoxicity is the potential of a substance to harm or kill organisms in the environment, including aquatic organisms and wildlife (Hauschild et al., 2018).

In addition, LCA studies should be complemented with other sustainability assessment tools, such as social Life Cycle Assessment (S-LCA) and techno-economic analysis, to provide a comprehensive evaluation of microalgae-based systems.

Overall, LCA studies have shown that water integration in microalgae-based systems can significantly impact the environmental and economic performance of the system. By using LCA to evaluate the environmental impact of different water management strategies, researchers and practitioners can identify opportunities for improvement and develop more sustainable and economically viable microalgae-based systems. However, it is important to note that the results of LCA studies may vary depending on the system boundary, assumptions, and data used. Therefore, it is important to use transparent and consistent methodology and data to ensure the accuracy and reliability of the results.

10.7.2 Effect of water integration in LCA of microalgae-based systems

Water integration in microalgae-based systems can also affect the environmental performance of the system. Therefore, LCA studies have been conducted to evaluate not only the economic viability but also environmental trade-offs of different water management strategies. For example, a study by Deprá et al. (2020) evaluated the environmental impact of commercial microalgae-based products, for species such as *Nitzschia laevis* (heterotrophic cultivation) and *Nannochloropsis oculata* (autotrophic cultivation) for the production of eicopentanoic acids (EPA). The authors report that the autotrophic and heterotrophic production has different impact regarding water footprint, particularly the PBRs showed values of 47.6 m^3 of water footprint per kg_{EPA}, whereas the heterotrophic production in fermenters led to a water footprint of 4.0 m^3 per kg_{EPA}.

Other study (Santos et al., 2020) has focused on establishing sustainability metrics for recovering energy and nutrients from wastewater, to produce bulk oil and lipid extracted algae, in an integrated process of agro-industrial wastewater treatment through microalgal heterotrophic bioreactors. In what concerns water integration, the integrated process has shown a reduction of 98% of the water footprint (gray water footprint as comparator). The authors recognize that using microalgae to mediate process integration is a suitable and innovative way to meet the demands of green engineering, particularly regarding nutrient cycling. Nonetheless, the primary hurdle will be incorporating these heterotrophic systems into process chains that are already in place for managing wastewater.

10.8 Water integration in microalgae-based systems: case studies

Several case studies have demonstrated the feasibility and benefits of water integration in microalgae-based systems. The following section presents a few case studies that illustrate different strategies for water integration in microalgae cultivation or downstream processing.

10.8.1 Water reuse in closed photobioreactor systems

In a study by Martins et al. (2018), the water footprint of microalgae cultivation in a closed pilot-scale multitubular PBR was assessed. The study took into consideration the entire life cycle of the PBR including construction and operation for microalgae cultivation. The findings indicated that the total water footprint ranges from 2.4 to 6.8 m^3/kg of dry biomass, with the PBR operation stage being responsible for over 60% of the water consumption. This intensive water consumption is mainly due to electricity and nutrient production for PBR operation. However, the direct water consumption for microalgae growth is relatively low, as almost all of the harvesting water is reused.

10.8.2 Microalgae cultivation using municipal wastewater

The production of microalgae with wastewater claims to require about 90% less freshwater. It is noteworthy to highlight that besides the integration of a substantial fraction of water, in parallel there is a considerable amount of mass integration, resulting in the reduction of the nitrogen requirement by up to 94%. This is because, depending on the composition of the effluent, the microalgae can remove approximately 85% of nitrates and 75% of ammonia, besides other nutrients such as phosphates and organic carbon demand (Rawat et al., 2013).

10.8.3 Microalgae cultivation using brackish water

Barahoei et al. (2021) have explored the potential of using *Chlorella vulgaris* as a new technique for desalinating brackish water. First, the researchers evaluated the adaptability of *Chlorella vulgaris* to saline water and then utilized the living microalgae cells to desalinate water in a bubble column PBR. The study investigated the effect of culture medium, time, salinity, and initial inoculum on microalgae growth and salinity removal. To ensure that the microalgae could consume sodium chloride (NaCl) in the water, the researchers modified the BG11 culture medium by substituting chloride and sodium-containing salts with nitrate, calcium, and potassium-containing minerals. The results indicated that using the modified-BG11 (MBG11) culture medium enhanced microalgae growth and salt removal efficiency. Using *Chlorella vulgaris* in the MBG11 culture medium, the researchers observed a decrease in brackish water electrical conductivity ranging from 80% to 40% for different NaCl concentrations between 1,000 and 5,000 ppm, respectively.

10.8.4 Recycling of flocculated medium for microalgal recultivation

Harvesting water recycling from microalgal-based production systems to re-grow microalgae is needed to save water resources together with the recovery of nutrients lost during the harvesting process due to changes in chemical composition of the spent water. Flocculation followed by sedimentation has been widely used for concentrating microalgae during the harvesting stage.

Zhu et al. (2018) compared and evaluated spent medium recycling after microalgal (*Chlorella vulgaris*) biomass harvesting using chitosan as a natural flocculant and aluminum sulfate as a traditional flocculant. The optimal doses for chitosan and aluminum sulfate to achieve more than 90% biomass recovery were 0.25 and 2.5 g/L, respectively, for a sedimentation time of 10 min.

After flocculation and sedimentation, the cultivation media was separated, the pH value was restored to the original level by HCl addition, and N and P concentrations were adjusted to fresh modified Bristol medium by $NaNO_3$ and KH_2PO_4 additions, respectively, toward the recultivation of the next batch of microalgal cells. The same authors proved successfully the spent medium recycling after chitosan flocculation for robust growth compared with the recycled medium from aluminum sulfate flocculation and at the same level compared with the conventional culture medium. Furthermore, the lag phase of microalgal growth was shortened in the treatments with spent media compared with the control since the recycled medium still contained some un-harvested microalgal cells, likely accelerating the growth of microalgae. Another possible explanation could be the fact that the unharvested microalgal cells had already undertaken an adaptation to the medium environment, facilitating microalgal cells to utilize nutrients available.

10.8.5 Cultivation water purification using ultrafiltration membrane

Wu et al. (2021) studied the reuse of water for (re)cultivating microalgae (*Euglena gracilis*) after a purification step of cultivation water using an ultrafiltration membrane (UFM). Several parameters have been evaluated such as the biomass productivity, biochemical composition, appearance, and accumulation of the growth inhibitors. *E. gracilis* grew well through two growth cycles with water that has been filtered and reused. Significant inhibition was achieved when the water was utilized a third time. Another interesting finding was, as the number of reused water cycles increased, an accumulation of Cl^- was obtained (up to five-fold compared to the control) in the cultivation water, exceeding the osmolality tolerance range. The authors suggested a solution, replacing NH_4Cl with urea as the source of N in the growth medium.

10.8.6 Overview of the analyzed case studies

These studies show the potential of integrating water management into microalgae-based systems for sustainable and economical production of microalgae-based products. Different water sources were used, including wastewater and brackish water, reducing environmental impact, and improving economic feasibility. Operating parameters were optimized for higher productivity and yield. Challenges remain, such as selecting appropriate microalgae strains and developing cost-effective water treatment and nutrient recovery technologies. Further research is needed to fully realize the potential of water integration in microalgae-based systems for sustainable and economical production.

10.9 Conclusions

Microalgae-based systems have the potential to provide sustainable and renewable sources of food, feed, fuel, and chemicals. However, the water use and environmental impact of microalgae-based systems can be significant and require careful consideration and optimization. Water integration is a promising approach to improve the water use efficiency and environmental impact of microalgae-based systems. LCA is a powerful tool to assess the environmental impact of microalgae-based systems and identify opportunities for improvement and optimization, particularly the effect of water integration. Case studies of open pond and closed PBR systems illustrate the potential benefits and challenges of water integration in microalgae production systems. The use of nonfreshwater sources, the use of water-saving technologies, the integration of microalgae production with other processes, and the use of the microalgae biomass for other value-added products are some of the strategies that can be used to improve the sustainability and economic viability of microalgae-based systems and simultaneously cope with the water use concerns.

References

Acién, F. G., Molina, E., Fernández-Sevilla, J. M., Barbosa, M., Gouveia, L., Sepúlveda, C., Bazaes, J., & Arbib, Z. (2017). Economics of microalgae production. *Microalgae-Based Biofuels and Bioproducts: From Feedstock Cultivation to End-Products*, 485–503. https://doi.org/10.1016/B978-0-08-101023-5.00020-0

Acién Fernández, F. G., Gómez-Serrano, C., & Fernández-Sevilla, J. M. (2018). Recovery of nutrients from wastewaters using microalgae. *Frontiers in Sustainable Food Systems*, *2*. https://doi.org/10.3389/fsufs.2018.00059

Acién Fernández, F. G., Reis, A., Wijffels, R. H., Barbosa, M., Verdelho, V., & Llamas, B. (2021). The role of microalgae in the bioeconomy. *New Biotechnology*, *61*, 99–107. https://doi.org/10.1016/j.nbt.2020.11.011

Ali, E. M. (2022). Life cycle assessment for microalgae-derived biofuels. *Handbook of Algal Biofuels*, 523–545. https://doi.org/10.1016/B978-0-12-823764-9.00012-1

Barahoei, M., Hatamipour, M. S., & Afsharzadeh, S. (2021). Direct brackish water desalination using *Chlorella vulgaris* microalgae. *Process Safety and Environmental Protection*, *148*, 237–248. https://doi.org/10.1016/j.psep.2020.10.006

Batan, L., Quinn, J. C., & Bradley, T. H. (2013). Analysis of water footprint of a photobioreactor microalgae biofuel production system from blue, green and lifecycle perspectives. *Algal Research*, *2*(3), 196–203. https://doi.org/10.1016/J.ALGAL.2013.02.003

Chu, W.-L. (2017). Strategies to enhance production of microalgal biomass and lipids for biofuel feedstock. *European Journal of Phycology*, *52*(4), 419–437. https://doi.org/10.1080/09670262.2017.1379100

Chua, S. Y., Cheng, Y. W., Lam, M. K., Dasan, Y. K., Kadir, W. N. A., Rosli, -S.-S., Lim, J. W., Tan, I. S., & Lim, S. (2022). Microalgae Cultivation for Sustainable Biofuel Production. In: Value-Chain of Biofuels. Elsevier, pp. 137–158. https://doi.org/10.1016/B978-0-12-824388-6.00006-3

Ciardi, M., Gómez-Serrano, C., Lafarga, T., González-Céspedes, A., Acién, G., López-Segura, J. G., & Fernández-Sevilla, J. M. (2022). Pilot-scale annual production of *Scenedesmus almeriensis* using diluted pig slurry as the nutrient source: Reduction of water losses in thin-layer cascade reactors. *Journal of Cleaner Production*, *359*, 132076. https://doi.org/10.1016/J.JCLEPRO.2022.132076

Costa, J. A. V., & de Morais, M. G. (2014). An Open Pond System for Microalgal Cultivation. In: Biofuels from Algae. Elsevier, pp. 1–22. https://doi.org/10.1016/B978-0-444-59558-4.00001-2

Costa, J. A. V., Freitas, B. C. B., Santos, T. D., Mitchell, B. G., & Morais, M. G. (2019). Open Pond Systems for Microalgal Culture. In: Biofuels from Algae. Elsevier, pp. 199–223. https://doi.org/10.1016/B978-0-444-64192-2.00009-3

Daiek, C., Liao, W., & Liu, Y. (2022). Effects of water recirculation on microalgae assemblage and corresponding sustainability of the photobioreactor cultivation system. *Biomass and Bioenergy*, *157*, 106326. https://doi.org/10.1016/J.BIOMBIOE.2021.106326

Dash, A., & Banerjee, R. (2017). Enhanced biodiesel production through phyco-myco co-cultivation of *Chlorella minutissima* and *Aspergillus awamori*: An integrated approach. *Bioresour Technology*, *238*, 502–509. https://doi.org/10.1016/j.biortech.2017.04.039

Dębowski, M., Zieliński, M., Kazimierowicz, J., Kujawska, N., & Talbierz, S. (2020). Microalgae cultivation technologies as an opportunity for bioenergetic system development – Advantages and limitations. *Sustainability*, *12*(23), 9980. https://doi.org/10.3390/su12239980

Deprá, M. C., Severo, I. A., Dos Santos, A. M., Zepka, L. Q., & Jacob-Lopes, E. (2020). Environmental impacts on commercial microalgae-based products: Sustainability metrics and indicators. *Algal Research*, *51* (August), 102056. https://doi.org/10.1016/j.algal.2020.102056

Santos, A. M., Deprá, M. C., dos Santos, A. M., Cichoski, A. J., Zepka, L. Q., & Jacob-Lopes, E. (2020). Sustainability metrics on microalgae-based wastewater treatment system. *Desalination and Water Treatment*, *185*, 51–61. https://doi.org/10.5004/dwt.2020.25397

Fayyaz, M., Chew, K. W., Show, P. L., Ling, T. C., Ng, I. S., & Chang, J. S. (2020). Genetic engineering of microalgae for enhanced biorefinery capabilities. *Biotechnology Advances*, *43*, 107554. https://doi.org/10.1016/J.BIOTECHADV.2020.107554

Figueroa-Torres, G. M., Pittman, J. K., & Theodoropoulos, C. (2021). Optimisation of microalgal cultivation via nutrient-enhanced strategies: The biorefinery paradigm. *Biotechnology for Biofuels*, *14*(1), 64. https://doi.org/10.1186/s13068-021-01912-2

Gerbens-Leenes, W., Vries, G. J. D., & Xu, L. (2013). The Water Footprint of Biofuels from Microalgae. In: Dallemand, J. F., & Gerbens-Leenes, P. W. (Eds.), Bioenergy and Water. Publications Office of the European Union, pp. 191–200. https://doi.org/10.2790/94637

Gnansounou, E., Pandey, A., Gnansounou, E., & Raman, J. K. (2017). Chapter 7 – Life cycle assessment of algal biorefinery In: Life-Cycle Assessment of Biorefineries, pp. 199–219. https://doi.org/10.1016/B978-0-444-63585-3.00007-3

Guieysse, B., Béchet, Q., & Shilton, A. (2013). Variability and uncertainty in water demand and water footprint assessments of fresh algae cultivation based on case studies from five climatic regions. *Bioresource Technology*, *128*, 317–323. https://doi.org/10.1016/J.BIORTECH.2012.10.096

Hauschild, M. Z., Rosenbaum, R. K., & Olsen, S. I. (Eds.) (2018). Life Cycle Assessment: Theory and Practice. Springer International Publishing. https://doi.org/10.1007/978-3-319-56475-3

Leite, G. B., Abdelaziz, A. E. M., & Hallenbeck, P. C. (2013). Algal biofuels: Challenges and opportunities. *Bioresource Technology*, *145*, 134–141. https://doi.org/10.1016/J.BIORTECH.2013.02.007

Li, G., Hu, R., Wang, N., Yang, T., Xu, F., Li, J., Wu, J., Huang, Z., Pan, M., & Lyu, T. (2022a). Cultivation of microalgae in adjusted wastewater to enhance biofuel production and reduce environmental impact: Pyrolysis performances and life cycle assessment. *Journal of Cleaner Production*, *355*, 131768. https://doi.org/10.1016/J.JCLEPRO.2022.131768

Li, S., Qu, W., Chang, H., Li, J., & Ho, S.-H. (2022b). Microalgae-driven swine wastewater biotreatment: Nutrient recovery, key microbial community and current challenges. *Journal of Hazardous Materials*, *440*, 129785. https://doi.org/10.1016/j.jhazmat.2022.129785

Martins, A. A., Marques, F., Cameira, M., Santos, E., Badenes, S., Costa, L., Vieira, V. V., Caetano, N. S., & Mata, T. M. (2018). Water footprint of microalgae cultivation in photobioreactor. *Energy Procedia*, *153*, 426–431. https://doi.org/10.1016/J.EGYPRO.2018.10.031

Mata, T. M., Martins, A. A., & Caetano, N. S. (2010). Microalgae for biodiesel production and other applications: A review. *Renewable and Sustainable, Energy Reviews*, *14*(1), 217–232. https://doi.org/10.1016/J.RSER.2009.07.020

Mohsenpour, S. F., Hennige, S., Willoughby, N., Adeloye, A., & Gutierrez, T. (2021). Integrating micro-algae into wastewater treatment: A review. *Science of the Total Environment*, *752*, 142168. https://doi.org/10.1016/j.scitotenv.2020.142168

Molazadeh, M., Ahmadzadeh, H., Pourianfar, H. R., Lyon, S., & Rampelotto, P. H. (2019). The use of microalgae for coupling wastewater treatment with CO_2 biofixation. *Frontiers in Bioengineering and Biotechnology*, *7*(Mar). https://doi.org/10.3389/fbioe.2019.00042

Nagappan, S., Devendran, S., Tsai, P. C., Dahms, H. U., & Ponnusamy, V. K. (2019). Potential of two-stage cultivation in microalgae biofuel production. *Fuel*, *252*, 339–349. https://doi.org/10.1016/J.FUEL.2019.04.138

Narala, R. R., Garg, S., Sharma, K. K., Thomas-Hall, S. R., Deme, M., Li, Y., & Schenk, P. M. (2016). Comparison of microalgae cultivation in photobioreactor, open raceway pond, and a two-stage hybrid system. *Frontiers in Energy Research*, *4*. https://doi.org/10.3389/fenrg.2016.00029

Ng, I., Tan, S., Kao, P., Chang, Y., & Chang, J. (2017). Recent developments on genetic engineering of microalgae for biofuels and bio-based chemicals. *Biotechnology Journal*, *12*(10), 1600644. https://doi.org/10.1002/biot.201600644

Norsker, N.-H., Barbosa, M. J., Vermuë, M. H., & Wijffels, R. H. (2011). Microalgal production – A close look at the economics. *Biotechnology Advances*, *29*(1), 24–27. https://doi.org/10.1016/j.biotechadv.2010.08.005

Nur, M. M. A., & Buma, A. G. J. (2019). Opportunities and challenges of microalgal cultivation on wastewater, with special focus on palm oil mill effluent and the production of high value compounds. *Waste Biomass Valorization*, *10*(8), 2079–2097. https://doi.org/10.1007/s12649-018-0256-3

Plöhn, M., Spain, O., Sirin, S., Silva, M., Escudero-Oñate, C., Ferrando-Climent, L., Allahverdiyeva, Y., & Funk, C. (2021). Wastewater treatment by microalgae. *Physiologia Plantarum*, *173*(2), 568–578. https://doi.org/10.1111/ppl.13427

Rawat, I., Ranjith Kumar, R., Mutanda, T., & Bux, F. (2013). Biodiesel from microalgae: A critical evaluation from laboratory to large scale production. *Applied Energy*, *103*, 444–467. https://doi.org/10.1016/j.apenergy.2012.10.004

Reijnders, L. (2020). Life Cycle Assessment of Microalgae-based Processes and Products. In: Handbook of Microalgae-Based Processes and Products: Fundamentals and Advances in Energy, Food, Feed, Fertilizer, and Bioactive Compounds. Academic Press, pp. 823–840. https://doi.org/10.1016/B978-0-12-818536-0.00030-0

Santos, C. A., Caldeira, M. L., Lopes da Silva, T., Novais, J. M., & Reis, A. (2013). Enhanced lipidic algae biomass production using gas transfer from a fermentative *Rhodosporidium toruloides* culture to an autotrophic chlorella protothecoides culture. *Bioresource Technology*, *138*, 48–54. https://doi.org/10.1016/j.biortech.2013.03.135

Santos, C. A., & Reis, A. (2014). Microalgal symbiosis in biotechnology. *Applied Microbiology and Biotechnology*, *98*(13), 5839–5846. https://doi.org/10.1007/s00253-014-5764-x

Seckler, D., Barker, R., & Amarasinghe, U. (1999). Water scarcity in the twenty-first century. *International Journal of Water Resources Development*, *15*(1–2), 29–42. https://doi.org/10.1080/07900629948916

Severo, I. A., Deprá, M. C., & Jacob-Lopes, E. (2020). Process Integration Applied to Microalgae-based Systems. In: Handbook of Microalgae-Based Processes and Products: Fundamentals and Advances in Energy, Food, Feed, Fertilizer, and Bioactive Compounds. Academic Press, pp. 709–735. https://doi.org/10.1016/B978-0-12-818536-0.00026-9

Sialve, B., Bernet, N., & Bernard, O. (2009). Anaerobic digestion of microalgae as a necessary step to make microalgal biodiesel sustainable. *Biotechnology Advances*, *27*(4), 409–416. https://doi.org/10.1016/J.BIOTECHADV.2009.03.001

Siddiki, S., A., Y., Mofijur, M., Kumar, P. S., Ahmed, S. F., Inayat, A., Kusumo, F., Badruddin, I. A., Khan, T. M. Y., Nghiem, L. D., Ong, H. C., & Mahlia, T. M. I. (2022). Microalgae biomass as a sustainable source for biofuel, biochemical and biobased value-added products: An integrated biorefinery concept. *Fuel*, *307*, 121782. https://doi.org/10.1016/j.fuel.2021.121782

Statista. (2022). Where Water Stress Will Be Highest by 2040. https://www.statista.com/chart/26140/water-stress-projections-global/

Sydney, E. B., Sydney, A. C. N., De carvalho, J. C., & Soccol, C. R. (2019). Microalgal strain selection for biofuel production. In: Biomass, Biofuels, Biochemicals: Biofuels from Algae (Second Edition), pp. 51–66. https://doi.org/10.1016/B978-0-444-64192-2.00003-2

Tzanakakis, V. A., Paranychianakis, N. V., & Angelakis, A. N. (2020). Water supply and water scarcity. *Water (Switzerland)*, *12*(9). https://doi.org/10.3390/w12092347

UNESCO. (1998). World Water Resources: A New Appraisal and Assessment for the 21st Century. https://unesdoc.unesco.org/ark:/48223/pf0000112671

United Nations. (2022). Goal 6: Ensure access to water and sanitation for all. The Sustainable Development Goals Report 2022. https://www.un.org/sustainabledevelopment/water-and-sanitation/

Vieira de Mendonça, H., Assemany, P., Abreu, M., Couto, E., Maciel, A. M., Duarte, R. L., Barbosa Dos Santos, M. G., & Reis, A. (2021). Microalgae in a global world: New solutions for old problems? *Renewable Energy*, *165*, 842–862. https://doi.org/10.1016/j.renene.2020.11.014

Wan Alwi, S. R., & Abd Manan, Z. (2023). Water integration and water pinch analysis. In: Handbook of Process Integration (PI), pp. 391–417. https://doi.org/10.1016/B978-0-12-823850-9.00035-9

Wang, D., Hubacek, K., Shan, Y., Gerbens-Leenes, W., & Liu, J. (2021). A review of water stress and water footprint accounting. *Water (Switzerland)*, *13*(2). https://doi.org/10.3390/w13020201

Wu, M., Du, M., Wu, G., Lu, F., Li, J., Lei, A., Zhu, H., Hu, Z., & Wang, J. (2021). Water reuse and growth inhibition mechanisms for cultivation of microalga *Euglena gracilis*. *Biotechnology for Biofuels*, *14*(1), 132. https://doi.org/10.1186/s13068-021-01980-4

Yen, H.-W., Hu, I.-C., Chen, C.-Y., Ho, S.-H., Lee, D.-J., & Chang, J.-S. (2013). Microalgae-based biorefinery – From biofuels to natural products. *Bioresource Technology*, *135*, 166–174. https://doi.org/10.1016/j.biortech.2012.10.099

Zhu, L., Li, Z., & Hiltunen, E. (2018). Microalgae chlorella vulgaris biomass harvesting by natural flocculant: Effects on biomass sedimentation, spent medium recycling and lipid extraction. *Biotechnology for Biofuels*, *11*(1), 183. https://doi.org/10.1186/s13068-018-1183-z

Akhil Rautela, Shweta Rawat, Indrajeet Yadav, Agendra Gangwar, Sanjay Kumar*

Chapter 11 Process integration opportunities applied to microalgae biomass production

Abstract: Microalgae-assisted, carbon-neutral, green approaches for biofuel production have become more demanding in recent years. Energy transition policies from fossil fuel to renewable energy and global concern for minimization of greenhouse gas emissions are key drivers for using biomass as a feedstock for biofuel production, along with the significant applications as dietary supplements with high protein content and nutritive value, bioactive secondary metabolites, and other value-added products. However, the high cost of microalgae and microalgae-derived products are governed by several challenging issues such as higher recovery cost, lower lipid fraction of dried microalgae biomass, and other complexities of mass cultivation in open and closed growth systems. In this context, the present paper focuses on low-cost technologies to cultivate microalgae by utilizing wastewater from different sources such as industrial water, produced water, pharmaceutical water, and food industry water. Integrating microalgae cultivation with wastewater treatment is a key strategy to recover nutrients from waste. Further, microalgae-assisted anaerobic digestion or co-digestion to produce biofuel is also addressed. In brief, the present study provides comprehensive detail concerning integral approaches for bioremediation with low-cost microalgae biomass production as a sustainable and renewable feedstock for biofuel production with current pilot-scale harvesting challenges and future perspectives.

Keywords: microalgae-assisted, greenhouse gas, low-cost technologies, renewable feedstock, pilot-scale

11.1 Introduction

As the world population is increasing, so is the energy demand and a quest for sustainable and renewable organic feedstock. Microalgae species rise up to be the savior as they are photosynthetic in nature (convert solar energy to chemical energy) and give high biomass yields. Microalgae feedstocks have the potential to produce cosmet-

***Corresponding author: Sanjay Kumar**, Assistant Professor, School of Biochemical Engineering, IIT BHU, Varanasi 221005, Uttar Pradesh, India, e-mail: sanjaykr.bce@iitbhu.ac.in
Akhil Rautela, Shweta Rawat, Indrajeet Yadav, Agendra Gangwar, School of Biochemical Engineering, IIT BHU, Varanasi, Uttar Pradesh, India

https://doi.org/10.1515/9783110781267-011

ics, nutraceuticals, and biofuels (Nicoletti 2016; Spolaore et al., 2006; Suali and Sarbatly, 2012). The use of microalgae, *Nostoc commune,* and *Nostoc flagelliforme* as food dates back 2,000 years (Gao 1998; Roney et al., 2009). Being a potential source of protein and antibiotics, their consumption increased when famine struck after the Second World War. Microalgae are rich in lipid, protein, and carbohydrate, which forms the basis for the products produced from them. *Spirulina*, *Dunaliella*, and *Chlorella* biomass are used as supplements. *Spirulina* and *Chlorella* are available as single-cell protein in the market. *Spirulina* contains up to 70% protein and was first successfully used by NASA as a dietary supplement (Karkos et al., 2011). *Chlorella* was the first microalgae whose mass production plant was set up at the Massachusetts Institute of Technology in 1951. Like humans, biomass can also be used as a feed additive for animals and fishes.

Unlike plants and animals, microalgae are able to synthesize polyunsaturated fatty acids (PUFAs). Apart from fishes, microalgae are an excellent source of PUFAs. γ-Linolenic acid (GLA), Arachidonic acid (AA), Eicosapentaenoic acid (EPA), and Docosahexaenoic acid (DHA) are some PUFAs synthesized by them, out of which only DHA is commercialized (Figure 11.1). DHA is an omega-3 fatty acid essential for eye and brain functioning. According to estimates, the global DHA market will increase to 8.91 billion euros by 2025 (Molino et al., 2020). Synthetic pigments are banned in many countries due to their adverse health effects. Therefore, natural or nature-identical pigments are preferred. Since plants alone cannot meet the demand, pigments of microalgae come to the rescue. Pigments are synthesized in the form of carotenoids. Some prime carotenoids used commercially are β-carotene, astaxanthin, and lutein. β-carotene has multiple applications of being used as a food pigment, provitamin A, and antioxidant. *Dunaliella salina* is the preeminent producer of β–carotene, with up to 14% of its dry weight (Pourkarimi et al., 2020a). Astaxanthin-rich feed in aquaculture can enhance the immunity of animals. Astaxanthin costs about $2,000 per kg (Onorato and Rösch, 2020). Microalgal production of astaxanthin is commercially not feasible since the natural producer, *Haematococcus pluvialis* accumulates only 1.5–3% astaxanthin of dry cell weight (Lu et al., 2021). Apart from these, microalgae are used as a source of stable isotopes widely used in Nuclear Magnetic Resonance (NMR) and Mass Spectrometry (MS). The ability of microalgae to grow in stringent conditions like heavy water, $^{13}CO_2$, and $^{15}NO_3$ leads to the synthesis of stable isotopes (Cox et al., 1988). Different products from microalgae sources and their applications are summarized in Table 11.1.

Growing microalgae for biomass production in a defined media is not cost-effective at the industrial level. Moreover, using freshwater whose demand is already high is not a wise decision for microalgae cultivation. The wastewater is enriched in C, N, and P, which aids in microalgae growth. Microalgae utilize and remove (remediation) these nutrients and increase their biomass. Further, this biomass and treated water can be used in a number of ways, as shown in Figure 11.2. Therefore, integration of microalgae biomass generation with wastewater treatment is the need of the hour. In this chapter, different cultivation systems for microalgae and factors af-

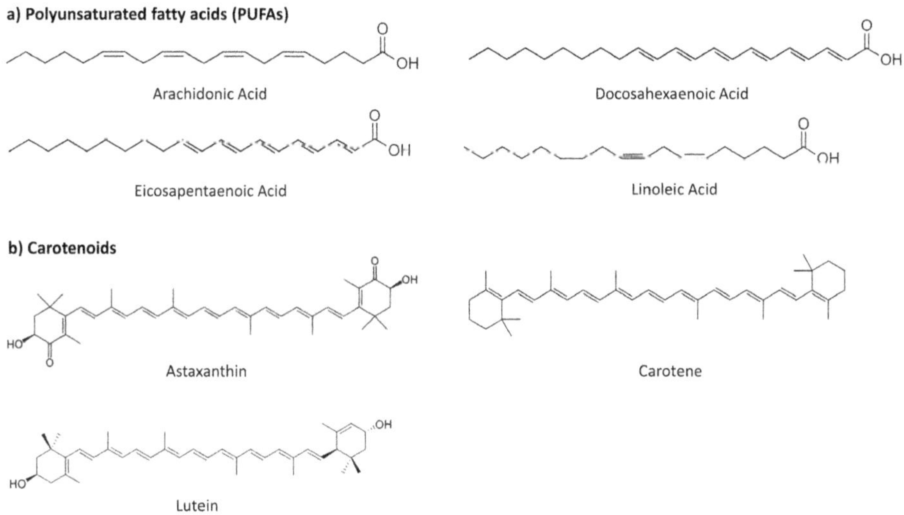

Figure 11.1: Chemical structures of a) PUFAs and b) Carotenoids.

fecting microalgae growth are discussed to describe microalgae culturing briefly. Further, the strategies for high cell cultivation using low-cost feedstock are summarized with a two-stage cultivation strategy. In addition, wastewater integration with microalgae cultivation is shown for biomass and biodiesel production. Lastly, the challenges and limitations of mass cultivation of microalgae are discussed, focusing on challenges with the strain selection, media (wastewater) selection, biomass harvesting, and techno-economical analysis of the process.

Table 11.1: Products available from microalgae sources with their applications.

Product	Microalgae used	Application	References
β-carotene	*Dunaliella salina*	Used as food coloring agent, vitamin A precursor, and potential antioxidant	(Pourkarimi et al., 2020)
Astaxanthin	*Haematococcus pluvialis*	Potential antioxidant, used in aquaculture, nutraceuticals, cosmetics, and pharmaceuticals industry	(Shah et al., 2016)
Zeaxanthin	*Scenedesmus almeriensis*	Food, pharmaceutical, nutraceutical, and treats age-related conditions	(Granado-Lorencio et al., 2009)
Lutein	*Scenedesmus almeriensis*, *Dunaliella tertiolecta*	Used as pharmaceutical agent to prevent cataract	(Granado-Lorencio et al., 2009; Kim et al., 2017)

Table 11.1 (continued)

Product	Microalgae used	Application	References
Eicosapentaenoic acid	*Nannochloropsis* sp.	Dietary supplement, omega-3 fatty acid, essential for brain	(Gu et al., 2021)
Docosahexaenoic acid	*Schizochytrium* sp.	Dietary supplement, omega-3 fatty acid, essential for brain	(Wang et al., 2021)
Squalene	*Chlamydomonas reinhardtii*	Cosmetics and pharmaceutical industry	(Potijun et al., 2021)
Polyhydroxyalkanoate	*Botryococcus braunii*	Biopolymer	(Costa et al., 2019)

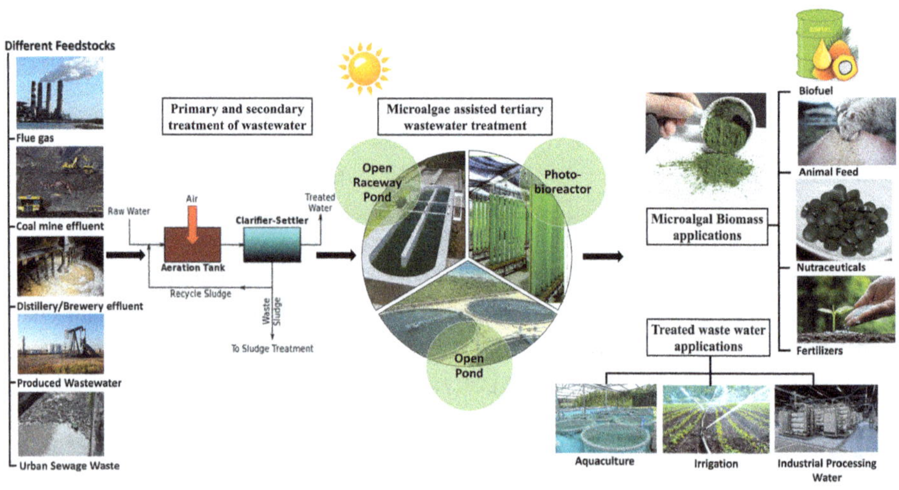

Figure 11.2: Schematic representation of microalgae-assisted wastewater treatment with its applications.

State of the art

In the current situation, the production and availability of fresh water have become a key concern. The conventional physicochemical water treatment approaches like reverse osmosis, electrodialysis, nanofiltration, and distillation are costly and not available to all. Apart from the water scarcity, the energy crisis is also one of the global issues to be resolved. Nonrenewable energy sources are depleting, and there is an earnest need to find sustainable alternate fuel resources. New energy policies are committed to energy transition from fossil fuels to bioenergy systems. Therefore, considering the importance of water and energy together for future generations, microalgae-assisted wastewater technologies offer potential solutions.

The enormous volume of wastewater with high levels of chemical oxidation demand, total dissolved solids, total solids, and total suspended solids with high concentrations of nitrogen and phosphorous offers suitable media for microalgae growth. In replacement of costly defined media, wastewater provides a low-cost growth system for microalgae on an industrial scale. The generated biomass is a value-added product and can be applied in versatile ways, notably for biofuel production. The treated water can be further used for fisheries, cooling water, and irrigation. Besides this, microalgae-based integral approaches can be applied to the bioremediation of agro-waste, sewage sludge, and activated sludge to produce biomethane and biohydrogen. Based on the evaluation, the present chapter focuses on different types of wastewater (dairy effluent, brewery/distillery effluent, produced water, municipal wastewater) as feedstock, low-cost microalgae cultivation, and limitation and challenges concerned with the process.

11.2 Algae cultivation systems

Open and closed systems are the two main cultivation systems for microalgae. Each system has its pros and cons. The open system requires lower investment and offers less control over algal production, while the closed system requires high investment and helps better control over cultivation conditions. The selection of the cultivation system is mainly subject to strain type, source of nutrients, and investment cost. Table 11.2 summarizes different cultivation systems for different strains with biomass and lipid produced.

11.2.1 Open system

Open ponds, shallow ponds, circular ponds, tanks, and raceway ponds are the most widely used open systems (Ugwu et al., 2008). Low expenditure in constructing the open system is the major advantage of this system. However, this system can be easily contaminated, and maintaining the axenic strain in the system is difficult. Moreover, the culture conditions in these systems are determined by the environment, which cannot be controlled. A detailed discussion of a few open systems is given below.

11.2.1.1 Open pond system

The open pond was the first algae cultivation system proposed and is still the most widely accepted technique. Generally, an open pond system includes an impermeable oval pond made of cement or compacted earth. An open pond system requires low energy, low construction cost, and low operation cost than a closed system. While the control over the culturing conditions and contaminations is limited, the open system is advised for those

Table 11.2: Different cultivation systems used for different species of microalgae.

Cultivation system		Microalgae culture	Biomass production (g/L)	Comment	References
Open culture system	Open pond	*Scenedesmus dimorphus*	1.34	When the pond was saturated with CO_2	(Liu et al., 2018)
	Circular pond	*Chlorella* sp.	1.1	At 2-cm culture depth	(Liang et al., 2013)
	Raceway pond	*Scenedesmus obliquus (Turpin) Kützing* GA 45	1.15 ± 0.08	Maximum growth in the winter season	(Bagchi et al., 2019)
		Scenedesmus accuminatus	1.21	Maximum growth in the winter season	(Koley et al., 2019)
		Nannochloropsis salina	1.25	Maximum growth at 23rd day	(Mohan et al., 2021)
Closed culture system	Flat-panel photobioreactor	*Micractinium* sp.	0.43 ± 0.03	12th day	(Piligaev et al., 2018)
	Stirred-tank photobioreactor	Desmodesmus communis	1.78	25 °C and 300 μmol m^2/s	(Vanags et al., 2015)
	Tubular photobioreactor	*Spirulina*	5.82	Helical tubular photobioreactor	(Travieso et al., 2001)
	Plastic bag Photobioreactor	*Nannochloropsis oceanica* CY2	3.5	When operated in semi-batch mode	(Chen et al., 2018)
	Membrane Photobioreactor	*Chlorella vulgaris*	0.251	At the 14th day, when operated in batch flow mode	(Gao et al., 2014)

algae species that are capable of sustaining in extreme environmental conditions. Other disadvantages linked to the open pond system are ineffective mixing, low mass transfer coefficient, low biomass productivity, uncontrollable light intensity, temperature fluctuation, and the need for a high amount of sunlight (Kiran et al., 2014). An open pond system can be classified into shallow lagoons and ponds, inclined (cascade) systems, circular central-pivot ponds (circular ponds), mixed ponds, and raceway ponds.

The largest shallow pond system (up to about 200 ha each) is used to cultivate *Dunaliella salina* for β-carotene in two plants located at Hutt Lagoon, Western Australia, and Whyalla, South Australia, owned and operated by BASF (Borowitzka and Hallegraeff, 2007). An inclined pond system is a type of culture system in which microalgae culture flows down the inclined surface and is collected at the bottom and recirculated via the pump. Chini Zittelli and coworkers first developed such a type of culture system in the 1960s at Trebon in the Czech Republic (Chini et al., 2013). Mixed ponds are the simplest type of 50–80 cm-deep pond with aeration from the base of the pond, which helps in providing some mixing. These ponds are usually used to produce alga for aquaculture feed (Borowitzka and Reza Moheimani, 2013).

11.2.1.2 Raceway pond

A raceway pond is a close loop, race track-like, oval-shaped shallow pond system used for algal production. The depth of the raceway pond should be approximately 3 meters to ensure proper light penetration and subsequently good microalgae growth rate (Tan et al., 2020). The raceway ponds are constructed using concrete or compressed earth, or sometimes coated with white plastic and featured with a paddlewheel to maintain the hydrodynamics of the pond. This paddlewheel also helps in gas mixing and avoids sedimentation of cells. Nutrients and microalgae inoculum are introduced in front of the paddlewheel, and harvesting is performed in the rear of the paddlewheel after circulation through the complete loop (Faried et al., 2017). The world's largest raceway pond microalgae cultivation system is operating in Calipatria, CA (USA) to produce *Spirulina* and *Spirulina*-based products, residing in an area of 44,000 m^3 (Kiran et al., 2014).

11.2.1.3 Circular pond

As its name suggests, the circular pond consists of a circular-shaped culture tank fitted with a rotating agitator in the middle of the pond. This rotating agitator ensures efficient mixing and protects cells from sedimentation. A circular pond typically has a depth and width of 30–70 cm and 50 meters, respectively (Meng et al., 2015). However, larger circular ponds of 1,000 m^2 are avoided due to inefficient mixing (Chhandama et al., 2021). Such large ponds are not preferred for commercial purposes. In addition to the high construction costs, they require a high amount of energy for mixing. This type of cultivation is used in Japan to cultivate *Chlorella* sp. for consumption purposes (Borowitzka, 1999).

11.2.2 Closed system

Close cultivation systems are usually photobioreactors (PBRs). In contrast to the open system, PBRs are controllable in terms of light, air, nutrients, and space. This gives growth conditions to specific strains and avoids contamination while maintaining axenic culture. All these advantages come at the expense of capital cost and expertise required to maintain the reactor. Many types of PBRs are available now, which are briefly discussed here.

11.2.2.1 Stirred-tank PBR

It is a PBR system in which agitation is offered manually using impellers at the top of the vessel (Aslanbay Guler et al., 2020). It is an up-gradation of stirred-tank bioreactor equipped with a light source. Mixing is an essential parameter in PBR because it aids in keeping the algal cells moving and provides improved mass transfer efficiency inside the system; for this purpose, a mechanical agitator is used (Benner et al., 2022). Baffles are usually used to reduce vortex formation. A system for CO_2 transfer is installed at the bottom to supply a carbon source for algae growth. An external light (LED and CFL) source can be installed to avoid low yield due to the fluctuation of sunlight intensity (Singh et al., 2015).

11.2.2.2 Flat-panel PBR

A flat-panel PBR is a common type of PBR used for medium-scale studies. It is made of two plates joined in a way that creates equal space and provides a high surface-area-to-volume ratio (Faried et al., 2017). Flat-panel photobioreactors are constructed of transparent or semitransparent materials like glass, plexiglass, polycarbonate, and plastic bags (Suparmaniam et al., 2019). It offers a large surface for light radiation, and therefore light usage capacity is high and dark zone can be avoided in the algal suspension. Flat-panel PBR is classified as indoor (Artificial light) or outdoor (sunlight) based on light sources. Either air bubbles provide the agitation through a perforated tube or the mechanical rotation of a motor (Faried et al., 2017). It has been commercially used by the Algamo company located at Krkonoše, Czech Republic, for astaxanthin production (Tan et al., 2020).

11.2.2.3 Tubular PBR

A tubular PBR,consists of a long transparent tube of glass or transparent plastic tubes. Based on tube configurations, tubular PBR is classified as vertical tubular PBR, horizontal PBR, helix tubular PBR, and slanted tubular PBR (Mishra et al., 2019). These transparent tubes are known as solar collectors, which assist in solar light collection. The diameter of the solar collectors should be less than 1 meter to enable proper light penetration to maximize microalgae production. Academic researchers and commercial producers have used tubular photobioreactors or their variants for over 50 years (Tan et al., 2018). In these PBRs, microalgae are circulated with the help of a mechanical pump or airlift. A tubular photobioreactor is commercially being used for the large-scale cultivation of *Chlorella* and *Haematococcus* in Israel and Germany, respectively (Tan et al., 2018).

11.2.2.4 Plastic bag PBR

A plastic bag PBR is attracting the attention of researchers due to its simplicity and cost-effectiveness. This photobioreactor is made of transparent plastic material like polyethylene (Huang et al., 2017). These plastic bags can be installed with an aerator to provide proper mass transfer to promote cell yield. Plastic bags can be arranged in different patterns such as horizontal, vertical, and helical according to need. For example, Chen and coworkers used the plastic bag photobioreactor to cultivate *Nannochloropsis oceanica* for eicosapentaenoic acid production (Chen et al., 2018).

11.2.2.5 Membrane PBR

Membrane PBR (MPBR) is an integrated approach in which photobioreactors with membrane filtration processes are used for microalgae cultivation (Zou et al., 2022). MPBR is a promising technology for simultaneous microalgae cultivation and nutrient removal. In these systems, the cost of nutrients required for microalgae cultivation is reduced by using sewage, and the biomass produced can be used for biofuel production. This type of bioreactor is the best fit for use in wastewater treatment with microalgae cultivation and helps decrease the CO_2 load from the environment (Amini et al., 2022). The performance of MPBR is highly influenced by operating conditions such as light intensity, hydraulic retention time (HRT), and solids retention time (SRT) (Luo et al., 2017). Among all these factors, SRT is the most critical factor that significantly influences the microalgae productivity, biomass concentration, and nutrient removal in MPBR (Maity et al., 2014).

11.3 Factors affecting microalgae cultivation

The growth of microalgae is affected by physiochemical factors such as light intensity, mixing, gas transfer, photoperiod, temperature, pH, nutrient concentration, and salinity. One will get the maximum yield with respect to either biomass or product, if proper conditions are provided for the microalgae species. Like plants, microalgae are autotrophs and use environmental CO_2 and water to generate glucose and O_2 and help decrease CO_2 load on the environment. Some of the factors that affect the growth of microalgae are discussed below.

11.3.1 Temperature

Temperature is a crucial environmental factor that affects the growth of microalgae. Similar to the effect of light intensity, as the temperature increases, the growth rate of microalgae increases first. After reaching optimum temperature, microalgae growth starts decreasing if further temperature increases, and cells start to decline (or increase in death) in culture (Chhandama et al., 2021). In contrast, temperature below optimum or above the frozen temperature does not kill the cell but inhibits microalgae growth. Most often, the growth of different microalgae occurs in temperatures ranging from 20–30°C (Singh and Singh, 2015).

11.3.2 Light intensity

Microalgae, being photoautotrophic, require light to grow. During the light reaction of photosynthesis, the light energy gets converted into ATP and NADPH, and that energy is utilized by the dark reaction to produce glucose and O_2 as a by-product (Metsoviti et al., 2019). The biomass yield of microalgae is affected by the incident light intensity, light period (photoperiod), and wavelength. As the light intensity increases, the growth rate and lipid accumulation will increase until it reaches a maximum value termed light saturation (at saturation, there is an equilibrium between photorespiration and photoinhibition) (Maltsev et al., 2021). If the light intensity is more than saturation, photoinhibition will be visible by a decrease in growth rate and less efficient CO_2 fixation (Dickinson et al., 2017). Photosynthesis and biomass growth are strongly affected by photoperiod (light and dark cycle). If a proper light and dark cycle is provided to microalgae, it will increase yield and low production cost. Microalgae use visible light spectra of light for photosynthesis, a wavelength ranging from 400–700 nm, and are called photosynthetically active regions (Zarmi et al., 2020).

11.3.3 pH

Microalgae are sensitive to changes in pH as it affects different processes such as metabolism, enzyme activity, membrane permeability, protein, and function of cell organelles (Brindhadevi et al., 2021). The requirement of optimum pH for microalgae is strain- or species-specific, and the most optimum pH is in the range of 7.5–8.5. The optimum pH for marine microalgae *Nannochloropsis salina* is around 9.0, while for *Dunaliella acidophila* optimum pH is 1.0 (Gimmler et al., 1991; Kumar, S., 2018). According to research by Song et al. highest carbon sequestration for *Chlorella* sp. L38 takes place at pH 8, and the toxic effect of ammonia is significantly less in this pH range because less ammonia will be converted to free ammonia (Song et al., 2019).

11.3.4 Nutrient content

Microalgae require inorganic nutrients such as calcium, potassium, nitrogen, sulfur, sodium, phosphorous, iron, and vitamins (B6 and B12) and some trace elements (iron, cobalt, nickel, zinc, magnesium) for their growth (Hossain and Mahlia, 2019). However, the cultivation is mostly affected by the concentration of nitrogen and phosphorous (Zhuang et al., 2018). Phosphorous and nitrogen, respectively, are part of DNA and protein. This makes them necessary macronutrients for microalgae development. High nitrogen and phosphorous concentrations will increase the protein, nucleic acid, and lipid content of up to 50% of the dry biomass (Rehman et al., 2022). In the study conducted by Mishra and the group, the growth of Isochrysis galbana decreased at low nitrogen concentration, and at 80 mg/L nitrate concentration they showed optimum growth (Mishra et al., 2019). Magnesium is a constituent of chlorophyll and that is why magnesium concentration directly affects chlorophyll biogenesis and photosynthesis (Polat et al., 2020).

11.3.5 Mixing

After adding the culture medium and inoculum, proper mixing is essential for providing equal nutrients and light to microalgae. Good mixing is essential for maximum microalgae growth. Low mixing should be avoided because low mixing creates an anaerobic zone in the culture system, leading to deterioration. If high mixing is provided to the culture system, it will damage the cell due to the shear stress (Wang and Lan, 2018). Therefore, a certain extent of mixing should be given to the culture. Different mixing systems are applied to the culture system, such as mechanical stirrer, sparger, and magnetic stirrer.

11.4 Integration of microalgae cultivation with industrial wastewater treatment

Microalgae provide a sustainable solution to reduce environmental pollutants from agricultural waste, sewage waste, industrial waste, and coal mine wastewater. Microalgae biomass conversion to lipid and further fermentative products bioethanol, biobutanol, biodiesel, and biohydrogen production is coupled with waste management and water treatment. Recently, two microalgae species *Neochloris oleoabundans* and *Chlorella vulgaris* have been used to monitor sludge waste feedstock with biomass and lipid content (Altunoz et al., 2020). An integrated approach toward bioremediation and biofuel production provides an exciting opportunity towards high-value metabolite production, biomass yield, and biofuel production (Altunoz et al., 2017). Microalgal bioremediation has a wide range of applications for purifying contaminated wetlands, which provides multiple advantages: obtaining high-value metabolites, biofuels, or high-value biomass yield, along with performing waste management and environmental stabilization (Cheah et al., 2016). Integration of microalgal biomass production with wastewater treatment covers a broad domain of industrial waste remediation. The growth of microalgae in different types of wastewater is summarized in Table 11.3.

Table 11.3: Summary of microalgae growth study in different types of wastewater.

Type of water	Strain	Biomass production (g/L)	Lipid productivity	References
Dairy effluent	*Scenedesmus* sp.	1.22	0.031 g/L/d	(Pandey et al., 2019)
Hydrothermal carbonization process water	*Chlorella minutissima*	0.130	–	(Tarhan et al., 2020)
Reverse osmosis reject water	*Chlorella pyrenoidosa*	1.37	0.45 g/L	(Bhandari and Prajapati, 2022)
Municipal waste water	*Chlorella pyrenoidosa*	0.35	0.136 g/L	(Zhou et al., 2020)
Starch processing water	*Chlorella pyrenoidosa*	2.26	0.305 g/L	(Chu et al., 2015)
Sewage sludge waste	*Chlorella vulgaris*	–	0.069 g/L	(Altunoz et al., 2020)
Centrate waste water	*Chlorella vulgaris*	0.72	0.02 g/L/d	(Ge et al., 2018)
Piggery waste water	*Scenedesmus* sp.	1.1	–	(Prandini et al., 2016)
Waste water with waste glycerol	*Chlorella vulgaris*	2.92	0.163 g/L/d	(Ma et al., 2016)
Paddy-soaked waste water	*Chlorella pyrenoidosa*	1.56	0.428 g/L	(Umamaheswari et al., 2020)
Biochar-supplemented aqueous dye solution	*Chlorella pyrenoidosa*	0.229	0.02 g/L	(Behl et al., 2022)

11.4.1 Produced water treatment

In hydrothermal carbonization (HTC), wet biomass is converted to solid, carbon-enriched hydrochar, mainly utilized as solid biofuel with an aqueous phase as a by-product of the HTC process (Leng et al., 2018). *Botryococcus braunii* and *Chlorella minutissima* were reported to grow in HTC process water with the highest growth rates of 0.078–0.093 g/L/day and 0.108–0.128 g/L/day, respectively, with a minimum doubling time of 5.33–6.41 day (*C. minutissima*) and 7.45–8.89 day (*B. braunii*). The combination of wastewater pretreatment with microalgae cultivation may be considered a sustainable, low-cost approach to cultivate microalgae. In this direction, one pilot study reported upflow anaerobic sludge blanket digestion (UASB)-treated piggery waste water as a nutrient resource for *C. sorokiniana* with a production rate of 1 g/L and average removal of ammonia, orthophosphate, and inorganic carbon in the range of 100%, 40–60%, and 46–56% (Leite et al., 2019).

Produced water (PW) co-extracted with oil and gas provides an exciting opportunity for microalgae cultivation with biofuel production. In this direction, *Dunalliella, Nannochloropsis, Scenedesmus rotundus, Chaetoceros gracilis,* and many other unknown strains are reported to be cultivated in PW with high productivity and lipid yield (Sullivan Graham et al., 2017). However, shifting PW from oil and gas extraction sites to microalgae cultivation sites is economically challenging. Microalgae cultivation-assisted coal seam water treatment may be considered a novel water management strategy with high yield biofuel production (Millar et al., 2016) Bioremediation of petroleum industries-derived PW effluent was analyzed by using five different microalgae strains of *Chlorella, Neochloris, Scenedesmus* and *Dictyosphaerium* species resulting in high biomass density with significant nutrient removal efficiency (Hakim et al., 2018). Further, *Dictyosphaerium* species is reported to remove various metals and phosphorus up to 88.83%.

Oilfield-produced water may be considered as an unclaimed cheap wastewater resource to cultivate microalgae with significant pollutant removal efficiency (Gillard et al., 2021). As a dominant microalga sp. of phytoplankton communities, *Phaeodactylum tricornutum* is reported as a superior lipid producer with CO_2 trapping efficiency of up to 25%. Therefore, *P. tricornutum* is considered as a potential microalgal organism to treat wastewater sources with improved lipid production (Hakim et al., 2018).

11.4.2 Industrial wastewater treatment

Considering the significance of microalgal biomass cultivation and lipid production, microalgae-assisted tertiary wastewater treatment (polishing) is an integrated approach to minimizing pollutants and reforming waste residues for final disposal with clean, sustainable technology. Mixotrophic microalgae cultivation in pharmaceutical wastewater is reported to cultivate 2.8 g/L biomass with 73% carbon removal and 62% nitrates removal (Hemalatha and Venkata Mohan, 2016). Further, as a value-added product, the total lipid content of 17.2% in light and 15.8% in dark condition is reported. In the direction of distillery wastewater treatment, algal treatment coupled with advanced chemical process is gaining interest for organic pollutants and dyes removal with commercially valuable biomass production and high-yield lipid production.

Cultivation of microalgae, *Chlorella* sp. and *Chlorococcum* sp., by utilizing nutrients from industrial wastewater and CO_2 from coal-fired flue gas is represented as an integrated mode of process development by nutrient removal (nitrate, phosphate, ammonium, and chemical oxygen demand (COD)), CO_2 fixation, and lipid production for sustainable biorefinery application (Yadav et al., 2020) Further, bacterial–algal coupling system was reported to treat high strength wastewater with low-cost biomass production. In this direction, microalgae *Chlorella vulgaris* co-cultured with *Bacillus* sp. was reported to remove COD by 94.4% and NH_4^+-N by 68.8% (Zhang et al., 2021). Further, integration of sequence batch reactor (SBR), bioelectrochemical treatment, and microalgal treatment was reported as an enhanced treatment approach with significant microalgae contribution of 56% COD removal and 44% nitrate removal and value-added biomass cultivation (Hemalatha et al., 2017).

However, microalgae utilization for primary wastewater treatment suffers from a number of limitations : (i) high turbidity creates problems in light penetration, (ii) inefficient in removing high COD level, and (iii) low production of lipids and pigment due to growth inhibition (da Silva et al., 2021). Microalgae cultures co-digested with sewage sludge are reported to produce CH_4 yields of 408 ± 16 N cm^3g/VS, which shows significant biochemical methane potential (Olsson et al., 2014). In a similar direction of biomethane production, *Nannochloropsis salina* and *Chlorella* sp. are reported to have high biomethane yield (Schwede et al., 2013). Recently, a microalgae–bacteria consortium system has been applied to treat wastewater with 15% better removal efficiency than the pure microalgae system. Therefore, microalgae-assisted wastewater treatment explores the low-cost cultivation of biomass for producing biofuel, biofertilizers, animal feed, and value-added biochemicals with the potential to treat industrial, agro-industrial, domestic, and landfill leachate waste water, as tertiary operation (Aditya et al., 2022).

11.5 High cell density cultivation strategies

Being the promising feedstock for the production of biofuels like biodiesel and bioethanol, microalgae have the potential to overcome the limitations of petroleum-based fuels. Microalgae are the best cellular factories for producing high lipid content by utilization and conversion of environmental CO_2. However, large-scale cultures are unable to accumulate high-density biomass and high volumetric productivities. Low cell density cultures cannot be implemented at large-scale productions due to requirements of high energy input and the high cost of downstream processing. When the microalgae cultures are supplemented with organic carbon sources, the biomass as well as lipid yield was enhanced (Coelho et al., 2014). Using only a single culture mode, either autotrophic or heterotrophic, cannot impart high cell density in comparison to other heterotrophs such as bacteria and yeast. Combining the low-cost phototrophic cultivation and achieving high cell density using heterotrophic culture mode has been the major strategy for the efficient production of biofuels by microalgae. Economically feasible ultra-high cell density culture techniques are needed for the sustainable production of biofuels from microalgae (Jin et al., 2020). To achieve high cell density scalable to large scales, various strategies have been followed by the biotechnology community. Another approach to biofuel production in which lipid is extracted for biodiesel production and further residual biomass is used for the fermentative production of biohydrogen and methane in a process called bio-hythane (Bauer et al., 2021).

11.5.1 Low-cost nutrient resources

Recently, microalgae have emerged as a sustainable source of bioenergy production. *Chlorella* strains have been reported to accumulate up to 50% lipid component of the total biomass, which makes it a suitable organism for biodiesel production. Basic nutritional requirements of microalgae include solar energy, CO_2, water, and some elemental nutrients in the form of salts. Artificial media containing elemental salts increase the cost of the microalgae culture process, reducing its sustainability. Alternatively, low-cost nutrient sources can be used for economical and sustainable cultivation of microalgae (Granado-Lorencio et al., 2009). Additionally, microalgae assist in organic load or nutrient removal from the waste stream of a particular industry effluent while growing to higher cell densities and producing the desired biofuels.

11.5.1.1 Wastewater as feedstock

Different wastewaters are potential feedstock for microalgae culture. Brewery wastewater is one of the low-cost nutrient resources loaded with high content of organic matter and could be the best choice for the growth of microalgae. The brewery industry uses a huge amount of water in the brewery production process and later releases almost 70% of it as wastewater effluent. The brewery effluent is rich in nitrogen, phosphorous, COD, and many other organic components; when discharged without treatment into the environment it could cause environmental pollution (Table 11.4). Microalgae utilize the organic matter for their growth and simultaneously reduce the organic load converting them into biomass from the brewery wastewater. Thus, the biomass produced can be used for lipid extraction, biofertilizer, animal feed, and biofuels (Amenorfenyo et al., 2019). Simulated brewery wastewater was used to grow *Scenedesmus obliquus* to produce biomass, and nutrient removal was assayed at varying light exposure, light intensity, and aeration rates. A biomass yield of 0.9 g of dry cell weight per liter of culture was obtained in nine days (Mata et al., 2012). In another study, *Scenedesmus obliquus* was grown in a bubble column photobioreactor using brewery wastewater operated in batch and continuous culture modes, attaining the biomass productivity of 0.2 g dry cell weight per day while removing nitrogen, COD, and phosphorus of 97%, 74%, and 23% respectively (Marchão et al., 2018).

Table 11.4: Brewery wastewater characteristics (Amenorfenyo et al., 2019).

Parameter	Unit	Value
pH	–	3–12
COD	mg/L	2,000–6,000
BOD	mg/L	1,200–3,600
Phosphate	mg/L	10–15
Nitrate	mg/L	25–80
Total solids	mg/L	500–8,750
Total dissolved solids	mg/L	2,020–5,940
Total suspended solids	mg/L	2,901–3,000
Volatile fatty acids	mg/L	1,000–2,500

COD, chemical oxygen demand; BOD, biological oxygen demand.

Aquaculture wastewater is a promising cost-effective feedstock for the cultivation of microalgae. Integrated microalgae biomass production as well as wastewater bioremediation, can be done in an economically favorable process. Organic and inorganic components of aquaculture wastewater are in the range of COD 100–150 mg/L, nitrates 2–110 mg/L, ammonia 3–7 mg/L, and phosphates 2–50 mg/L (Ansari et al., 2017). Aquaculture wastewater stream can be connected with the microalgae production process, making a promising biorefinery that could produce microalgal biomass utilizing aquaculture wastewater. Due to the high nutritional value of aquaculture wastewater, dependency on high-cost artificial nutrients will decrease for microalgal biomass production. Further, the microalgal biomass can be used to produce biodiesel and other high-value-added products. Egloff et al. (2018) cultivated *Chlorella vulgaris* in an open thin-layer photobioreactor system achieving a maximum biomass yield of 20 g/L. When the *Platymonas subcordiformis* was cultivated using aquaculture wastewater from different stages having different nutritional values, biomass yield and nitrogen and phosphorus removal efficiencies were studied. Biomass yield increases 8.9-fold, and nitrogen and phosphorus removal efficiency was observed up to 95% and 99%, respectively (Guo et al., 2013).

Dairy effluents, having a high organic carbon load, is one of the most important low-cost feedstocks for microalgae cultivation. The dairy industry produces a huge amount of effluents (0.2–10 L effluent per liter of milk processed) that are disposed of in the environment without any treatment, causing environmental damage (Ummalyma and Sukumaran, 2014). Alternatively, dairy effluents could be utilized for the cultivation of microalgae for biofuel production. Dairy effluents are rich in organic carbon, phosphates, nitrates, COD, and biological oxygen demand (BOD). Microalgae cultivation using dairy effluents not only produces biomass, but bioremediation also occurs by removing the organic load from the effluent. Oleaginous microalgae *Chlorococcum* sp. RAP13 was cultivated in a mixotrophic mode using dairy effluents and supplied with waste glycerol from the biodiesel industry; maximum biomass yield was observed up to 1.94 g/L, and lipid accumulated up to 42% (Ummalyma and Sukumaran, 2014). Microalgal biomass and lipid production were studied using a novel isolate of *Chlorella sp.* utilizing dairy effluents. The maximum biomass yield was found to be 1.37 g/L (Choi et al., 2018).

11.5.1.2 Anaerobic digestate

Anaerobic digestate or anaerobic effluents rich in organic content is a great alternative low-cost nutrient to artificial mineral nutrient media for microalgae cultivation, increasing economic viability and sustainability for biofuel production. Anaerobic digestate is produced during biogas production as a by-product of anaerobic digestion processes. Depending on the source of the digestate, a pretreatment step such as centrifugation, filtration, sterilization, hydrogen peroxide oxidation, or thermochemical treatment is required for contaminant removal prior to microalgae cultivation. Nutrients are optimized by supplementing organic carbon and inorganic carbon in low-carbon effluents. The transparency of the digestate is maintained such that light can pass through it. Other factors such as microbial contamination, metal toxicity, and the presence of ammonia in the digestate affect the growth of microalgae adversely. It has been observed that microalgae growth rates on digestate are comparable to that cultivated using artificial media (Bauer et al., 2021; Chong et al., 2022). Microalgae *Chlorella vulgaris* was cultivated using liquid digestate from an agricultural-based biogas plant at laboratory-scale bubble column photobioreactor and pilot-scale thin layer photobioreactor. Cultures were grown to achieve a high cell density of up to 14 g/L in 21 days (Pulgarin et al., 2021).

11.5.2 Two-stage cultivation strategy

Microalgae can accumulate high lipid content when grown in nitrogen and phosphorus stress conditions. A high cell density biomass algal culture corresponds to high lipid production. Further, transesterification and hydrogenation reactions are done and lipid is converted to biodiesel. Two-stage cultivation has been the major strategy of biotechnologists for enhanced biodiesel and other chemical production. The first stage provides optimal culture conditions; nitrogen and phosphorus are supplemented in sufficient amounts to achieve the highest cell density. In the second-stage, nitrogen and phosphorus supply is limited to provide nutritional stress conditions for directing the metabolic pathway of microalgae toward lipid synthesis (Benasla and Hausler, 2021). Once the maximum cell density is achieved in the first stage, the biomass is harvested by centrifugation and transferred to a new medium lacking or having low concentrations of nitrogen and phosphorus. After culturing for a few days, the culture is harvested and lipid is extracted. In a study, microalgae *Raphidocelis subcapitata* immobilized in alginate gel were initially cultured in a complete media and after attaining the highest biomass cells were transferred to nitrogen- and phosphorus-deficient media for high lipid accumulation. Lipid content was achieved up to 31.7% and 19.4% under nitrogen- and phosphorus-deficient conditions, respectively (Benasla and Hausler, 2021).

Zheng et al. (2012) grew a heterotrophic first-stage culture of microalgae *Chlorella sorokiniana* biomass. In the second stage, culture was shifted to phototrophic mode. A higher yield was obtained in the heterotrophic culture mode; growth rate, cell density, and productivity were found to be 3,0-, 3.3-, and 7.4-fold respectively higher than in the phototrophic culture mode. Another approach of two-stage cultivation is the utilization of wastewater to culture microalgae and further- the culture medium is recycled in the second stage. *Chlorella* sp. was grown in poultry wastewater in the first stage until the highest biomass was achieved. Further, chlorella biomass was harvested by centrifugation, and the recycled media was used for the cultivation of *Spirulina platensis* in the second stage. *Chlorella* sp. and *Spirulina platensis* biomass yield was attained up to 0.39 g/L and 3.4 g/L, respectively (Wang et al., 2018). A two-stage cultivation process was used for the production of polyhydroxybutyrate (PHB) by *Cupriavidus necator* DSM 545 from CO_2. Microalgal biomass was grown in the heterotrophic growth conditions supplemented with two carbon sources, namely glucose and waste glycerol, in the first stage. In the second stage, nitrogen and oxygen stress conditions were applied for PHB biosynthesis in autotrophic growth conditions resulting in the highest PHB yield of 28 g/L (Garcia-Gonzalez et al., 2015).

11.6 Limitations and challenges concerned with large-scale cultivation

The sections above discussed different cultivation systems, integration approaches related to microalgae, and the production of different value-added products. Despite the adequacy shown by microalgae culture systems, several challenges are associated with it. The major drawback is the scale-up of the lab-scale process to the pilot-scale. Another primary concern is the economics of the process. Limitations of the strain selection, biomass harvesting, integration process linked to wastewater remediation, and techno-economic analysis (TEA) are discussed in short, here.

Algal strains such as *Botryococcus braunii* Kutzing (70% lipid content) having higher lipid content are preferred. In addition to this, the strain should also be able to withstand fluctuating environmental conditions and contamination. Native strains are favored so that they can grow effortlessly. Lipid is the most useful by-product of the algae culture, which can be further used in the form of biodiesel. However, favorable conditions for increased lipid content decrease biomass concentration and vice versa. Isolation of indigenous strains and then selecting them under the stress conditions like nitrogen deprivation is one of the approaches (Rodolfi et al., 2009). For instance, *Nannochloropsis* sp. F&M-M24 is able to reach about 60% lipid content under nitrogen deprivation conditions. The strain grew outdoors, giving an annual production of 20 tons of lipids per hectare. Isolating strains and genetically modifying them can be a tedious and cumbersome task. To avoid this, adaptive laboratory evolution (ALE) is applied in which the strain is allowed to grow under stress conditions, and with the subsequent generations, the strain achieves tolerance to the stress (Zhao and Huang, 2021). ALE strains could help in environmental remediation as they can grow in high salt concentrations, varying temperatures, and wastewater. However, these conditions can lead to reactive oxygen species generation, which can damage the photosystem of algae. A number of cycles are required to generate the desired strains; for example, 31 cycles of ALE gave two *Chlorella* sp. AE10 and AE20, which can tolerate 10% and 20% CO_2, respectively (Li et al., 2015).

After selecting the suitable algae for biomass generation, the media in which it will grow plays a pivotal role. Usually, feigned media BG-11 is used widely to grow algae systems; nevertheless, using this at a pilot scale is not a sustainable approach. Initially, the algae were used for wastewater bioremediation, reducing up to 85% BOD in the wastewater. Nonetheless, wastewater cannot be used directly as it contains multiple micropollutants like heavy metals, drugs, medicines, microorganisms, soaps, and detergents. Many techniques such as filtration, adsorption, coagulation, and flocculation are used to remove heavy metals but are not feasible. Algae strains such as *Cladophora fracta* can remove 85–99% of Cu, Zn, Cd, and Hg (Ji et al., 2012). However, this is not consistent with the conditions provided for every metal. Chlorination is the method of choice to remove microorganisms from wastewater but can lead to the generation of carcinogenic by-products (Zarpelon et al., 2016). Moreover, the dead and decaying algae in wastewater can increase the load of microorganisms.

Another concern after the successful growth of microalgae is biomass harvesting. It has been estimated that 30% of the total expense is biomass harvesting cost (Molina Grima et al., 2003). Several processes for harvesting are being used, such as filtration, centrifugation, gravity sedimentation, chemical flocculation, dissolved air floatation, and bio-flocculation. All these processes have their pros and cons. For instance, the filtration and submerged membrane filtration process has the advantage of low cost but is slow and shows the problem of membrane fouling. In contrast to filtration, centrifugation is fast and efficient but is energy-intensive and adds to the economics of the process. Similarly, bio-flocculation and electrolytic flocculation are high energy requirement processes but are highly efficient. Gravity sedimentation does not require energy but is slow and only relevant for algae species with high sedimentation rates and density (Collet et al., 2011). Out of these processes, the chosen one should not hamper microalgae's biomass and lipid quality. Since the process selection criteria vary from species to species, maintaining a single species in open cultivation systems is challenging.

A process has to undergo rigorous assessments, and therefore, the majority of microalgae integrated processes are still in the nascent stage in several developing countries. Cost is the major factor in any process and determines its commercial feasibility. TEA assesses the process's cost and feasibility by studying different factors and costs in each phase of research and development to scale up. TEA was done on different systems, primarily on open raceway ponds, as cultivation in them is much cheaper. A plethora of literature is available for TEA of microalgal biofuel from the number of feedstocks ranging from whole cells to wood. The cost of biofuel estimated from TEA has a broad range due to the fact that different researchers use different assumptions, methods, cultivation systems, and parameters (Ranganathan and Savithri, 2019). Various cultivation systems like high volume V-shape Pond (HVVP), open raceway pond (OP), closed tubular photobioreactor (PBR), algal turf scrubbers, and many more are used for the evaluation purposes (Davis et al., 2011; Hoffman et al., 2017; Kumar et al., 2020). For instance, Vázquez-Romero and colleagues did TEA for the production of *Phaeodactylum tricornutum* in tubular PBR, acquiring 29.48 tons of biomass per hectare per year, whose cost was estimated to be 108.26 or 44 €/kg of dry weight (on 1- and 100-hectare scale, respectively) (Vázquez-Romero et al., 2022). The source of light is crucial for the growth of microalgae, and countries that see less sunlight require an artificial source of light, which increases the process cost by 95%, and its absence decreases biomass production. Reverse results were obtained when conditions were favorable for the growth. This was seen in Spain, where conditions were pleasing, yielding biomass of 39.02 tons per hectare per year with the cost of 23.08 €/kg dry weight on a 100-hectare scale.

11.7 Conclusion and recommendations

To meet the growing population's energy demand, an energy source distinct from the natural one must be identified as the nonrenewable source/s may be able to fulfill the demand for the next ~50–60 years. In this regard, microalgal biofuel is considered to be potential replacement for petroleum products. Biofuel generated from microalgal biomass is considered to be a win-win situation as biomass cultivation can be integrated with wastewater treatment, and further, the biomass can be accumulated for lipid extraction and hence, biodiesel. This removes the nutrients from the effluent and generates biomass. Additionally, there are several benefits to growing microalgae in wastewater as the growth medium, which include less aeration requirement, greater P consumption than in biological treatment, and the microalgae's ability to biofix CO_2.

Based on the critical analysis of various microalgae-assisted cultivation technologies for wastewater treatment, the following major points are recommended:

- Microalgae cultivation in wastewater can contribute to simultaneous wastewater treatment and biological CO_2 fixation. Since the wastewater has high concentrations of C, N, and P, it can be employed as a microalagal cultivation media.
- Although the integral microalgae cultivation approaches have little impact in the lab- scale, they have a significant role in the pilot/industrial level to make the process sustainable.
- Anaerobic digestate, activated sludge, and municipal solid waste are investigated as potential substrates for biomass and biofuel (biomethane, biohydrogen) production.
- Moreover, the treated water can be used for several purposes like aquaculture, irrigation, and industrial processing.

Hence these integral technologies will lead toward sustainable biofuel production coupled with wastewater treatment.

References

Aditya, L., Mahlia, T. M. I., Nguyen, L. N., Vu, H. P., & Nghiem, L. D. (2022). Microalgae-bacteria consortium for wastewater treatment and biomass production. *Science of the Total Environment, 838*, 155871. https://doi.org/10.1016/J.SCITOTENV.2022.155871

Altunoz, M., Allesina, G., Pedrazzi, S., & Guidetti, E. (2020). Integration of biological waste conversion and wastewater treatment plants by microalgae cultivation. *Process Biochemical, 91*, 158–164. https://doi.org/10.1016/J.PROCBIO.2019.12.007

Altunoz, M., Pirrotta, O., Forti, L., Allesina, G., Pedrazzi, S., Obali, O., Tartarini, P., & Arru, L. (2017). Combined effects of LED lights and chicken manure on neochloris oleoabundans growth. *Bioresour Technology, 244*, 1261–1268. https://doi.org/10.1016/J.BIORTECH.2017.04.094

Amenorfenyo, D. K., Huang, X., Zhang, Y., Zeng, Q., Zhang, N., Ren, J., & Huang, Q. (2019). Microalgae brewery wastewater treatment: Potentials, benefits and the challenges. *International Journal of Environmental Research and Public Health, 2019, 16*(11), 1910. https://doi.org/10.3390/IJERPH16111910

Amini, M., Mohamedelhassan, E., & Liao, B. (2022). The biological performance of a novel electrokinetic-assisted membrane photobioreactor (EK-MPBR) for wastewater treatment. *Membranes, 2022, 12*(6), 587. https://doi.org/10.3390/MEMBRANES12060587

Ansari, F. A., Singh, P., Guldhe, A., & Bux, F. (2017). Microalgal cultivation using aquaculture wastewater: Integrated biomass generation and nutrient remediation. *Algal Research, 21*, 169–177. https://doi.org/10.1016/j.algal.2016.11.015

Aslanbay Guler, B., Deniz, I., Demirel, Z., Oncel, S. S., & Imamoglu, E. (2020). Computational fluid dynamics modelling of stirred tank photobioreactor for haematococcus pluvialis production: Hydrodynamics and mixing conditions. *Algal Research, 47*. https://doi.org/10.1016/J.ALGAL.2020.101854

Bagchi, S. K., Patnaik, R., Sonkar, S., Koley, S., Rao, P. S., & Mallick, N. (2019). Qualitative biodiesel production from a locally isolated chlorophycean microalga scenedesmus obliquus (turpin) kützing GA 45 under closed raceway pond cultivation. *Renewable Energy, 139*, 976–987. https://doi.org/10.1016/J.RENENE.2019.02.115

Bauer, L., Ranglová, K., Masojídek, J., Drosg, B., & Meixner, K. (2021). Digestate as sustainable nutrient source for microalgae – Challenges and prospects. *Applied Sciences (Switzerland)*, *11*(3), 1–21. MDPI AG. https://doi.org/10.3390/app11031056

Behl, K., Jaiswal, P., Nigam, S., Prasanna, R., Abraham, G., & Singh, P. K. (2022). Treatment of Textile Waste Effluents Using Microalgae: A Suitable Approach for Wastewater Remediation and Lipid Production, pp. 103–137. https://doi.org/10.1007/978-981-19-0793-7_5

Benasla, A., & Hausler, R. (2021). A two-step cultivation strategy for high biomass production and lipid accumulation of raphidocelis subcapitata immobilized in alginate gel. *Biomass*, *1*(2), 94–104. https://doi.org/10.3390/biomass1020007

Benner, P., Meier, L., Pfeffer, A., Krüger, K., Oropeza Vargas, J. E., & Weuster-Botz, D. (2022). Lab-scale photobioreactor systems: Principles, applications, and scalability. *Bioprocess and Biosystems Engineering, 2022*, *45*(5), 791–813. https://doi.org/10.1007/S00449-022-02711-1

Bhandari, M., & Prajapati, S. K. (2022). Use of reverse osmosis reject from drinking water plant for microalgal biomass production. *Water Research*, *210*. https://doi.org/10.1016/J.WATRES.2021.117989

Borowitzka, M. A. (1999). Commercial production of microalgae: Ponds, tanks, tubes and fermenters. *Journal of Biotechnology*, *70*(1–3), 313–321. https://doi.org/10.1016/S0168-1656(99)00083-8

Borowitzka, M. A., & Reza Moheimani, N. (2013). Open pond culture systems. *Algae for Biofuels and Energy*, 133–152. https://doi.org/10.1007/978-94-007-5479-9_8/COVER/

Borowitzka, M., & Hallegraeff, G. (2007). *Economic importance of algae – Open Access Repository*. Algae of Australia: Introduction, Algae of Australia, CSIRO Publishing / Australian Biological Resources, Canberra/Collingwood. https://eprints.utas.edu.au/5904/

Brindhadevi, K., Mathimani, T., Rene, E. R., Shanmugam, S., Chi, N. T. L., & Pugazhendhi, A. (2021). Impact of cultivation conditions on the biomass and lipid in microalgae with an emphasis on biodiesel. *Fuel*, *284*, 119058. https://doi.org/10.1016/J.FUEL.2020.119058

Cheah, W. Y., Ling, T. C., Show, P. L., Juan, J. C., Chang, J. S., & Lee, D. J. (2016). Cultivation in wastewaters for energy: A microalgae platform. *Applied Energy*, *179*, 609–625. https://doi.org/10.1016/J.APENERGY.2016.07.015

Chen, C. Y., Nagarajan, D., & Cheah, W. Y. (2018). Eicosapentaenoic acid production from nannochloropsis oceanica CY2 using deep sea water in outdoor plastic-bag type photobioreactors. *Bioresource Technology*, *253*, 1–7. https://doi.org/10.1016/J.BIORTECH.2017.12.102

Chhandama, M. V. L., Satyan, K. B., Changmai, B., Vanlalveni, C., & Rokhum, S. L. (2021). Microalgae as a feedstock for the production of biodiesel: A review. *Bioresource Technology Reports*, *15*, 100771. https://doi.org/10.1016/J.BITEB.2021.100771

Chini Zittelli, G., Rodolfi, L., Bassi, N., Biondi, N., & Tredici, M. R. (2013). Photobioreactors for microalgal biofuel production. *Algae for Biofuels and Energy*, 115–131. https://doi.org/10.1007/978-94-007-5479-9_7

Choi, Y. K., Jang, H. M., & Kan, E. (2018). Microalgal biomass and lipid production on dairy effluent using a novel microalga, chlorella sp. isolated from dairy wastewater. *Biotechnology and Bioprocess Engineering, 2018*, *23*(3), 333–340. https://doi.org/10.1007/S12257-018-0094-Y

Chong, C. C., Cheng, Y. W., Ishak, S., Lam, M. K., Lim, J. W., Tan, I. S., Show, P. L., & Lee, K. T. (2022). Anaerobic digestate as a low-cost nutrient source for sustainable microalgae cultivation: A way forward through waste valorization approach. *Science of the Total Environment*, *803*, 150070. https://doi.org/10.1016/J.SCITOTENV.2021.150070

Chu, H. Q., Tan, X. B., Zhang, Y. L., Yang, L. B., Zhao, F. C., & Guo, J. (2015). Continuous cultivation of chlorella pyrenoidosa using anaerobic digested starch processing wastewater in the outdoors. *Bioresource Technology*, *185*, 40–48. https://doi.org/10.1016/J.BIORTECH.2015.02.030

Coelho, R. S., Vidotti, A. D. S., Reis, É. M., & Franco, T. T. (2014). High cell density cultures of microalgae under fed-batch and continuous growth. *Chemical Engineering Transaction*, *38*, 313–318. https://doi.org/10.3303/CET1438053

Collet, P., Hélias Arnaud, A., Lardon, L., Ras, M., Goy, R. A., & Steyer, J. P. (2011). Life-cycle assessment of microalgae culture coupled to biogas production. *Bioresource Technology*, *102*(1), 207–214. https://doi.org/10.1016/j.biortech.2010.06.154
Costa, S. S., Miranda, A. L., De morais, M. G., Costa, J. A. V., & Druzian, J. I. (2019). Microalgae as source of polyhydroxyalkanoates (PHAs) – A review. *International Journal of Biological Macromolecules*, *131*, 536–547. https://doi.org/10.1016/J.IJBIOMAC.2019.03.099
Cox, J., Kyle, D., Radmer, R., & Delente, J. (1988). Stable-isotope-labeled biochemicals from microalgae. *Trends in Biotechnology*, *6*(11), 279–282. https://doi.org/https://doi.org/10.1016/0167-7799(88)90125-4
da Silva, T. L., Moniz, P., Silva, C., & Reis, A. (2021). The role of heterotrophic microalgae in waste conversion to biofuels and bioproducts. *Processes, 2021*, *9*(7), 1090. https://doi.org/10.3390/PR9071090
Davis, R., Aden, A., & Pienkos, P. T. (2011). Techno-economic analysis of autotrophic microalgae for fuel production. *Applied Energy*, *88*(10), 3524–3531. https://doi.org/10.1016/j.apenergy.2011.04.018
Dickinson, S., Mientus, M., Frey, D., Amini-Hajibashi, A., Ozturk, S., Shaikh, F., Sengupta, D., & El-Halwagi, M. M. (2017). A review of biodiesel production from microalgae. *Clean Technologies and Environmental Policy*, *19*(3), 637–668. https://doi.org/10.1007/S10098-016-1309-6/FIGURES/6
Egloff, S., Tschudi, F., Schmautz, Z., & Refardt, D. (2018). High-density cultivation of microalgae continuously fed with unfiltered water from a recirculating aquaculture system. *Algal Research*, *34*, 68–74. https://doi.org/10.1016/j.algal.2018.07.004
Faried, M., Samer, M., Abdelsalam, E., Yousef, R. S., Attia, Y. A., & Ali, A. S. (2017). Biodiesel production from microalgae: Processes, technologies and recent advancements. *Renewable and Sustainable, Energy Reviews*, *79*, 893–913. https://doi.org/10.1016/J.RSER.2017.05.199
Gao, F., Yang, Z. H., Li, C., Wang, Y. J., Jin, W. H., & Deng, Y. B. (2014). Concentrated microalgae cultivation in treated sewage by membrane photobioreactor operated in batch flow mode. *Bioresource Technology*, *167*, 441–446. https://doi.org/10.1016/J.BIORTECH.2014.06.042
Gao, K. (1998). Chinese studies on the edible blue-green alga, nostoc flagelliforme: A review. *Journal Applied Phycology*, *10*, 37–49.
Garcia-Gonzalez, L., Mozumder, M. S. I., Dubreuil, M., Volcke, E. I. P., & de Wever, H. (2015). Sustainable autotrophic production of polyhydroxybutyrate (PHB) from CO2 using a two-stage cultivation system. *Catalysis Today*, *257*(Part 2), 237–245. https://doi.org/10.1016/j.cattod.2014.05.025
Ge, S., Qiu, S., Tremblay, D., Viner, K., Champagne, P., & Jessop, P. G. (2018). Centrate wastewater treatment with chlorella vulgaris: Simultaneous enhancement of nutrient removal, biomass and lipid production. *Chemical Engineering Journal*, *342*, 310–320. https://doi.org/10.1016/J.CEJ.2018.02.058
Gillard, J. T. F., Hernandez, A. L., Contreras, J. A., Francis, I. M., & Cabrales, L. (2021). Potential for biomass production and remediation by cultivation of the marine model diatom phaeodactylum tricornutum in oil field produced wastewater media. *Water, 2021 13*(19), 2700. https://doi.org/10.3390/W13192700
Gimmler, H., Treffny, B., Kowalski, M., & Zimmermann, U. (1991). The resistance of dunaliella acidophila against heavy metals: The importance of the zeta potential. *Journal of Plant Physiology*, *138*(6), 708–716. https://doi.org/10.1016/S0176-1617(11)81320-9
Granado-Lorencio, F., Herrero-Barbudo, C., Acién-Fernández, G., Molina-Grima, E., Fernández-Sevilla, J. M., Pérez-Sacristán, B., & Blanco-Navarro, I. (2009). In vitro bioaccesibility of lutein and zeaxanthin from the microalgae scenedesmus almeriensis. *Food Chemistry*, *114*(2), 747–752. https://doi.org/10.1016/J.FOODCHEM.2008.10.058
Gu, W., Kavanagh, J. M., & McClure, D. D. (2021). Photoautotrophic production of eicosapentaenoic acid. *Critical Reviews in Biotechnology*, *41*(5), 731–748. https://doi.org/10.1080/07388551.2021.1888065
Guo, Z., Liu, Y., Guo, H., Yan, S., & Mu, J. (2013). Microalgae cultivation using an aquaculture wastewater as growth medium for biomass and biofuel production. *Journal of Environmental Sciences*, 85–88. www.jesc.ac.cnAvailableonlineatwww.sciencedirect.com

Hakim, M. A. A., Al-Ghouti, M. A., Das, P., Abu-Dieyeh, M., Ahmed, T. A., & Aljabri, H. M. S. J. (2018). Potential application of microalgae in produced water treatment. *Desalination and Water Treatment, 135*, 47–58. https://doi.org/10.5004/DWT.2018.23146

Hemalatha, M., Sravan, J. S., Yeruva, D. K., & Venkata Mohan, S. (2017). Integrated ecotechnology approach towards treatment of complex wastewater with simultaneous bioenergy production. *Bioresource Technology, 242*, 60–67. https://doi.org/10.1016/J.BIORTECH.2017.03.118

Hemalatha, M., & Venkata Mohan, S. (2016). Microalgae cultivation as tertiary unit operation for treatment of pharmaceutical wastewater associated with lipid production. *Bioresource Technology, 215*, 117–122. https://doi.org/10.1016/J.BIORTECH.2016.04.101

Hoffman, J., Pate, R. C., Drennen, T., & Quinn, J. C. (2017). Techno-economic assessment of open microalgae production systems. *Algal Research, 23*, 51–57. https://doi.org/10.1016/j.algal.2017.01.005

Hossain, N., & Mahlia, T. M. I. (2019). Progress in physicochemical parameters of microalgae cultivation for biofuel production. *Critical Reviews in Biotechnology, 39*(6), 835–859. https://doi.org/10.1080/07388551.2019.1624945

Huang, Q., Jiang, F., Wang, L., & Yang, C. (2017). Design of photobioreactors for mass cultivation of photosynthetic organisms. *Engineering, 3*(3), 318–329. https://doi.org/10.1016/J.ENG.2017.03.020

Ji, L., Xie, S., Feng, J., Li, Y., & Chen, L. (2012). Heavy metal uptake capacities by the common freshwater green alga cladophora fracta. *Journal of Applied Phycology, 24*(4), 979–983. https://doi.org/10.1007/s10811-011-9721-0

Jin, H., Zhang, H., Zhou, Z., Li, K., Hou, G., Xu, Q., Chuai, W., Zhang, C., Han, D., & Hu, Q. (2020). Ultrahigh-cell-density heterotrophic cultivation of the unicellular green microalga scenedesmus acuminatus and application of the cells to photoautotrophic culture enhance biomass and lipid production. *Biotechnology and Bioengineering, 117*(1), 96–108. https://doi.org/10.1002/BIT.27190

Karkos, P. D., Leong, S. C., Karkos, C. D., Sivaji, N., & Assimakopoulos, D. A. (2011). Spirulina in clinical practice: Evidence-based human applications. *Evidence-based Complementary and Alternative Medicine, 2011*. https://doi.org/10.1093/ecam/nen058

Kim, M., Ahn, J., Jeon, H., Jin, E. S., Cutignano, A., & Romano, G. (2017). Development of a dunaliella tertiolecta strain with increased zeaxanthin content using random mutagenesis. *Marine Drugs*, 2017 *15*(6), 189. https://doi.org/10.3390/MD15060189

Kiran, B., Kumar, R., & Deshmukh, D. (2014). Perspectives of microalgal biofuels as a renewable source of energy. *Energy Conversion and Management, 88*, 1228–1244. https://doi.org/10.1016/J.ENCONMAN.2014.06.022

Koley, S., Mathimani, T., Bagchi, S. K., Sonkar, S., & Mallick, N. (2019). Microalgal biodiesel production at outdoor open and polyhouse raceway pond cultivations: A case study with scenedesmus accuminatus using low-cost farm fertilizer medium. *Biomass and Bioenergy, 120*, 156–165. https://doi.org/10.1016/J.BIOMBIOE.2018.11.002

Kumar, A. K., Sharma, S., Dixit, G., Shah, E., & Patel, A. (2020). Techno-economic analysis of microalgae production with simultaneous dairy effluent treatment using a pilot-scale high volume V-shape pond system. *Renewable Energy, 145*, 1620–1632. https://doi.org/10.1016/j.renene.2019.07.087

Kumar, S. (2018). Effect of salinity and ph ranges on the growth and biochemical composition of marine microalga- nannochloropsis salina. *International Journal of Agriculture, Environment and Biotechnology, 11*(4). https://doi.org/10.30954/0974-1712.08.2018.6

Leite, L. D. S., Hoffmann, M. T., & Daniel, L. A. (2019). Coagulation and dissolved air flotation as a harvesting method for microalgae cultivated in wastewater. *Journal of Water Process Engineering, 32*. https://doi.org/10.1016/J.JWPE.2019.100947

Leng, L., Li, J., Wen, Z., & Zhou, W. (2018). Use of microalgae to recycle nutrients in aqueous phase derived from hydrothermal liquefaction process. *Bioresource Technology, 256*, 529–542. https://doi.org/10.1016/J.BIORTECH.2018.01.121

Li, D., Wang, L., Zhao, Q., Wei, W., & Sun, Y. (2015). Improving high carbon dioxide tolerance and carbon dioxide fixation capability of chlorella sp. by adaptive laboratory evolution. *Bioresource Technology*, *185*, 269–275. https://doi.org/10.1016/j.biortech.2015.03.011

Liang, F., Wen, X., Geng, Y., Ouyang, Z., Luo, L., & Li, Y. (2013). Growth rate and biomass productivity of chlorella as affected by culture depth and cell density in an open circular photobioreactor. *Journal of Microbiology and Biotechnology*, *23*(4), 539–544. https://doi.org/10.4014/JMB.1209.09047

Liu, W., Chen, Y., Wang, J., & Liu, T. (2018). Biomass productivity of scenedesmus dimorphus (chlorophyceae) was improved by using an open pond–photobioreactor hybrid system. *European Journal of Phycology*, *54*(2), 127–134. https://doi.org/10.1080/09670262.2018.1519601

Lu, Q., Li, H., Zou, Y., Liu, H., & Yang, L. (2021). Astaxanthin as a microalgal metabolite for aquaculture: A review on the synthetic mechanisms, production techniques, and practical application. *Algal Research*, *54*. Elsevier B.V. https://doi.org/10.1016/j.algal.2020.102178

Luo, Y., Le-Clech, P., & Henderson, R. K. (2017). Simultaneous microalgae cultivation and wastewater treatment in submerged membrane photobioreactors: A review. *Algal Research*, *24*, 425–437. https://doi.org/10.1016/J.ALGAL.2016.10.026

Ma, X., Zheng, H., Addy, M., Anderson, E., Liu, Y., Chen, P., & Ruan, R. (2016). Cultivation of chlorella vulgaris in wastewater with waste glycerol: Strategies for improving nutrients removal and enhancing lipid production. *Bioresource Technology*, *207*, 252–261. https://doi.org/10.1016/J.BIORTECH.2016.02.013

Maity, J. P., Bundschuh, J., Chen, C. Y., & Bhattacharya, P. (2014). Microalgae for third generation biofuel production, mitigation of greenhouse gas emissions and wastewater treatment: Present and future perspectives – A mini review. *Energy*, *78*, 104–113. https://doi.org/10.1016/J.ENERGY.2014.04.003

Maltsev, Y., Maltseva, K., Kulikovskiy, M., & Maltseva, S. (2021). Influence of light conditions on microalgae growth and content of lipids, carotenoids, and fatty acid composition. *Biology*, *10*(10), 1060. https://doi.org/10.3390/BIOLOGY10101060

Marchão, L., da Silva, T. L., Gouveia, L., & Reis, A. (2018). Microalgae-mediated brewery wastewater treatment: Effect of dilution rate on nutrient removal rates, biomass biochemical composition, and cell physiology. *Journal of Applied Phycology*, *30*(3), 1583–1595. https://doi.org/10.1007/S10811-017-1374-1/TABLES/3

Mata, T. M., Melo, A. C., Simões, M., & Caetano, N. S. (2012). Parametric study of a brewery effluent treatment by microalgae scenedesmus obliquus. *Bioresource Technology*, *107*, 151–158. https://doi.org/10.1016/J.BIORTECH.2011.12.109

Meng, C., Huang, J., Ye, C., Cheng, W., Chen, J., & Li, Y. (2015). Comparing the performances of circular ponds with different impellers by CFD simulation and microalgae culture experiments. *Bioprocess and Biosystems Engineering*, *38*(7), https://doi.org/10.1007/S00449-015-1376-9

Metsoviti, M. N., Papapolymerou, G., Karapanagiotidis, I. T., & Katsoulas, N. (2019). Effect of light intensity and quality on growth rate and composition of chlorella vulgaris. *Plants*, *9*(1), 31. https://doi.org/10.3390/PLANTS9010031

Millar, G. J., Couperthwaite, S. J., & Moodliar, C. D. (2016). Strategies for the management and treatment of coal seam gas associated water. *Renewable and Sustainable Energy Reviews*, *57*, 669–691. https://doi.org/10.1016/J.RSER.2015.12.087

Mishra, A. K., Kaushik, M. S., & Tiwari, D. N. (2019). Nitrogenase and hydrogenase: Enzymes for nitrogen fixation and hydrogen production in cyanobacteria. *Cyanobacteria: From Basic Science to Applications*, 173–191. https://doi.org/10.1016/B978-0-12-814667-5.00008-8

Mishra, N., Srivastava, R., Tripathi S., & Mishra N. (2019). Variability in biochemical composition of microalgae Isochrysis galbana under nitrate deprivation. *Plant archives*, *19*(2), 173–175

Mohan, N., Rao, P. H., Boopathy, A. B., Rengasamy, R., & Chinnasamy, S. (2021). A sustainable process train for a marine microalga-mediated biomass production and CO2 capture: A pilot-scale cultivation of nannochloropsis salina in open raceway ponds and harvesting through electropreciflocculation. *Renewable Energy*, *173*, 263–272. https://doi.org/10.1016/J.RENENE.2021.03.147

Molina Grima, E., Belarbi, E.-H., Acién Fernández, F. G., Robles Medina, A., & Chisti, Y. (2003). Recovery of microalgal biomass and metabolites: Process options and economics. *Biotechnology Advances*, *20*(7), 491–515. https://doi.org/https://doi.org/10.1016/S0734-9750(02)00050-2

Molino, A., Lovinea, A., Leone, G., Di Sanzo, G., Palazzo, S., Martino, M., Sangiorgio, P., Marino, T., & Musmarra, D. (2020). Microalgae as alternative source of nutraceutical polyunsaturated fatty acids. *Chemical Engineering Transactions*, *79*, 277–282. https://doi.org/10.3303/CET2079047

Nicoletti, M. (2016). Microalgae nutraceuticals. *Foods*, *5*(3), 1–13. https://doi.org/10.3390/foods5030054

Olsson, J., Feng, X. M., Ascue, J., Gentili, F. G., Shabiimam, M. A., Nehrenheim, E., & Thorin, E. (2014). Co-digestion of cultivated microalgae and sewage sludge from municipal waste water treatment. *Bioresource Technology*, *171*(1), 203–210. https://doi.org/10.1016/J.BIORTECH.2014.08.069

Onorato, C., & Rösch, C. (2020). Comparative life cycle assessment of astaxanthin production with haematococcus pluvialis in different photobioreactor technologies. *Algal Research*, 50. https://doi.org/10.1016/j.algal.2020.102005

Pandey, A., Srivastava, S., & Kumar, S. (2019). Isolation, screening and comprehensive characterization of candidate microalgae for biofuel feedstock production and dairy effluent treatment: A sustainable approach. *Bioresource Technology*, 293. https://doi.org/10.1016/J.BIORTECH.2019.121998

Piligaev, A. V., Sorokina, K. N., Samoylova, Y. V., & Parmon, V. N. (2018). Lipid production by microalga micractinium sp. IC-76 in a flat panel photobioreactor and its transesterification with cross-linked enzyme aggregates of burkholderia cepacia lipase. *Energy Conversion and Management*, *156*, 1–9. https://doi.org/10.1016/J.ENCONMAN.2017.10.086

Polat, E., Yüksel, E., & Altınbaş, M. (2020). Mutual effect of sodium and magnesium on the cultivation of microalgae auxenochlorella protothecoides. *Biomass and Bioenergy*, *132*, 105441. https://doi.org/10.1016/J.BIOMBIOE.2019.105441

Potijun, S., Jaingam, S., Sanevas, N., Vajrodaya, S., & Sirikhachornkit, A. (2021). Green microalgae strain improvement for the production of sterols and squalene. *Plants 2021*, *10*(8), 1673. https://doi.org/10.3390/PLANTS10081673

Pourkarimi, S., Hallajisani, A., Nouralishahi, A., Alizadehdakhel, A., & Golzary, A. (2020). Factors affecting production of beta-carotene from Dunaliella salina microalgae. *Biocatalysis and Agricultural Biotechnology*, 29. Elsevier Ltd. https://doi.org/10.1016/j.bcab.2020.101771

Prandini, J. M., da Silva, M. L. B., Mezzari, M. P., Pirolli, M., Michelon, W., & Soares, H. M. (2016). Enhancement of nutrient removal from swine wastewater digestate coupled to biogas purification by microalgae scenedesmus spp. *Bioresource Technology*, *202*, 67–75. https://doi.org/10.1016/J.BIORTECH.2015.11.082

Pulgarin, A., Kapeller, A. G., Tarik, M., Egloff, S., Mariotto, M., Ludwig, C., & Refardt, D. (2021). Cultivation of microalgae at high-density with pretreated liquid digestate as a nitrogen source: Fate of nitrogen and improvements on growth limitations. *Journal of Cleaner Production*, *324*, 129238. https://doi.org/10.1016/J.JCLEPRO.2021.129238

Ranganathan, P., & Savithri, S. (2019). Techno-economic analysis of microalgae-based liquid fuels production from wastewater via hydrothermal liquefaction and hydroprocessing. *Bioresource Technology*, *284*, 256–265. https://doi.org/10.1016/j.biortech.2019.03.087

Rehman, M., Kesharvani, S., Dwivedi, G., & Gidwani Suneja, K. (2022). Impact of cultivation conditions on microalgae biomass productivity and lipid content. *Materials Today: Proceedings*, *56*, 282–290. https://doi.org/10.1016/J.MATPR.2022.01.152

Rodolfi, L., Zittelli, G. C., Bassi, N., Padovani, G., Biondi, N., Bonini, G., & Tredici, M. R. (2009). Microalgae for oil: Strain selection, induction of lipid synthesis and outdoor mass cultivation in a low-cost photobioreactor. *Biotechnology and Bioengineering*, *102*(1), 100–112. https://doi.org/10.1002/bit.22033

Roney, B. R., Renhui, L., Banack, S. A., Murch, S., Honegger, R., & Cox, P. A. (2009). Consumption of fa cai nostoc soup: A potential for BMAA exposure from nostoc cyanobacteria in china? *Amyotrophic Lateral Sclerosis*, *10*(SUPPL. 2), 44–49. https://doi.org/10.3109/17482960903273031

Schwede, S., Kowalczyk, A., Gerber, M., & Span, R. (2013). Anaerobic co-digestion of the marine microalga nannochloropsis salina with energy crops. *Bioresource Technology*, *148*, 428–435. https://doi.org/10.1016/J.BIORTECH.2013.08.157

Shah, M. M. R., Liang, Y., Cheng, J. J., & Daroch, M. (2016). Astaxanthin-producing green microalga haematococcus pluvialis: From single cell to high value commercial products. *Frontiers in Plant Science*, *7*(APR2016), 531. https://doi.org/10.3389/FPLS.2016.00531/BIBTEX

Singh, R. N., Sharma, S., Singh, A. K., & Srivastava, N. (2015). Design and development of a simple stirred tank photobioreactor for algal production. *Journal of Solar Energy Research Updates*, *2*(2), 24–26. https://doi.org/10.15377/2410-2199.2015.02.02.1

Singh, S. P., & Singh, P. (2015). Effect of temperature and light on the growth of algae species: A review. *Renewable and Sustainable Energy Reviews*, *50*, 431–444. https://doi.org/10.1016/J.RSER.2015.05.024

Song, C., Qiu, Y., Xie, M., Qi, Y., Li, S., & Kitamura, Y. (2019). Novel bio-regeneration concept via using rich solution as nutrition resource for microalgae cultivation: Effect of ph and feeding modes. *ACS Sustainable Chemistry and Engineering*, *7*(17), 14471–14478. https://doi.org/10.1021/ACSSUSCHEMENG.9B01839/ASSET/IMAGES/LARGE/SC9B01839_0008.JPEG

Spolaore, P., Joannis-Cassan, C., Duran, E., & Isambert, A. (2006). Commercial applications of microalgae. *Journal of Bioscience and Bioengineering*, *101*(2), 87–96. https://doi.org/10.1263/jbb.101.87

Suali, E., & Sarbatly, R. (2012). Conversion of microalgae to biofuel. *Renewable and Sustainable Energy Reviews*, *16*(6), 4316–4342. https://doi.org/10.1016/j.rser.2012.03.047

Sullivan Graham, E. J., Dean, C. A., Yoshida, T. M., Twary, S. N., Teshima, M., Alvarez, M. A., Zidenga, T., Heikoop, J. M., Perkins, G. B., Rahn, T. A., Wagner, G. L., & Laur, P. M. (2017). Oil and gas produced water as a growth medium for microalgae cultivation: A review and feasibility analysis. *Algal Research*, *24*(B), 492–504. https://doi.org/10.1016/J.ALGAL.2017.01.009

Suparmaniam, U., Lam, M. K., Uemura, Y., Lim, J. W., Lee, K. T., & Shuit, S. H. (2019). Insights into the microalgae cultivation technology and harvesting process for biofuel production: A review. *Renewable and Sustainable Energy Reviews*, *115*, 109361. https://doi.org/10.1016/J.RSER.2019.109361

Tan, J. S., Lee, S. Y., Chew, K. W., Lam, M. K., Lim, J. W., Ho, S. H., & Show, P. L. (2020). A review on microalgae cultivation and harvesting, and their biomass extraction processing using ionic liquids. *Bioengineered*, *11*(1), 116. https://doi.org/10.1080/21655979.2020.1711626

Tan, X. B., Lam, M. K., Uemura, Y., Lim, J. W., Wong, C. Y., & Lee, K. T. (2018). Cultivation of microalgae for biodiesel production: A review on upstream and downstream processing. *Chinese Journal of Chemical Engineering*, *26*(1), 17–30. https://doi.org/10.1016/J.CJCHE.2017.08.010

Tarhan, S. Z., Koçer, A. T., Özçimen, D., & Gökalp, İ. (2020). Cultivation of green microalgae by recovering aqueous nutrients in hydrothermal carbonization process water of biomass wastes. *Journal of Water Process Engineering*, *40*, 101783. https://doi.org/10.1016/J.JWPE.2020.101783

Travieso, L., Hall, D. O., Rao, K. K., Benítez, F., Sánchez, E., & Borja, R. (2001). A helical tubular photobioreactor producing spirulina in a semicontinuous mode. *International Biodeterioration and Biodegradation*, *47*(3), 151–155. https://doi.org/10.1016/S0964-8305(01)00043-9

Ugwu, C. U., Aoyagi, H., & Uchiyama, H. (2008). Photobioreactors for mass cultivation of algae. *Bioresource Technology*, *99*(10), 4021–4028. https://doi.org/10.1016/J.BIORTECH.2007.01.046

Umamaheswari, J., Kavitha, M. S., & Shanthakumar, S. (2020). Outdoor cultivation of chlorella pyrenoidosa in paddy-soaked wastewater and a feasibility study on biodiesel production from wet algal biomass through in-situ transesterification. *Biomass and Bioenergy*, *143*, 105853. https://doi.org/10.1016/J.BIOMBIOE.2020.105853

Ummalyma, S. B., & Sukumaran, R. K. (2014). Cultivation of microalgae in dairy effluent for oil production and removal of organic pollution load. *Bioresource Technology, 165*(C), 295–301. https://doi.org/10.1016/j.biortech.2014.03.028

Vanags, J., Kunga, L., Dubencovs, K., Galvanauskas, V., & Grigs, O. (2015). Influence of light intensity and temperature on cultivation of microalgae desmodesmus communis in flasks and laboratory-scale stirred tank photobioreactor. *Latvian Journal of Physics and Technical Sciences, 52*(2), 59–70. https://doi.org/10.1515/LPTS-2015-0012

Vázquez-Romero, B., Perales, J. A., De vree, J. H., Böpple, H., Steinrücken, P., Barbosa, M. J., Kleinegris, D. M. M., & Ruiz, J. (2022). Techno-economic analysis of microalgae production for aquafeed in Norway. *Algal Research, 64*, 102679. https://doi.org/https://doi.org/10.1016/j.algal.2022.102679

Wang, C., & Lan, C. Q. (2018). Effects of shear stress on microalgae – A review. *Biotechnology Advances, 36*(4), 986–1002. https://doi.org/10.1016/J.BIOTECHADV.2018.03.001

Wang, Q., Han, W., Jin, W., Gao, S., & Zhou, X. (2021). Docosahexaenoic acid production by schizochytrium sp.: Review and prospect. *Food Biotechnology, 35*(2), 111–135. https://doi.org/10.1080/08905436.2021.1908900

Wang, X., Lin, L., Lu, H., Liu, Z., Duan, N., Dong, T., Xiao, H., Li, B., & Xu, P. (2018). Microalgae cultivation and culture medium recycling by a two-stage cultivation system. *Frontiers of Environmental Science and Engineering 2018, 12*(6), 1–10. https://doi.org/10.1007/S11783-018-1078-Z

Yadav, G., Meena, D. K., Sahoo, A. K., Das, B. K., & Sen, R. (2020). Effective valorization of microalgal biomass for the production of nutritional fish-feed supplements. *Journal of Cleaner Production, 243*. https://doi.org/10.1016/J.JCLEPRO.2019.118697

Zarmi, Y., Gordon, J. M., Mahulkar, A., Khopkar, A. R., Patil, S. D., Banerjee, A., Reddy, B. G., Griffin, T. P., & Sapre, A. (2020). Enhanced algal photosynthetic photon efficiency by pulsed light. *IScience, 23*(5), 101115. https://doi.org/10.1016/J.ISCI.2020.101115

Zarpelon, F., Galiotto, D., Aguzolli, C., Carli, L. N., Figueroa, C. A., Baumvol, I. J. R., Machado, G., Crespo, J. D. S., & Giovanela, M. (2016). Removal of coliform bacteria from industrial wastewaters using polyelectrolytes/silver nanoparticles self-assembled thin films. *Journal of Environmental Chemical Engineering, 4*(1), 137–146. https://doi.org/10.1016/j.jece.2015.11.013

Zhang, Z., Guo, L., Liao, Q., Gao, M., Zhao, Y., Jin, C., She, Z., & Wang, G. (2021). Bacterial-algal coupling system for high strength mariculture wastewater treatment: Effect of temperature on nutrient recovery and microalgae cultivation. *Bioresource Technology, 338*, 125574. https://doi.org/10.1016/J.BIORTECH.2021.125574

Zhao, Q., & Huang, H. (2021). Adaptive evolution improves algal strains for environmental remediation. Trends in Biotechnology, 39(2), 112–115. Elsevier Ltd. https://doi.org/10.1016/j.tibtech.2020.08.009

Zheng, Y., Chi, Z., Lucker, B., & Chen, S. (2012). Two-stage heterotrophic and phototrophic culture strategy for algal biomass and lipid production. *Bioresource Technology, 103*(1), 484–488. https://doi.org/10.1016/j.biortech.2011.09.122

Zhou, X., Jin, W., Wang, Q., Guo, S., Tu, R., Han, S. F., Chen, C., Xie, G., Qu, F., & Wang, Q. (2020). Enhancement of productivity of chlorella pyrenoidosa lipids for biodiesel using co-culture with ammonia-oxidizing bacteria in municipal wastewater. *Renewable Energy, 151*, 598–603. https://doi.org/10.1016/J.RENENE.2019.11.063

Zhuang, L. L., Azimi, Y., Yu, D., Wu, Y. H., & Hu, H. Y. (2018). Effects of nitrogen and phosphorus concentrations on the growth of microalgae scenedesmus. LX1 in suspended-solid phase photobioreactors (sspbr). *Biomass and Bioenergy, 109*, 47–53. https://doi.org/10.1016/J.BIOMBIOE.2017.12.017

Zou, H., Rutta, N. C., Chen, S., Zhang, M., Lin, H., & Liao, B. (2022). Membrane photobioreactor applied for municipal wastewater treatment at a high solids retention time: Effects of microalgae decay on treatment performance and biomass properties. *Membranes 2022, 12*(6), 564. https://doi.org/10.3390/MEMBRANES12060564

Part III: **Process intensification applied to microalgae-based processes**

Carlos Eduardo Guzmán-Martínez, Juan Manuel Vera-Morales,
Efraín Quiroz-Pérez, Araceli Guadalupe Romero-Izquierdo,
Claudia Gutiérrez-Antonio*

Chapter 12
Process intensification applied to bioreactor design

Abstract: Microalgae are microorganisms of great relevance towards sustainable development, due to their elevated growth rate without the requirement of fertile lands; they can also absorb carbon dioxide and other pollutants during their cultivation. Moreover, all the fractions of microalgae can be transformed to produce chemical compounds, biofuels, as well as other value-added products. In the cultivation of microalgae, the bioreactor design is a key element. In this case, the bioreactor must enable the reception of the highest values of photosynthetically active radiation and it has to promote good availability of nutrients for allowing good microalgae culture; also, this system must reduce both energy and water flow requirements in order to ensure the positive energetic balance and reduce the operating costs. In this context, process intensification is a powerful tool that allows improving the design of the bioreactors since both mass and heat transfer rates are increased; in some cases, it is possible to combine two individual unit operations or reduce the equipment size. Therefore, in this chapter, a review of the available literature regarding the intensification of bioreactors for microalgae cultivation will be presented. The outline of the chapter includes information about the most relevant microalgae species and their bioproducts, as also the conventional bioreactors used for their cultivation. Later, the intensification strategies reported in the literature will be presented, based on which the future trends will be discussed.

Keywords: microalgae, process intensification, bioreactors

Acknowledgments: Financial support provided by CONACyT given to A. G. Romero-Izquierdo through a scholarship for the realization of her postgraduate studies is gratefully acknowledged.

***Corresponding author: Claudia Gutiérrez-Antonio**, Facultad de Ingeniería, Universidad Autónoma de Querétaro, Campus Amazcala, El Marqués-Querétaro 76265, México, e-mail: claugtez@gmail.com
Carlos Eduardo Guzmán-Martínez, Juan Manuel Vera-Morales, Efraín Quiroz-Pérez, Araceli Guadalupe Romero-Izquierdo, Facultad de Ingeniería, Universidad Autónoma de Querétaro, Campus Amazcala, El Marqués-Querétaro 76265, México

https://doi.org/10.1515/9783110781267-012

12.1 Introduction

Nowadays, society faces big challenges. The world population is growing, and the trend keeps increasing; as a consequence, the demand for goods and services will also increase, as will the energy consumption. Regarding this point, over the past 10 years, oil is the most used energy source for the transport sector, power generation, and chemical industries all over the world (Cherubini, 2010); it is important to mention that many chemical products are generated from oil refineries; almost 4% of oil is used for plastic and chemical production. Moreover, the intensive use of fossil fuel combustion, along with the land-use change, has mainly contributed to the problem of climate change, which began because of the increase of carbon dioxide concentration in the atmosphere. In this context, society has aligned its efforts to soften climate change and to decrease the dependence on oil. For this, it is recognized that there is a set of solutions that can contribute to get closer to the goal by working in coordinated way; for instance, improvement of vehicle technologies, design of sustainable processes, optimization of existing processes to decrease wastes, as well as the development of renewable raw material sources for sustainable processes are some alternatives.

As renewable raw material source, biomass has been gained recognition globally. Biomass can be defined as plant-based raw materials, which has the potential to substitute fossil resources as feedstock for industrial productions. This includes two fronts: the energy and non-energy sectors. The latter implies chemical and high valued materials. The strength of biomass lies in its characteristic of being a carbon-rich source available on the earth, unlike other energy sources such as wind, water, sun, and fossil sources. However, the sustainable generation of biomass is a key point, especially due to a possible competition for fertile land for food production. The sustainable biomass production from regional to global levels allows generating bioenergy, biofuels, and biochemical products; in this sense, it is possible to stand up to climate change, to develop rural communities, and to contribute to energy security.

There are several sources to obtain biomass, which is a criterion to classify biomass. Thus, it is possible to mention 4 generations:

– First-generation biomass

First-generation biomass includes raw materials that are used in food industries. Due to this, the use of this biomass for energy and other non-energy uses drives ethical, political and environmental concerns due to the competition aforementioned. First-generation raw materials are sugar, starch, edible oils, or animal fats. The feedstock includes mainly seeds and grains such as wheat, corn, soybean, rapeseed, palm oil, and so on. Biochemical methods, like fermentation or hydrolysis, are employed to convert them to biofuels or high-value products. The main advantage of these materials is their easy conversion due to their high oil or sugar content; however, for the same reason, it continues to be responsible for the food vs. fuel debate (Cherubini, 2010).

– Second-generation biomass
Second-generation biomass are raw materials that come from several nonfood crops; besides, this category includes lignocellulosic materials, among which residues from agriculture, industry, forestry and lignocellulosic nonedible crops are included. Unlike first-generation raw materials, the second generation ones do not work so easily due to their chemical composition. Cellulose, hemicellulose and lignin, which compound this feedstock, must be pretreated in order to be transformed to less complex molecules such as monosaccharides; this is the only way that microorganisms can carry out a biochemical transformation. The conversion routes, as pretreatment, are thermochemical processes, flash pyrolysis, and enzymatic digestion. The main advantages of these materials are that they are widespread and their relatively cheap cost. In some cases, such as agricultural residues, the feedstock for biorefineries used to be a debated issue because it can be used as fodder for livestock.

– Third-generation biomass
The third-generation biomass is derived from microorganisms, mainly, micro and macro algae biomass. This kind of biomass, microalgae specifically, has attracted attention as an alternative raw material because its cultivation does not compete with land utilization for agricultural activities and also does not require fresh water. Besides, the growth rate of microalgae is significantly higher than crops cultivated on land. These features highlight the wide potential to mitigate climate change that this feedstock has, if phototrophic nature is added (Tan et al., 2020). The base of phototrophic nature is photosynthesis; this is defined as a biochemical process which uses solar energy, atmospheric CO_2, and water to produce polysaccharides (cellulose and hemicellulose) and monosaccharides (glucose). Obtained carbohydrates are transformed to high value products by the biorefinery concept. Similar to the second-generation raw material, pretreatments are necessary if polysaccharides such as cellulose are used; however, these (thermochemical processes or enzymatic saccharification) are not as aggressive as the ones used for lignin degradation (Tan et al., 2020). Even though third-generation raw materials results are attractive to supply biorefineries, they have their own challenges, for instance, the need for extensive downstream processing such as dewatering requirement for biofuels. This fact can increase energy demand, and hence, increase in both investment and operation costs (Tan et al., 2020).

– Fourth-generation biomass
Derived from the advantages shown by third-generation raw materials, the fourth-generation ones are represented by microorganism, mainly algae; however, the difference is that microorganisms are modified via genetic engineering to alter their properties and cellular metabolism (Sikarwar et al., 2017). Through genetic engineering, the microorganisms are capable of enhancing their product yields, CO_2 capture ability, production rate, and tolerance to extreme environmental conditions. Based on the better properties and advantages shown by fourth-generation raw materials as com-

pared to third-generation ones, one might think there is nothing wrong with their use. However, initial investment is high though it is economical in the long run (Sikarwar et al., 2017).

As mentioned, there are many kinds of feedstock for biorefineries; however, it is important to highlight the most promising one, which are microalgae. On deep analysis, the employment of microalgae for production of high-value products has the following advantages over higher plants:

(1) Microalgae produce and amass big amounts of lipids (20–50% dry weight of biomass) and they have elevated growth rates.
(2) Microalgae can be cultivated all year; therefore, the oil yield per area of microalgae cultures surpasses the ones reported for even the best oilseed crops.
(3) Microalgae require less water than land crops, consequently decreasing the load on freshwater sources.
(4) Microalgae cultivation does not need the application of herbicides or pesticides.
(5) Microalgae absorb carbon dioxide during their cultivation from flue gases released from power plants that utilize fossil fuels and other sources.
(6) Microalgae can be used in wastewater bioremediation for the removal of NH^{+4}, NO^{-3}, and PO_4^{-3} from several types of wastewater sources (e.g., agricultural, animal feed operations, industrial and municipal wastewaters).
(7) Microalgae can be cultivated in saline/brackish water/coastal seawater on nonarable land; they also can grow under aggressive conditions with small needs for nutrients, and do not compete for resources with traditional farming systems.
(8) Microalgae species contains other compounds that can also be extracted; these compounds include polyunsaturated fatty acids, natural dyes, polysaccharides, pigments, antioxidants, high-value bioactive compounds, and proteins.

As biological organisms, the taxonomy of macroalgae and microalgae is so extensive; thus, it is important to define which species will be used in order to fit the desired bioproduct. After selecting the microalgae strain, a bioprocess needs to be proposed to enable its viable commercialization. Therefore, the design and optimization of bioreactors for microalgae cultivation is a key step in generating a marketable product. In spite of all potential applications, only a few species of algae are cultivated for commercial purposes; this is because bioreactor technology for microalgae is still scarce. From a commercial perspective, a microalgae cultivation system must have high area productivity, high volumetric productivity, reduced investment and maintenance costs, ease of control of the operating parameters, and reliability (Dragone et al., 2010). Different designs of cultivation systems try to fulfill these characteristics through several approaches. Thus, the term "photobioreactor" (PBR) is applied to open ponds and channels; however, some phycologists indicated a difference between open-air systems and PBRs (devices that allow monoseptic culture) (Dragone et al., 2010).

In PBRs, the regulation and control of all important operating parameters is a distinctive characteristic. Moreover, in this equipment, there is reduced contamination risk, no CO_2 losses, reproducible cultivation conditions, and flexible technical design. PBRs receive sunlight directly, through transparent container walls, or indirectly, via light fibers or tubes that channel it from sunlight collectors. In spite of this relative success, microalgae cultivation requires closed systems since many of the high-value products must be grown free of pollution and potential contaminants; this is due to the type of applications. Several designs have been developed, which include: (1) tubular (e.g., helical, manifold, serpentine), (2) flat (e.g., glass plates), and (3) column (e.g., bubble columns and airlift). A great amount of work has been carried out to optimize different PBR systems for microalgae cultivation (Dragone et al., 2010).

There are challenges around PBRs, which are directly related with some design parameters and algae nature; for example, surface-volume ratio, which has a high impact on light supply, CO_2/O_2 exchange, cell density, shear stress, and others. However, process intensification is a strong tool that can contribute, in an effective way, to offer a solution to the problems in algae production. Process intensification can be considered as any activity that integrates one or more of the following (Gómez et al., 2019):

- Smaller size equipment for a specific throughput
- Higher throughput value for a given size equipment or production process
- Fewer materials in inventory required for the processing of materials, considering the same throughput
- Fewer requirements of utility materials and feedstock for a specific throughput or size equipment
- Major performance for a specific size equipment or production process

To recapitulate, the main objective of process intensification is the development of cleaner, safer, smaller, and high energy-efficient processes. In algae production, technically, this implies the improvement or increase of the heat and/or mass transfer rates between the culture media and microorganism; also, it must be supplied a safety light source through a combination of different methods, processes, and equipment, without compromising the integrity of microorganism (Jacob et al., 2020).

Therefore, this chapter provides an overview of microalgae biotechnology. This chapter includes basic concepts related with microalgae and its cultivation systems; also, the main challenges in biomass production and improvements achieved by process intensification are discussed. Finally, the future trends in microalgae process intensification and use as feedstock in biorefineries are presented.

12.2 Microalgae

Microalgae are unicellular photosynthetic microorganisms that can grow in sea and fresh water. Microalgae species vary in intracellular composition, growing behavior, nutrient requirements, and other characteristics; and they can be taxonomically classified according to their metabolism, shape, color, and general morphology. Microalgae cells are mainly composed of three kinds of molecules, carbohydrates, proteins, and lipids, in different proportions depending on the species. Besides these macro components, different species of microalgae have adapted to a variety of harsh environments producing certain metabolites such as polyunsaturated fatty acids (PUFAs), antioxidants, released polysaccharides, and silicates, among others (Vuppaladadiyam et al., 2018).

Commercial microalgae cultures can generate different substances. Macro compounds such as lipids for biofuels, proteins for nutrition, and carbohydrates for fermentation are typically the main interest. Specific compounds can also be obtained from different species, PUFAs, amino acids, antioxidants, diatoms, etc. Table 12.1 shows some examples of species and the compounds of interest that they produce. Species with an industrial interest can be classified by the metabolites of interest. Microalgae production is a growing industry due to the variety of compounds produced by these microorganisms; however, there are still several obstacles to achieving a sustainable and economical production (Usher et al., 2014).

Table 12.1: Some species of microalgae and their significance for industry (Vieira et al., 2020).

Microalgae class	Species	Metabolite	Significance
Chlorophyceae	*Chlorella sp.*	Lipids, PUFAs	Biofuels, antioxidant
	Haematococcus pluvialis	Astaxanthin	Antioxidant
	Dunaliela salina	B-carotene	Antioxidant
Cyanophyceae	*Arthrospira platensis*	Proteins, c-phycocyanine	Nutrition, antioxidant
Porphyridiophyceae	*Porphyridium sp.*	Phycoerythrin, PUFAs	Imunomodulator, antioxidant
Bacillariophyceae	*Phaeodactylum tricornutum*	PUFAs, Fucoxanthin	Antioxidant, anti-inflammatory

Microalgae cultures are basic systems with fundamental components: a flow-like water medium, a light source, a carbon source, and mineral nutrients. Besides this fundamental composition, a specific culture generally requires particular conditions to obtain a determined metabolite. Fundamentally, microalgae require an aquatic medium to grow. This medium can be fresh water or sea water, depending on the species; in addition, nutrients such as nitrogen and phosphates are required, and frequently a moderately acid pH. Fresh water can be replaced with residual water from sources such as greenhouse drainage, as long as there are no toxic components. This alternative can also help consider microalgae culture as an option for bioremediation of certain industry wastes (Lu et al., 2020).

Moreover, since microalgae are photosynthetic, a light source is required. Sunlight is the preferred source for its renewable nature; nevertheless, additional light can improve biomass growth, and spectra-specific light can also increase specific compound production such as antioxidants. Therefore, the culture system is often designed to receive the maximum usable light, since photoperiod can also be relevant to the production of certain metabolites.

Another fundamental nutrient is a source of carbon, whose most typical source is atmospheric CO_2, due to the interest in using the microalgae production as a carbon capture technique, and its relatively cheap cost. Carbon dioxide can also be obtained from an industrial waste airflow such as cement production or yeast fermentations. The carbon dioxide source airflow can ingress to the medium through simple surface liquid-gas interchange or can be injected through pressure flow into the water medium to increase availability across the culture. Since oxygen is also produced by microalgae respiration, the culture requires adequate turbulence to allow the oxygen liberation. Other sources of carbon can be used, such as sugars, or glycerol, turning the culture into a mixotrophic system, which can improve the biomass growth. These alternative carbon sources can come from residual waters, but this usually carries the risk of contamination by other microorganisms.

Microalgae cultures also require a series of nutrients for optimal growth. Typical macronutrients for plant production are used for microalgae production. Nitrogen is fundamental for amino acids and protein synthesis, phosphates are involved in lipid production, and others such as potassium and sodium are needed for intra and extra cell molecule transportation. These nutrient requirements can be an economical obstacle for microalgae production. However, the use of wastewaters can reduce the dependency of added nutrients. Other biomolecules are sometimes needed to achieve optimal growing conditions such as amino acids or vitamins, plus some biocides can be required to control antagonist microorganisms.

Once the microalgae culture reaches a determined concentration and/or maturity, a harvesting process is required to process the biomass and obtain the compounds of interest. This process benefits from a flocculation stage promoted either by physical, chemical, or biological mechanisms. Once the biomass is separated from the water medium, microalgae cells require a cell wall rupture process, to allow the separation of all its intracellular contents. This process must not damage the different compounds and ideally be 100% effective. Once the different compounds are free from the cell wall, the actual compound can be obtained by an effective process to separate the different products without damaging the other parts.

Adequate growing conditions require the flow of the medium, in order to augment the exposition of the microalgae to the light, nutrients, and air. Cultures can be developed in raceway ponds, closed systems, hybrid systems, or other ways. Conditions can be controlled by automation or grown in acceptable weather. The whole process of obtaining different products from a microalgae culture requires the design of a multistage unit that is able to harvest the biomass, eliminate the excess of water, break the cell wall, separate the different compounds, purify the ones that can be used in its natural form, and even process the ones that can be transformed into a different final form. This unit, a bioreactor, requires a total adaptation for the cultures intended to be processed.

12.3 Conventional bioreactors

12.3.1 Bioreactors: an overview

A bioreactor can be defined as an equipment in which a set of biochemical reactions are carried out by living cells, which includes bacteria, fungi, protozoan, plant and mammalian cells, or by enzymes. An example of an important biochemical process performed within bioreactors is fermentation. Fermentation is an alternating metabolic pathway when oxidative phosphorylation cannot be performed in facultative microorganisms due to oxygen deficiency. There are many kinds of fermentations, classified by final products (such as ethanol, butyric acid, lactic acid, acetic acid, and propionic acid, among others) and undertaken by different microorganisms. It is important to highlight that although these biochemical processes share the name "fermentation,", the only feature in common is their anaerobic condition. Commonly, people used to call bioreactors as fermenters, even if fermentation does not take place within bioreactor. In this sense, industrial fermentation can include both aerobic and anaerobic large-scale cultivation of microorganisms; therefore, and due to industrial trends, the concepts of fermenter and bioreactor are the same, although just enzymes are involved (Dutta R., 2008).

Bioreactors are employed for manufacturing different biological products. Their main objective is to provide a controlled environment to ensure optimal growth and/or product formation in the cell system. Their yield, selectivity, and performance are a function of several variables, such as biomass concentration, sterile conditions, agitations, aeration, product removal, nutrient supply, product inhibition, microbial activities, shear conditions, heat removal / supply, presence of toxic agents, and special requirements such as light source (Najafpour G., et al., 2015).

On an industrial scale, bioreactors can be grouped into three sets: stirred and aerated ones, non-stirred and aerated ones, non-stirred and non-aerated ones, the last set being the most employed (Najafpour G., et al., 2015). The conventional bioreactors involved in biotechnological industry are (Najafpour G., et al., 2015):

- Batch bioreactor: As a batch reactor, it has neither inflow nor outflow of culture media, microorganisms, or product while the bioreaction is being carried out. If

the mixture is perfectly mixed, the rate of reaction throughout the reactor volume is the same. Since this is a volume-limited closed system, its resources are finite.
- Continuous stirred-tank bioreactor: Its main feature is the feed flow rate that is the same as product flow rate; thus, it provides a constant product without variation in concentrations across time if all process variables keep constant. Culture media is added by feed stream and a mixture of products, microorganisms, and culture media is obtained from the product stream. Due to the supplied agitation, the composition inside the bioreactor is the same as output stream composition.
- Airlift pressure cycle bioreactor: A pressurized air flow is added to the bioreactor to provide a constant oxygen transfer.
- Loop bioreactor: It is like an airlift bioreactor; however, a pump feeds the air and liquid through an internal/external loop in the reactor.
- Immobilized system: Enzymes or microorganisms are fixed on a solid surface for making a film where culture media or substrates will flow over it.
- Fluidized bed: Packed beds are used in an up-flow mode. The bed spreads out at elevated flow rates, and it is desirable to avoid the channeling and clogging of the bed. A typical application of this reactor type is in wastewater biological treatment.
- Trickle-bed: It is a different kind of packed bed. Fluid is sprinkled onto the top of the packing and trickles down through the bed. On the other hand, air is fed at the base, since liquid does not flow continuously throughout the column, while air moves easily around the wetted packing media. This type of bioreactor is widely used for aerobic wastewater treatment as well as nitrification and denitrification of wastewater.

12.3.1.1 Bioreactor for microalgae

Microalgae can be produced through several ways, which vary from laboratory methods, where all variables are controlled, to less restrained ones in outdoor tanks. Indoor culture allows control of nutrients, temperature, illumination, chemical or biological contamination and microalgae competition, while outdoor ponds do not; although the latter are cheaper, it is very hard to grow specific cultures for long periods in them.

Outdoor ponds

Several types of ponds are designed for microalgae cultivation; these ponds differ in size, shape, materials used for construction, and mixing device. Big outdoor ponds can be unlined, with a natural bottom, or lined, with inexpensive materials, such as clay, brick, or cement, or expensive plastics such as polyethylene, PVC sheets, glass fiber, or polyurethane. Unlined ponds exhibit silt suspension, percolation, and heavy contamination; indeed, their use is constrained to some algal species, as well as to particular soil and environmental conditions (Barsanti & Gualtieri 2005).

Natural bioreactors, such as lakes or small natural reservoirs, can be employed for microalgae production, since they offer adequate climatic conditions and sufficient nutrients. Some cases (Barsanti & Gualtieri 2005) are:

- *Arthrospira sp.* is cultivated as monoculture within temporary or permanent lakes along the northeast border of Lake Chad, and it is consumed by the Kanembou people who inhabit that place.
- *Arthrospira sp.* naturally blooms in old volcanic craters that contain alkaline waters in the Myanmar region. The production of these microalgae began in Twin Taung Lake in 1988, and by 1999, increased to 100 tons per year; 60% of this production is harvested from boats on the lake, and the rest is grown in outdoor ponds alongside the lake. During the blooming season, the cyanobacterium forms thick mats on the lake, so people in boats, collect them in buckets. *Arthrospira* is harvested on parallel inclined filters, washed with fresh water, dewatered, and pressed again. This paste is extruded into noodle-like filaments that are sun dried on transparent plastic sheets. Dried chips are moved to a pharmaceutical factory in Yangon and pasteurized, where they are pressed into tablets ready to be sold.
- *Aphanizomenon flos-aquae* are harvested from Upper Klamath Lake, Oregon, since they were used as food and health supplements. In 1998, the market for A. flos-aquae had a production close to 106 kg (dry weight). The harvested biomass is screened and centrifuged, where the algal concentrate is separated in cells and colonies, eliminating about 90% of the remaining water. Once concentrated, the product is cooled to 2 °C and stored in boxes, before they are shipped to the freezer facility for storage. The final product is converted into capsules or tablets.

In natural ponds, the environmental control is the least and there are no mixing requirements; due to this, natural ponds are used for extensive cultivation systems, for example:

- The biggest natural ponds used for commercial production of microalgae are *Dunaliella salina* lagoons in Australia. In this country, Western Biotechnology Ltd. operates 250 ha. of ponds (semi-intensive cultivation) at Hutt Lagoon.
- In South Australia, Betatene Ltd., a division of Henkel Co. (Germany), operates 460 ha. unmixed ponds (extensive cultivation) at Whyalla for the production of biomass for β-carotene.
- In Hawaii and Earthrise farms in California, raceway culture ponds are operated by Cyanotech Co. for the production of *Haematococcus* and *Artrosphira* biomass. The utilized raceway ponds have sizes from 1,000 to 5,000 m^2, where mixing is performed by one large paddle wheel per pond.
- Raceway ponds are also employed for intensive cultivation of *D. salina* by Nature Beta Technologies Ltd. in Israel.

The nutrient medium for outdoor cultures is based on agricultural-grade fertilizers instead of laboratory-grade reagents. In addition, natural blooms are kept at a reasonable cell density throughout the year, and the ponds are flushed with oceanic water whenever required. Culture depths are typically 0.25–1 m. Cultures from indoor production may serve as inoculum for monospecific cultures.

Algal production in outdoor ponds is relatively inexpensive; however, it cannot be maintained for long periods of time. Indeed, this kind of production is only suitable for a few fast-growing species due to contamination problems. Moreover, outdoor production is often characterized by poor consistency between production batches as well as unpredictable culture crashes caused by changes in weather, sunlight, or water quality. At present, large-scale commercial production of microalgae biomass is limited to *Dunaliella, Haematococcus, Arthrospira,* and *Chlorella,* which are cultivated in open ponds at farms located in Australia, Israel, Hawaii, Mexico, and China. These algae are a renewable source for carotenoids, pigments, proteins, and vitamins, which can be used for the production of nutraceuticals, pharmaceuticals, animal feed additives, and cosmetics. Mass algal cultures in outdoor ponds are implemented in Taiwanese shrimp hatcheries, where *Skeletonema costatum* is produced successfully in rectangular outdoor concrete ponds of 10–40 tons of water with depth of 1.5–2 m (Barsanti & Gualtieri, 2005).

Photobioreactors (PBRs)

PBRs are an alternative to open ponds for the production of microalgae biomass. A PBR is a closed system where there is no direct exchange of gases or contaminants between the contained algal culture and the atmosphere. These devices provide a protected environment for cultivated species, which is also relatively safe from contamination; moreover, in a PBR, the operating conditions can be better controlled. The operation costs are also reduced in PBRs, since they avoid evaporation, reduce water use, and lower CO_2 losses due to outgassing; also, they allow higher cell concentration, which increases productivity. Nevertheless, this type of cultivation system is more expensive to construct and operate than ponds, due to the need for cooling, strict control of oxygen accumulation, and biofouling; hence, their use is limited to the production of very high-value compounds from algae, which cannot be cultivated in open ponds. Different configurations of PBRs exist (González and Muñoz, 2017); among them, the main ones are:

– Tubular PBRs: They are the most popular design of closed systems that are used on a large scale for microalgae cultivation (Torzillo & Zittelli, 2015). These systems are usually fabricated of transparent glass or plastic tubes of 0.1 m diameter; inside them, the culture is moved by pumps or air streams (airlift). Tubular PBRs have surface-volume ratios up to 80 m^2/m^3, which allow having high biomass concentration cultures. The dimensions of the tubes are designed to avoid O_2 accumulation and to reduce PBR head loss. In this type of cultivation system, temperature control is imperative since considerable amounts of heat can be absorbed by the culture. For this, PBR cooling is performed by spraying water on the tube surface, shading (e.g., by overlapping of tubes), and immersion of the photostage in a water bath with temperature control and heat exchangers (González and Muñoz 2017).

The tubular reactor is divided in two sections and each one must be accurately designed: (1) photostage loop and (2) mixing (retention) tank (González and Muñoz, 2017). The photostage loop is the main section of the reactor, where photosynthesis and biomass growth occurs; on the other hand, the mixing tank allows removal of oxygen and the control of cultivation variables. In the photostage loop, the tube diameter is determined by the irradiance on the reactor surface and photosynthetic efficiency of the strain used, the latter ranging from 0.03–0.09 m. After diameter is calculated, the total length of the loop (L) is estimated in order to prevent inhibition of dissolved oxygen concentrations evolved as a function of photosynthesis. Moreover, the circulation of liquid along the tubes can be performed with mechanical or airlift systems. The required power is mainly used to overpass the head loss by friction inside the tubes. Regardless of the selected method for culture circulation, cell damage must be avoided; this aspect highlights the importance of an adequate selection of the pumping device (centrifugal, peristaltic, airlift, etc.) (González and Muñoz, 2017).

In general, tubular PBRs can be subdivided into three categories: (1) serpentine, (2) manifold, and (3) helical. Serpentine and manifold PBRs can have a horizontal, vertical, inclined, or conical arrangement (Zittelli et al., 2013).

Serpentine reactors, the oldest closed systems developed (Burlew, 1953), consist of straight tubes connected by U-bends to form a flat loop (the photostage), which can be arranged either vertically or horizontally. Gas exchange and nutrient addition are normally carried out in a separate vessel, and culture circulation (at flow rates between 20 and 30 cm/s) is achieved through a pump or an airlift.

In manifold PBRs, a series of parallel tubes are plugged in at the extremes by two manifolds, one for distribution and the other for collection of the culture suspension. The main advantages of manifold systems over serpentine loop reactors are the decrement of head losses and lower oxygen concentrations; these factors are key to making easier scale-up to industrial size bioreactors.

On the other hand, helical PBRs consist of small-diameter flexible tubes coiled around an upright supporting structure. The main advantage of helical-type systems is that they allow the use of relatively long tubes on a small land area, as compared to other PBR types. The cleaning problems and hydrodynamic stress are still not easy to solve, the main aspects that influence the complexity of these issues being the tube diameter, flow rate, and microalgae species. To the best of our knowledge, no commercial plant of this design has been operated to date.

It is important to mention that tubular PBRs are used to produce biomass of elevated quality for the production of high-value applications, mainly related to human consumption and the production of sensible strains.

– Flat-plate PBRs: Flat-plate reactors are built of a transparent material for optimal use of solar radiation. They are usually integrated by narrow panels to reach high area-to-volume ratios (usually from 16–80 m^2/m^3) as well as high volumetric biomass productivities (sometimes above 2 g/L//d). The basic design is integrated by two parallel panels with a thin-layer of microalgae suspension flowing in between. Between the two transparent panels there is a separation of a few centimeters, which allows efficient light transfer; these panels are usually fabricated of PVC, polycarbonate, polymethyl methacrylate, glass, or polyethylene. The main advantage of this design is the widespread illumination surface with respect to the volume of required culture medium for biomass production. The flat chamber is the simplest configuration (basic flat-plate design), but alveolar reactors are also proposed where flat-panels (sheets) are partitioned into a series of internal rectangular channels (namely alveoli). Alveolar reactors have been proposed for the higher structural rigidity, more efficient culture flow, increased versatility, and commercial options availability with lower building costs (González and Muñoz 2017).

Mixing can be achieved through pump-assisted circulation; in this case, the culture is circulated from an open gas exchange unit (open headspace) for improved gas transfer and better oxygen clearance. This open zone can compromise sterility; nevertheless, several parallel panels placed horizontally are proven to be efficient in overpassing the problem of oxygen buildup or by gas bubbling (González and Muñoz 2017). Indeed, air supply is the operation which contributes most to energy consumption in flat-plate PBRs, governing the mass transfer capacity at the same time.

Flat-plate PBRs can be collocated vertically or tilted at any angle (inclination) to the horizontal; this inclination allows optimization of solar energy capture. These reactors can even be oriented toward the sun, thus conceptually enabling better efficiency in terms of energy absorbed (González and Muñoz 2017).

– Thin-layer systems. This type of systems employs low-depth/thin-layer cultures, which enables augmenting biomass concentration and maximizing light utilization efficiency. In spite of the specific system design, the operation conditions – suitable biomass density, culture layer, cell movement patterns, mixing, and gas exchange – must be developed to optimize the use of high photon flux densities (PFDs) (González and Muñoz 2017).

The thin-layer systems are divided in two sections: (1) the surface where photosynthesis is performed, and (2) the retention tank where the culture is managed. In these systems, the culture depth is the main design variable, which is calculated using the relative roughness and slope of the surface used. Once the water depth is determined, the total length of the channel is calculated as in tubular PBRs, considering the maximum dissolved oxygen concentration admissible (<250%Sat.) (González and Muñoz 2017).

At the moment, different configurations of closed PBRs are evaluated for microalgae cultures, among which it are: (1) tubular systems (glass, plastic, bags), (2) flattened, plate-type systems, and (3) ultrathin immobilized configurations. Vertical arrangements of horizontal running tubes or plates are preferred due to their light distribution and appropriate flow. Table 12.2 shows a brief comparison between open ponds and closed algal cultivation devices.

In all types of bioreactors, the culture vessels must be nontoxic (chemically inert), transparent to light, easily cleaned and sterilized, and they need to provide a large surface-to-volume ratio. The most used materials include polycarbonate, polystyrene, and borosilicate glass (Barsanti & Gualtieri 2005).

Table 12.2: Advantages and disadvantages of open and closed algal cultivation devices (Pulz O., 2001).

Parameter	Open ponds (raceway ponds)	Closed systems (PBR systems)
Contamination risk	Extremely high	Low
Space required	High	Low
Water losses	Extremely high	Almost none
CO_2 losses	High	Almost none
Biomass quality	Not susceptible	Susceptible
Variability as to cultivatable species	Not given, cultivation possibilities are restricted to a few algal varieties	High, nearly all microalgae varieties may be cultivated
Flexibility of production	Change of production between the possible varieties nearly impossible	Change of production without any problems
Reproducibility of production parameters	Not given, dependent on exterior conditions	Possible within certain tolerances
Process control	Not given	Given
Standardization	Not possible	Possible
Weather dependence	Absolute, production impossible during rain	Insignificant, because closed configurations allow production also during bad weather.
Period until net production is reached after start or interruptions	Long, approx. 6–8 weeks	Relatively short, approx. 2–4 weeks
Biomass concentration during production	Low, approx. 0.1–0.2 g/l	High, approx. 2–8 g/l
Efficiency of treatment processes	Low, time-consuming, large volume flows due to low concentrations	High, short-time, relatively small volume flows

The use of vessels for algal cultivation dates back to the late 1940s, due to the research on the principles of photosynthesis realized with *Chlorella*. From the first vertical tubular reactors set up in the 1950s for the cultivation of *Chlorella* under artificial light and sunlight, several types of PBRs have been designed and tested. Most of these equipment are small-scale systems, in which experimentation was conducted mainly indoors, and only few of them have been scaled up. Significantly higher photosynthetic efficiencies and a higher degree of system reliability have been reached in recent years, due to the progress in understanding of the growth dynamic and requirements of microalgae under mass cultivation conditions. Nevertheless, there are only few examples of PBR technology that have reached the market (Barsanti & Gualtieri 2005). Two of the largest commercial systems in operation at present are:

- The Klötze plant in Germany for the production of *Chlorella* biomass. This plant consists of compact and vertically arranged horizontal running glass tubes with a total length of 500,000 m and a total PBR volume of 700 m^3.
- The Algatechnologies plant in Israel for the production of *Haematococcus* biomass.

Both plants use tubular, pump-mixed, single phase PBRs. Other industrial plants actually operating are:

- A plant built in Maui, Hawaii (USA) by Micro- Gaia Ltd. (now BioReal, Inc. a subsidiary of Fuji Chemical Industry Co., Ltd.), which is based on a rather complex design called BioDomeTM for the cultivation of *Haematococcus.*
- The rigid plastic tubes PBR of AAPS (Addavita Ltd., UK).
- The flexible plastic tubes PBR of the Mera Growth Module (Mera Pharmaceuticals, Inc., USA).

Biotechnological challenges for microalgae culture through photobioreactors

In open ponds, natural life conditions of microalgae can be present; for example, maximum cell densities of 10^3 cells/ml, average distance between cells of 1,350 μm (250 times the cell diameter), vertical or horizontal displacements of 5×10^{-3} to 3×10^{-5} m/s, adequate PFD, light supply available to daytime rhythm, CO_2 and nutrient conditions, prolonged stability of pH value, ion concentrations, and temperature. On the other hand, for PBRs the latter parameters are cell densities up to 10^8 cells/ml, average distance between cells reduced to 60 μm (10 times the cell diameter), spatial displacements ranging from 0.3–1.2 m/s, turbulence-conditioned PFD variations with frequencies of 0.1–1,000 s, optimum nutrient and CO_2 supply, pH values, and temperatures undergoing completely non-physiological variations, and continuous mechanical stress on the cell walls and the cells themselves (Pulz, 2001). Based on these differences, it is possible to know the challenges.

It is common that light source becomes the main challenge in photobiotechnology due to photoautotrophic life requirements. As regards photosynthesis, its rate is directly proportional to light intensity as long as the light compensation point is overcome; however, there is a light saturation point, which if surpassed, illumination intensity will damage the photosynthetic receptor system in few minutes (photoinhibition). Photosynthesis is saturated at about 30% of the total terrestrial solar radiation, i.e., 1,700–2,000 $\mu E/m^2/s$.

Several submersed light sources in illuminated stirred fermenters drive an average productivity in the range of 100–1,000 mg DW/day, as long as the surface-to-volume ratio of 2–8 m^2/m^3 for this illumination design has been reached. In tubular or plate-type PBRs, surface-to-volume ratios of 20–80 m^2/m^3 and light incidence values (PAR) up to 1,150 $\mu E/(m^2\ s)$ are achieved. At a layer thickness of up to 5 mm, PBRs could have a productivity of 2–5 g DW/day.

Many laboratory bioreactors are illuminated stirred fermenters that have been used to carry out physiological studies, mainly for testing and establishment miniature ecosystems and they have been successful if the surface-to-volume ratio is greater than 2 m^2/m^3. Unfortunately, several of them represent an important challenge to be scaled up due to surface-to-volume ratio decreases.

The surface-to-volume ratio is a very important light supply criterion for determination of maximum productivity of microalgae. If this parameter is too small and cell density is high, it is possible to create highly illuminated or dark zones where neither microalgae are photo-inhibited nor can the light compensation point be reached (Kroumov et al., 2013).

As it is known, oxygen is one among different products obtained from photosynthesis. Unfortunately, a high oxygen concentration inside the culture can become toxic for microalgae because:

- Oxygen reacts with photons from light source to generate superoxide ion, which can damage the cell membrane, and therefore cause cell disruption.
- Excess oxygen reduces CO_2 mass transfer from media to chloroplasts, so the carboxylating enzyme, Rubisco, cannot fix it in Calvin cycle. Thus, photosynthesis is inhibited.

In high-cell-density cultures of microalgae with optimum growth, species-specific O_2 evolution rates between 28 and 120 mg $O_2/(g\ DW\ h^{-1})$ have been reached. Thus, the challenge in PBR design is to achieve an efficient CO_2/O_2 balance. Sometimes, the procedures used to eliminate O_2 are: increasing the turbulence, where is important to not increase cell stress through shear forces, O_2 stripping with air, injecting pure CO_2, and although the methodology is still in experimental phase, by using membranes suitable for gas exchange. Another aspect to take care is temperature. The higher it is, the less CO_2 is solved in media, so the efficiency of photosynthesis decreases.

Photoautotrophic organisms live in natural environment at a density of 10^3 cells/ml and distances of more than 1,000 µm between cells. Thus, in high-cell-density microalgae cultures of up to 10^9 cells/ml, natural conditions are not suitable for high productivity; thus, the turbulence represents a challenge for PBR operation.

12.4 Intensified bioreactors for microalgae production

Process intensification (PI) was originally developed to make a drastic reduction in the volume of chemical plants (ideally between 100- and 1,000-fold) by replacement of the conventional unit operations with novel and very compact designs, often by combining two or more operations into one hybrid unit (Boodhoo and Harvey, 2013). In this sense, one of the first established definitions of PI was: "*PI consists in the physical miniaturization of process equipment while retaining throughput and performance*" (Cross and Ramshaw, 1986).

Over the last two decades, this basic definition of PI has evolved into a more complete one: "*PI consists in the development of innovative apparatus and techniques that offer drastic improvements in chemical manufacturing and processing, substantially decreasing equipment volume, energy consumption, or waste formation, and ultimately leading to cheaper, safer, sustainable technologies*" (Stankiewicz and Moulijn, 2000).

The reduction in the plant scale achieved by PI has many advantages in chemical engineering operations. First of all, the reduced path lengths of the diffusion/conduction interfaces enable a decrease in the mass and heat transfer resistances; this, coupled with a more intense fluid dynamic in the enhanced equipment, allows chemical reactions to proceed at their inherent rates. In this sense, the increase in mixing rates produced by the low reaction volumes maximizes both conversion and selectivity. In addition, PI promotes a reduction in the residence times of different processes, which brings beneficial consequences for energy consumption and process safety (Boodhoo and Harvey, 2013).

PI includes a wide range of processing equipment types and methodologies based on the use of multifunctional reactors, hybrid separations, and alternative sources of energy. Many of the equipment classified as intensified technologies, such as structured packed columns, compact heat exchangers, and static mixers have long been implemented in the chemical industry (Boodhoo and Harvey, 2013). In this sense, the implementation of some PI strategies on PBRs for the production of microalgae has been proposed as a promising alternative to improve the performance of these systems.

12.4.1 Intensified PBRs for the production of microalgae

The most common methods for microalgae production can be classified into open cultivation systems (open ponds, tanks, and raceway ponds), and controlled closed cultivation systems (closed PBRs). Open cultivation systems offer some interesting advantages, such as minimal capital and operating costs, as well as a lower energy requirement for culture mixing. Nevertheless, open systems often require large areas to scale up and they are sensitive to contamination and adverse weather conditions. On the other hand, closed PBRs are more efficient in terms of quality, since these systems can be operated at controlled conditions; therefore, they can overcome the drawbacks of open cultivation systems (Narala et al., 2016). Even though closed PBRs have some disadvantages including bio-fouling, overheating, cleaning issues, and high build-up of dissolved oxygen resulting in growth limitation, most of the research on the intensified production of microalgae has been focused on these systems, especially on finding solutions to the aforementioned problems; this, in order to provide more efficient PBRs designs.

12.4.2 Membrane PBRs

A typical membrane PBR is a continuously-operated system that combines an enclosed PBR with a submerged or side-stream membrane filtration system using micro- or ultrafiltration membranes for solid–liquid separation (Luo et al., 2017). In this sense, some studies have focused on membrane PBRs for the intensified production of microalgae in wastewater (Table 12.3). As it can be seen, one of the main advantages of the use of membrane modules on PBRs is the increase in the concentration of microalgae biomass in comparison with that in conventional PBRs. In addition, membrane modules have some other interesting advantages in the performance of PBRs, such as a significant increase in the fixation of CO_2 in microalgae and a reduction in the concentration of dissolved oxygen (Cheng et al., 2006), as well as a more efficient removal of N, P, and unionized ammonia in wastewater (Chang et al., 2016; Gao et al., 2016). Despite the aforementioned advantages, membrane modules in PBRs exhibit some important limitations that must be resolved. These limitations include fouling and pressure resistance, which can reduce mass transfer rates during the operation of these systems; the high cost of some specific membranes as well as their limited permeability are other factors that may limit the use of membrane modules in PBRs.

12.4.3 Use of internal structures in PBRs

It has been reported that internal structures can change the mixing conditions in PBRs; therefore, a very common method to improve the performance of these systems consists of setting or optimizing such structures (Degen et al., 2001; Zhang et al., 2013). Internal structures, such as baffles and static mixers can enhance the mixing in PBRs, thus allowing the existence of light/dark cycles, which can be beneficial for microalgae growth (Ugwu et al., 2002).

In this case, several studies have proposed the use of different baffles designs in flat-plate PBRs, including inclined baffles (Zhang et al., 2013), ladder-type crossed baffles (Zhao et al., 2018), and paddlewheels (Xu et al., 2020a). The use of helical baffles in tubular PBRs for intensification in the production of microalgae has also been re-

Table 12.3: Main aspects of the reviewed works on membrane PBRs for the intensified production of microalgae.

Technology	Microalgae/ culture medium	Advantages	Limitations	Reference
A cylindrical glass PBR with an external hollow fiber membrane module.	*Chlorella vulgaris* Filtered seawater with added minerals	The CO_2 fixation capacity in the system was enhanced from 80–260 mg/L/h, the retention time of the gas bubbles increased from 2 s to more than 20 s, and the dissolved oxygen dropped by a factor of 30.	Membrane modules must resolve fouling and pressure resistance to ensure high mass transfer rates for extended periods of operation.	Cheng et al. (2006)
Two hollow concentric cylinders of polymethyl methacrylate: the smaller cylinder being the wastewater chamber and the space between cylinders being the microalgae chamber. Cation and anion exchange membranes adhered onto four holes evenly spaced on the wall of the small cylinder.	*Chlorella vulgaris* Wastewater	Microalgae biomass concentrations were significantly improved from 2.34, 2.15, and 0 g/L to 4.24, 3.13 and 2.04 g/L. Removal efficiencies of N and P in wastewater were also greatly improved.	Ion-exchange-membranes (IEMs) can be expensive. In addition, some organic matters in wastewater can hardly permeate through the IEMs in the proposed intensified technology.	Chang et al. (2016)
A cylindrical PBR with a hollow-fiber membrane module submerged in the middle of it.	*Chlorella vulgaris, Scenedesmus obliquus* Aquaculture wastewater	An increase of 5.8-fold in the average volumetric biomass productivity. An average reduction in N and P of 86.1% and 82.7%, respectively. Effective removal of unionized ammonia.	Preferably should be applied only to diluted wastewaters.	Gao et al. (2016)

cently proposed (Xu et al., 2020b) (Table 12.4). As shown in Table 12.4, the presence of baffles in flat-plate and tubular PBRs tend to promote a significant increase in the microalgae biomass, referred to as biomass productivity (~25.2–29.94%), biomass concentration (62.3%), and biomass accumulation (40.8%), compared to PBRs without baffles. It is important mentioning that even though baffles can enhance the performance of PBRs in terms of the production of microalgae biomass, more studies on some important aspects are required for the development of these systems on a commercial scale. Such aspects include mixing, mass transfer, and the effects of shear stress generated by some internal structures on microalgae cells.

Table 12.4: Main aspects of the reviewed works on the use of internal structures in PBRs for the intensified production of microalgae.

Technology	Microalgae/ culture medium	Advantages	Limitations	References
A flat-plate PBR with inclined baffles attached perpendicularly to each horizontal surface. Each baffle is set with an angle of inclination with respect to the flow direction.	*Chlorella sp.* BG11 medium	An increase of 29.94% in the maximum biomass productivity compared to that in a PBR without baffles. The baffles cause an enhancement of the mixing effects.	Being an outdoor technology, the effects of mixing on growth rate are significant only on sunny days, given the high light intensity in these days.	Zhang et al. (2013)
A thin-film flat-plate PBR consisting of 16 shunt-wound channels with inclined baffles attached perpendicularly to the film surface. Each baffle is set with an angle of inclination with respect to the flow direction. A pilot-scale PBR system with 48 thin-film PBRs (as described above).	*Chlorella sp.* *Scenedesmus dimorphus.* BG11 medium	The biomass productivity of *Chlorella sp.* in the system was 25.2% higher than that in a PBR without baffles. The maximum biomass productivity of *Chlorella sp.* and *Scenedesmus dimorphus* was 11.0 and 21.9 g/m^2/day, respectively, in the pilot-scale system. The maximum cell concentration of *Chlorella sp.* and *Scenedesmus dimorphus* was 1.83 and 2.23 g/L, respectively, in the pilot-scale system.	More research on the processing technology of the photobioreactor and the factors influencing microalgae productivity is required for the scaling of the system.	Yan et al. (2016)
A flat-plate PBR with three different designs of a built-in horizontal ladder-type cross baffle.	*Chlorella vulgaris* BG11 medium	The growth rate of *Chlorella vulgaris* reached 0.783 g/L/d with a total biomass of 7.21 g/L within six days, which are higher than those of a PBR without baffles.	More research on mixing and mass transfer is required for the development of the proposed system on a commercial scale.	Zhao et al. (2018)

Table 12.4 (continued)

Technology	Microalgae/ culture medium	Advantages	Limitations	References
A flat-plate PBR with two paddlewheels placed inside of it, each one consisting of a central spindle and four plexiglass blades.	*Chlorella* PY-ZU1 Culture medium not specified	The microalgae biomass in the proposed system was 1.543 g/L, which was 62.3% higher than that in the PBR without paddlewheels.	More research regarding the effects of shear stress generated by the paddlewheels on microalgae cells is required.	Xu et al. (2020a)
A horizontal tubular PBR with a helical baffle and a central hollow tube situated in the CO_2 dissolver.	*Chlorella* PY-ZU1 An optimized SE culture medium	An increase in microalgae biomass accumulation of 40.8% with 15% CO_2 in the proposed system compared to that in a PBR without a helical baffle.	More research on mixing and mass transfer is required for the development of the proposed system on a commercial scale.	Xu et al. (2020b)

12.4.4 PBRs with nonconventional designs

One of the most common drawbacks in some designs of PBRs is the low efficient use of light intensity and an insufficient synergistic interaction between the flow field and light field. This, in turn, can affect the performance of PBRs for microalgae cultivation in terms of biomass productivity (Zhao et al., 2019). Therefore, some studies have recently focused on the effects of the geometrical shape and structure of PBRs on the spatial distribution of light and the production of microalgae biomass (Table 12.5). In this sense, it has been established that the surface area-to-volume ratio (SVR) is one key parameter in the geometric design of PBRs, as it determines the light intensity captured per unit volume and plays an important role in the photosynthetic growth of microalgae (Posten, 2009).

An interesting PBR design with a high SVR ratio has been recently proposed by Zhao et al. (2019). In this case, the authors developed a fractal tree-like PBR and used for the production of *Chlorella vulgaris*. It was found that the proposed PBR significantly increases both the CO_2 capture efficiency and the growth rate of microalgae biomass. In a recent study, the authors analyzed the effect of gas bubble size on the performance of their proposed PBR (Zhao et al., 2021). It has been reported that micro bubbles tend to enhance gas-liquid mixing by generating a more homogeneous dispersed flow; this, in turn, increases the microalgae growth rate. As indicated by the authors, the pumping cost of the proposed fractal tree-like PBR can be high. Therefore, further studies focused on strategies for energy reduction in the proposed system are required.

Table 12.5: Main aspects of the reviewed works on PBRs with nonconventional designs for the intensified production of microalgae.

Technology	Microalgae/ culture medium	Advantages	Limitations	Reference
A fractal tree-like PBR with a high surface area to volume (SVR) ratio.	*Chlorella vulgaris* A modified BG11 medium	The high SVR in the proposed system promotes an improvement in the CO_2 capture efficiency and the growth rate of biomass, compared with those in a multitubular PBR and a serpentine PBR.	The proposed system has the potential problem of high pumping cost, so more research on the reduction of energy consumption is required.	Zhao et al. (2019)
A fractal tree-like PBR with a high surface area-to-volume ratio (SVR) with two different gas distributors to generate micro bubbles (MBs) and large bubbles (LBs).	*Chlorella vulgaris* A BG11 culture medium	MBs generate a homogeneous dispersed flow pattern in the proposed system, which intensifies residence time, gas–liquid contact area, and multiphase mixing. The average specific growth rate in the proposed system is higher than that in conventional PBRs (flat-plate, tubular and column).	More research on the reduction of energy consumption due to pumping is required.	Zhao et al. (2021)

12.4.5 PBRs with a microbubbles generation system

As mentioned before, microbubbles (MBs) have some interesting beneficial effects on the performance of PBRs. In this sense, some studies have focused on the development and implementation of systems for the generation of MBs in conventional PBRs (Table 12.6). Cheng et al. (2019) proposed a plate PBR with four symmetric nozzles for the generation of a swirling flow of MBs. In this case, the authors observed a considerable decrement in the average bubble size with a subsequent increase in the gas-liquid mass transfer coefficient. This, in turn, caused an increase of 49.4% in the dry weight of microalgae biomass compared to that in a conventional PBR. The positive effect of MBs on microalgae production in PBRs has also been observed by Barahoei et al. (2020). In this case, when the authors used a gas sparger that generates MBs, they reported an increase of 34% in the dry weight of microalgae biomass, in comparison with the experiments when they used a sparger that generates large bubbles (LBs). The biomass productivity observed in the PBR with MBs was also higher compared to that observed in the PBR with LBs. Despite the benefits of MBs in the aforementioned PBRs, it is important to mention that there is need for more research on the effects of microbubbles on mixing, mass transfer, biomass growth rate, and biomass productivity for PBRs on a commercial scale.

Table 12.6: Main aspects of the reviewed works on PBRs with a microbubbles generation system for the intensified production of microalgae.

Technology	Microalgae/ culture medium	Advantages	Limitations	Reference
A plate PBR with four centrally symmetric nozzles. Each jetflow coming from the nozzles is tangent to a tangential circle that drove the surrounding microalgae solution in a vertical vortex movement.	*Chlorella* PY-ZU1 An modified SE culture medium	An increase in the biomass dry weight by 49.4% compared to that in a traditional airlift plate PBR. The average bubble diameter decreased by 80.2% and the mass transfer coefficient increased 4.6 times to 48.9/h.	More research on the enhancement of mixing and mass transfer caused by the vortex flow field of the proposed system on a commercial scale is required.	Cheng et al. (2019)
A bubble column PBR with two air spargers with different aperture size.	*Chlorella vulgaris* A BG11 culture medium	The biomass dry weight for the small bubbles at a CO_2 concentration of 7% was nearly 34% higher than that obtained with the LBs. The biomass productivity in the case of small bubbles was 48, 68, 58, 59, and 56% greater than that obtained with LBs at a CO_2 concentration of 3, 5, 7, 9, and 11%, respectively.	More research on the effects of bubble size on mass transfer, biomass growth rate and biomass productivity for the proposed system on a commercial scale is required.	Barahoei et al. (2020)

12.4.6 PBRs with an oscillatory/ultrasound system

As it has been mentioned, gas-liquid mixing is a key parameter that positively influences microalgae production in PBRs. In this sense, mixing can be achieved by aeration, mechanical agitation, or a combination of these. Nevertheless, some microalgae cells are sensitive to the hydrodynamic stresses produced by the aforementioned mixing mechanisms, which leads to a decrease in the biomass growth rate (Suh and Lee, 2003). In this sense, some studies have proposed the use of nonconventional systems for the gas–liquid mixing in PBRs as an alternative to traditional mixing mechanisms, Table 12.7.

Abbott et al. (2015) have proposed the use of a tubular PBR with periodically spaced orifice baffles for the intensified production of microalgae. In this case, orifice baffles generate the oscillatory motion of the fluid system thus promoting an enhanced gas-liquid mixing. The authors reported an increment of 95% in the maximum biomass growth rate compared to that in control cultures in T-flasks. Despite these results, there is need for studies on the proposed PBR system in order to achieve exponential growth and higher cell concentrations, as suggested by the authors. In a different study, the incorporation of a resonant ultrasound-field system to a PBR has been proposed (Lee and Li, 2017). In this case, the authors reported an increase in the volumetric biomass productivity of 2.6-fold with respect to that in a PBR without an ultrasound system. Given that the experiments on the proposed PBR system have been carried out on a laboratory scale, more research on the ultrasound effects on the productivity of microalgae biomass should be done in a system on a commercial scale.

12.5 Future trends

Microalgae cultures require a sustainable source of water, light, and nutrients. To maintain adequate growing conditions, a source of energy is required, and ideally this source needs to be sustainable on its own. The subsequent processes to obtain different products need to be scalable and safe for all compounds of interest. A low-cost raceway is exposed to weather and low production, while a more productive tubular system has high cost and high energy consumption. Secondary metabolites such as antioxidants have the highest market value, but in order to achieve sustainability, all possible compounds must be used – protein and carbs for nutrition, lipids and carbs for energy production, fermentation, etc.

Based on the advantages and limitations presented for each one of the intensified technologies for the production of microalgae, it can be noticed that more research on the performance of most of these PBRs on a commercial scale is required. In this sense, some intensified PBRs exhibit additional important limitations that should be overcome to make this possible; for instance, the high cost of membrane modules and

Table 12.7: Main aspects of the reviewed works on PBRs with an oscillatory/ultrasound system for the intensified production of microalgae.

Technology	Microalgae/ culture medium	Advantages	Limitations	Reference
A tubular PBR with a piston located at one end to generate oscillatory motion of internal material through periodically spaced orifice plates (baffles).	*Chlamydomonas reinhardtii* CCAP 11/32C Sueoka high salt medium	An increase of 95% in the average maximum growth rate compared to control cultures in T-flasks.	More research is required to achieve exponential growth and higher cell concentrations for the efficient mixing and enhanced mass transfer provided by the proposed system.	Abbott et al. (2015)
A dynamic PBR with an incorporated resonant ultrasound field system.	*Nannochloropsis oculata* Walne medium	An increased volumetric productivity in terms of the yields of biomass, which exhibited a 2.6-fold increase with respect to that in a system without ultrasound treatment.	More research on the ultrasound effects on biomass productivity for the proposed system on a commercial scale is required.	Lee and Li (2017)

reduced permeability in membrane reactors, high pumping costs in fractal PBRs, and potential high energy costs in PBRs with an ultrasound system. Taking these points into account, internal structures in PBRs represent one of the most promising strategies for the intensified cultivation of microalgae on a commercial scale. In this case, internal structures do not represent additional energy costs, and they can be fabricated in a great variety of sizes and geometric shapes. In addition, the performance of different designs of baffles and static mixers could be tested by using computational techniques, such as Computational Fluid Dynamics (CFD), in order to analyze their behavior in high volume PBRs. In this sense, internal structures with a gas distribution system, such as nozzles or orifices, could enhance the gas-liquid mixing in the PBRs through the generation of microbubbles.

12.6 Concluding remarks

As it was aforementioned, the third and fourth generations of raw materials (being specific microalgae) have big potential to make a positive impact on environmental issues and human quality life through biorefinery schemes. Even though process intensification has supported the development of new technologies along all processes (algae cultivation by using photobioreactors, pretreatments, and downstream processes) and overcome the cited challenges, there is need to join multidisciplinary efforts.

In order to improve algal technology and to decrease process costs for achieving viable and sustainable processes, the support of different science fields is necessary, such as: genetic engineering by modifying genetically microorganisms and making them more resistant to aggressive environmental condition, mechanical engineering by designing more electrically efficient pumps and static mixers, material engineering by developing cheaper material for membranes without lose effectiveness, and chemical engineering by proposing new intensified and controllable operations; finally, all of them to maximize shapes and configurations to supply, in an efficient way, light sources.

References

Abbott, M. S., Brain, C. M., Harvey, A. P., Morrison, M. I., & Perez, G. V. (2015). Liquid culture of microalgae in a photobioreactor (PBR) based on oscillatory baffled reactor (OBR) technology – A feasibility study. *Chemical Engineering Science*, *138*, 315–323.

Barahoei, M., Hatamipour, M. S., & Afsharzadeh, S. (2020). CO2 capturing by chlorella vulgaris in a bubble column photo-bioreactor; Effect of bubble size on CO2 removal and growth rate. *Journal of CO2 Utilization*, *37*, 9–19.

Barsanti, L., & Gualtieri, P. (2005). Algae: Anatomy, Biochemistry, and Biotechnology. CRC Press, Boca Raton.

Boodhoo, K., & Harvey, A. (2013). Process Intensification Technologies for Green Chemistry: Engineering Solutions for Sustainable Chemical Processing. John Wiley & Sons, UK.

Burlew, J. S. (ed.) Algal Culture from Laboratory to Pilot Plant. (Carnegie Institution of Washington, Publication 600) Washington, 1953.

Cross, W., & Ramshaw, C. (1986). Process intensification: Laminar flow heat transfer. *Chemical Engineering Research and Design*, *64*(4), 293–301.

Chang, H.-X., Fu, Q., Huang, Y., Xia, A., Liao, Q., Zhu, X., Zheng, Y.-P., & Sun, C.-H. (2016). An annular photobioreactor with ion-exchange-membrane for non-touch microalgae cultivation with wastewater. *Bioresource Technology*, *219*, 668–676.

Cheng, J., Lai, X., Ye, Q., Guo, W., Xu, J., Ren, W., & Zhou, J. (2019). A novel jet-aerated tangential swirling-flow plate photobioreactor generates microbubbles that enhance mass transfer and improve microalgal growth. *Bioresource Technology*, *288*, 121531.

Cheng, L., Zhang, L., Chen, H., & Gao, C. (2006). Carbon dioxide removal from air by microalgae cultured in a membrane-photobioreactor. *Separation and Purification Technology*, *50*(3), 324–329.

Cherubini, F. (2010). The biorefinery concept: Using biomass instead of oil for producing energy and chemicals. *Energy Conversion and Management*, *51*(7), 1412–1421.

Degen, J., Uebele, A., Retze, A., Schmid-Staiger, U., & Trösch, W. (2001). A novel airlift photobioreactor with baffles for improved light utilization through the flashing light effect. *Journal of Biotechnology*, *92*(2), 89–94.

Dragone, G., Fernandes, B. D., Vicente, A. A., & Teixeira, J. A. (2010). Third generation biofuels from microalgae, https://repositorium.sdum.uminho.pt/handle/1822/16807?mode=full.

Dutta, R. (2008). Fundamentals of Biochemical Engineering (No. 660.63 D8F8). Ane Books India, New Delhi.

Gao, F., Li, C., Yang, Z.-H., Zeng, G.-M., Feng, L.-J., Liu, J.-Z., Liu, M., & Cai, H.-W. (2016). Continuous microalgae cultivation in aquaculture wastewater by a membrane photobioreactor for biomass production and nutrients removal. Ecological Engineering, *92*, 55–61.

Ghasem D. N. (2015). Biochemical Engineering and Biotechnology (Second Edition), Elsevier, ISBN 9780444633576, DOI https://doi.org/10.1016/C2013-0-09819-2

Gómez-Castro, F. I., & Segovia-Hernández, J. G. (Eds.) (2019). Process Intensification: Design Methodologies. Berlin, Boston: De Gruyter, 2019. https://doi.org/10.1515/9783110596120.

Gonzalez-Fernandez, C., & Muñoz, R. (2017). Microalgae-based Biofuels and Bioproducts. Woodhead Publishing (Elsevier), Kindlington, United Kingdom.

Jacob-Lopes, E., Maroneze, M. M., Queiroz, M. I., & Zepka, L. Q. (Eds.) (2020). Handbook of Microalgae-based Processes and Products: Fundamentals and Advances in Energy, Food, Feed, Fertilizer, and Bioactive Compounds. Academic Press, London, UK, San Diego, US, Cambridge, US, Oxford, UK.

Kroumov, A., Gacheva, G., Iliev, I., Alexandrov, S., Pilarski, P., & Petkov, G. (2013). Analysis of sf/v ratio of photobioreactors linked with algal physiology. *Genetics and Plant Physiology*, *3*(1–2), 55–64.

Lee, Y.-H., & Li, P.-H. (2017). Using resonant ultrasound field-incorporated dynamic photobioreactor system to enhance medium replacement process for concentrated microalgae cultivation in continuous mode. Chemical Engineering Research and Design, *118*, 112–120.

Lu Z., L., Loftus, S., Sha, J., Wang, W., Park, M. S., Zhang, X., Johnson, Z. I., & Hu, Q. (2020). Water reuse for sustainable microalgae cultivation: Current knowledge and future directions. *Resources, Conservation and Recycling*, *161*, 104975. https://doi.org/10.1016/j.resconrec.2020.104975

Luo, Y., Le-Clech, P., & Henderson, R. K. (2017). Simultaneous microalgae cultivation and wastewater treatment in submerged membrane photobioreactors: A review. *Algal Research*, *24*, 425–437.

Narala, R. R., Garg, S., Sharma, K. K., Thomas-Hall, S. R., Deme, M., Li, Y., & Schenk, P. M. (2016). Comparison of microalgae cultivation in photobioreactor, open raceway pond, and a two-stage hybrid system. *Frontiers in Energy Research*, *4*, 29.

Posten, C. (2009). Design principles of photo-bioreactors for cultivation of microalgae. *Engineering in Life Sciences*, *9*(3), 165–177.

Pulz, O. (2001). Photobioreactors: Production systems for phototrophic microorganisms. *Applied Microbiology and Biotechnology*, *57*(3), 287–293.

Sikarwar, V. S., Zhao, M., Fennell, P. S., Shah, N., & Anthony, E. J. (2017). Progress in biofuel production from gasification. *Progress in Energy and Combustion Science*, *61*, 189–248.

Stankiewicz, A. I., & Moulijn, J. A. (2000). Process intensification: Transforming chemical engineering. *Chemical Engineering Progress*, *96*(1), 22–34.

Suh, I. S., & Lee, C.-G. (2003). Photobioreactor engineering: Design and performance. *Biotechnology and Bioprocess Engineering*, *8*(6), 313–321.

Tan, I. S., Lam, M. K., Foo, H. C. Y., Lim, S., & Lee, K. T. (2020). Advances of macroalgae biomass for the third generation of bioethanol production. *Chinese Journal of Chemical Engineering*, *28*(2), 502–517.

Torzillo, G., & Zittelli, G. C. (2015). Tubular Photobioreactors. In: Prokop, A., Bajpai, R. K., & Zappi, M. E. (Eds.), Algal Biorefineries. Products and Refinery Design, vol. 2. Springer, Switzerland, pp. 187–212.

Ugwu, C., Ogbonna, J., & Tanaka, H. (2002). Improvement of mass transfer characteristics and productivities of inclined tubular photobioreactors by installation of internal static mixers. *Applied Microbiology and Biotechnology*, *58*(5), 600–607.

Usher, P. K., Ross, A. B., Camargo-Valero, M. A., Tomlin, A. S., & Gale, W. F. (2014). An overview of the potential environmental impacts of large-scale microalgae cultivation. *Biofuels*, *5*(3), 331–349. 10.1080/17597269.2014.913925

Vieira, M. V., Pastrana, L. M., & Fuciños, P. (2020). Microalgae encapsulation systems for food, pharmaceutical and cosmetics applications. *Marine Drugs*, *18*, 644. https://doi.org/10.3390/md18120644

Vuppaladadiyam, A. K., Prinsen, P., Raheem, A., Luque, R., & Zhao, M. (2018). Microalgae cultivation and metabolites production: A comprehensive review. *Biofuels, Bioproducts and Biorefining*, *12*, 304–324. https://doi.org/10.1002/bbb.1864

Xu, J., Cheng, J., Lai, X., Zhang, X., Yang, W., Park, J.-Y., Kim, H., & Xu, L. (2020a). Enhancing microalgal biomass productivity with an optimized flow field generated by double paddlewheels in a flat plate photoreactor with CO2 aeration based on numerical simulation. *Bioresource Technology*, *314*, 123762.

Xu, J., Cheng, J., Xin, K., Xu, J., & Yang, W. (2020b). Developing a spiral-ascending CO2 dissolver to enhance CO2 mass transfer in a horizontal tubular photobioreactor for improved microalgal growth. *ACS Sustainable Chemistry and Engineering*, *8*(51), 18926–18935.

Yan, C., Zhang, Q., Xue, S., Sun, Z., Wu, X., Wang, Z., Lu, Y., & Cong, W. (2016). A novel low-cost thin-film flat plate photobioreactor for microalgae cultivation. *Biotechnology and Bioprocess Engineering*, *21*(1), 103–109.

Zhang, Q. H., Wu, X., Xue, S. Z., Wang, Z. H., Yan, C. H., & Cong, W. (2013). Hydrodynamic characteristics and microalgae cultivation in a novel flat-plate photobioreactor. *Biotechnology Progress*, *29*(1), 127–134.

Zhao, L., Peng, C., Zhang, J., & Tang, Z. (2021). Synergistic effect of microbubble flow and light fields on a bionic tree-like photobioreactor. *Chemical Engineering Science*, *229*, 116092.

Zhao, L., Tang, Z., Gu, Y., Shan, Y., & Tang, T. (2018). Investigate the cross-flow flat-plate photobioreactor for high-density culture of microalgae. *Asia-Pacific Journal of Chemical Engineering*, *13*(5), e2247.

Zhao, L., Zeng, G., Gu, Y., Tang, Z., Wang, G., Tang, T., Shan, Y., & Sun, Y. (2019). Nature inspired fractal tree-like photobioreactor via 3D printing for CO_2 capture by microalgae. *Chemical Engineering Science*, *193*, 6–14.

Zittelli, G. C., Biondi, N., Rodolfi, L., & Tredici, M. R. (2013). Photobioreactors for Mass Production of Microalgae. In: Richmond, A., & Qiang, H. (Eds.), Handbook of Microalgal Culture: Applied Phycology and Biotechnology. Wiley-Blackwell, Hoboken, NJ, pp. 225–266.

Priya Yadav, Parag R. Gogate*

Chapter 13 Process intensification approaches applied to the downstream processing of microalgae production

Abstract: Microalgae are unicellular organisms that carry out photosynthesis and are found in fresh as well as marine environment. Microalgae contribute to the production of various products such as biofuels, food supplements, cosmetics, and bio-based chemicals and also help in environmental cleaning by utilizing CO_2. High production costs and scale-up problems for microalgae-based products are key challenges in commercializing these products. To overcome these hurdles, process intensification approaches can be applied at various production steps. Process intensification (PI) is an important aspect to improve the processing and is looked upon with great interest. Process intensification, through the revolutionary development of new products and processes that ensure reduced material usage and energy consumption, and reduced environmental impacts while offering greater flexibility in the scale of operations, is the need of the hour. Broadly speaking, process intensification is the modification of the original equipment or replacement by new equipment, which enhances the process rate by improving mass and heat transfer rates. In the current chapter, various intensification approaches applied to downstream steps of microalgae-based products will be discussed. Initially, a brief description about microalgae will be provided, followed by their cultivation, harvesting techniques, and extraction of the products, mainly focusing on the intensification aspects. Overall, this chapter is expected to critically analyze the various approaches and present the benefits due to the involvement of process intensification in various downstream steps of production.

Keywords: microalgae productivity, extraction, ultrasound, process intensification, biofuels

13.1 Introduction

Microalgae are unicellular organisms found in seawater, hyper saline habitat, and freshwater as well as in rocks and moist soils (Dittami et al., 2017). Microalgae are a phylogenetically diverse group, encircling different classes and phyla (Hayes et al.,

Corresponding author: Parag R. Gogate, Chemical Engineering Department, Institute of Chemical Technology, Matunga, Mumbai 400019, Maharashtra, India, e-mail: pr.gogate@ictmumbai.edu.in
Priya Yadav, Chemical Engineering Department, Institute of Chemical Technology, Matunga, Mumbai 400019, Maharashtra, India

https://doi.org/10.1515/9783110781267-013

2017). Due to their high photosynthetic efficiency and the presence of a variety of components (fats, proteins, nucleic acids, and carbohydrates) as compared to other crops, microalgae can be used for various commercial and industrial purposes. Microalgae offer various applications such as biofuel production (Lum et al., 2013), synthesis of high-value specialty molecules, formulation of food (Gouveia et al., 2006; Sousa et al., 2008), cosmetics (Wang et al., 2015), health products (Hamed et al., 2015), and fertilizers (Uysal et al., 2016) as represented in Figure 13.1. In addition to these applications, microalgae are also used to make the environment clean and safer by reducing greenhouse gases by performing carbon dioxide fixation from the atmosphere.

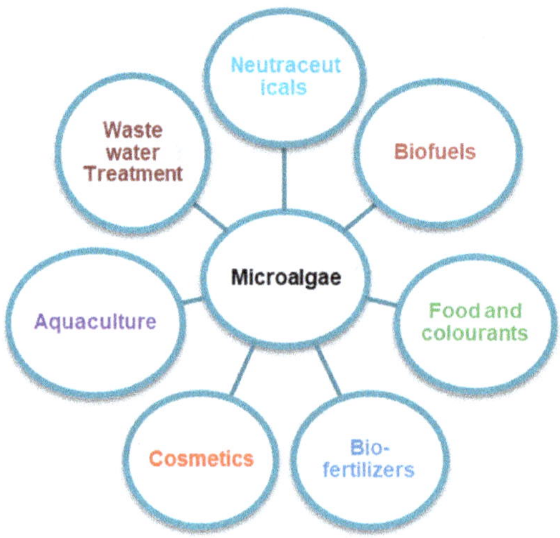

Figure 13.1: Applications of microalgae.

The composition of the different compounds in the microalgae may vary according to the microalgae species, depending on the habitat or surrounding conditions such as pH, nutrients, light intensity, and temperature. Microalgae can grow in marine as well as in fresh water. Freshwater microalgae grow on rocks, in the mud of river, and in streams. The growth of microalgae is found more in stagnant water as compared to flowing water. The problem associated with algae in fresh water is contamination of fresh water due to the massive algal growth. Cultivation of marine algae can boost economics as they can be grown in salt marshes, brackish water, and near coastal areas. However, marine algae cultivation imparts some adverse effects on marine flora, and there is a problem of premature rupturing of algal cells because of the high salinity of marine water and the necessity for marine water pretreatment, which negatively affects the economic and financial feasibility. So, it is very important to have a deeper insight before choosing the cultivation source and the production approach for microalgae cultivation.

Production of microalgae biomass provides several advantages over other biomass feedstocks and plants, such as less doubling time, high production rate and utilization of nonarable ground, coupled with feasible cultivation using waste or brackish water. Microalgae can survive in environments with limited nutrients and can also be more efficient in producing higher quantum of products under stress. A large amount of biologically active compounds in the form of pigments, vitamins, minerals, proteins, carbohydrates, and polyunsaturated fatty acids have been observed in the biomass of microalgae as shown in Figure 13.2 (Gouveia et al., 2006). Compounds obtained from microalgal biomass have been used for curing human health issues and may find application as nutraceuticals, based on their antioxidant, anti-inflammatory, antimicrobial, anti-obesity, and anticancer properties. The compounds can also be used as nutritious feed mainly in aquaculture to enhance the immune response of marine organisms.

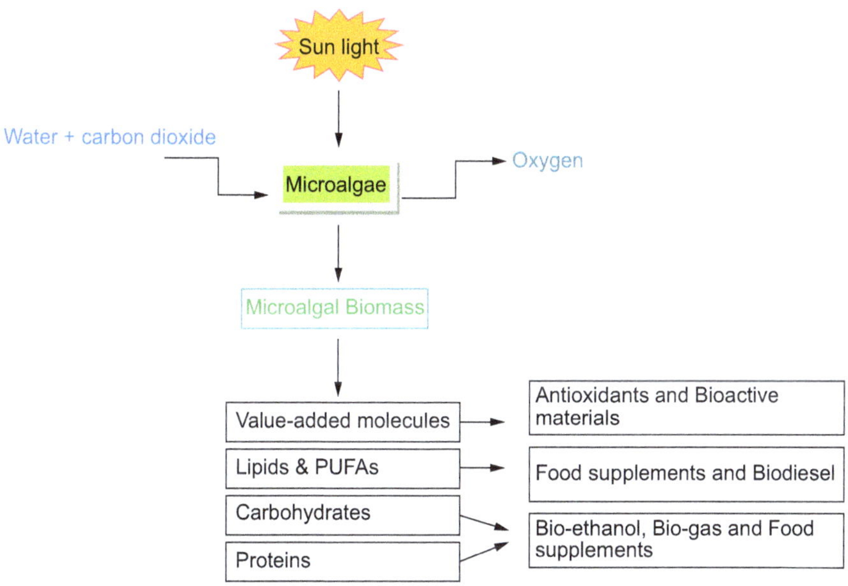

Figure 13.2: Compounds found in algal biomass and their applications.

The increasing trend of using microalgae-based products and their diverse range of applications force industries to take active efforts to manufacture microalgae-based products with excellent nutritional value, safety and texture. Despite promising research activities on microalgae-based products, the current products in the market are limited and costly. These products also face technological challenges in terms of high energy requirements, cost, and waste generated during the manufacturing process (Patil et al., 2020). These challenges can be overcome by applying process intensification approaches during the upstream and downstream stages of production. Process intensification (PI) is receiving increased attention and importance nowadays because of its potential to offer innovative and sustainable process design alternatives in different fields such as in chemical, biochemical, and bioprocess engineering. The main objectives of process intensification approaches are to build up safer, smaller, cleaner, and more energy-efficient processes (Vaghari et al., 2015). The current chapter provides information about the various intensification approaches that can be applied to downstream stages during the production of various microalgae-based products. In the initial section, a short description about the various PI strategies involved in microalgae synthesis steps such as cultivation, harvesting, and extraction are also described. Subsequently, the chapter reviews the different process intensification methods applied during the manufacturing of microalgae-based products.

13.2 Implementation of process intensification approaches in microalgae biomass production

13.2.1 Selection of microalgae strain

Considering the enormous potential of microalgae in different areas to produce value-added molecules, the significance of microalgae as cellular factories is increasing. However, due to the high cost of microalgae (requirement of fertilizers, fresh water, harvesting, and extraction cost) scale–up at the commercial level is challenging. In order to achieve efficient production of microalgae, it is necessary to cultivate the microalgae strain with a higher productivity of the desired constituents, which will facilitate considerable enhancement in subsequent product yield, with lower costs at different stages of production. The microalgae strain should have a high rate of photosynthesis, high rate of carbon dioxide consumption, and a high rate of growth (Patil et al., 2020). The strain should also be flexible and robust to tolerate different environmental stresses as given in Table 13.1. In order to get high productivity with cost-effective benefits, genetic manipulation of the strain can be performed. Genetic manipulation of microalgae could propose various advantages to the biotech industry and could be considered as an important tool for biotechnology and synthetic biology. The modified strain can provide better productivity with new cultivation and harvesting protocols, reduced energy input and consumption, and lower extraction cost (Mata et al., 2010). In recent times, significant research has been conducted on the genetic manipulation of microalgae to enhance their photosynthetic rate, carbon dioxide consumption, capability to grow in severe environment conditions, and also to achieve preferred buildup of desired compounds in large amounts (Wan et al., 2014). Most manipulation techniques involve a strategy such as adaptation that includes the application of stress conditions such as light or irradiations, nutrient limitations, pH, temperature, and others. The main goal for the development of super strain through genetic engineering is to enhance the production process profitability with an eco-friendly and greener approach.

Table 13.1: Different PI strategies applied for strain modifications.

PI approach	Microalgae strain	Improvement in performance
New strain **Indigenous microalgae strain isolated from the natural habitat (Mutanda et al., 2011)**	*Botryococcus braunii*	25% to 75% enhanced accumulation of oil content
Nutrient stress **Nitrogen stress (Chu, 2017)**	*Tetradesmus (Scenedesmus) obliquus*	Enhancement in lipid content from 26 to 43%
Nutrient stress **Phosphorous stress (Xin et al., 2010 algal book)**	Scenedesmus sp.	Enhancement in lipid content up to 53%
Nutrient stress **All element stress (Chu, 2017)**	*Parachlorella kessleri*	Enhancement in lipid content from 0% to 29%

13.2.2 Cultivation of microalgae

A very early example of microalgae cultivation dates back to the World War II, where microalgae were utilized as food supplements. After industrialization, microalgae cultivation was done for carbon dioxide abatement (Burlew 1953). In the 1970s, mass cultivation of microalgae was performed for obtaining healthy foods. The application of microalgae in wastewater treatment was reported in United States, where methane (source of energy) was produced from the recovered biomass (Burlew 1953). Apart from these initial applications, mass cultivation of microalgae also became important for aquaculture (Muller-Feuga, 2000), fine chemical production (Lorenz and Cysewski, 2000), and as animal and human supplements (Dallaire et al., 2007). Microalgae also offered new applications as immobilization systems in the production of extracellular compounds (Chetsumon et al., 1994), fixation of harmful gases (Chae et al., 2006), and heavy metal biosorption (Karthikeyan et al., 2007). Considering the importance of microalgae, the cultivation needs to be improved to also obtain good quality biomass with higher productivity. The control on growth and quality of microalgae is quite complex due to the requirement of optimum nutrient/substrate concentrations, varied environmental conditions, and the complex nature of photosynthesis. At the cultivation stage, information regarding growth requirements in relation to temperature, light, and nutrients is very important and proper analysis of these issues can lead to employing correct process intensification strategies for obtaining maximum biomass productivity.

Cultivation of microalgae can be done in open or closed systems. Lagoons and open ponds are considered as open systems. Different types of open-pond systems are circular ponds, raceway ponds, and unstirred ponds. Mixing in the open systems is carried out using rotating arms, impellers, and paddle wheels. Open systems have some advantages such as easy maintenance and construction, and low operating cost making them suitable for large-scale cultivation. Open systems have some drawbacks as well such as poor control on culture conditions, necessity of a larger space, prone to contamination with predators and microbes, water loss due to evaporation, and carbon dioxide loss into the environment as well as limited mass transfer rates. Closed systems offer some advantages over open systems. For example, closed photobioreactors have good control over the cultivation conditions, provide high biomass production, and can help in prevention from contamination as compared to open photobioreactors. Different types of closed systems are flat-plate, tubular, and vertical-column photobioreactors as listed in Table 13.2; the table also lists their advantages and limitations.

Table 13.2: Advantages and limitations of closed culture systems.

Cultivation systems	Advantages	Limitations
Flat-plate photobioreactors	High photosynthetic efficiency, large illumination surface, high productivity of algal biomass, suitable for algae immobilization, easy cleaning of reactor after cultivation, and relatively economical	Difficulty in scale-up due to requirements of various support materials and compartments, appearance of wall growth, and difficult to control temperature
Vertical-column photobioreactor	Good mass transfer and mixing, consumption of less energy, high scalability potential, and low photoinhibition	Less illumination surface and requirement of sophisticated materials for construction
Tubular photobioreactors	large illumination surface, good algal biomass productivity, and relatively economical	Appearance of wall growth, fouling, and requires large land area for cultivation

To improve biomass productivity, a good understanding of the hydrodynamic conditions and mass transfer is necessary. Most of the time, carbon dioxide is used as a nutrient. The introduction of carbon dioxide also helps in mixing to enhance biomass productivity. The use of PI approaches can further help in improving the mixing, say based on the use of ultrasound that can generate turbulence in the liquid. The main aim of the process intensification approach is to optimize the design and operation so that energy consumption can be reduced. A summary of the different process intensification strategies applied has been listed in Table 13.3; it also gives the observed performance improvements. It can be said that process intensification strategies mainly focus on improved gas exchange, i.e., removal of oxygen and supply of carbon dioxide, uniform supply of nutrients, optimization of the configuration of the reactor for best performance and easy scale-up, and also for achieving an efficient distribution of light within the photobioreactor.

Table 13.3: Intensification strategies applied to cultivation systems.

Microalgae species	Details of the process intensification technique / method	Process conditions / equipment details for intensification	Performance improvement	Reference
Chlorella vulgaris	Light emitting diode-based bubble column photobioreactors	Light emitting diode (LED) is used as a source of light.	Improvement in biomass productivity	(Fu et al., 2012)
Nannochloris atomus	Floating horizontal bioreactor	Combines the advantages of both open ponds and photobioreactors; system can be used in ground as well as in water; a pump is used to transfer liquid through silicone tubing into or out of the bioreactor	Indoor - high density cultivation of microalgae Outdoor - high density and also higher microalgae productivity, without contamination	(Dogaris et al., 2015)
Dunaliella salina	Airlift loop photobioreactor with the addition of fluidic oscillator system	Fluidic oscillator used to generate monodispersed, uniformly spaced and regular size microbubble cloud, which influences the mass transfer and mixing efficiencies positively	Prevention of dead zone with 100% survivability at pilot scale; high density of microalgae produced	(Zimmerman et al., 2011)

Table 13.3 (continued)

Microalgae species	Details of the process intensification technique / method	Process conditions / equipment details for intensification	Performance improvement	Reference
Chlorella vulgaris	Hybrid photobioreactor (stirred airlift photobioreactors)	Hybrid photobioreactor used to promote mass transfer, gas hold-up and medium homogenization; to prevent vortex formation, four baffles were used inside the draft tube	High productivity of carbohydrates and lipids at low rotational speed, which reduced the power consumption and made the process economically viable	(de Jesus and Filho et al., 2017)
Scenedesmus Species	Split internal-loop photobioreactor	This type of photobioreactor has four regions - riser, downcomer, upper, and bottom sections. Sparger was introduced in the riser section to introduce gas into the reactor. To provide light, eight cool fluorescent lamps were used	Increase in gas hold-up, cell movement, turbulent kinetic energy, and liquid velocity were observed; gas-liquid mixing was also improved	(Sabri et al., 2021)
***Tetraselmis* sp. KCTC12236BP**	Semipermeable membrane photobioreactor	Semipermeable membranes were used to transfer the dissolved nutrients in sea water to the packed bed reactors (PBRs)	High fatty acid production in all SPM-PBRs, but high biomass productivity only in the SPM-PBRs, which has high permeability of the membrane	(Kim et al., 2015)
***Tetraselmis* sp.**	Floating Photobioreactor with installed internal partitions	To improve mixing efficiency and mass transfer, ocean waves were used.	Enhancement in biomass and FAME productivity	(Kim et al., 2016)

Table 13.3 (continued)

Microalgae species	Details of the process intensification technique / method	Process conditions / equipment details for intensification	Performance improvement	Reference
Haematococcuspluvialis	Submerged light photobioreactor	Light emitting diode (LED) is used as a light source	Enhancement in light delivery by 5 times; reduction in dark zone	(Murray et al., 2017)
Haematococcuspluvialis	Flashing light photobioreactor	Photosynthesis efficiency increased by the activation of photosystems I and II	Enhancement in astaxanthin production by four times	(Abu-Ghosh et al., 2016)
Chlorellasorokiniana IAM-C212	Stirred tank photobioreactor	For improving mixing, a modified mechanical agitator was used; gas exchange by using improved sparger	Increase in biomass production	(Singh and Sharma, 2012)

13.2.3 Harvesting of microalgae

Microalgae harvesting is the process of detachment or separation of microalgae from the culture so as to translate it into valuable products. A harvesting process is ideal when it is effective for most microalgae strains and allows recovery of high biomass concentrations with moderate operation, low maintenance, and reduced energy costs (Danquah et al., 2009). Harvesting process is a sequential operation that includes three major steps - recovery of biomass, dewatering, and drying. The harvesting methods depend on the microalgae strain chosen, size and density of microalgae cell, scope for recyclability of nutrient medium, and final product specifications. Harvested biomass may be processed further to obtain the desired product; so the harvested process should not be toxic or inducing contamination. For microalgae harvesting, there are two different methodologies generally applied – a two-step process and a one-step process. In the two step process, diluted suspension of microalgae is thickened or concentrated to a slurry of about 2–7% of the total suspended solids (TSS), and then dewatered to 15–25% of the total suspended solids (Brennan et al., 2010). To overcome the disadvantages of the two-step method, the single step dynamic technique has been developed. In recent times, microalgae harvesting is done by biological, mechanical, electrical, and chemical based methods as summarized in Figure 13.3. Mechanical-based methods are mostly used for microalgae harvesting. Combining the mechanical methods with flocculation and coagulation makes the process more efficient in terms of efficiency of harvesting, maintenance, and operations cost. Techniques such as flocculation, coagulation under chemical methods, bio-flocculation, under biological methods, and flotation, and gravity sedimentation under physical methods are the improved techniques for harvesting that make microalgae more concentrated. A brief discussion about each technique is presented in the upcoming sections, while the advantages and disadvantages of each technique are summarized in Table 13.4 along with a specific recovery reported in the study, for a better comparison.

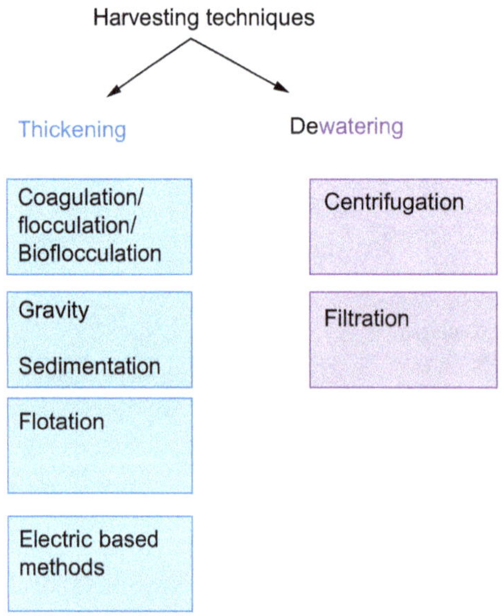

Figure 13.3: Microalgae harvesting techniques.

13.2.3.1 Coagulation/flocculation

Chemical coagulation/flocculation is an important approach that can lead to an economic optimization of the processes for harvesting of microalgae. Coagulation is based on pH adjustment and addition of electrolytes whereas flocculation involves the addition of cationic polymers. Microalgae cells have negative charge on their surface, their density is close to the nutrient medium, and in the dispersed stage; they keep the microalgae in natural sedimentation. Chemical flocculants are used to neutralize the negative charge present on the microalgae cells and allow agglomeration, leading to better separations. Flocculent use for harvesting can be organic/polyelectrolyte and inorganic. The harvesting approach based on flocculation typically concentrates the microalgae suspension 20–100 times (Vandamme et al., 2013). This type of harvesting technique can enhance the effective particle size before the dewatering step so that the energy requirement can be reduced. For a cost-effective harvesting, coagulation/flocculation is usually followed by gravity sedimentation (Smith and Davis, 2012).

13.2.3.2 Gravity sedimentation

Solids and liquids are separated from each other using gravitational forces. The concept of different densities of microalgae is the basis for separation using gravity sedimentation. This method of harvesting is well suited for applications related to synthesis of products like biofuels. Normally, coagulation/flocculation is used prior to gravity sedimentation to increase the settling of microalgae and also to reduce the energy demand (Al Hattab et al., 2015).

13.2.3.3 Flotation

Flotation is based on use of external gas bubbles or air to bring the suspended materials to the top of the liquid surfaces from where they can be collected using a skimming process (Singh et al., 2011). For microalgae species having low density and self-floating features, this harvesting technique can be relatively faster and more effective as compared to gravity sedimentation (Singh et al., 2011). Flotation has been used efficiently for both fresh and marine microalgae harvesting. For successful harvesting using flotation, the microalgae cell should have a hydrophobic nature and a high molecular weight. Based on the air or gas bubble size, flotation process can be classified into dissolved air flotation, for a size range between 10–100 µm, and dispersed air flotation for a size range of 700–500 µm (Patil et al., 2020). Ozone gas is also another flotation media used and has gained significant importance as compared to atmospheric air, as it can produce charged bubbles, which give a better separation efficiency and the use of ozone oxidizes the organic matter present in the liquid medium.

13.2.3.4 Electrical energy-based methods

Electrical energy-based intensification methods are highly versatile due to their applicability to a wide range of microalgae species and because of their eco-friendly nature. In this approach, cells from the culture broth are separated with the application of an electric field to the culture medium (Uduman et al., 2010). During this process, hydrogen bubbles formed by water electrolysis stick to the microalgae flocs and bring them to the surface. This process of separation can also be described as electro-coagulation-flotation. If the cells precipitate on the electrode, it is termed as electrophoresis, and if cells accumulate on the base of the container, it is termed as electro-flocculation. In this process, two types of electrodes can be used:

I) Sacrificial electrodes – In this type of electrode, metal ions are released into the culture broth, depending on the electricity passed through the electrolyte solution. The dissolution of the metal ions released from the anode causes the formation of a coagulant, which in turn destabilizes the algal suspension. After this, microalgae form aggregates when they come in contact with positively charge metal ions (Uduman et al., 2010).
II) Non-sacrificial electrodes – This process is based on the movement of negatively charged microalgae cells toward the anode. The microalgae cells, upon reaching the anode, lose their charge and form aggregates. This type is very much prone to fouling as compared to the chemical methods (Uduman et al., 2010).

Electricity-based processes are influenced by the material from which the electrodes are made. Iron and aluminum are widely used metals, with aluminum being more effective. For example, during a 45 min harvesting period, an electrode made up of aluminum demonstrated a removal efficiency of 100% while an electrode with iron showed only 78.9% removal efficiency; this is attributed to the fact that iron electrodes have very less current efficiency compared to aluminum electrodes. The current density influences the reaction kinetics and hence the harvesting efficiency. Typically, an increment in current density offers shorter harvesting periods. Implementation of mixing and settling after electro-flocculation reduces the energy consumption, though it also means a higher operation time with a low concentration slurry. Electro-flocculation, followed by flotation, offers a high concentrated slurry and low energy consumption within a short processing period. Scale up of this process is difficult and requires that all parameters like temperature, pH, and type of electrodes are in an optimum range, clearly necessitating a detailed study.

13.2.3.5 Centrifugation

Centrifugation is the fastest method to harvest microalgae. However, due to the high energy requirement, it is the most expensive method and thus its application is limited to only high-value products. In this process, the separation of microalgae cells is induced by a centrifugal force. There are different types of centrifuges applied, such as the basket centrifuge, decanter-perforated basket centrifuge, disk stack centrifuge, and imperforated hydro cyclones. The disc stack and decanter type centrifuge are most widely used for microalgae harvesting (Patil et al., 2020). Centrifugation provides a high rate of recovery and a chemical-free microalgae biomass. Slurry concentration may be increased with less energy requirement if coagulation/flocculation is applied prior to centrifugation. Centrifugation also has its disadvantages in that high gravitational force and shear forces may result in cell damage. In addition, for large-scale operations, treatment time, capital cost, maintenance, and energy consumption can be very high.

13.2.3.6 Filtration

In filtration, the algae culture medium is passed through filters of a specific pore size under pressure. Vacuum or gravity holds the microalgae to the membrane in a thick paste and let the water pass through the membranes pore. Filtration is generally performed to harvest large colonies that form microalgae. Harvesting of small-sized microalgae cannot be effectively performed with conventional filtration and hence membrane filtration is preferred. Filtration through membranes offers good quality in terms of the harvested biomass as compared to others due to the absence of chemical use and very less cell disruption. Filtration is classified on the basis of the pore size of the membrane as microfiltration, macrofiltration, ultrafiltration, and reverse osmosis. Microfiltration uses a pore size between 0.1 and 10 μm. It is preferred for fragile small microalgae harvesting and allows very high initial flux, but clogging takes place very easily. Macrofiltration uses pore size of 10 μm, ultrafiltration uses pore size between 0.005 and 0.1 μm and reverse osmosis uses pore size of 0.001 μm. Membrane filtration can be used to harvest microalgae with low density (Patil et al., 2020). An important problem with membrane-based separations is that membranes, especially those used in microfiltration and ultrafiltration, need to be replaced frequently and it makes the process more costly and also energy-intensive. The membrane-based harvesting technique is also not suitable for large-scale application due to very high pumping, replacement, and operational cost.

Application of high-gradient magnetic filtration is another promising harvesting technique for microalgae. Biosorption of submicron magnetic particles offers good results for harvesting of microalgae. These magnetic particles must be reusable, of low cost, offer high capacity of adsorption, and should be chemically stable.

A comparison of different processes allows us to understand that for volumes less than 2 m^3/d, the application of microfiltration can be cost-effective as compared to centrifugation. On the other hand, for volumes greater than 20 m^3/d, the application of centrifugation may be more cost-effective (Grima et al., 2003). Membrane filtration techniques are energy-intensive and account for approximately 20–30% of the total processing cost. To save energy consumption and processing cost, preconcentration operations, such as sedimentation, flotation, or flocculation, can be performed. As a specific case study, application of preconcentration using flocculation reduced energy consumption by 10 folds as compared to only centrifugation (AL Hattab et al., 2015).

Table 13.4: Comparison of different microalgae harvesting methods.

Harvesting techniques	Advantages	Disadvantages	Recovery (%)
Coagulation/ flocculation (Papazi et al., 2010)	Simple, fast, can be used for large scale needs, less energy requirement, applied to a wide range of microalgae species	Chemicals may be expensive, flocculants may be toxic to microalgae biomass, face difficulty in separation of the coagulate from the harvested biomass	60
Bioflocculation (Zhou et al., 2013)	Inexpensive, nontoxic to microalgae biomass, recycling of nutrient medium	Risk of microbiological contamination, changes in cellular composition	63
Gravity sedimentation (Harith et al., 2009)	Relatively simple and inexpensive method	Risk of microalgae biomass deterioration, time-consuming, requirement of high cell density, low final concentration	91
Flotation (Liu et al., 1999)	Suitable for large-scale operations, requires less space and operating time	Requires chemical flocculants, not feasible for marine microalgae, algae-specific	90
Centrifugation (Ahmad et al., 2014)	Feasible for almost all microalgae species, rapid method with high recovery efficiency	Requires high energy, feasible only for the recovery of high value-added products, risk of cell damage due to high shear, time consuming, expensive for large-scale applications	100

13.2.4 Recovery of microalgae-based products

After the dewatering of microalgae, there is still very low content of dry solid biomass, and to avoid processing problems, drying needs to be applied. There are various drying methods for microalgae cells, such as spray drying, solar drying, freeze drying, vacuum shelf drying, incinerator drying, cross-flow drying and convective drying, based on the requirement or application of the final product. The drying processes should be aimed largely at minimizing or avoiding deterioration of microalgae cells in terms of their biochemical properties. Among all the drying techniques, solar drying is the most suitable and affordable, especially for the production of biofuels. Drying of high-value micro-algal products can be achieved using spray drying so as to obtain a dark green color powder of dry microalgae (Oliveira et al., 2009). Products obtained using spray drying method retain higher nutrients while convective drying leads to a loss of 10–20% proteins. Some other drying technologies like roller/drum and fluidized-bed can also be used for getting dry biomass of microalgae with better properties. Roller/drum was indeed reported for *Scenedesmus sp.*, mainly for the sterilization of the samples and disruption of microalgae cell wall, simultaneously, with good maintenance of properties. Drying of microalgae *Dunaliella salina* using fluidized bed was also investigated and it was reported that the total loss of carotenoids was between 13 and 20%, which was lesser than that from the spray drying method of microalgae from *Spirulina sp.* (Leach et al., 1998). Thus, the final form and market choice for the desired end product and their related cost constraints definitely influence the choice of drying method.

After drying, the dried microalgae biomass is disrupted using various disruption methods. Disruption methods are applied based on the nature of the desired end product and the strength of cell wall of microalgae. Disruption can be by mechanical methods such as homogenizer, bead mill, ultrasound, autoclave, and spray drying as well as non-mechanical by using osmotic shock, treatment using enzyme, alkali, acid, or organic solvent. Ultrasound and microwave-assisted cell disruption have acquired significant attention due to their intensification benefits and economic feasibility. For example, Prabakaran and Ravindran (2011) reported that the application of sonication and microwave enhance the extent of cell disruption, in comparison to other techniques. Ultrasound is a sound wave that is pitched above the human hearing range of 20 Hz to 20 kHz. Ultrasound is passed through the liquid medium using transducers that convert electrical energy into vibrational sound energy. Ultrasonic transducers are available in a range of shapes, size, and frequencies to be selected, depending on the application. The effect of ultrasound is based on the cavitation phenomenon, which is described as the formation, growth, and sudden collapse of bubbles (cavities) with a release of energy, leading to conditions of high localized temperatures and pressure as well as liquid microcirculation and local turbulence. These cavitational effects results in the intensification of the physical and chemical changes. In the case of microalgae cell disruption, sonication disrupts the cell wall of the microalgae due to the generation of shear forces and microscale turbulence. Cell disruption using sonication offers advantages in terms of time reduction or higher yield. For example, in the study of (Wiltshire et al., 2000), it was reported that application of ultrasound on *Scenedesmus sp.* Biomass increased the recovery of pigment by 22 times compared to the control. Ultrasound also lowers the overall consumption of solvent and offers better diffusion of solvents in the cells. However, excess use of ultrasound may be detrimental to heat-sensitive components as the generation of local energy can deteriorate these components. Thus, controlling the temperature during sonication is required so as to maintain or improve the quality of the product. To cut down the overall cost and energy consumption, ultrasound can be integrated with other methods of cell disruption.

Another effective method of cell disruption is microwave. Microwave works by generating local heating as a result of frictional forces from intramolecular and intermolecular movements, which causes cell lysis and then the liberation of cell components into the liquid medium. Dvoretsky et al. (2017) reported that microwave-assisted cell disruption was effective at 2,450 MHz frequency. Gudhe et al. (2014) investigated the effect of the sonication and microwave on the disruption of *Scenedesmus sp.* Cell wall for the recovery of lipid. According to the results, lipid yield was more in microwave-assisted disruption as compared to sonication-assisted disruption. Rakesh et al. (2015) studied the effect of microwave irradiation on the recovery of lipid yield and reported that by using microwave, there was an enhancement in the lipid yield by 2 factors as compared to the control.

A summary of the different process intensification strategies for recovery has been given in Table 13.5, which establishes that trends are specific to the species of microalgae, mainly attributed to the differences in the cell wall characteristics and the effects of inducing the irradiations.

Table 13.5: Comparison of different intensified cell disruption methods for product recovery.

Method applied	Microalgae species	Efficient method	Lipid extraction (%)	References
Ultrasound **Autoclave** **Microwave** **Osmotic shock**	*Botryococcus sp.*	Microwave	28.6	(Lee et al., 2010)
Ultrasound **Autoclave** **Microwave** **Osmotic shock**	*Chlorella sp.*	Ultrasound	20.1	(Prabakaran and Ravindran, 2011)
French press-homogenization **Ultrasound**	*Schizochytrium*	Ultrasound	34.5	(Yel et al., 2017)
Water bath **Microwave**	*Scenedesmus obliquus*	Microwave	77	(Balasubramanian et al., 2011)
Microwave **Grinding** **ultrasound**	*Scenedesmus sp.*	Ultrasound	90.8	(Kim et al., 2017)

13.3 Implementation of process intensification approaches in the extraction of microalgae-based products

13.3.1 Bioethanol

In recent years, there is a spike in energy consumption due to the rapid increase of the world population and expansion of the emerging economies. The dominant source of energy generation is fossil fuels, which are not greener. To reduce the consequences of using fossil fuels in energy generation, several countries are making efforts in terms of using environment-friendly sources for energy generation. One economic and environment friendly source is lignocellulosic biomass, which is used to produce biofuels such as bioethanol (John et al., 2011). At present, bioethanol is produced from starch crops such as corn and sugarcane as well as from lignocellulosic materials such as waste papers, saw dust, switch grass, rice straw etc. High content of lignin in the biomass of these materials hinders the process, requiring multiple steps and hence the cost of bioethanol production is much higher. Production of biofuels using microalgae has gained much attention due to their high content of carbohydrate in terms of cellulose and starch (Nigam and Singh, 2011; Brennan and Owende, 2010). The advantage of using microalgae for the production of bioethanol is their fast growth and their ability to fix carbon dioxide at a much higher rate as compared to terrestrial plants and presence the of carbohydrates in the form of cellulose and starch (absence of lignin), which makes the saccharification process easier as compared to using lignocellulosic biomass as substrate. The other advantage of using microalgae for the production of bioethanol is their capacity to store starch within their cells and their ability to extract starch to get reducing sugars by disrupting the cells. Starch extraction is done in solvent or water after disrupting the microalgae cells. For hydrolysis, there are several methods available such as alkaline hydrolysis, acid hydrolysis, and enzymatic hydrolysis. Hydrolysis using acid is easier, cheaper, and faster but may generate unwanted substances by decomposing the reducing sugars, which may hinder the fermentation process (Gírio et al., 2010). Enzymatic hydrolysis is quite slow and expensive as compared to acid hydrolysis, but offers a greener approach and higher reducing sugar yield without producing inhibitory compounds during fermentation. Onay (2019) reported the use of a microalgae strain named *Hindakia H. tetrachotoma ME03*, pretreated with acid and followed by enzymatic hydrolysis, which produced bioethanol, with a yield of 11.2 g/L. Guo et al. (2013) compared the efficacy of enzymatic hydrolysis and acid hydrolysis on the microalgae strain named *S. abundans*. Hydrolysis with enzymes (β-glucosidase/cellulose) resulted in 92% saccharification, similar to acid hydrolysis (3% v/v), but finally, acid hydrolysis resulted in 52% higher bioethanol. In another study, a comparison was made between the pretreatment methods (enzymatic, alkaline, and acid) in terms of their reducing sugar production using a culture of mixed microalgae. The results showed that there was an improvement in the reducing sugar production when a combination of magnesium sulphate and dilute sulfuric acid was used as compared to hydrolysis with only enzyme and sulfuric acid (Shokrkar et al., 2017). It is also important to note that the production of bioethanol can also be enhanced by using genetic engineering methods.

13.3.2 Biogas/biohydrogen

Microalgae are composed of carbohydrate, proteins, and lipids, along with phosphorous and nitrogen as well as micronutrients such as zinc, iron, and cobalt. $CO_{0.48}H_{1.83}N_{0.11}P_{0.01}$ represents the chemical composition of microalgae (Grobbelaar, 2004). The compounds present in the microalgae could be transformed into biogas/biohydrogen using anaerobic digester. Biogas production using microalgae requires the same design of the anaerobic digester reactor as required by other biomasses. In the anaerobic digester (AD), organic materials are degraded by microalgae using oxygen-deprived conditions for biogas production. An ideal AD reactor must protect from corrosive gases, UV light, water, chemicals, and water. Reactors for AD can differ in terms of their size, design, mode of operation, temperature, construction materials, and total solid load. Suitable reactors can be selected by taking some aspects into consideration, such as feasibility in terms of maintenance, availability, and cost-effectiveness (Kuria and Maringa, 2008). There are various type of AD reactors used for biogas production, such as plastic biogas reactor, fixed-dome reactor, textile bioreactor, and floating-drum reactor (Patinvoh et al., 2017). Among these AD reactors, textile bioreactors are novel for biogas production. The economic viability and technical feasibility of textile bioreactors were analyzed by Patinvoh et al. (2017) using manure bedded with straw. Microalgae does not contain lignin in their structure, thus diminishing the recalcitrance index of algal biomass as compared to lignocellulosic biomass. However, the presence of pectin and cellulose in the cell wall of microalgae make it difficult for hydrolysis, by resisting the accessibility of the enzymes. In addition to these, there is the presence of polyphenol in brown algae and aliphatic biopolymers in green algae that hamper the hydrolysis of the biomass due to lower activity of alginate lyase. So, biodegradability of biomass is also a serious issue in terms of getting high conversion efficiency and biogas yield. Recalcitrance of biomass can be overcome by pretreating the biomass to reduce the crystallinity index and particle size, and also to break the cell wall of the microalgae. Microalgae pretreatment methods include thermal (steam explosion and hydrothermal), mechanical (microwave, ultrasound, and shaking), biological (enzymatic, fungi, and bacteria) and chemical (alkali and acid) as reported by Passos et al. (2018). Each of the applied methods for pretreatment can yield a different degree of improvement in the final product. For example, Passos and Ferrer (2015) studied the effect of thermal pretreatment on the production of biogas and reported a marginal enhancement in biogas generation at 75–95 °C with a pretreatment time of 10 h. On the other hand, there was a significant increase in biogas production (by 108%) at 120 °C with the pretreatment time of 2 h. Some new emerging technologies (microwave- and ultrasound-assisted intensification) have also been reported for the production of biogas as these techniques produce the desired output in a lesser pretreatment time as well as through lesser energy. Thermal and mechanical pretreatment methods are found to be most effective as compared to other pretreatment methods in terms of biodegradability of the microbial biomass and biogas production. A combination of different methods can also be effective. For example, pretreatment of microbial biomass with acid/alkali was found to be more effective when applied in combination with thermal pretreatment (Mendez et al., 2013). Bohutskyi et al. (2014) investigated the effect of pretreatment methods (single and in combination) on five microalgae strains. According to the results, no single pretreatment approach was most effective. According to the study, thermal and mechanical pretreatment were found to enhance biogas yield but their energy balance was not acceptable. On the contrary, energy balance for alkali pretreatment was satisfactory but biogas yield was not as effective as with mechanical and thermal pretreatment (Cho et al., 2013). When a combined pretreatment of thermo-alkali was introduced, there was an enhancement in the methane yield by 30–40% at manageable energy requirements.

13.3.3 Biodiesel

Microalgae are known to be accumulators of lipids in large quantity and hence considered as renewable feedstock for the production of biodiesel. Metabolites (carbohydrate, lipid, and protein) present in microalgae play a vital role in the production of biofuels. The lipid content of microalgae has acquired significant attention, especially for biodiesel production. The chemical identity of biodiesel is fatty acid methyl ester and it can be synthesized by performing acid catalyst-assisted esterification of triglycerides present in lipids, followed by transesterification in the basic environment. The production of biodiesel in an economical manner requires certain major optimization steps that include strain selection for cultivation, providing and maintaining appropriate conditions at the time of cultivation for maximum growth, and harvesting of algal biomass, followed by extraction of lipid from the biomass and transesterification of the lipid into biodiesel. A summary of the intensification approaches applied and the obtained benefits during the different steps has been presented in Table 13.6.

To achieve a high production of biodiesel, the very first step of microalgae strain selection must be handled wisely. The microalgae strain should contain a high lipid concentration. Generally, there is a 20–50% lipid content in a majority of the microalgae species. Some examples of microalgae strain, with commercial applications, having high growth rate and lipid content are *Ankistrodesmus, Tetradesmus (Scenedesmus) obliquus, and C. vulgaris.* Selection of microalgae strain could depend on the desired environment conditions such as tolerance to high salinity or temperature. After selecting the microalgae with potentially higher lipid content, further enhancement in microalgae lipid could be obtained by manipulating the environmental conditions. Controlling the exposure of light at the time of cultivation can provide enhancement in lipid content. A suitable exposure to light may be achieved by applying new process intensification strategies such as using paints, light emitting diodes, and dyes to get the desired light wavelength and intensity. The photosynthetic efficiency is controlled by the light intensity that affects the growth of microalgae during cultivation (Stockenreiter et al., 2013). Maximum growth of microalgae is obtained at the saturated light intensity and starts declining as the light intensity moves away from saturation (Sorokin and Krauss, 1958). Temperature is another crucial factor that affects their survival, growth, apart from the biochemical composition of the microalgae, including the lipid content and type. Microalgae can survive in a temperature range of 15 to 40 °C. In the majority of the microalgae, unsaturation of fatty acid enhances as the temperature decreases and on the other hand, saturation of fatty acid enhances as the temperature increases. Another strategy for enhancing the lipid content is by managing the nutrient stress. The growth and lipid content of microalgae is affected by the availability of nutrients in the culture medium. Nutrient stress conditions in growth medium declines the rate of cell division, although biosynthesis of fatty acid is not affected under adequate availability of carbon dioxide and light (Thompson, 1996). Being a crucial micronutrient, nitrogen affects the lipid content and growth of the microalgae. Limitation of nitrogen at the time of cultivation allows the accumulation of lipid in the cellular content of the microalgae (Table 13.6). Nitrogen stress reduces the cellular fraction of the thylakoid membrane, induces activation of acyl hydrolase, and stimulates phospholipid hydrolysis, which in turn enhances the intracellular proportion of the fatty acid acyl-CoA. Another important nutrient is phosphorus, which affects the respiration, energy transfer, cellular metabolic pathway mediated by signal transduction, and photosynthesis. The limitation of phosphorous impairs cell division, which leads to the arrest of cell division and enhances the synthesis of phospholipids. The effect of accumulation of high lipid content under phosphorus stress in various species of microalgae is given in Table 13.6. The biochemical and lipid content of microalgae can also be enhanced by metabolic and genetic engineering. With a modified microalgae strain, one can enhance the production of recombinant proteins, photosynthetic activities, and metabolism. After the selection of microalgae strain, the most important aspect is selection of suitable cultivation system. Different reactors for the cultivation of microalgae are discussed in Section 13.2.2. After cultivation, the harvesting of microalgae biomass is done using different methods. The most commonly used method for biomass harvesting is flocculation from the cultivation medium in which algae starts aggregating, thus enhancing the efficiency of the processes like sedimentation, centrifugation, and filtration. After harvesting the biomass, lipid extraction is the next step and this extraction step accounts for 40–60% of the total production cost. Microalgae oil can be extracted after performing cell disruption or solvent extraction, followed by dewatering and drying. After the extraction of lipids, transesterification of the extracted lipids is performed to obtain biodiesel. Biodiesel production can be performed by two methods; one is the two-stage approach, which includes drying of the biomass (Guldhe et al. 2014), extraction of the lipid, and transesterification, followed by purification steps. Another approach is direct transesterification, in which extraction of the lipid and transesterification take place concurrently. In a study, both approaches for transesterification were investigated for the Microalgae strain *S. limacinum.* According to results, 57% biodiesel and 66.37% FAME were obtained from the two-stage approach and on the other hand, from direct transesterification, 66% crude biodiesel % and 66.46% FAME were obtained (Johnson and Wen, 2009).

Table 13.6: Process intensification applied at different steps for biodiesel production from microalage.

Process intensification Strategy	Strain	Performance enhancement	References
Strain selection Indigenous strain chosen from the natural habitat	*Botryococcus braunii*	25–75% enhanced accumulation of oil content	(Mutanda et al., 2011)
Nutrient stress Phosphorous stress	*Tetradesmus obliquu*	30% increase in lipid content	(Chu et al., 2017)
Nutrient stress Nitrogen stress	*Nannochloropsis salina*	Enhancement in lipid content, up to 56.1% of dry weight	(Fakhry and El Maghraby, 2015)
Nutrient stress Nitrogen stress	*Nannochloropsis oculata*	Enhancement in lipid content, up to 49.7% of dry weight	(Millán-Oropeza et al., 2015)
Genetic engineering Alteration in gene and lipid metabolism	*Chlorella sorokiniana*	Improvement in lipid content by 51%	(Wan et al.,2014)
Cultivation system Hybrid system (combination of open pond and PBR)	*Arthrospira platensis*	lipid content improved from 19% to 36.6% of dry cell weight	(Chu et al., 2017)
Cultivation system Hybrid photobioreactor (stirred airlift photobioreactors)	*Chlorella vulgaris*	High productivity of carbohydrates and lipids at low rotational speed, which reduced the power consumption, making it economically viable	(de Jesus and Filho et al., 2017)
Harvesting Flocculation by magnetic particles	*Chlorella ellipsoidea*	Recovery of 95% cells	(Wang et al., 2014)
Harvesting Autoflocculation	*Chaetoceros calcitrans*	Recovery of 98% cells	(Harith et al., 2009)

Table 13.6 (continued)

Process intensification Strategy	Strain	Performance enhancement	References
Extraction and transesterification Cell disruption by hydrodynamic cavitation and lipid extraction	*Nannochloropsis salina*	Improvement in lipid yield (25.9–99%)	(Lee and Han, 2015)
Extraction and transesterification Cell disruption by Hydrodynamic cavitation and lipid extraction	*Nannochloropsis sp.*	Improvement in lipid yield (5.4–26.9%)	(Seo et al., 2016)
Extraction Ultrasound	*Chlorella vulgaris*	lipid recovery of 31.7%	(Lee et al., 2017)
Cell disruption and lipid extraction Ultrasound	*Chlorella vulgaris*	about 52.5% lipid yield obtained	(Araujo et al., 2013)
Cell disruption and lipid extraction Microwave	*Nannochloropsis gaditana*	Lipid yield obtained was 40%	(Bermúdez Menéndez et al., 2014)

13.4 Conclusions

Microalgae can be used effectively to produce value-added products in various industries, including biofuels as a substitute to fossil fuels. Despite being an effective alternate for many products, there are no available technologies that can be implemented at commercial scale. Implementation of process intensification approaches is hence needed at different stages of the microalgae-based product development to derive benefits in terms of energy efficiency, reduced time, and scope of commercialization. Microalgae biomass formation in sufficient quantity during cultivation is the main requirement in the development of microalgae-based value-added products and biofuels, with economic feasibility. Strain selection is the key aspect in terms of getting a high growth rate at the time of cultivation. To improve the efficiency and biomass from microalgae strain, approaches such as limitation of nutrients, manipulation of environmental conditions, genetic engineering, and co-cultivation could be applied. Cultivation and harvesting of microalgae from the culture media also play critical roles in getting a high product yield with less capital cost. So, appropriate process intensification approaches, such as the use of hybrid photobioreactors during cultivation or some modification in the existing technology can be implemented. There are various new process intensification methods available, such as ultrasound and microwave, to efficiently extract the lipid for further applications with high yield, less time, energy, and temperature requirement. Overall, it can be said that by applying process intensification approaches, microalgae can be used as an efficient feedstock/biomass for the generation of value-added products, including biofuels, with higher yield in less time, energy, and cost.

References

Abu-Ghosh, S., Fixler, D., Dubinsky, Z., & Iluz, D. (2016). Flashing light in microalgae biotechnology. *Bioresource Technology*, *203*, 357–363.

Ahmad, A. L., Yasin, N. M., Derek, C. J. C., & Lim, J. K. (2014). Comparison of harvesting methods for microalgae chlorella sp. and its potential use as a biodiesel feedstock. *Environmental Technology*, *35*(17), 2244–2253.

Al Hattab, M., Ghaly, A., & Hammouda, A. (2015). Microalgae harvesting methods for industrial production of biodiesel: Critical review and comparative analysis. *Journal of Fundamentals of Renewable Energy and Applications*, *5*(2), 1000154.

Araujo, G. S., Matos, L. J., Fernandes, J. O., Cartaxo, S. J., Gonçalves, L. R., Fernandes, F. A., & Farias, W. R. (2013). Extraction of lipids from microalgae by ultrasound application: Prospection of the optimal extraction method. *Ultrasonics Sonochemistry*, *20*(1), 95–98.

Balasubramanian, S., Allen, J. D., Kanitkar, A., & Boldor, D. (2011). Oil extraction from scenedesmus obliquus using a continuous microwave system–design, optimization, and quality characterization. *Bioresource Technology*, *102*(3), 3396–3403.

Bermúdez Menéndez, J. M., Arenillas, A., Menéndez Díaz, J. Á., Boffa, L., Mantegna, S., Binello, A., & Cravotto, G. (2014). Optimization of microalgae oil extraction under ultrasound and microwave irradiation. *Journal of Chemical Technology and Biotechnology*, *89*(11), 1779–1784.

Bohutskyi, P., Betenbaugh, M. J., & Bouwer, E. J. (2014). The effects of alternative pretreatment strategies on anaerobic digestion and methane production from different algal strains. *Bioresource Technology*, *155*, 366–372.

Brennan, L., & Owende, P. (2010). Biofuels from microalgae – A review of technologies for production, processing, and extractions of biofuels and co-products. *Renewable and Sustainable, Energy Reviews*, *14*(2), 557–577.

Burlew, J. S. (1953). Algal culture. *From Laboratory to Pilot Plant, Carnegie Institution of Washington Publication*, *600*(1), 375.

Chae, S. R., Hwang, E. J., & Shin, H. S. (2006). Single cell protein production of euglena gracilis and carbon dioxide fixation in an innovative photo-bioreactor. *Bioresource Technology*, *97*(2), 322–329.

Chetsumon, A., Maeda, I., Umeda, F., Yagi, K., Miura, Y., & Mizoguchi, T. (1994). Antibiotic production by the immobilized cyanobacterium, Scytonema sp. TISTR 8208, in a seaweed-type photobioreactor. *Journal Applied Phycology*, *6*(5), 539–543.

Cho, S., Park, S., Seon, J., Yu, J., & Lee, T. (2013). Evaluation of thermal, ultrasonic and alkali pretreatments on mixed-microalgae biomass to enhance anaerobic methane production. *Bioresource Technology*, *143*, 330–336.

Chu, W. L. (2017). Strategies to enhance production of microalgae biomass and lipids for biofuel feedstock. *European Journal of Phycology*, *52*(4), 419–437.

Dallaire, V., Lessard, P., Vandenberg, G., & De La Noüe, J. (2007). Effect of algal incorporation on growth, survival and carcass composition of rainbow trout (oncorhynchus mykiss) fry. *Bioresource Technology, 98*(7), 1433–1439.

Danquah, M. K., Ang, L., Uduman, N., Moheimani, N., & Forde, G. M. (2009). Dewatering of microalgae culture for biodiesel production: Exploring polymer flocculation and tangential flow filtration. *Journal of Chemical Technology & Biotechnology: International Research in Process, Environmental & Clean Technology, 84*(7), 1078–1083.

de Jesus, S. S., & Filho, R.M. (2017). Potential of algal biofuel production in a hybrid photobioreactor. *Chemical Engineering Science, 171*, 282–292.

Dittami, S. M., Heesch, S., Olsen, J. L., & Collén, J. (2017). Transitions between marine and freshwater environments provide new clues about the origins of multicellular plants and algae. *Journal of Phycology, 53*(4), 731–745.

Dogaris, I., Welch, M., Meiser, A., Walmsley, L., & Philippidis, G. (2015). A novel horizontal photobioreactor for high-density cultivation of microalgae. *Bioresource Technology, 198*, 316–324.

Dvoretsky, D., Dvoretsky, S., Temnov, M., Akulinin, E., & Zuorro, A. (2017). The effect of the complex processing of microalgae chlorella vulgaris on the intensification of the lipid extraction process. *Chemical Engineering Transaction, 57*, 721–726.

Fakhry, E. M., & El Maghraby, D. M. (2015). Lipid accumulation in response to nitrogen limitation and variation of temperature in nannochloropsis salina. *Botanical Studies, 56*(1), 1–8.

Fu, W., Gudmundsson, O., Feist, A. M., Herjolfsson, G., Brynjolfsson, S., & Palsson, B. Ø. (2012). Maximizing biomass productivity and cell density of chlorella vulgaris by using light-emitting diode-based photobioreactor. *Journal of Biotechnology, 161*(3), 242–249.

Gírio, F. M., Fonseca, C., Carvalheiro, F., Duarte, L. C., Marques, S., & Bogel-Łukasik, R. (2010). Hemicelluloses for fuel ethanol: A review. *Bioresource Technology, 101*(13), 4775–4800.

Gouveia, L., Raymundo, A., Batista, A. P., Sousa, I., & Empis, J. (2006). Chlorella vulgaris and haematococcus pluvialis biomass as colouring and antioxidant in food emulsions. *European Food Research and Technology, 222*(3), 362–367.

Grima, E. M., Belarbi, E. H., Fernández, F. A., Medina, A. R., & Chisti, Y. (2003). Recovery of microalgae biomass and metabolites: Process options and economics. *Biotechnology Advances, 20*(7–8), 491–515.

Grobbelaar, J. U. (2004). Algal nutrition: Mineral nutrition. In: Handbook of Microalgae Culture: Biotechnology and Applied Phycology, Ed. A. Richmond, Blackwell Publishing Ltd, Hoboken, New Jersey. DOI: 10.1002/9780470995280, pp. 97–115.

Guldhe, A., Singh, B., Rawat, I., Ramluckan, K., & Bux, F. (2014). Efficacy of drying and cell disruption techniques on lipid recovery from microalgae for biodiesel production. *Fuel, 128*, 46–52.

Guo, H., Daroch, M., Liu, L., Qiu, G., Geng, S., & Wang, G. (2013). Biochemical features and bioethanol production of microalgae from coastal waters of pearl river delta. *Bioresource Technology, 127*, 422–428.

Hamed, I., Özogul, F., Özogul, Y., & Regenstein, J. M. (2015). Marine bioactive compounds and their health benefits: A review. *Comprehensive Reviews in Food Science and Food Safety, 14*(4), 446–465.

Harith, Z. T., Yusoff, F. M., Mohamed, M. S., Shariff, M., Din, M., & Ariff, A. B. (2009). Effect of different flocculants on the flocculation performance of flocculation performance of microalgae, chaetoceros calcitrans, *cells. African Journal of Biotechnology, 8*(21), e-article.

Hayes, M., Skomedal, H., SkjDnes, K., Mazur-Marzec, H., Toruńska-Sitarz, A., Catala, M., Isleten Hosoglu, M., & García-Vaquero, M. (2017). Microalgae Proteins for Feed, Food and Health. In: Gonzalez-Fernandez, C., & Munoz, R. (Eds.), Microalgae-based Biofuels and Bioproducts. From Feedstock Cultivation to End Products (First Edition). Woodhead Publishing Series in Energy, Cambridge, UK, pp. 347–368.

John, R. P., Anisha, G. S., Nampoothiri, K. M., & Pandey, A. (2011). Micro and macroalgal biomass: A renewable source for bioethanol. *Bioresource Technology, 102*(1), 186–193.

Johnson, M. B., & Wen, Z. (2009). Production of biodiesel fuel from the microalga schizochytrium limacinum by direct transesterification of algal biomass. *Energy and Fuels*, *23*(10), 5179–5183.
Karthikeyan, S., Balasubramanian, R., & Iyer, C. S. P. (2007). Evaluation of the marine algae ulva fasciata and sargassum sp. for the biosorption of cu (II) from aqueous solutions. *Bioresource Technology*, *98*(2), 452–455.
Kim, M. G., Hwang, H. W., Nzioka, A. M., & Kim, Y. J. (2017). Enhanced lipid extraction from microalgae in biodiesel production. *Hemijska Industrija*, *71*(2), 167–174.
Kim, Z. H., Park, H., Hong, S. J., Lim, S. M., & Lee, C. G. (2016). Development of a floating photobioreactor with internal partitions for efficient utilization of ocean wave into improved mass transfer and algal culture mixing. *Bioprocess and Biosystems Engineering*, *39*(5), 713–723.
Kim, Z. H., Park, H., Ryu, Y. J., Shin, D. W., Hong, S. J., Tran, H. L., Lim, S.-M., & Lee, C. G. (2015). Algal biomass and biodiesel production by utilizing the nutrients dissolved in seawater using semi-permeable membrane photobioreactors. *Journal of Applied Phycology*, *27*(5), 1763–1773.
Kuria, J., & Maringa, M. (2008). Developing simple procedures for selecting, sizing, scheduling of materials and costing of small bio-gas units. *international. Journal for Service Learning in Engineering, Humanitarian Engineering and Social Entrepreneurship*, *3*(1), 9–40.
Leach, G., Oliveira, G., & Morais, R. (1998). Production of a carotenoid-rich product by alginate entrapment and fluid-bed drying of dunaliella salina. *Journal of the Science of Food and Agriculture*, *76*(2), 298–302.
Lee, I., & Han, J. I. (2015). Simultaneous treatment (cell disruption and lipid extraction) of wet microalgae using hydrodynamic cavitation for enhancing the lipid yield. *Bioresource Technology*, *186*, 246–251.
Lee, J., Yoo, Y., Jun, C., Ahn, C. Y., S. Y., & Oh, H. M. (2010). Comparison of several methods for effective lipid extraction from microalgae. *Bioresource Technology*, *101*(1), S75–S77.
Lee, S. Y., Cho, J. M., Chang, Y. K., & Oh, Y. K. (2017). Cell disruption and lipid extraction for microalgae biorefineries: A review. *Bioresource Technology*, *244*, 1317–1328.
Liu, J. C., Chen, Y. M., & Ju, Y. H. (1999). Separation of algal cells from water by column flotation. *Separation Science and Technology*, *34*(11), 2259–2272.
Lorenz, R. T., & Cysewski, G. R. (2000). Commercial potential for haematococcus microalgae as a natural source of astaxanthin. *Trends in Biotechnology*, *18*(4), 160–167.
Lum, K. K., Kim, J., & Lei, X. G. (2013). Dual potential of microalgae as a sustainable biofuel feedstock and animal feed. *Journal of Animal Science and Biotechnology*, *4*(1), 1–7.
Mata, T. M., Martins, A. A., & Caetano, N. S. (2010). Microalgae for biodiesel production and other applications: A review. *Renewable and Sustainable, Energy Reviews*, *14*(1), 217–232.
Mendez, L., Mahdy, A., Timmers, R. A., Ballesteros, M., & González-Fernández, C. (2013). Enhancing methane production of chlorella vulgaris via thermochemical pretreatments. *Bioresource Technology*, *149*, 136–141.
Millán-Oropeza, A., Torres-Bustillos, L. G., & Fernández-Linares, L. (2015). Simultaneous effect of nitrate (NO3-) concentration, carbon dioxide (CO2) supply and nitrogen limitation on biomass, lipids, carbohydrates and proteins accumulation in nannochloropsis oculata. *Biofuel Research Journal*, *2*(1), 215–221.
Muller-Feuga, A. (2000). The role of microalgae in aquaculture: Situation and trends. *Journal Applied Phycology*, *12*(3), 527–534.
Murray, A. M., Fotidis, I. A., Isenschmid, A., Haxthausen, K. R. A., & Angelidaki, I. (2017). Wirelessly powered submerged-light illuminated photobioreactors for efficient microalgae cultivation. *Algal Research*, *25*, 244–251.
Mutanda, T., Ramesh, D., Karthikeyan, S., Kumari, S., Anandraj, A., & Bux, F. (2011). Bioprospecting for hyper-lipid producing microalgae strains for sustainable biofuel production. *Bioresource Technology*, *102*(1), 57–70.
Nigam, P. S., & Singh, A. (2011). Production of liquid biofuels from renewable resources. *Progress in Energy and Combustion Science*, *37*(1), 52–68.

Oliveira, E. G. D., Rosa, G. S., Moraes, M. A., & Pinto, L. A. A. (2009). Characterization of thin layer drying of spirulina platensis utilizing perpendicular air flow. *Bioresource Technology*, *100*(3), 1297–1303.

Onay, M. (2019). Bioethanol production via different saccharification strategies from H. tetrachotoma ME03 grown at various concentrations of municipal wastewater in a flat-photobioreactor. *Fuel*, *239*, 1315–1323.

Papazi, A., Makridis, P., & Divanach, P. (2010). Harvesting chlorella minutissima using cell coagulants. *Journal Applied Phycology*, *22*(3), 349–355.

Passos, F., & Ferrer, I. (2015). Influence of hydrothermal pretreatment on microalgae biomass anaerobic digestion and bioenergy production. *Water Research*, *68*, 364–373.

Passos, F., Mota, C., Donoso-Bravo, A., Astals, S., Jeison, D., & Muñoz, R. (2018). Biofuels from Microalgae: Biomethane. In: Energy from Microalgae. Ed. E. Jacob-Lopes, Springer, Cham, pp. 247–270.

Patil, R. A., Kausley, S. B., Joshi, S. M., & Pandit, A. B. (2020). Process Intensification Applied to Microalgae-based Processes and Products. In: Handbook of Microalgae-Based Processes and Products. Ed. E. Jacob-Lopes, M. M. Maroneze, M.I. Queiroz, L.Q. Zepka, Academic Press, Cambridge, USA, pp. 737–769.

Patinvoh, R. J., Arulappan, J. V., Johansson, F., & Taherzadeh, M. J. (2017). Biogas digesters: From plastics and bricks to textile bioreactor-a review. *Journal of Energy and Environmental Sustainability*, *4*, 31–35.

Prabakaran, P., & Ravindran, A. D. (2011). A comparative study on effective cell disruption methods for lipid extraction from microalgae. *Letters in Applied Microbiology*, *53*(2), 150–154.

Rakesh, S., Dhar, D. W., Prasanna, R., Saxena, A. K., Saha, S., Shukla, M., & Sharma, K. (2015). Cell disruption methods for improving lipid extraction efficiency in unicellular microalgae. *Engineering in Life Sciences*, *15*(4), 443–447.

Sabri, L. S., Sultan, A. J., & Al-Dahhan, M. H. (2021). Split internal-loop photobioreactor for Scenedesmus sp. microalgae: Culturing and hydrodynamics. *Chinese Journal of Chemical Engineering*, *33*, 236–248.

Seo, Y. H., Sung, M., Oh, Y. K., & Han, J. I. (2016). Lipid extraction from microalgae cell using persulfate-based oxidation. *Bioresource Technology*, *200*, 1073–1075.

Shokrkar, H., Ebrahimi, S., & Zamani, M. (2017). Bioethanol production from acidic and enzymatic hydrolysates of mixed microalgae culture. *Fuel*, *200*, 380–386.

Singh, A., Nigam, P. S., & Murphy, J. D. (2011). Mechanism and challenges in commercialisation of algal biofuels. *Bioresource Technology*, *102*(1), 26–34.

Singh, R. N., & Sharma, S. (2012). Development of suitable photobioreactor for algae production – A review. *Renewable and Sustainable, Energy Reviews*, *16*(4), 2347–2353.

Smith, B. T., & Davis, R. H. (2012). Sedimentation of algae flocculated using naturally-available, magnesium-based flocculants. *Algal Research*, *1*(1), 32–39.

Sorokin, C., & Krauss, R. W. (1958). The effects of light intensity on the growth rates of green algae. *Plant Physiology*, *33*(2), 109.

Sousa, I., Gouveia, L., Batista, A. P., Raymundo, A., & Bandarra, N. M. (2008). Microalgae in Novel Food Products. In: Papadopoulos, K. N. (Ed.), Food Chemistry Research Developments. Nova Science Publishers, New York, USA, pp. 75–112.

Stockenreiter, M., Haupt, F., Graber, A. K., Seppälä, J., Spilling, K., Tamminen, T., & Stibor, H. (2013). Functional group richness: Implications of biodiversity for light use and lipid yield in microalgae. *Journal of Phycology*, *49*(5), 838–847.

Thompson, G. A., Jr (1996). Lipids and membrane function in green algae. *Biochimica Et Biophysica Acta (Bba)-lipids and Lipid Metabolism*, *1302*(1), 17–45.

Uduman, N., Qi, Y., Danquah, M. K., Forde, G. M., & Hoadley, A. (2010). Dewatering of microalgae cultures: A major bottleneck to algae-based fuels. *Journal of Renewable and Sustainable Energy*, *2*(1), 012701.

Uysal, O., Uysal, O., & Ek, K. (2016). Determination of fertilizing characteristics of three different microalgae cultivated in raceways in greenhouse conditions. *Iucrari Stiintifice Seria Agronomie*, *59*, 15–18.

Vaghari, H., Eskandari, M., Sobhani, V., Berenjian, A., Song, Y., & Jafarizadeh-Malmiri, H. (2015). Process intensification for production and recovery of biological products. *American Journal of Biochemistry and Biotechnology*, *11*(1), 37.

Vandamme, D., Foubert, I., & Muylaert, K. (2013). Flocculation as a low-cost method for harvesting microalgae for bulk biomass production. *Trends in Biotechnology*, *31*(4), 233–239.

Wan, M., Jin, X., Xia, J., Rosenberg, J. N., Yu, G., Nie, Z., Oyler, G. A., & Betenbaugh, M. J. (2014). The effect of iron on growth, lipid accumulation, and gene expression profile of the freshwater microalga chlorella sorokiniana. *Applied Microbiology and Biotechnology*, *98*(22), 9473–9481.

Wang, H. M. D., Chen, C. C., Huynh, P., & Chang, J. S. (2015). Exploring the potential of using algae in cosmetics. *Bioresource Technology*, *184*, 355–362.

Wang, S. K., Wang, F., Hu, Y. R., Stiles, A. R., Guo, C., & Liu, C. Z. (2014). Magnetic flocculant for high efficiency harvesting of microalgae cells. *ACS Applied Materials and Interfaces*, *6*(1), 109–115.

Wiltshire, K. H., Boersma, M., Möller, A., & Buhtz, H. (2000). Extraction of pigments and fatty acids from the green alga scenedesmus obliquus (chlorophyceae). *Aquatic Ecology*, *34*(2), 119–126.

Xin, L., Hong-Ying, H., Ke, G., & Ying-Xue, S. (2010). Effects of different nitrogen and phosphorus concentrations on the growth, nutrient uptake, and lipid accumulation of a freshwater microalga scenedesmus sp. *Bioresource Technology*, *101*(14), 5494–5500.

Yel, N. V., Yelboğa, E., Tüter, M., & Karagüler, N. G. (2017). Comparison of cell disruption and lipid extraction methods for improving lipid content of schizochytrium sp. S31. *Journal of Molecular Biology and Biotechnology*, *1*(1), 9–12.

Zhou, W., Min, M., Hu, B., Ma, X., Liu, Y., Wang, Q., Shi, J., Chen, P., & Ruan, R. (2013). Filamentous fungi assisted bio-flocculation: A novel alternative technique for harvesting heterotrophic and autotrophic microalgae cells. *Separation and Purification Technology*, *107*, 158–165.

Zimmerman, W. B., Zandi, M., Bandulasena, H. H., Tesař, V., Gilmour, D. J., & Ying, K. (2011). Design of an airlift loop bioreactor and pilot scales studies with fluidic oscillator induced microbubbles for growth of a microalgae Dunaliella salina. *Applied Energy*, *88*(10), 3357–3369.

Part IV: Process integration and process intensification strategies applied to microalgae-based products

Florin Barla, Massimiliano Lega, Sandeep Kumar*

Chapter 14
Process integration opportunities applied to microalgae biomass production

Abstract: The rapid population growth and remarkable development of modern society has resulted in an increased demand for energy, and consequently, an intensified use of fossil fuels reserves, compromising the energy sector sustainability and accumulation of greenhouse gasses in atmosphere that are associated with climate change. Microalgae are photosynthetic microorganisms that have a higher growth rate than terrestrial plants and are capable of capturing CO_2, contributing to the planet's sustainability primarily by transforming CO_2 into O_2. They are considered biomass producers for aquatic systems. Microalgae biotechnology is presently contributing to the global bioeconomy producing valuable biomass for a variety of applications such as biofuels, bioplastics, pharmaceuticals, cosmetics, food, and feed. They can be cultivated in wastewater that is the most favorable process from the viewpoint of a green and circular economy. This chapter aims to summarize the recent studies regarding the integration of microalgae cultivation for biomass with wastewater treatment, biofuels production, as well as other value-added bioproducts based on microalgae.

Keywords: microalgae, biomass, process integration, sustainability, wastewater recycling, biofuels, HTL, bioplastic, circular economy

14.1 Introduction

Due to their inherent qualities, such as ability to grow on nonarable land, their high biomass productivity, and lipid yield, algae biomass has been extensively utilized to produce fuels, chemicals, and other type of valuable materials (Feng et al., 2022). Microalgae also gained great attention as a ground-breaking resource with the ability to generate large amounts of organic molecules such as proteins, carbohydrates, and lipids (Mustapha et al., 2022). Chlorophyll, lutein, astaxanthin, β-carotene, protein, polysaccharides, and polyunsaturated fatty acids (PUFAs) are some of the bioactive metabolites produced by algae that have been extensively investigated. Biofuels are sustainable

***Corresponding author: Sandeep Kumar**, Department of Civil and Environmental Engineering, Old Dominion University, Norfolk, VA 23529, USA, e-mail: skumar@odu.edu
Florin Barla, Circ INC, Danville, VA 24540, USA
Massimiliano Lega, Dipartimento di ingegneria, Universita degli Studi di Napoli Parthenope, 80143 – Napoli, Italy

https://doi.org/10.1515/9783110781267-014

fuels and are strategically important alternatives to fossil fuels, contributing significantly to reducing the greenhouse gasses emission, and therefore improving air quality (Milano et al., 2016; Khoo et al., 2020; Nigam and Singh, 2011). Typically, 1 kg of microalgae biomass can fix about 1.83 kg of CO_2, and their cultivation does not require large areas of land or freshwater as compared to lignocellulosic biomass; thus, microalgae are promising candidates for biofuels and bioproducts and contribute to reducing the impact of climate change and global warming (Wu and Chang, 2019). Microalgae can be cultivated on polluted municipal or industrial wastewater with lower nutrient requirement as compared to terrestrial crops this being the most feasible process from the perspective of a green and circular economy (Moreno-Osorio et al., 2021). The research and practice in the field are oriented to ensure the sustainability of wastewater treatment in terms of economic feasibility, environmental impacts, and resource recovery. Among the several proposed solutions, the integration of a microalgae-based treatment within the conventional wastewater treatment layout is often recommended due to the many potential synergies; the wastewater provides all the resources required for the algae to grow (Tua et al., 2021). There is a great potential for cultivation of microalgae as biofilms; this could tremendously increase production efficiency of microalgae biomass for industrial applications. The primary advantage of microalgal biofilms for biomass production is increased biomass concentration (e.g., many cells per unit volume) and the simple separation of attached cells from the surrounding growth medium without the use of extra energy or chemical procedures (Moreno-Osorio et al., 2021). Furthermore, biofilms make harvesting and dewatering easier, since the cells are simple scraped off the surface without requiring costly separation techniques such as filtration or centrifugation, and there is no need to flocculate and settle the biomass. Microalgae have the capacity to produce large quantities of lipids, proteins, and carbohydrates, some of which are mentioned below, and are the most prevalent components of bio-based products such bioplastics, biopolymers, and bio-based polyurethane. Such materials known as biopolymers have unique qualities including biodegradability, biocompatibility, and renewability. Growing interest is being shown in bio-based polymers for a variety of uses, including food packaging, medical equipment, electronics, and energy (Madadi et al., 2021). In addition, microalgae biomass can act as potential filler and reinforcement in the enhancement of bioplastic, either blending with conventional bioplastic or synthetic polymer (Chong et al., 2022). The categorization of different biofuels from microalgae biomass in the biomass conversion process is shown in Figure 14.1. Bio-oil and syngas can be produced by thermochemical conversion (pyrolysis, gasification, and hydrothermal liquefaction) processes, biodiesel can be produced by chemical conversion (transesterification) process, bio alcohols and biogas can be produced by biochemical conversion (fermentation and anaerobic digestion) process.

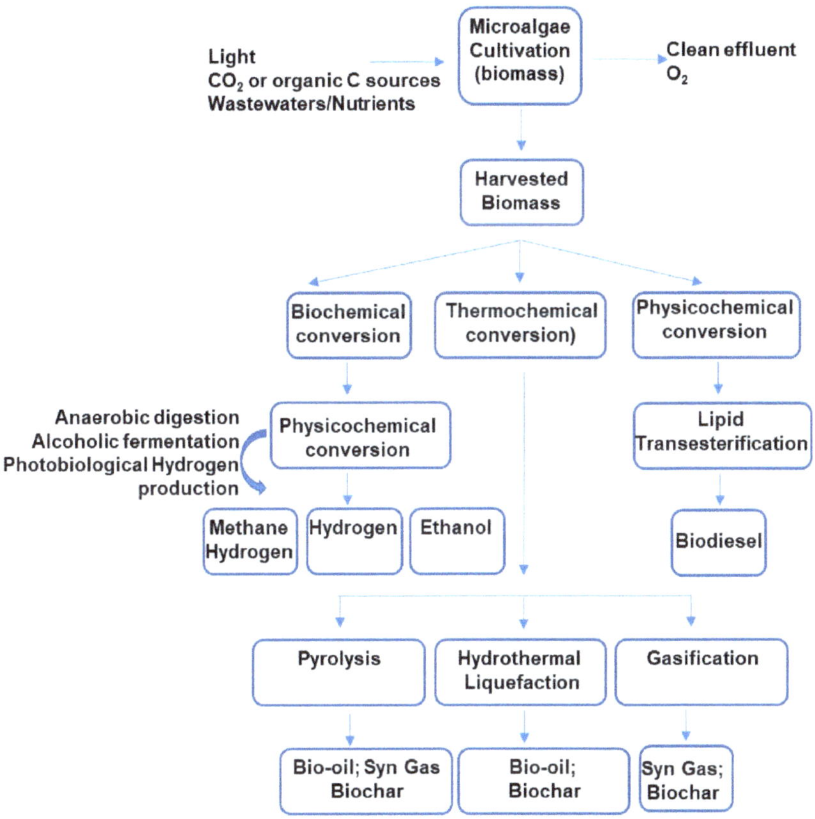

Figure 14.1: Technologies used for biofuels production based on microalgae biomass.

14.2 Microalgae derivatives for biofuels and bioactive compounds

Historically, the implementation and production of microalgae-based biofuel production was proposed as early as the 1950s and it was later extensively approached and investigated by researchers (Chew et al., 2017). As mentioned, algae are well-known to contain lipids, carbohydrates, proteins, and phytonutrients and these compounds can also be used as feedstock in the process of obtaining a variety of high-value products (Takagi et al., 2000; Lee et al., 2021). Lipids extracted from algae biomass can be refined into fatty acids, which can then be transesterified to produce biodiesel (Goncalves et al., 2013). Microalgae anaerobic digestion could produce biogas, which is a mixture of methane and carbon dioxide gas (Zhang et al., 2016). Gasification is a process that generates syngas by converting carbon-rich raw materials through partial oxidation within a limited oxygen supply range (Sikawar et al., 2017).

Although 350 crop species have been investigated and found suitable for oil production, their sustainability is challenged due to food crises around the world (Patel et al., 2015). The first- and second-generation biodiesels are derived from edible and nonedible plant oils, while the microbial oils are primarily used to produce the third-generation or advanced biodiesel. Currently, biodiesel is produced from plant oils by transesterification with an alcohol (methanol or ethanol); for cost reasons, methanol is used most frequently, in the presence of a base, an acid, or an enzyme as a catalyst (Figure 14.2). The transesterification is performed with methanol and triacylglycerides extracted from dried microbial biomass; however, the lipids transesterification can be done also via direct alcholysis of dried microbial biomass without previous lipid extraction (Figure 14.3). Biodiesel production from microbial lipids is presently receiving increasing attention as a cost-effective, sustainable alternative.

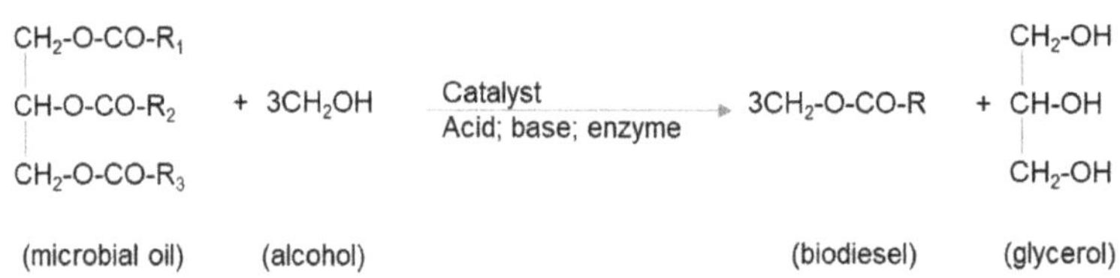

Figure 14.2: Biodiesel synthesis by transesterification reaction.

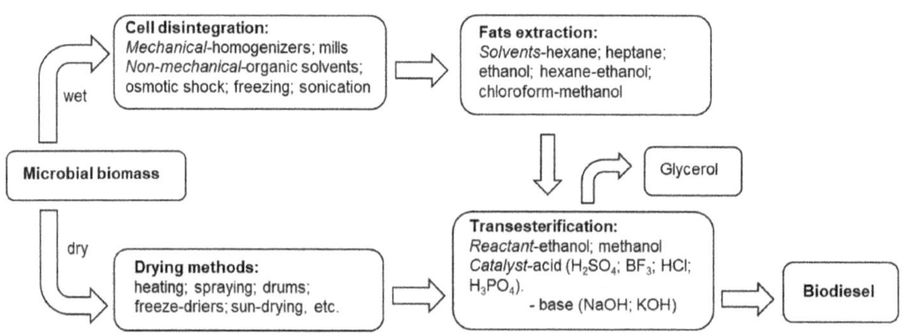

Figure 14.3: Summary of biodiesel production from oleaginous microorganisms via conventional transesterification with extracted lipids vs. direct transesterification of microbial biomass.

Since centuries, vegetable oils have been consumed as human food but more recently, it also finds applications in biofuels production. Lipids are now getting increasing attention for their applications, especially in the biofuels industry and food supplements industry. The importance of such lipid resources intensified and steadily increased from 1960s, when such products began to be used to produce laundry-grade lipases in combination with detergents and other cleaning agents (Colin and James, 2002). In the 1980s, microbial oil was used for the first time as a food supplement (gamma-linolenic acid supplement)

(Colin and James, 2002). Algae are considered important cell factories and they can produce bioactive compounds such as natural pigments, which can be used in food, animal feed, cosmetics, and pharmaceutical industries (Guedes et al., 2011). Currently, microorganisms, including microalgae, are referred to as sources of long-chain PUFAs being used as high-grade nutraceuticals for human and animal consumption. In addition, microorganisms may utilize low-cost substrates like agricultural wastes or industrial by-products. Studies indicate that microbial oils could be used in green-fuels production, pharmaceutical and cosmetic applications, food additives, or biopolymer production (Bharathiraja et al., 2017). Oils obtained from microalgae have better quality of PUFAs compared to oils derived from plants, and the unsaturated and fatty acids of more than four bonds can be reduced through partial catalytic hydrogenation of the oil. The oils can be obtained from microalgae biomass and converted into bio-oils and bio diesel via different ways to extract lipids from microalgae biomass such as liquid–liquid extraction, oil press, sub- and supercritical fluid extraction and ultrasound-assisted extraction (Chia et al., 2020). In oleaginous microorganisms, the lipid accumulation starts when the carbon source is present in excess and an element in the growth media becomes limiting. Also, studies showed that isocitrate dehydrogenase, ATP-citrate lyase, and malic enzyme are the three key enzymes in lipid accumulation (Christophe et al., 2012). Some microorganisms can produce PUFAs such as omega-3 and omega-6 series, known for their benefits on human health. PUFAs play a substantial role as precursors of eicosanoids and structural components of membrane phospholipids (Sakuradani et al., 2009). PUFAs are components of thrombocytes, neuronal and muscle cells, cerebral cortex, as well as immunocompetent cells. The potential demand for omega-3 PUFAs (based on 500 mg/day) is approximately 1.27 million tons and the supply by fish is about 0.84 million tons, so the gap is 0.43 million tons that should be obtained from other sources. Conjugated linoleic acid, which is a mixture of isomers of linoleic acid, also possesses well-known health benefits such as anticarcinogenic, antiobesity, antidiabetic, antihypertensive, antiatherogenic, immunomodulatory, and osteosynthetic properties (Vela-Gurovic et al., 2014; Beligon et al., 2016). On the other hand, the same kind of oils were investigated on their potential to be used as possible alternatives to vegetable oils for biodiesel production avoiding competition with human food (Christophe et al., 2012). Nowadays, PUFAs production using various microbial strains is an industrial reality. Some PUFAs, such as arachidonic acid (ARA) and docosahexaenoic acid (DHA) are now produced by various microorganisms at a commercial level. These types of oils are extensively used as dietary supplements in infant formulations (Beligon et al., 2016). Paik, (Paik et al., 2017) showed that *C. reinhardtii* cultivated under special conditions (perfusion microfluidic chip) under continuous feed system, showed a 2.4-fold increase in triacylglyceride production compared to the level obtained under an only ammonia-depleted condition. Also, the produced triacylglyceride that is a bio-derived neutral lipid has been regarded as a precursor for biodiesel fuel as well as for ricinoleic acid that has healthcare properties and industrial uses (Kajikawa et al., 2016). Numerous studies investigated the potential of various microorganisms such as microalgae, to produce oils. Some of these microalgae and their protein, lipids, and carbohydrates content are shown in Table 14.1

Table 18.1: Growth conditions and biochemical composition of various microalgae species (Mustapha et al., 2021; Wu and Chang, 2019).

Microlagae species	Microalgae source/ nutrient medium	Lipid (wt.%)	Carbohydrate (wt.%)	Protein (wt.%)
Anabena cylindrical	/	4-7	25-30	43-56
Chlorella vulgaris	Shandong Biotech Co., Ltd; medium not specified	15.6	13.2	47.4
Chlorella vulgaris	cultivated with standard nitrogen (0.150 kg/m3)	49.5	19.7	29
Chlorella vulgaris	low nitrogen (0.049 kg/m3)	43	51	6
Chlorella vulgaris	bold's basal mwdium at low temperature	15.67	20.99	41.51
Chlorella vulgaris	commercial sources /medium not specified	25	9	55
Chlorella pyrenoidosa	Wudi Iv Qi Bioeng. Co., Ltd	10.6	23.7	57.6
Chlorella protothecoides	cultivated in artificial sea water	14.57	10.62	52.64
Calothrix	/	/	7-9	21-28
Chlamydomonas	/	10-20	12-53	37-47
Dunaliella salina	Shandong Firstspirulina Biotech Co., Ltd	10.5	11.9	58.8
Dunaliella tertiolecata	Tanjing Microalgae Biotechnologies Co., Ltd	2.87	21.69	61.32
Euglena gracilis		14-20	14-18	39-61
Golenkinia sp.	NLP Co., Ltd Hadong Korea / medium not specified	17.3	26.7	45.1
Lyngbya	/	/	5-9	18-28
Microcystis aeruginosa	cultivated in artificial sea water	12.5	11.6	30.8
Nannochloropsis sp.	cultivated with BG-11 medium	30	21	44
Nannochloropsis oceanica	NLP Co., Ltd Hadong Korea / medium not specified	30.2	24.4	36.3
Nannochloropsis oculata	commercial sources /medium not specified	32	8	57
Nannochloropsis	Yantai Hearol Biotechnologies Co., Ltd	24.8	22.7	19.1
Oscillatoria	/	/	9-12	40-52
Pavlova sp.	Varicon Aqua Solutions / medium not specified	20	26	43
Porphyridium cruentum	commercial sources /medium not specified	8	40	43
Phormidium	/	/	15-17	49-64
Prymnesium parvum	/	22-38	25-33	28-45
Porphyridium cruentum	/	9-14	40-57	28-39
Spirulina	Xindaze Co., Ltd	7.4	19.3	65.4
Scenedesmus	Institute of Bioscience and Biotechnology of Daejeon	19.5	29.3	36.4
Scenedesmus sp.	Ingrepro B.V. /medium not specified	5.6	48.1	46.3
Spirogyra sp.	/	11-21	33-64	6-20

14.3 Microalgae

Microalgae are microscopic algae, photoautotrophic, microorganisms that can convert CO_2 to lipids and long-chain hydrocarbons in the presence of sunlight and nutrient sources. The lipids can be particularly used to produce biodiesel and, further, the biomass can be fermented to produce other fuels (Carioca, 2010; Beer et al., 2009). Numerous studies pointed out the advantages of using microalgae as feedstock for biofuels such as: relatively higher lipid content, usually ranging from 20–60% by weight dry biomass (Shi et al., 2011) under specific conditions and fast growth – some can double the biomass in 3.5 h and could be harvested daily (Christi, 2007; Kapdan and Kargi, 2006). For biodiesel production, lipids with long carbon chain are used. Microalgae mainly produce unsaturated fatty acids such as: palmitoleic (C16:1); oleic (C18:1); linoleic (C18:2); and linolenic (C18:3) as well as saturated fatty acids such as: palmitic (C16:0) and stearic (C18:0) (Shi et al., 2011). Some species can synthesize PUFAs, but the biodiesel produced from such kinds of compounds oxidizes faster than petroleum products. The faster the microalga grows, the higher the oils productivity, and therefore higher the biodiesel productivity (Shi et al., 2011). Lipid production and accumulation in various microalgae are shown in Table 14.2.

Table 14.2: Lipid accumulation in different microalgae.

Microalgae	Total fats content (% as dry weight)	References
N. laevis	69	(Chen et al., 2008)
C. emersonii	63	(Illman et al., 2000)
P. incisa	62	(Solovcenko et al., 2008)
C. minutissima	57	(Illman et al., 2000)
C. vulgaris	57	(Liu and Zhao, 2007)
N. oleoabundans	54	(Metting, 1996)
C. vulgaris	40	(Illman et al., 2000)
M. subterraneus	39	(Khozin and Cohen, 2006)
C. protothecoides	23	(Illman et al., 2000)
C. sorokiniana	22	(Illman et al., 2000)

Microalgae growth and lipid accumulation under phototrophic conditions are influenced by several factors: intensity of light, pH, dissolved oxygen, and nutrients concentration such as: nitrogen (Rodolphi et al., 2009), phosphorus (Wu et al., 2010), silicon (Griffiths and Harrison, 2009) iron (Liu et al., 2008), and salinity (Rao et al., 2007). High lipid content in microalgae is achieved under environmental stress (Li et al., 2008) that can be caused by the various factors as mentioned above. Studies showed that nutrient deprivation in a batch process stimulates microorganisms to produce secondary metabolites, resulting in limitation of cellular growth. Recently, a new continuous perfusion system design was tested to culture *Chlamydomonas reinhardtii*, a microalga. The nutrient was kept under constant slightly lower concentration (near the limit of deplete), and for this particular case, when it was cultured in 7.5%/7.5% of NH_4 +/PO_4^{2-}, showed a 2.4-fold increase in triglycerides production and a 3.5-fold increase in biomass production, as compared to levels obtained under an only NH_4^+-depleted conditions (Paik et al., 2017). This type of reactor with a steady continuous nutrient flow can be used to optimize the conditions for secondary metabolite production, at the same time increasing the biomass production. This design can induce starvation but also encourage the cells to grow at a minimum growth rate (Paik et al., 2017). Most of the lipids produced by microalgae are polyunsaturated fatty acids; therefore the auto-oxidation of such lipids during storage becomes a challenge when used for biofuels production. Also, artificial lightening and long fermentation period may raise technical and economic problems (Christophe et al., 2012). Some microalgae are considered potential candidates in PUFAs production. These polyunsaturated fatty acids, which are important since mammals cannot synthesize them, should be found in food or be available as nutritional supplements. They are crucial to healthy growth because of their advantages highlighted in many works. It has been reported that the microalgae N. *oceanic*, P. *cruentum*, P. *tricornutum*, and C. *calcitrans* are able to synthesize eicosapentaenoic acid (EPA). Several microalgae have been shown to include DHA-producing bacteria, including I. *galbana* and *Crypthecodinium* spp. The effectiveness of light penetration is a frequent issue while creating a photobioreactor system for growing microalgae (Beligon et al., 2016). Single cell oils are the lipids produced by microalgae as lipid droplets, and they are present in chloroplasts. Specific enzymes are involved in the process. Briefly, the oleic acid can be converted successively into linoleic acid (LA) and then into alpha-linoleic acid (ALA). Furthermore, from LA under enzymatic catalysis, ARA is biosynthesized, and similarly, from ALA, EPA and DHA could be biosynthesized (Beligon et al., 2016). EPA and DHA play an important role against cardiovascular disease, reduce arterial LDL-cholesterol, and participate in the inflammatory response (Ward and Singh, 2005).

14.4 Methods of lipids extraction

The production of microbial lipids is directly correlated with a high conversion of the substrate into intracellular lipids combined with high extraction efficiency (Bharathiraja et al., 2017). Usually a form of biomass pretreatment, such as cell disruption, is required to remove the protective cell walls of microorganisms to make the intracellular lipids more accessible to solvent extraction (Figure 14.4).

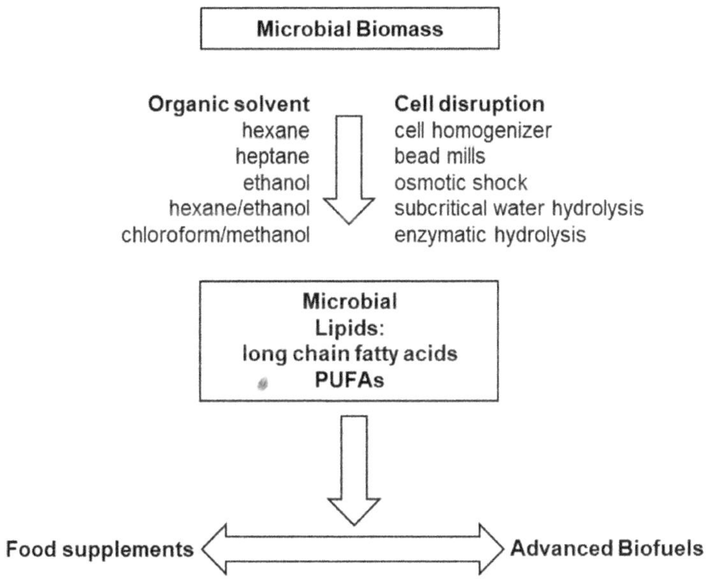

Figure 14.4: Microbial lipids extraction and utilization potential.

14.5 Organic solvent extraction

While the oil press approach yields 75% oil extraction, it is less effective and takes longer to complete the extraction process. In liquid–liquid extraction, organic solvents such cyclohexane, hexane, benzene, acetone, and chloroform are added to speed up the extraction process. The addition of the organic solvent aids in weakening the algal cell walls and oil is subsequently extracted from the liquid phase since the chemical characteristics of the oil have a greater solubility than water. Based on its maximum extraction yield and lowest extraction cost, hexane appears to be the more trustworthy substance. Using ethanol for lipid extraction and hexane for purification, the yield of the lipid extractions is increased to 80% (Khoo et al., 2020).

Soxhlet extraction is one of the most used methods in lipid extraction based on organic solvents such as hexane, where basically, the lipids extraction is a continuous process in which the sample is washed with solvent in a cyclic manner. The viability of microbial oil production is correlated with high conversion of the substrate into lipids, combined with high extraction efficiency at low energy input. Conventional processes of oil extraction are based on hexane or a mixture of other organic solvents. When organic solvents are used in oil extraction, some constituents of microbial oils may be degraded; also some other cell wall components may be co-extracted. Various other pretreatments that are done in order to ease the extraction process were evaluated such as: base addition, acid, and enzyme addition (Tsakona et al., 2014), sonication and microwaves application (Bharathiraja et al., 2017). According to several publications, microbial oil extraction processes are simpler and less labor-intensive compared to plant oil extraction procedures that require processing of the specific plant tissue (e.g., drying the seeds) followed by mechanical methods to grind the seeds and extract the oils. Microbial oils can be extracted in the aqueous phase or with the aid of simple cell disruption techniques like ultra-sonication to remove or weaken the protective cell walls of microorganisms, to make the intracellular lipids more accessible in solvent extraction (Jin et al., 2013; Zhang et al., 2014). Lipids should have a high partition coefficient in the selected solvent; hydrophobic lipids will relatively easily partition into nonpolar solvents, but the polar lipids are not easily extracted into nonpolar solvents (Dong et al., 2016). Cosolvents are usually used to break the linkages between the polar lipids and biomass; in Table 14.3, some organic solvents usually used in lipid extraction, the method, and applicability are summarized. Due to chloroform and methanol toxicity, various combinations of less toxic cosolvents such as hexane or short chain alcohols are used often in lipid extraction. The lipid extraction efficiency is affected by limited lipid accessibility. The biomass pretreatment or cell disruption prior to wet extraction will open the cellular wall, exposing the lipids to organic solvent and will increase lipid accessibility, thus improving mass transfer and reducing emulsion formation. There are many pretreatment methods described in the literature such as: high-pressure homogenization, bead milling, ultrasound, pulsed electric field, osmotic shock, microwave assistance, subcritical water hydrolysis, enzymatic hydrolysis, and chemical hydrolysis (Dong et al., 2016), which could be categorized basically into mechanical, enzymatic, and chemical pretreatment methods.

Table 14.3: Solvents used in lipid extraction and methods (Colin and James, 2002).

Organic solvent	Extraction method	Notes
$CHCl_3$: MeOH [2:1]	Folch's method	For intact cells
$CHCl_3$: MeOH [1:1]	Pedersen method	For freeze-dried cells
$CHCl_3$: MeOH: water [2:2:1.8]	Bligh and Dyer method	Acid pre-treated cells only
Hexane	Homogenization and pretreatment	Less toxic than $CHCl_3$/MeOH
Supercritical carbon dioxide	Gas chromatography	For enzyme-treated cells
EtOH (96%)	Microwave-assisted extraction	Clean and sustainable

$CHCl_3$ – Chloroform; MeOH – methanol; EtOH – ethanol

14.6 Protein extraction

Combining electrolysis cell disruption technique with alcohol/salt-based liquid biphasic flotation for extraction of protein from *Chlorella sorokiniana* CY-1 was also investigated (Sankaran et al., 2018). The impact of salt type/concentration, algae biomass loading, air flowrate of liquid biphasic electric flotation (LBEF), floatation time, and voltage of electrolysis and position of electrode immersed in LBEF was examined. Optimized conditions were found to be those of 60%(v/v) of 1-propanol as top phase, 250 g/L of dipotassium hydrogen phosphate as bottom phase, 0.1 g crude microalgae loading, 150cc/min air flowrate, 10 min flotation time, 20 V electrolysis voltage, and electrodes' tip touching the top phase of LBEF. After optimization, it was discovered that the protein recovery and separation efficiency were, respectively, 23.4 ± 1.25% and 173 ± 4.47%. A comparison of liquid biphasic flotation (LBF) and LBEF was also conducted, and it was shown that the protein recovery improved with the help of an electric supply. This work proved that LBF combined with in situ cell disruption is an effective strategy for protein extraction. The advantages of this unique technology include a reduction in the number of stages required to extract all the protein from the microalgae, as well as time and money savings in the bioprocessing sector. The potential for applying the LBEF system in downstream processing for various biomolecules in industries proved quite attractive (Sankaran et al., 2018).

14.7 Subcritical water hydrolysis of microalgae biomass

Subcritical water, commonly referred to as pressurized hot water, is liquid water that is heated to a temperature below the critical water temperature of 374 °C and above the ambient boiling point of 100 °C. By exerting pressure to keep the water in liquid condition, the subcritical state allows liquid water to be in equilibrium with vapor at saturated vapor pressure. There is an increase in the water's ability to self-ionize (acidify), solvating power, and diffusivity, when these circumstances are present. In contrast, a reduction in viscosity, surface tension, and its polarity could be observed. Water is a "green" solvent since it is harmless. Recently, subcritical water technology has been selected to extract a variety of active compounds from biomass, to remove non-oil biomass, leaving an oil-rich biomass to produce biodiesel (Levine et al., 2010), or to release nitrogen-containing compounds (amino acids and peptides) producing an oil-rich biomass residue (Levine et al., 2010; Moscoso-Garcia et al., 2013). Subcritical water extraction was also used to recover polysaccharides and protein-derived products from microalgae and yeast biomass (Chakraborty et al., 2012; Moscoso-Garcia et al., 2015), or it could be used to produce bio-crude, which can then be upgraded to liquid hydrocarbons (Moscoso-Garcia et al., 2015). When microalgae are flash-hydrolyzed in subcritical water, the proteins are released as water-soluble peptides and amino acids. The leftover material, which is made up of lipids and algal cells, may then be filtered and utilized to produce biofuel. (Kumar et al., 2014; Thakkar et al., 2021). Hydrothermal liquefaction (HTL) is an intriguing technique that uses the entire biomass of microalgae in a wet state to produce bio-oil, hydrochar, water-soluble compounds, and some gases (Guo et al., 2015; Chiaramonti et al.,

2015). Despite the benefits, bio-oils produced by hydrothermal microalgae liquefaction contain high levels of nitrogen and oxygen compounds that must be removed before they can be used as transportation fuel (Gollakota et al., 2017). In terms of their contribution to HTL bio-oil yield, many authors have shown that microalgae composition follows the trend lipid > protein > carbohydrate and are the only components that benefit from utilizing a catalyst (Biller and Ross, 2011). Intriguingly, a study found that, when heated to 310 ° C, microalgae biomass with a low nitrogen level (1.92%) produced HTL bio-oil with a significantly lower nitrogen percentage of about 0.915% (Guo et al., 2015). The HTL bio-oil that is generated has a low nitrogen level (0.91%), which is within the range for petroleum crude (0.1–2%) and is of excellent quality. This suggests that a microalgae biomass's initial nitrogen content may have a significant impact on the quality of the final product. (Mustapha et al., 2022). The bio-oil yield is a function of the characteristics of microalgae species, the hydrothermal conditions, and the catalyst used as shown in Table 14.4. Several recent studies on hydrothermal carbonization (HTC) of algal biomass are summarized in Table 14.5. Heilmann (Heilmann et al., 2010) showed that the resulting hydrochar from an HTC of algal biomass had a distinct composition and energy density comparable to that of bituminous coal. Hydrochar is formed via HTC of feedstock slurry, while biochar is produced from slow pyrolysis of dried feedstock. Hydrochar is a C-rich material and its properties have been investigated in various fields such as wastewater treatment for contaminants removal, soil amendment by enhancing nutrients and water holding capacity, and solid fuel. HTC may be used as a pretreatment, much like torrefaction, to upgrade biomass by eliminating moisture and contaminants. The enhanced feedstock

Table 14.4: Hydrothermal liquefaction of algae biomass to bio-oil under different conditions (Mustapha et al., 2022; Duan and Savage, 2011).

Microalgae species	Catalyst	Liquefaction conditions (wt% active metal in catalyst)	Bio-oil yield
Nannochloropsis sp.	Pd/C	350 °C, 60 min, 50 wt. % catalyst loading	57
Nannochloropsis sp.	Pt/C	350 °C, 60 min, 50 wt. % catalyst loading	49
Nannochloropsis sp.	Ru/C	350 °C, 60 min, 50 wt. % catalyst loading	50
Nannochloropsis sp.	Ni/SiO_2-Al_2O_3	350 °C, 60 min, 50 wt. % catalyst loading	50
Nannochloropsis sp.	CoMo/Al_2O_3	350 °C, 60 min, 50 wt. % catalyst loading	55
Nannochloropsis sp.	zeolite	350 °C, 60 min, 50 wt. % catalyst loading	48
C. pyrenoidosa	Ce/H-ZSM-5	300 °C, 20 min, 5 wt. % catalyst loading	52
B. braunii	Na_2CO_3	300 °C, 60 min, 5 wt. % catalyst loading	64
Chorella vulgaris	non-catalytic	350 °C, 15 min	36
B. braunii	non-catalytic	310 °C, 15 min	68
Nannochloropsis	non-catalytic	320 °C, 30 min, 5 wt. % solid loading	57
Scenedesmus	non-catalytic	320 °C, 30 min, 5 wt. % solid loading	55
E. prolifera	Na_2CO_3	300 °C, 30 min, 5 wt. % catalyst loading	23
Desmodesmus sp.	non-catalytic	375 °C, 5 min	49
Nannochloropsis sp.	non-catalytic	350 °C, 60 min	35
Chorella sp.	non-catalytic	320 °C, 30 min, 5 wt. % catalyst loading	40

can then be co-fired with coal to produce heat and electricity in a typical coal-fired power plant.

Table 14.5: Studies on HTC of algae showing the hydrochar caloric value (Feng et al., 2022).

Algae	Temp. (°C)	Residence time (min)	Caloric value (MJ/kg)
Algae in high-rate ponds	130–170	10–50	16.99–17.47
Scenedesmus (untreated)	180–260	240	7.14–26.64
Algae	180–240	60	/
H. reticulatum	150–270	60	18.29–25.4
C. vulgaris	150–270	60	24.78–34.53
C. vulgaris	180–240	30	21.9–27.74
S. horneri	180–210	120–960	19–25.1
Algae in municipal wastewater	180–240	60–240	/
Sea lettuce	150–220	30–120	20.23
C. vulgaris	180–250	30–240	21–25.3
S. platensis	180–300	30	/

14.8 The potential of microalgae in wastewater treatment

Water is a natural resource that is abundant around the world, but access to clean water has been a problem in many emerging nations with an industrial basis. Global water use has grown by 700% in the last century and is still rising, which is making water shortages worse in many parts of the world (Wada et al., 2016). In the past, most efforts on wastewater treatments plants were successfully focused on obtaining an effluent of high quality, while nowadays, research and practice in the field are oriented to ensure the sustainability of wastewater treatment plants in terms of economy feasibility, environmental impacts, and resource recovery (Tua et al., 2021). In the context of circular economy and sustainability, the recovery of resources from wastewater is equally crucial. Industrial, municipal, and agricultural wastewaters with significant amount of organic and inorganic contaminants are the main causes of water pollution. In addition to industrial and household trash dumped into rivers, the majority of wastewater sources contain heavy metals, xenobiotics, high concentrations of nitrates, phosphates, and carbon compounds as pollutants (Kim et al., 2016; de Oliviera et al., 2020). It is also recognized that agricultural runoff or leaching releases large amounts of pesticide into aquatic ecosystems and that some of these organic pollutants may be present there. Aquatic microalgae production may lower biological oxygen consumption, total suspended particles, pathogen organisms, and nitrogen and phosphorus concentrations in wastewater (Khoo et al., 2020). Table 14.6 highlights the characteristics of microalgae that address the key wastewater treatment goals.

Table 14.6: The impact of microalgae application on BOD, nutrient, heavy metals, and pathogen removal in wastewater treatment (Peter et al., 2021).

Criteria	Microalgae application
BOD (biochemical oxygen demand) removal	Microalgae in the treatment of brewery effluent and its biomass production have been evaluated. The maximum dry weight biomass reaches 0.917 g/L with 91.43% of BOD reduction.
Nutrient removal	Phytoremediation technology using microalgae indicating an excellent nutrient removal efficiency with low implementation and maintenance cost compared to other conventional physicochemical methods such as ion-exchange membrane separation, reverse osmosis.
Heavy metal removal	Macroalgae can absorb a large variety of heavy metals depending on the biomass growth conditions. Reports indicate that grown in photoautotrophic conditions, they showed the ideal biosorption characteristics.
Pathogen removal	Microalgae's photosynthesis process enhances the dissolved oxygen produced, which increases the pH value. This mechanism contributes to the reduction of pathogen bacteria present in the effluent.

Microalgae are used in wastewater remediation primarily because of their capacity to directly absorb or consume contaminants in order to grow. It is well-known that microalgae require nitrogen and phosphate in the cultivation process, which may be obtained directly during wastewater treatment; therefore, the primary targeted pollutants are nitrogen and phosphorus (Zhang et al., 2020). Microalgae have been found to be efficient at removing nutrients from wastewater as well as bacteria, heavy metals, and biological oxygen consumption. (Abdel-Raouf et al., 2012). Microalgae-based wastewater treatment has several benefits over standard microalgae culture and conventional wastewater treatment procedures; therefore, it is increasingly being used to treat wastewater, and these processes have the advantage of being green. High nutrient, CO_2, and energy needs are linked to the cultivation of microalgae, and these factors are the main barriers to commercial microalgae production. When microalgae are grown on wastewater, the costs of fertilizer, external CO_2 supply, and freshwater can all be decreased. Microalgae are diverse photosynthetic microorganisms and can be cultivated on nutrient-rich wastewater to produce food, feed, biofertilizer, biofuels, biopolymers, and many other high value products, although there are concerns for human consumption (Chaudry, 2021). In contrast to conventional wastewater treatment processes, which require a significant amount of energy (0.5 kWh/m^3), processes using wastewater treatment and microalgae to remove nutrients use a lot less energy (0.2 kWh/m^3) while simultaneously producing valuable biomass (Molazadeh et al., 2019). In a particular study, the cost of biomass production was found ranging from \$ 0.39 to \$ 0.92/kg, with a minimum cost for the anaerobically digested domestic effluent and centrate. The cost of wastewater treatment was found to be ranging from \$0.18–1.69/m^3, with minimum for the sewage (Chaudry, 2021). The estimated costs for wastewater treatment with microalgae for aerobically domestic effluent and centrate were \$1.69/m^3 and \$0.53/m^3, respectively. This wastewater treatment process produced 4.3 kg/m^3 of algal biomass from anaerobically digested domestic effluent and 1.34 kg/m^3 of centrate. According to the cost of biomass production, any credit above \$0.39/kg produced by biomass will counterbalance the cost of anaerobic wastewater and centrate treatment via algae cultivation. The estimated cost of sewage treatment, on the other hand, was \$0.18/m^3, which generates 0.24 kg of algal biomass/m^3 of wastewater. Any credit generated by the biomass above \$0.73/kg would offset the sewage treatment cost. (Chaudry, 2021). In order to determine the integrated microalgae cultivation's actual potential, the wastewater treatment process should be planned and studied on an individual basis, taking into account the geographic location, economic restrictions, land availability, and the localized end-use of biomass. Generally, the microalgae cultivation system technology could be done in open systems (non-stirred ponds and stirred ponds) and closed systems such as tubular photobioreactor, flat plate photobioreactor, and plastic bag photobioreactor, (Brennan and Owende, 2010). The best method depends a lot on the objective, scale, type of microalgae, species, and cost. Compared to photobioreactors, open ponds seem to be more cost-effective. Microalgae have furthermore been demonstrated to be quite effective in treating wastewater, in addition to CO_2 biofixation. Compared to electromechanical wastewater treatment systems, high-rate algal ponds can offer effective tertiary-level wastewater treatment at a significantly cheaper cost. Although microalgae may be used independently for the processes of CO_2 biofixation, wastewater treatment, or biofuel generation, integrating all of these processes would be more effective. The combination of wastewater treatment, biofuel generation, and CO_2 fixation based on microalgae might be viewed as multifaceted ways to managing significant environmental concerns. It is a sustainable and environmentally friendly method of lowering the price of producing biodiesel, fertilizers for farms, sewage treatment, and fuel gas treatment. Eventually, it may result in an effective decrease in greenhouse emissions. Future large-scale initiatives based on algae might reduce the existing technological and financial obstacles to the production of microalgae-based biofuels (Molazadeh et al., 2019). The qualities of the effluent (flowrate and nutrient concentration) are crucial to the cost-benefit analysis of wastewater treatment combined with microalgae production. Both the price of labor and the quantity of nutrients that must be removed are significant cost-inflationary factors. Reducing biomass production and nutrient removal can lower the cost of microalgae-based wastewater treatment and the quantity of land needed. On the other hand, the external supply of limiting nutrients can reduce the cost of algal biomass production on wastewater. Microalgae-based wastewater treatment has the potential to be particularly successful in developing nations when compared to traditional wastewater treatment methods due to the much cheaper labor cost. Integrated processes are more suited to wastewater with higher nutrient concentrations, such as centrate in domestic wastewater, piggeries effluent, and anaerobically digested abattoir effluent. However, there would be a trade-off between wastewater treatment (nutrient removal) and the land requirement for a microalgae-based wastewater treatment process in the case of extremely high nutrient concentrations (Chaudry, 2021).

14.9 Microalgae as potential feedstock for biopolymers and other valuable compounds production

Approximately one-third of the more than 400 million tons of plastic produced annually around the world ends up as plastic waste in landfills, freshwater lakes, rivers, and oceans (Modolo et al., 2010). Petroleum-based plastics and polymers cannot biodegrade and can have detrimental effects on the ecosystem (Horton et al., 2017). The circular economy concept has received a lot of attention as a solution to the present problems associated with the growth of consumption and production demands. This approach is primarily focused on the resource, recovery, and recycling strategy to limit raw material usage and natural resource consumption.

Global demand for plastic materials has considerably harmed the environment and especially the marine sea life; consequently, bioplastics have emerged as environmentally friendly alternatives, decreasing the fossil-based plastic usage. Plants, animals, agricultural wastes, and microorganisms can all be used to produce biopolymers. Bioplastics are currently originated from terrestrial crops such as maize and potatoes, competing with food sources in terms of enormous land areas, water, and nutrients (Zeller et al., 2013). Microalgae and cyanobacteria are considered promising sources of polyhydroxyalkanones (PHAs), cellulose, carbohydrates, and proteins as main components of microalgae used in bioplastics production process (Madadi et al., 2021). The capacity to produce bioplastics increased from 2 million tons in 2014 to about 6.7 million tons in 2018, primarily using starch and PLA-based polymers (Cinar et al., 2020). Bioplastics can be blended or combined with synthetic polymers to increase their physical and durability features (Rahman and Miller, 2017). The tensile strength for petroleum-based plastics, such as HDPE (high density polyethylene), LDPE (low density polyethylene), PVC (polyvinyl chloride), PP (polypropylene), and PLA (poly-lactic acid), is shown in Table 14.7.

The tensile strength of PHBs (polyhydroxybutyrates) is 40 MPa, which is equivalent to the tensile strengths of synthetic PLA, PP, and HDPE, which are 47, 30.6, and 10–43 MPa, respectively. As a result of having substantially lower tensile strength and elongation values than synthetic polymer and bioplastic blends, *chlorella* and *spirulina* biomass are not suitable for commercialization. The notion of integrating microalgae biomass with synthetic polymers has resulted in satisfactory outcomes or even superior mechanical qualities when compared to certain of the synthetic polymers described in Table 14.7. The lowest and highest bioplastic blends have elongation qualities that are lower than those of synthetic polymers, ranging from 1.4–26.3%

Table 14.7: Mechanical strength comparison of bioplastic from microalgae and synthetic polymer (Chong et al., 2022).

Microalgae source	Synthetic polymer	Plastic blend	Ratio algae/plastic	Tensile strength MPa	Elongation (%)
None	HDPE	/	0/100	10–43	>100
None	LDPE	/	0/100	3–56	>100
None	PP	/	0/100	30.6	147
None	PVC	/	0/100	55	2–100
None	PLA	/	0/100	47	3.2
Chlorella	/	/	100/0	5.7	3.4
Spirulina	/	/	100/0	3	1.4
Chlorella	/	Modified PE	10/90	22	14.1
Chlorella	/	Unmodified PE	10/90	20	26.3
Chlorella	/	PP with MPP	10/90	32	11
Chlorella	/	PP without MPP	10/90	28	7
Chlorella	/	PVC	10/90	41	1.9
Chlorella	/	PVC	20/80	32	1.9
Green algae	/	PLA	20/80	45	3
Algae	/	PHBs	100/0	40	5

(Chong at al., 2022). Microalgae can readily thrive in most lakes, seas, and rivers, and they can produce bioactive chemicals such as lipids, protein, and carbs. PHA, fatty acids, and cellulose may be used in a variety of bioplastic applications. Chew et al. (2017), proposed incorporating the concept of microalgae biorefinery by producing multiple high-value products such as proteins, pigments, carbohydrates, vitamins, biofuel, and biogas via several conversion processes (thermochemical, biochemical, transesterification, photosynthetic microbial fuel cell), with the remaining residual biomass being processed into bioplastic. PHAs, which are biodegradable biopolymers made by various microorganisms, can replace petroleum-based plastics because of the mechanical quality similarities they have with synthetic polymers. When diverse microalgae species are pushed to produce PHAs and lipids for survival due to environmental stress or nutrient deficiency, PHAs are created from these microalgae species. PHAs are used for a number of different things, including packaging films for shopping bags, containers, fiber and foam materials, medical implants, medication delivery systems, cups, and more (Cheng et al., 2011). In some algae strains, roughly 30% of the microalgae biomass products are cellulose-based plastics, which are regarded as traditional types of bioplastics. The microalgae's cell wall is strengthened by polysaccharide matrix, which is followed by a microfibril matrix in the inner membrane, which contains some cellulose and hemicellulose (Khoo et al., 2020). Microalgal cellulose is fibrous, rigid, highly crystalline, insoluble in water, biodegradable, and biocompatible. It is frequently used as a filler and reinforcement for bioplastic (Chong et al., 2022). The cellulose concentration of filamentous green algae is found to be in the 20–30 wt% range, and the green algae *C. glomerata* can accumulate cellulose up to 45 wt%, making it a good reinforcement material for bioplastics. *Lyngby* sp. and *U. fasciata* can produce 14–17% cellulose. As algae-derived cellulose is not suitable for bioplastics due to its lack of thermal stability, these issues are addressed by incorporating various forms of cellulose such as nanofibers, cellulose nanofibers, and microcrystalline cellulose. Microalgae-derived cellulose nanofibers can be used as reinforcement in polyurethane foams to improve mechanical tensile strength, biodegradability, thermal resistance, and elasticity. Lactic acid is an organic acid that is used as a food preservative. It is an important intermediate feedstock in the manufacture of polylactic acids. Lactic acid is produced by chemical synthesis and microbial fermentation and is categorized as L-lactic acid and D-lactic acid; both forms are safe to use as food additives. however D-lactic acid is unsafe for human consumption. Nguyen et al. (2012) produced lactic acid from microalgae *H. reticulum* via the fermentation process of *Lactobacillus paracasei* LA104, yielding 97.3% pure lactic acid. In three days, the production of pure D-lactic acid from microalgae *Nannochlorum* sp.26A4 converted 70% of the consumed starch into 70% D-lactic acid (Hirayama and Ueda, 2004). In another study, lipid-free microalgae hydrolysate is used as feedstock in a fermentation process to produce lactic acid. After the lipid extraction, the lipid-free hydrolysate is deionized to maintain lactic acid productivity of 90% or higher. *Lactobacillus pentosus* was used to produce lactic acid under anaerobic conditions, yielding 92.8%. Polymerization can be used to convert lactic acid into polylactic acid. Polylactic acid can be produced using traditional methods such as lactic acid polycondensation and lactide ring-opening polymerization. Direct polycondensation polymerization method has complications in producing high molecular weight polymer with good mechanical properties, whereas polycondensation ring-opening polymerization by lactide can synthesize high molecular polymer but requires high temperature and pressure, driving up the production costs.

Starch is an abundant polysaccharide, a natural component of most plants, which can be used in the food, textile, paper, and packaging sectors. It is also easily digestible and biodegradable, being an important part of a balanced diet. Little starch granules (0.5–2.1 mm) produced by microalgae are regarded as new feedstock for the manufacture of starch-based bioplastics. Microalgae have the benefit of being lignin-free sources of carbohydrates like starch and cellulose. The following three methods can be used to incorporate starch into polymeric materials: using starch as a raw material to create low molecular weight hydroxylated compounds (dextrins and glycolyzed products are two examples of polymers used in polyurethane formulations), or directly converting starch into the monomers needed to synthesize polymers such as poly lactic acid from lactic acid, polyethylene from ethylene prepared by ethanol dehydration from or even PHAs from ethylene. Using the starch from microalgae, research was done on the creation of starch-based bioplastics (Mathiot et al., 2019). After 20 days of sulfur deprivation (under autotrophic conditions), *Chlamydomonas reinhardtii* 11-32A strain produced the most starch, up to 49% w/w, at a concentration of 5.07 g/L in flasks. Starch-enriched macroalgae biomass was successfully plasticized with glycerol using twin-screw extrusion. In a different study, the starch content of the algal biomass and the suitability of the starch from the marine microalga *Klebsormidium flaccidum* for the production of starch-based bioplastics were evaluated. To ensure the sustainability of starch produced from marine microalgae in bioplastic formulations, starch properties including granule size, amylose/amylopectin content, swelling power, solubility, and turbidity were examined (Ramli et al., 2020). The results were compared with cornstarch, commonly used in the production of bioplastic and demonstrated the high viability of the approach using starch from the microalgae (Ramli et al., 2020). Polymer blending is a straightforward process that results in better polymeric materials; numerous studies on PHA blends have been carried out to enhance their qualities and lower their production costs. Biodegradable polymers like cellulose, lignin, amylose, amylopectin, polylactic acid, and polycaprolactone are frequently combined with PHAs to enhance their characteristics. It was discovered that cellulose derivatives have an excellent compatibility with PHAs, including ethyl cellulose, cellulose propionate, and cellulose acetate butyrate. PHA and plasticizer blending is a well-known method for improving the performance of most bio-based polymers. The best plasticizer should be nonvolatile, stable, nontoxic, and biodegradable. Glycerol, oxypropylated glycerin, glycerol triacetate, polyethylene glycol, 4-nonylphenol, acetyl tributyl citrate, acetylsalicylic acid ester, salicylic ester, dioctyl phthalate, and dibutyl phthalate are the most prevalent plasticizers (Madadi et al., 2021). Compression molding is the most prevalent method for producing microalgae-polymer blends, in which a mixture of biomass, polymers, and additives is placed in a mold and squeezed at elevated pressure and temperature for a short period of time to make bio-composites. Temperature, pressure, and time all play a crucial role According to some reports, temperatures range from 130–160 °C, pressures range from 20 kPa–0 MPa, and molding times range from 3–20 min. Before the molding process, the mixes must be adequately homogenized. Figure 14.5 illustrates the bioplastic synthesis process from microalgae biomass.

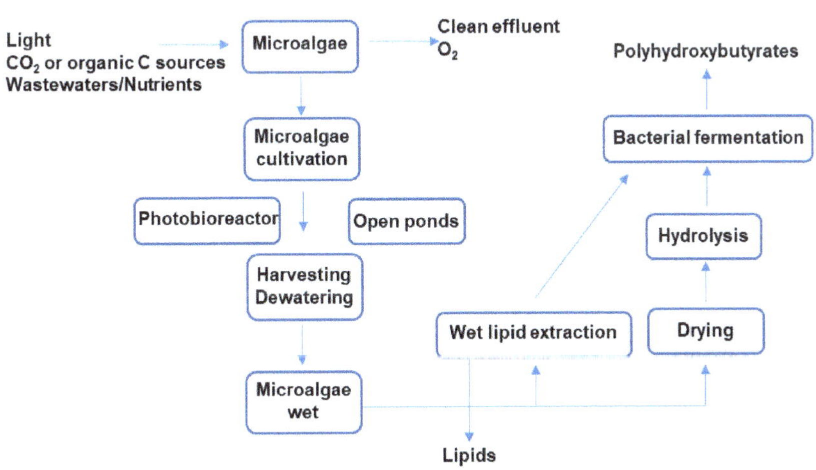

Figure 14.5: A schematic representation of bioplastic production process from microalgae.

The first step is microalgae cultivation (open ponds or photobioreactors) followed by harvesting, which is a fundamental step to collect microalgae biomass; this may include filtration, floatation centrifugation, and sedimentation. The most suitable is flocculation due to its low energy requirement; however, the optimum method should be selected based on microalgae types and the final product characteristics. As shown in Figure 14.5, two distinct techniques can be used to produce PHB from microalgae, following culture and harvesting. The dried microalgae are further hydrolyzed to produce fermentable sugars in the first method, which could then be used for *Escherichia coli* fermentation and PHB production. The second method involves a wet lipid extraction procedure, which has the potential to lower overall costs since it does not require a drying step (Cinar et al., 2020). Spirulina species can be used to make bioplastics. Moreover, Wang (Wang, 2014) suggested a novel mixing approach to enhance the performance of bioplastics traditionally made from 100% spirulina biomass. The system includes ethylene glycol plasticization, ultrahigh-molecular-weight polyethylene mixing, and polyethylene-graft-maleic anhydride compatibilization with the spirulina. To determine the protein concentration and a rough estimate of the molecular weight, two methods can be used: the cinchonic acid test assay and sodium dodecyl sulfate-polyacrylamide gel electrophoresis. Microalgae and Cyanobacteria contain valuable compounds for health and cosmetic applications; these include mainly bioactive carbohydrates, antioxidants, pigments, and fatty acids, among others. Carotenoids produced from microalgae are important for health protection. Beta-carotene from *Dunaliella salina* has proven efficient in preventing angiogenesis in laboratory studies, thus helping prevent the irregular form of new blood vessels in pathologies such as cancer (Guruvayoorappan and Kuttan, 2007). Astaxanthin from *Haematococcus pluvialis* has been reported to lower blood pressure; it was shown to act as an antihypertensive in spontaneously hypertensive rats, at a dosage of 50 mg/kg over 14 days (Hussein et al., 2005). Carotenoids have also been reported for their anticancer activity, especially contributing to the prevention of cancers such as breast, hepatic, intestinal, and prostate cancer as well as leukemia. Likewise, carotenoids have been reported for treating and controlling diabetes, with beta-carotene concentrations in the blood being inversely associated with fasting blood glucose levels and insulin resistance, respectively (Ylonen et al., 2003). In addition, carotenoids such as lycopene, lutein, and zeaxanthin protect against diabetic retinopathy. Regarding anti-obesity activity, the chemical structures of carotenoids are important for suppressing adipocyte differentiation, with fucoxanthin and neoxanthin exhibiting significant anti-obesity functions (Fernandez et al., 2021).

14.10 State-of-the-art

The durability of microalgae plastics can be effectively increased by blending them with a variety of petroleum-based plastics, including HDPE, LDPE, PVC, PP, and PLA (Chong et al., 2022). The following describes the state-of-the-art of producing bioplastics utilizing microalgae as feedstock: blending microalgae biomass synthetic polymers; microalgae biomass as feedstocks for PHAs; cellulose/lipid-based plastics; intermediate processing of PLA involving the production of lactic acid from microalgae and downstream state-of-the-art polymerization techniques; blending microalgae biomass with synthetic polymers; and blending microalgae biomass with plastics. Due to their high protein contents of about 51–58% and 46–63%, respectively, both chlorella and spirulina are recognized as being appropriate for the synthesis of bioplastics. Equally, chlorella and spirulina contain very small cells and high protein content, according to Zeller (Zeller et al., 2013). Besides, bioplastic conversion is feasible without any pretreatment step, resulting in cost-effective large-scale production and waste reduction. Concerns expressed by society and governments must be included in discussions on the impact of plastics. Furthermore, creating a circular economy for these products, across all commercial sectors including food, health, cosmetics, and nutraceutical industries and services, is a main priority. Bioplastics, as a component of the bioeconomy, are frequently referred to be the pinnacle of circularity since they not only recycle CO_2 but also utilize renewable raw materials to create more sustainable everyday items. Microalgae culture is regarded as a contemporary biotechnologist innovation that supports advanced biotechnology. According to estimates, 5,000 tons of dry matter of microalgae-based biomass might be produced annually by 2036, with a market value of $1,250 million (Khan et al., 2018). The atmospheric air balance, on the other hand, has been of concern, particularly for large-scale activities. Studies have also shown that absorbing CO_2 gas emitted by factories and industries could promote the growth of microalgae (Raja et al., 2014). When compared to land-based plants, aquatic microalgae have been found to grow more quickly due to their higher rates of carbon fixation (Khan et al., 2018). It was determined through research on an integrated algal biorefinery process that the combination of algae processing lowers the cost of biofuel by 9% while improving energy production to 126 gallons of gasoline equivalent per ton of biomass (Posada, 2016). The wastewater streams are mostly used in the microalgae process integration for nutrients. This is due to the fact that microalgae may absorb the nutrients found in the wastewater flow, acting as a treatment agent at the same time (Tao, 2016). In conclusion, the new microalgae-based technologies have proven to be viable alternatives. Furthermore, recovery of lipids and other beneficial bio-compounds, which can then be refined into biofuel, biodiesel, or pharmaceutical health supplements, is achievable. Microalgae can be used in the efforts of protecting the environment. The diverse application of microalgae has to be commercialized on a large-scale especially as biological assessment, as an alternative biological treatment for wastewater treatment. The biomass produced can also be used as an environmental regulator. Moreover, microalgae contribute to environmental sustainability by considering atmospheric balance, aquatic environmental sustainability, land, and water resources. Enhancing research on the implementation of eco-technology in process integration using microalgae is critical in producing a sustainable end-product that can meet energy demand without harming the global environment. Recently, a study indicated that there are several potential benefits of co-locating an oil refinery and an algae-based biorefinery and that the use of microalgae as feedstock is more beneficial than the use of microalgae from a system energy perspective (Andersson et al., 2020). It takes several days to produce biofuels using biochemical conversion techniques like fermentation, and it is still very difficult to achieve viability for commercialization (Nahak et al., 2011). The potential for microalgae to be valued and converted into biofuel and other useful byproducts is high. The emphasis is on finding ways to improve the quality and yield of bio-oils because the bio-oil product from the thermal conversion processes is still far from the desired quality when compared to petroleum-based fuel. Bio-oil with its high nitrogen content is still one of the most significant challenges affecting the fuel quality produced by thermal conversion processes, and study efforts should be directed toward creating strategies for improving fuel quality.

A state-of-the-art drying process for microalgae, as well as their integration with gasification and power generation, has been proposed by Aziz, (Aziz et al., 2014). An improved process integration methodology has also been proposed, in which the concepts of exergy recovery and process integration are used to achieve the lowest possible exergy destruction. As a result, greater energy efficiency may be realized. Due to the high moisture content and the presence of bound water within the microalgae, drying is an energy-intensive process. High-energy-efficiency technologies are urgently needed to improve the energy returned on energy invested (ERoEI) from the use of microalgae. Using the proposed process integration technology in microalgae could significantly reduce the energy required for drying. Depending on the target moisture content, the total energy input and compressor outlet pressure differ. The minimum total required work is approximately 28 kW (wet microalgae flow rate of 1 ton h^{-1}), which can be associated with drying to a moisture content of 10 wt.% (wet basis). Furthermore, for drying to a moisture content of 10 wt.% wet basis, the highest coefficient of drying performance that can be achieved is around 18.5. In this case, nearly all of the energy used in drying can be recovered with minimal exergy destruction. There are several methods, including thermochemical, chemical, and biological, for turning algae into fuels and chemicals. Their techno-economic viability is still unclear, though. However, integrating the manufacture of high-value products like bioactive compounds into the algal biorefinery system is one way to increase the economic feasibility of the process. Another strategy to increase the algal biorefinery's sustainability is to grow algae in wastewater. Moreover, the development of a sustainable algal biorefinery system may benefit from the utilization of CO_2-containing industrial flue gas to produce algae (Feng et al., 2022). Encouraging results have recently been obtained by boosting light use efficiency or by strengthening specific metabolic pathways. Additional research efforts and funding for implementing innovative biorefineries will realistically support progress toward next-generation algal biotechnology (Benedetti et al., 2018).

14.11 Recommendations

Microalgae cultivation provides opportunity for sequestering anthropogenic waste carbon dioxide into biomass feedstock, which can be used for producing biofuels and bioproducts. Microalgae contain biopolymeric compounds such as protein, lipids, and carbohydrates and high value carotenoid including β-carotene, lutein, and zeaxanthin. Traditionally, microalgae have been considered for biodiesel production using lipids and biogas production from residual biomass. This approach leads to low-value application of some of the important compounds such as carotenoid, protein, and carbohydrates. To utilize the full value of microalgae biomass, it is important to look at atom economy approach and try to utilize each component of microalgae biomass for high-value products. One of the challenges is how to fractionate multiple components without degrading some of the other fractions. The use of subcritical water in a continuous flow reactor provides a chemical free and environmentally friendly process for fractionating microalgae biomass components. The process can use microalgae slurry, thus eliminating the need for cost-intensive drying process. At low temperature (below 200 °C), with a very short residence time, high value carotenoids can be extracted without degrading the other components of biomass. With temperature in the range of 200–280 °C with a residence time of few seconds under subcritical water condition, protein and carbohydrates can be hydrolyzed without changing the quality of lipids. The residual solids, which contain all the lipids and unhydrolyzed compounds, can be used for biofuels production via transesterification or catalytic hydrotreating process. In this microalgae biorefinery method, high value compounds typically provide economic feasibility of producing biofuels matching the fossil fuels cost.

References

Abdel-Raouf, N., Al-Homaidan, A. A., & Ibraheem, I. B. M. (2012). Microalgae and wastewater treatment. *Saudi Journal of Biological Sciences*, *19*(3), 257–275.

Andersson, V., Heyne, S., Harvey, S., & Berntsson, T. (2020). Integration of algae-based biofuel production with an oil refinery: Energy and carbon footprint assessment. *International Journal of Energy Research*, *44*, 10860–10877.

Aziz, M., Takuya, O., & Takao, K. (2014). Integration of energy-efficient drying in microalgae utilization based on enhanced process integration. *Energy*, *70*, 307–316.

Beer, L., Boyd, E., & Peters, J. (2009). Engineering algae for biohydrogen and biofuel production. *Current Opinion in Biotechnology*, *20*, 264–271.

Beligon, V., Christophe, G., & Fontanille, P. (2016). Microbial lipids as potential source to food supplements. *Current Opinion in Food Science*, *7*, 35–42.

Benedetti, M., Vecchi, V., Barera, S., & Dall'Osto, L. (2018). Biomass from microalgae: The potential of domestication towards sustainable bio factories. *Microbial Cell Factories*, *17*, 173. 10.1186/s12934-018-1019-3

Bharathiraja, B., Sridharan, S., Sowmya, V., Yuvaraj, D., & Praveenkumar, R. (Jun 2017). Microbial oil – A plausible alternate resource for food and fuel application. *Bioresource Technology, 233*, 423–432. 10.1016/j.biortech.2017.03.006

Biller, P., & Ross, A. B. (2011). Potential yields and properties of oil from the hydrothermal liquefaction of microalgae with different biochemical content. *Bioresource Technology*, *102*(1), 215–225.

Brennan, L., & Owende, P. (2010). Biofuels from microalgae – A review of technologies for production, processing, and extractions of biofuels and co-products. *Renewable and Sustainable Energy Reviews, 14*, 557–577.

Carioca, J. B. (2010). Problems, challenges and perspectives. *Biotechnology Journal, 5*, 260–273.

Cheng, C. H., Du, T. B., Pi, H. C., Jang, S. M., Lin, Y. H., & Lee, H. T. (2011). Comparative study of lipid extraction from microalgae by organic solvent and supercritical CO2. *Bioresource Technology*, *102*(21), 10151–10153. 10.1016/j.biortech.2011.08.064

Chaudry, S. (2021). Integrating microalgae cultivation with wastewater treatment: A peek into economics. *Applied Biochemistry and Biotechnology, 193*, 3395–3406.

Chakraborty, M., Miao, C., & McDonald, A. (2012). Concomitant extraction of bio-oil and value-added polysaccharides from *chlorella sorokiniana* using a unique sequential hydrothermal technology. *Fuel, 95*, 63–70.

Chen, G. Q., Jiang, Y., & Chen, F. (2008). Variation of lipid class composition in nitzschia laevis as a response to growth temperature change. *Food Chemistry, 109*, 88–94.

Chew, K. W., Yap, J. Y., Show, P. L., Suan, N. H., Juan, J. C., Ling, T. C., Lee, D., & Chang, J. (2017). Microalgae biorefinery: High value products perspectives. *Bioresource Technology, 229*, 53–62.

Chiaramonti, D., Prussi, M., Buffi, M., Casini, D., & Rizzo, A. M. (2015). Thermochemical conversion of microalgae: Challenges and opportunities. *Energy Procedia, 75*, 819–826.

Chia, W. Y., D.Y.Y, T., Khoo, K. S., Lup, A. K., & Chew K, W. (2020). Nature's fight against plastic pollution: Algae for plastic biodegradation and bioplastics production. *Environmental Science and Ecotechnology, 4*, 100065. ISSN 2666-4984. https://doi.org/10.1016/j.ese.2020.100065

Chong, J. W. R., Tan, X., Khoo, K. S., & H.s, N. (2022). Microalgae-based bioplastics: Future solution towards mitigation of plastic waters. *Environmental Research, 206*, 112620–112635.

Christi, Y. (2007). Biodiesel from microalgae. *Biotechnology Advances, 25*, 294–306.

Christophe, G., Kumar, V., & Nouaille, R. (2012). Recent developments in microbial oils production: A possible alternative to vegetable oils for biodiesel without competition with human food? *Brazilian Archives of Biology and Technology*, *55*(1), 29–46.

Cinar, S. O., Chong, Z. K., Kucuker, M. A., Wieczorek, N., Cengiz, U., & Kuchta, K. (2020). Bioplastic production from microalgae: A review. *International Journal of Environmental Research and Public Health, 17*, 3842–3863.

Colin, R., & James, P. W. (2002). The Biochemistry and Molecular Biology of Lipid Accumulation in Oleaginous Miccroorganisms. Elsevier Science, USA.

de Oliveira, M., Frihling, F. J. C. M., Velasques, J., Filho, F., Cavalheri, P. S., & Migliolo, L. (2020). Pharmaceuticals residues and xenobiotics contaminants: Occurrence, analytical techniques and sustainable alternatives for wastewater treatment. *Science of the Total Environment, 25*(*705*), 135568. 10.1016/j.scitotenv.2019.135568

Dong, T., Knoshaug, E. P., Pienkos, P. T., & Laurens L, M. L. (2016). Lipid recovery from wet oleaginous microbial biomass for biofuel production: A critical review. *Applied Energy, 177*, 897–895.

Duan, P., & Sanvage, P. E. (2011). Hydrothermal liquefaction of a microalga with heterogeneous catalyst. *Industrial and Engineering Chemistry Research, 50*, 52–61.

Fernandez G, A. F., Reis, A., Wijffels, R. H., Barbosa, M., Verdelho, V., & Llamas, B. (2021). The role of microalgae in the bioeconomy. *New Biotechnology, 61*, 99–107.

Feng, S., Kang, K., Salaudeen, S., Ahmadi, A., He, S. Q., & Hu, Y. (2022). Recent advances in algae-derived biofuels and bioactive compounds. *Industrial and Engineering Chemistry Research, 61*, 1232–1249.

Guedes, A. C., Amaro, H. M., & Malcata, F. (2011). Microalgae as sources of carotenoids. *Marine Drugs*, *9*, 625–644.

Gollakota, A., Kishore, N., & Gu, S. (2017). A review on hydrothermal liquefaction of biomass. *Renewable Sustainable Energy Reviews*, *81*, 1378–1392.

Guo, Y., Yeh, T., Song, W., Xu, D., & Wang, S. (2015). A review of bio-oil production from hydrothermal liquefaction of algae. *Renewable and Sustainable Energy Reviews*, *48*, 776–790.

Goncalves, A. L., Pires, J. C. M., & Simones, M. (2013). Green fuel production: Processes applied to microalgae. *Environmental Chemistry Letters*, *11*, 315–324.

Griffiths, M. J., & Harrison, S. T. L. (2009). Lipid productivity as a key characteristic for choosing algal species for biodiesel production. *Journal of Applied Phycology*, *21*, 493–507.

Guruvayoorappan C. and Kuttan G., 2007. β-Carotene inhibit tumor-specific angiogenesis by altering the cytokine profile and inhibits the nuclear translocation of transcription factors ibn B16F-10 melanoma cells. Integr. *Cancer Ther.* 6, 258–70.

Heilmann, S. M., Davis H, T., Jader L, R., Lefebvre P, A. P., Sadowsky M, J., Schendel, F. J., Von keitz, M. G., & Valentas K, J. (2010). Hydrothermal carbonization of microalgae. *Biomass and Bioenergy*, *34*(6), 875–882.

Hirayama, S., & Ueda, R. (2004). Production of optically pure d-lactic acid by nannochlorum sp. 26ba4. *Applied Biochemistry and Biotechnology*, *119*, 71–77.

Hussein, G., Nakamura, M., Zhao, Q., Iguchi, T., Goto, H., & Sankawa, U. (2005). Antihypertensive and neuroprotective effects of astaxanthin in experimental animals. *Biological and Pharmaceutical Bulletin*, *28*, 47–52.

Horton, A. A., Walton, A., Spurgeon, D. J., Lahive, E., & Svendsen, C. (2017). Science of the total environment microplastics in freshwater and terrestrial environments: Evaluating the current understanding to identify the knowledge gaps and future research priorities. *Science of the Total Environment*, *586*, 127–141.

Illman, A. M., Scragg, A. H., & Shales, S. (2000). Increase in *chorella* strains calorific values when grown in low nitrogen medium. *Enzyme and Microbial Technology*, *27*, 631–635.

Jin, G., Zhang, Y., Shen, H., Yang, X., Xie, H., & Zhao, Z. K. (2013). Fatty acid ethyl esters production in aqueous phase by the oleaginous yeast *rhodosporium toruloides*. *Bioresource Technology*, *150*, 266–270.

Kajikawa, M. A., Tatsuki, K. F., & Ken-Ichi, Y. D. (2016). Production of ricinoleic acid containing monoestolide triacylglycerides in an oleaginous diatom *chaetoceros gracillis*. *Scientific Reports*, *6*, 36809. 10.1038/srep36809

Kapdan, I. K., & Kargi, F. (2006). Bio-hydrogen production from waste materials. *Enzyme and Microbial Technology*, *38*, 569–582.

Khan, M. I., Shin, J. H., & Kim, J. D. (2018). The promising future of microalgae: Current status, challenges, and optimization of a sustainable and renewable industry for biofuels, feed, and other products. *Microbial Cell Factories*, *17*, 1–21.

Khoo, K. S., Chew, K. W., Yew, G. Y., Leong, W. H., Chai, Y. H., Show, P. L., & Chen, W.-H. (2020). Recent advances in downstream processing of microalgae lipid recovery for biofuel production. *Bioresource Technology*, *304*, 122996.

Khozin-Goldberg, I., & Cohen, Z. (2006). The effect of phosphate starvation on the lipid and fatty acid composition of the fresh water eustigmatophyte *monodus subterraneus*. *Phyochemistry*, *67*, 696–701.

Kim, G., Mujtaba, G., & Lee, K. (2016). Effects of nitrogen sources on cell growth and biovhemical compositions of marine chlorophyte *tetraselmis* sp. for lipid production. *Algae*, *31*, 257–266.

Kumar, S., Hablot, E., Moscoso-Garcia, J., Obeid, W., Hatcher, P., DuQuette, B., Graiver, D., Narayan, R., & Balan, V. (2014). Polyurethanes preparation using proteins obtained from microalgae. *Journal of Materials Science*, *49*(22), 7824–7833.

Lee, S. Y., Khorioh, I., Vo, D. V., Ponnusamy, S. K., & Show, P. L. (2021). Techniques of lipid extraction from microalgae for biofuel production: A review. *Environmental Chemistry Letters*, *19*, 231–251.

Levine, R. B., Pinnarat, T., & Savage, P. E. (2010). Biodiesel production from wet algal biomass through in situ lipid hydrolysis and supercritical transesterification. *Energy and Fuels*, *24*, 5235–5243.

Li, Q., Du, W., & Liu, D. (2008). Perspectives of microbial oils for biodiesel production. *Applied Microbiology and Biotechnology*, *80*, 749–756.

Liu, Z. Y., Wang, G. C., & Zhou, B. C. (2008). Effect of iron on growth and lipid accumulation in *chlorella vulgaris*. *Bioresource Technology*, *99*, 4717–4722.

Liu, B., & Zhao, Z. B. (2007). Biodiesel production by direct methanolysis of oleaginous microbial biomass. *Journal of Chemical Technology and Biotechnology*, *82*, 775–780.

Madadi, R., Maljaee, H., Serafim, L. S., & Ventura, V. (2021). Microalgae as contributors to produce biopolymers. *Marine Drugs*, *19*, 466–493.

Mathiot, C., Ponge, P., Gallard, B., Sassi, J.-F., Delrue, F., & Le moigne, N. (2019). Microalgae starch-based bioplastics: Screening of ten strains and plasticization of unfractionated microalgae by extrusion. *Carbohydrate Polymers*, *208*, 142–151.

Metting, F. (1996). Biodiversity and application of microalgae. *Journal of Industrial Microbiology and Biotechnology*, *17*, 477–489.

Milano, J., Ong, H. C., Masjuki, H. H., Chong, W. T., Lam, M. K., Loh, P. K., & Vellayan, V. (2016). Microalgae biofuels as an alternative to fossil fuel for power generation. *Renewable and Sustainable Energy Reviews*, *58*, 180–197.

Molazadeh, M., Ahmadzadeh, H., Pourinfar H, R., Lyon, S., & Rampelotto, P. H. (2019). The use of microalgae for coupling wastewater treatment with CO_2 biofixation. *Frontiers in Bioengineering and Biotechnology*, *7*, 42. 10.3389/fbioe.2019.00042

Modolo, R., Benta, A., Ferreira, V. M., & Machado, L. M. (2010). Pulp and paper plant wastes valorisation in bituminous mixes. *Waste Management*, *30*, 685–696.

Moreno-Osorio, J. H., Pollio, A., Frunzo, L., Lens, P. N. L., & Esposito, G. (2021). A review of microalgal biofilm technologies: Definition, applications, settings and analysis. *Frontiers in Chemical Engineering*, *3*, 737710. https://www.frontiersin.org/article/10.3389/fceng.2021.737710; doi: 10.3389/fceng.2021.737710 ISSN=2673-2718

Moscoso-Garcia, J. L., Obeid, W., & Kumar, S. (2013). Flash hydrolysis of microalgae (*scenedesmus sp.*) for protein extraction and production of biofuels intermediates. *Journal of Supercritical Fluids*, *82*, 183–190.

Moscoso-Garcia, J. L., Teymouri, A., & Kumar, S. (2015). Kinetics of peptides and arginine production from microalgae (*scenedesmus sp.*) by flash hydrolysis. *Industrial and Engineering Chemistry Research*, *54*(7), 2048–2058.

Mustapha, S. I., Rawat, I., & Bux, F. (2022). Enhancing the efficiency of thermal conversion of microalgae: A review. *Biomass Conversion and Biorefinery*. https://doi.org/10.1007/s13399-021-02086-5

Nahak, S., Nahak, G., Pradhan, I., & Sahu, R. (2011). Bioethanol from marine algae: A solution to global warming problem. *Journal of Applied Environmental and Biological Sciences*, *1*, 74–80.

Nguyen, C. M., Kim, J. S., Hwang, H. J., Park, M. S., Choi, G. J., Choi, Y. H., Jang, K. S., & Kim, J. C. (2012). Production of l-lactic acid from a green microalga, *hydrodictyon reticulum*, by *lactobacillus paracasei* LA104 isolated from the traditional Korean food, makgeolli. *Bioresource Technology*, *110*, 552–559.

Nigam, P. S., & Singh, A. (2011). Production of liquid biofuels from renewable resources. *Progress in Energy and Combustion Science*, *37*, 52–68.

Paik, S. M., Sim, S. J., & Jeon, N. L. (2017). Microfluidic perfusion bioreactor for optimization of microalgal lipid productivity. *Bioresource Technology*, *233*, 433–437.

Patel, A., Shindhu, D. K., Arora, N., Singh, R. P., & Pruthi, P. A. (2015). Biodiesel production from non-edible lignocellulosic biomass of *cassia fistula* L. fruit pulp using oleaginous yeast *rhodosporidium kratochvilovae* HIMPA1. *Bioresource Technology*, *197*, 91–98.

Peter, A. P., Khoo, K. S., Chew, K. W., Ling, T. C., Ho, S. H., Chang, J. S., & Show, P. L. (2021). Microalgae for biofuels, wastewater treatment and environmental monitoring. *Environmental Chemistry Letters*, *19*, 2891–2904.

Posada, J. A. (2016). Conceptual design of sustainable integrated microalgae biorefineries parametric analysis of energy use, greenhouse gas emissions and techno-economics. *Algae Red*, *17*, 113–131.

Rahman A. and Miller C.D., 2017b. Microalgae as a Source of Bioplastics. Algal Green Chemistry, pp. 121–138.

Raja, R. (2014). Oceanography and marine research biomass from microalgae: An overview. *Journal of Oceanography and Marine Research*, *2*, 1–7.

Ramli, R. N., Lee, C. K., & Kassim, M. A. (2020). Extraction and characterization of starch from microalgae and comparison with commercial corn starch. *IOP Conference Series: Materials Science and Engineering*, *716*, 012012.

Rao, A. R., Dayananda, C., Sarada, R., et al. (2007). Effect of salinity on growth of green alga *botryococcus braunii* and its constituents. *Bioresource Technology*, *98*, 560–564.

Rodolphi, L., Zittelli, G. C., Bassi, N., et al. (2009). Microalgae for oil: Strain selection, induction of lipid synthesis and outdoor mass cultivation in a low-cost photobioreactor. *Biotechnology and Bioengineering*, *102*, 100–112.

Sakuradani, E., Ando, E., Ogawa, A., et al. (2009). Improved production of various polyunsaturated fatty acids through filamentous fungus *mortierellq alpine* breeding. *Applied Microbiology and Biotechnology*, *84*(1), 1–10.

Sankaran, R., Show, P. L., Cheng, Y. S., Tao, Y., Ao, X., Nguyen, D. P., & Quyen, D. V. (2018). Integration process for protein extraction from microalgae using liquid biphasic electric flotation (LBEF) system. *Molecular Biotechnology*, *60*, 749–761.

Sikarwar, V. S., Zhao, M., Fennell, P. S., et al. (2017). Progress in biofuel production from gasification. *Progress in Energy and Combustion Science*, *61*, 189–248.

Solovchenko, A. E., Khozin-Goldberg, I., & Didi-Cohen, S. (2008). Effects of light intensity and nitrogen starvation on growth, total fatty acids and arachidonic acid in the green microalga *P. incisa*. *Journal of Applied Phycology*, *20*, 245–251.

Shi, S., Valle-Rodriguez, J. O., Siewers, V., et al. (2011). Prospects for microbial biodiesel production. *Biotechnology Journal*, *6*, 277–285.

Tao, D. (2016). Combined algal processing: A novel integrated biorefinery process to produce algal biofuels and bioproducts. *Algal Research*, *19*, 316–323.

Takagi, M., Watanabe, K., Yamaberi, K., et al. (2000). Lilted feeding of potassium nitrate for intracellular lipid and triglyceride accumulation of *nannochloris* sp. UTEX LB1999. *Applied Microbiology and Biotechnology*, *54*, 112–117.

Thakkar, A., Barbera, E., Sforza, E., Bertucco, A., Davis, R., & Kumar, S. (2021). Flash hydrolysis of yeast (saccharomyces cerevisiae) for protein recovery. *Journal of Supercritical Fluids*, *173*, 105240. 10.1016/j.supflu.2021.105240

Tsakona, S., Kopsahelis, N., Chatzifragkou, A., et al. (2014). Formulation of fermentation media from flour-rich waste streams for microbial lipid production by lipomyces starkeyi. *Journal of Biotechnology*, *189*, 36–45.

Tua, C., Ficara, E., Mezzanotte, V., & Rigamonti, L. (2021). Integration of a side-stream microalgae into a municipal wastewater treatment plant: A life cycle analysis. *Journal of Environmental Management*, *279*, 111605–111615.

Vela-Gurovic, M. S., Gentili, A. R., Oliviera, N. L., et al. (2014). Lactic acid bacteria isolated from fish gut produce conjugated linoleic acid without the addition of exogenous substrate. *Process Biochemical*, *49*, 1071–1077.

Wada, Y., Florke, M., Hanasaki, N., Eisner, S., Fischer, G., Tramberend, S., Satoh, Y., Van Vliet, M. T. H., Yillia, P., Ringler, C., Burek, P., & Wiberg, D. (2016). Modeling global water use for the 21st century: The water futures and solutions (WFaS) initiative and its approaches. *Geoscientific Model Development*, *9*(1), 175–222.

Wang, K., 2014. Bio-plastic potential of spirulina microalgae. In: Proceedings of the Beijing Institute of Fashion Technology, Beijing, China, May 2014. https://getd.libs.uga.edu/pdfs/wang_kun_201405_ms.pdf

Ward, O. P., & Singh, A. (2005). Omega-3/6 fatty acids: Alternative sources of production. *Process Biochemical*, *40*, 3627–3652.

Wu, S., Hu, C., Jin, G., et al. (2010). Phosphate-limitation mediated lipid production by *rhodosporium toruloides*. *Bioresource Technology*, *101*, 6124–6129.

Wu, W., & Chang J, S. (2019). Integrated algal biorefineries from process systems engineering aspects: A review. *Bioresource Technology*, *291*, 121939.

Ylonen, K., Alfthan, G., Groop, L., Saloranta, C., Aro, A., & Virtanen, S. M. (2003). Dietary intakes and plasma concentrations of carotenoids and tocopherols in relation to glucose metabolism in subjects at high risk of type 2 diabetes: The botnia dietary study. b am. *Journal of Clinical Nutrition*, *77*, 1434–41.

Zeller, M. A., Hunt, R., Jones, A., & Sharma, S. (2013). Bioplastics and their thermoplastic blends from spirulina and chlorella microalgae. *Journal of Applied Polymer Science*, *130*, 3263–3275.

Zhang, C., Wang, X., Z., M., et al. (2020). Removal of phenolic substances from wastewater by algae. *A Review Environmental Chemistry Letters*, *18*, 377–392.

Zhang, X., Yan, S., Tyagi R., D., et al. (2014). Ultrasonication assisted lipid extraction from oleaginous microorganisms. *Bioresource Technology*, *158*, 253–261.

Zhang, Q., Hu, J., & Lee, D. J. (2016). Biogas from anaerobic digestion processes: Research updates. *Renew Energy*, *98*, 108–119.

Júlio Cesar de Carvalho*, Denisse Tatiana Molina Aulestia, Juliana Cardoso, Hissashi Iwamoto, Maria Clara Manzoki, Carlos R. Soccol

Chapter 15 Process integration opportunities applied to microalgae specialty chemicals production

Abstract: The growing demand for bioproducts from algae is being met by constant improvement in mass cultivation and downstream. However, there still is a gap between the perceived possibilities of microalgae products and the reality of feasible and profitable production. In order to reach economic production of microalgae bioproducts such as biofuels, nutritional ingredients, and specialty products, it is essential to reduce biomass processing costs – from microalgae harvesting and dewatering to pretreatment, fractionation, and purification. While the production routes for specialty microalgae products (e.g., carotenoids or lipids) may leave modest space for alternatives, utilizing other bioprocessing fractions can improve the process economy and sustainability, eventually leading to new, valuable coproducts. This chapter discusses the downstream processing of selected commercial algal products from the point of view of process integration and how to evaluate its multiple possibilities for biorefinery processing.

Keywords: microalgae, lipids, extraction, pigments, carotenoids, biorefinery, protein, valorization

15.1 Introduction

Microalgae products are decades-old ideas finally gaining traction. The enormous potential of these unicellular microorganisms for producing food, feed, biofuels, and biochemicals caused acceleration in research and patenting in the early eighties that oscillated throughout 40 years but has always been exponential, as can be seen from publication statistics. There has been a hype cycle: first, excitement about the fantastic prospects in producing biofuels and capturing carbon dioxide, fueled by the high productivity of microalgae in pure laboratory cultures. Then, techno-economic and life cycle analyses and difficulty in operating pilot plants showed the brutal reality of uneconomical processes. However, further research showed that the pathway to un-

***Corresponding author: Júlio Cesar de Carvalho,** Federal University of Paraná, UFPR, Curitiba, PR, Brazil, e-mail: jccarvalho@ufpr.br
Denisse Tatiana Molina Aulestia, Juliana Cardoso, Hissashi Iwamoto,Maria Clara Manzoki, Carlos R. Soccol, Federal University of Paraná, UFPR, Curitiba, PR, Brazil

https://doi.org/10.1515/9783110781267-015

leashing the microalgae potential might lie in producing high value-added products – and, in the process, perhaps producing biofuels, in a biorefinery approach.

We are living in this integration moment now: there are a few established commercial products of microalgae, dozens of companies worldwide, and much more knowledge about microalgae biotechnology, production, and fractionation. There is even a new wave of investigating wastewaters use in microalgae production, an old idea but *on par* with modern sustainable development goals. There is also much more knowledge about microalgae fractionation into bioproducts, leading to integration and biorefinery processing opportunities.

In theory, many microalgae components can be converted into chemicals: lipids can be converted into biodiesel or, trough metathesis, into drop-in hydrocarbon biofuels; starch granules can be recovered and converted into glucose and then into biofuel ethanol and all sorts of fermentation products, etc. However, the raw biomass of selected microalgae already has considerable value as nutraceuticals or ingredients, and the minimal fractionation of these biomasses can be used to produce even more valuable *specialty chemicals*.

Specialty chemicals are those produced in low volumes with high added value, compared to commodity or basic chemicals. Usually produced because of their unique properties, they may be the final consumer products but are, more commonly, ingredients or additives used in industry to prepare final formulations (ACC, 2021) and, in the case of microalgae, are the renewed focus of many initially biofuel-directed initiatives. Phycocyanin from *Spirulina*, carotenoids from *Dunaliella* and *Haematococcus*, or polyhydroxyalkanoates from many cyanobacteria are examples of alga-derived specialty chemicals.

Because economically proven production processes such as those for *Spirulina* and *Chlorella* are the most likely to be integrated into biorefineries, the following sections discuss the microalgae scene and the most important commercial strains, describing its downstream processing to identify by-products or routes to integration. These routes cover important compound classes (lipids, carotenoids, proteins, etc.) and can therefore serve as templates for other microalgae whose processing technology is developing.

15.1.1 Microalgae market evolution – Products, biomass, extracts

The microalgae market is growing at a CAGR (Compound Annual Growth Rate) of 5.4%. In terms of value, it is projected to rise from $977 million in 2020 to $1,485 million in 2028 (Kumar, 2021). When analyzing by applications, most microalgae production is destined for dietary supplements: the biomass is rich in minerals, vitamins, and proteins, among other components (Kumar, 2021), and there is a growing demand in the nutraceutical sector. Increase in consumers' awareness of links between nutrition and health leads to the search for more natural, healthy, vegan, and sustainable food, which favors microalgae commercialization (Meticulous Research, 2021). Besides, the growing demand for microalgae-derived natural pigments for food and beverages makes this segment expected to grow with an even higher CAGR during the following years (Meticulous Research, 2021). A similar trend is expected in the cosmetics sector, where the demand for microalgae products has increased in the last decade.

Spirulina sp. (*Arthrospira*) is the dominant microalga cultivated globally, followed by *Chlorella* sp. and *Dunaliella salina* (Kumar, 2021). *Spirulina* in powder form is expected to continue dominating the global market in value and volume, with a CAGR of 10.4% (Persistent Market Search, 2017); the *Chlorella* market is expected to reach $ 412.3 million by 2028, at a CAGR of 6.3% (Meticulous Research, 2022). North America dominates the overall microalgae market, followed by Asia-Pacific and Europe, although the last one is expected to experience the most significant growth during the next few years (Meticulous Research, 2021).

15.1.2 What is currently produced using microalgae?

Numerous bioproducts can be obtained from microalgae. After simple downstream processing, the biomass can be used as human food or in food supplements, commercialized in different forms such as powder, tablets, gel capsules, and liquids (Khan et al., 2018; Mobin et al., 2019). Microalgae have always been present in the nutrition of some populations, but as new species that have begun to be popularized; they have been studied and recognized by regulatory agencies all over the world as safe for consumption, although this process can demand many safety tests that consume time and resources (Torres-Tiji et al., 2020). In the case of the United States, for example, microalgae can be classified as Generally Recognized as Safe (GRAS) by the Food and Drug Administration (FDA): *Arthrospira platensis, Schizochytrium sp., Chlamydomonas reinhardtii, Auxenochlorella protothecoides, Chlorella vulgaris, Dunaliella bardawil,* and *Euglena gracilis* have GRAS-status (Torres-Tiji et al., 2020; de Carvalho et al., 2022). Biomass can also be used in feed for pets, farm animals, fish, and zooplankton, to improve their immune reaction and reproduction (Mobin et al., 2019).

More advanced biomass processing can generate bioproducts like polyunsaturated fatty acids (PUFAs), antioxidants, pigments, vitamins, and anticancer and antimicrobial substances. Those compounds can be afterward incorporated into other final products for different sectors. In the cosmetic industry, microalgae extracts are added to regenerating creams, lotions, body and sun protection creams, and shampoos (Mobin et al., 2019). In nutraceuticals, microalgae can be incorporated into pasta, biscuits, snacks, ice cream, etc., to enhance nutritional and health values, generating functional foods (Mobin et al., 2019). In the biopharmaceutical industry, microalgae pigments like chlorophyll, β-carotene, and other carotenoids, phycobiliproteins, and astaxanthin are promising in therapies for tumorigenesis, neuronal disorders, and optical diseases (Khan et al., 2018).

Furthermore, biomass can also be used to extract liquid and gaseous compounds for the biofuel industry, such as biochar, bioethanol, biodiesel, vegetable oils, biohydrogen, and biosyngas (Khan et al., 2018). There are ongoing pilot projects, but so far, no industries have produced biofuels from microalgae on a large scale due to the economic balance of the production process (de Carvalho et al., 2022).

15.1.3 Product classes: Definition

15.1.3.1 Whole biomass

Microalgae biomass, the integral microorganism cells, is produced through the many cell divisions during photosynthesis. This is achieved through carbon fixation by CO_2, generating the energy necessary for the growth of these organisms. Other nutrients such as phosphorus and nitrogen are vital for the nutritional composition of biomass, which can contain high values of carbohydrates, proteins, and lipids; so microalgae biomass is widely used commercially as animal feed and dietary supplements for humans. The culture conditions of microalgae, such as light, temperature, pH, and nutrients present in the medium, are critical factors for high biomass production (Su, 2021). *Spirulina* and *Chlorella* are the prototypical algae biomass products, currently available as capsules, tablets, and powders for final consumers. Raw biomass is also increasingly applied to novel food ingredients. The production of other fractions starts with biomass, and the opportunity for integration here is the use of residue-based culture media and flue gases rich in CO_2.

15.1.3.2 Protein fractions

The high amount of protein generated by microalgae allows for obtaining various products classified by their protein content and fractionation level. Whole-cell proteins correspond to the intracellular protein content, which is less susceptible to denaturation due to the protection of the cell wall. Isolated, concentrated, and purified proteins are part of another group of microalgae proteins. Protein hydrolysates can be obtained when these proteins are fragmented into smaller peptides using enzymes. Bioactive peptides are protein fragments, ordinarily inactive in the protein sequence but, once released, have several kinds of bioactivity in humans or towards other organisms (Soto-Sierra et al., 2018).

15.1.3.3 Phycocyanin

Phycocyanin is a blue-colored protein pigment, nontoxic, and highly water-soluble; It is widely used in the food, cosmetics, and pharmaceutical industries, and has biological activities such as anti-inflammatory, antioxidant, anticancer, and antiviral, so it has been the target of intensive research. This pigment, a phycobilin that is part of an antenna complex called *phycobilisome*, can be produced using many cyanobacteria. The most used organism is *Arthrospira* sp. Methods such as fractionated precipitation, chromatography, and dialysis are employed to separate and purify this pigment (Safaei et al., 2019; Peeters and Huyskens, 1993). An opportunity for integrating microalgae processes into biorefineries lies in using cyanobacterial extremophiles capable of growing in residues to produce biomass for phycocyanin production.

15.1.3.4 Biofuel products

Microalgae biomass can also be used as a feedstock for biodiesel production. Due to their high growth rate and efficiency in using solar energy, microalgae are promising sources for biofuel production. They are great alternatives to using fossil fuels or producing biofuel with edible feedstock. The lipids extracted from microalgae can be converted through transesterification into carbon fuels. In addition, microalgae carbohydrates can generate bioethanol and biobutanol through conversion to fermentable sugars by hydrolysis. Microalgae biomass can also be used as a substrate for anaerobic digestion, generating biogas (Ganesan et al., 2020; Delrue et al., 2017). Biofuels from algae have been proven to be technically possible, but they cannot economically compete with fossil fuels. One of the pathways for increasing the competitiveness of these advanced biofuels is lowering production costs by coupling microalgae production with agro-industrial residues treatment and valorization of the spent biomass meal.

15.1.3.5 Carotenoids

Carotenoids are a group of pigments widely distributed in nature. Their structure is based on C40 isoprene units (terpenoids) and is divided into two groups: carotenes and xanthophylls. They are lipophilic compounds, usually colored yellow, orange, or red (Gong and Bassi, 2016). Astaxanthin, β-carotene, lutein, lycopene, and canthaxanthin are some of the commercially valuable carotenoids produced from *Dunaliella salina*, *Chlorella* spp. and *Haematococcus pluvialis*. In microalgae, their main functions are photoprotection, oxidation-protective agents, and as part of light-harvesting complexes for photosynthesis. Due to anti-inflammatory and antioxidant properties, carotenoids are used in the food and feed industry. They have beneficial effects on human health, e.g., cancer prevention, lowering the risk of Type 2 diabetes, and improved eye health (de Carvalho et al., 2014). Similarly to phycocyanin, these high-added value products can improve the economics of existing processes via biomass fractionation.

15.1.3.6 Polysaccharides

Polysaccharides are polymers formed by monosaccharide chains, whose structure and composition depend on the type of monosaccharide and the bonds. Microalgae present two types of polysaccharides: (I) intracellular and (II) structural (including exopolysaccharides released into the medium, cell-bound polysaccharides, and cell wall polysaccharides). Their production and synthesis depend on the type of microalgae and the culture conditions, e.g., culture medium, salinity, irradiation, and temperature. As with lipids, the extraction of an intracellular polysaccharide fraction leaves protein-rich biomass that can be used to produce new product fractions, while extracellular polysaccharides are left in the spent media, and further use depends on a concentration process that can be resource-intensive. Some microalgae with a large amount of starch and cellulose are *Chlorella, Scenedesmus, Chlamydomonas, Tetraselmis,* etc.; this metabolite is the raw material for bioethanol production (Gong and Bassi, 2016; de Carvalho et al., 2014).

15.1.3.7 PHAs – (polyhydroxyalkanoates)

Polyhydroxyalkanoate (PHA) polyesters can be synthesized and accumulated in various microörganisms, including microalgae. They form intracellular inclusions produced during the stationary phase and under extreme conditions (biotic or abiotic stress factors). Their primary function is the storage of carbon and energy, and they can reach high intracellular concentrations without altering their osmotic state (Costa et al., 2019). *Synechococcus subsalsus* and *Spirulina* sp. have been reported as possible producers of PHAs under mixotrophic conditions and nitrogen and phosphorus deficiency. Two peculiarities that favor the integration of PHA production into existing processes are that it is not a food but a material and the possibility of using algal polycultures because the solvent fractionation used downstream can leave behind most contaminants.

15.1.3.8 Lipids

The production of lipids from microalgae is promising due to their potential use in the chemical, food, pharmaceutical, and cosmetic industries. Algae synthesize different lipids, e.g., triacylglycerols, phospholipids, glycolipids, or phytosterols. Based on their characteristics, it is possible to group them into two categories: (I) storage lipids (non-polar lipids) and (II) structural lipids (polar lipids). Triacylglycerides (TAGs) serve as energy storage and could be recovered by solvent extraction and become feedstock for biodiesel production. Polar (structural) lipids are important components of membranes and intermediates (or precursors) in cell signaling pathways, and play a role in response to environmental changes. The composition and amount of lipids depend on the species; the strains considered "oleaginous" can accumulate more than

20% of lipids on a dry basis. There are oleaginous strains of *Chlorella* sp., *Nannochloropsis* sp., *Scenedesmus* sp., and *Dunaliella* sp. (Sharma et al., 2012; De Luca et al., 2021), but most microalgae can have an enhanced lipid content in specific conditions. Despite the promise of algal lipids as biofuels, the most important strains are used today to produce dietary lipids, as is further discussed in this text.

15.2 Important algae products as cases

One of the best approaches to developing new processes is to study existing processes to identify winning strategies, improvement opportunities, and waste streams for integration into new processes. With that in mind, the most important microalgae products are discussed as cases in this section.

15.2.1 *Chlorella* as nutraceutical

Microalgae from the *Chlorella* genus are consumed as health food, used for livestock and aquaculture feeds, and in the drugs and cosmetics industries. Beijerinck first cultivated *Chlorella* in 1890–1893, and in 1894, Kruger established a pure culture (Ahmad et al., 2020). Many *Chlorella* species have a higher growth rate than land plants. This organism can be cultivated under high light intensities, allowing operational strategy implementation such as fed-batch and semicontinuous to avoid light limitation and enhance productivity (Chen et al., 2019).

Biomass is the main product of *Chlorella*; this rich source of protein, lipids, carbohydrates, and some pigments, makes it a supplement with high nutritional value (Table 15.1). From a nutraceutical point of view, *Chlorella* biomass may contain about 61.6% proteins, 12.5% fat, 13.7% carbohydrates, trace elements, vitamins, and their precursors, and a total fiber of 35%. The total fiber in untreated biomass consists of complex polysaccharides that can be less digestible by humans but can be considered for application in livestock feeding (Coronado-Reyes et al., 2022). It is necessary to have optimized microalgae cultivation to obtain a higher biomass concentration and productivity.

Many researchers showed that *Chlorella* is one of the best microalgae with easy adaptability and performance for the wide possibility of producing secondary metabolites based on the substrate they consume. Those metabolites have been extracted, characterized, and already applied in the formulation of food supplements. Regarding pigments, lutein and zeaxanthin have gained particular attention in *Chlorella*. As they are part of the photosynthetic system, there is a relationship between light intensity and total lutein content in the cell during the cultivation. Closed photobioreactors offer the best condition for production, considering the area, homogenization, and final concentration (Lin et al., 2015).

Table 15.1: Composition of dry *Chlorella vulgaris* biomass (adapted from Coronado-Reyes et al. (Coronado Reyes et al., 2022).

Component	(%)
Moisture	5.83
Dry matter	94.17
Crude protein	54.45
Carbohydrates	11.86
Crude lipids	12.18
Crude fiber	9.18
Total carotenoids, mg/g	1–2
Ash	9.50

There are two commonly used media for Chlorella sp. cultivation: BG11 (Blue Green medium) and BBM (Bold's Basal medium). There are many recipe modifications. Both media have nutrient profiles in different concentrations, which supply the macro- and micronutrients and metabolic cofactors (Safi et al., 2014; Mcclure et al., 2019). As with many other microalgae, *Chlorella* can also assimilate organic carbon, which can be a strategy to increase biomass production or modulate the cell composition using mixotrophic conditions. Table 15.2 presents the data on *Chlorella* biomass productivity (Table 15.2).

Table 15.2: Commonly cultivated medium for various *Chlorella* species biomass productivity (adapted from Mcclure et al., 2019).

Microalgae	Medium	*Carbon source	Productivity (g/L/d)
Chlorella minutissima	BBM	None	2.04
Chlorella protothecoides	BM and BG11	Glucose, lactose, galactose	0.59–2.47
Chlorella sorokiniana	BG11	Sodium acetate	0.48–1.28
Chlorella vulgaris	BBM and BG11	Glucose	0.13–1.76

*The original article has a detailed carbon source discussion.

Simultaneous recovery of various algae products can be carried out, allowing the reduced cost of production by an integrated process (Coronado-Reyes et al., 2022). Lin, Lee, and Chang (Su, 2021) have pointed out that microalgae can be considered a commercially viable lutein source. They have proposed an estimated production for lutein by microalgae and compared it to the current production via *Tagetes erecta* (flower).

In the conceptual process, based on (Chen et al., 2019; Lin et al., 2015), it is estimated that *C. sorokiniana* could reach a concentration of 3.51 g/L in a tubular photobioreactor. The culture medium is based on BG11 and 70% *shrimp culture wastewater*.

Daily production of 10 m^3 of microalgae suspension gives 31.6 kg of *Chlorella* biomass or, after lutein saponification and extraction, 22L of a lutein concentrate with 15,000 ppm of the pigment. This process produces 500L of an aqueous residue with a significant COD. An initial integration step can be neutralizing the residue and using it for the production of biogas or biohydrogen, since alkaline biomass pretreatment has proven feasible in microalgae biodigestion research (de Carvalho et al., 2020); further developments in extraction, such as using supercritical fluid extraction, could lead to less degraded biomass.

There are many possible applications for *Chlorella* besides biomass production, because of its fast growth rate, resistance, and nutritional composition. This can be explored for integration into processes as both a wastewater treatment option and a bioingredient source. The lipid content of selected strains makes them a potential source of biodiesel and biogas for energy production. As a cosmetic, the natural pigments can be applied as vitamin precursors (e.g., β-carotene), antioxidants (e.g., xanthophyll and chlorophyll), and ingredients for retinal degeneration supplements (e.g., lutein). This microalga can supply carbohydrates and mineral nutrition (Lin et al., 2015; Coronado-Reyes et al., 2022).

15.3 Haematococcus and Astaxanthin

Haematococcus sp. is a biflagellate Chlorophyte microalga that accumulates the reddish carotenoid astaxanthin (AX), recognized as one of the most potent carotenoid antioxidants. That accumulation is most evident when the green vegetative cells change to a palmella or aplanospore state after a few days under environmental stress conditions such as high salinity or light irradiation. The astaxanthin content in *H. pluvialis* can reach about 4%, making this microalga the most important source of natural astaxanthin.

There are three main classes of *Haematoccocus*-based products: biomass for feed, ingredients for functional foods and cosmetics, and nutritional supplements. Since natural astaxanthin has a considerably higher cost (and quality) than synthetic products, most of it goes to human use. Typical bulk products are dry biomass (with or without disruption to enhance absorption), oily extracts for use as an ingredient or in softgel capsules, and powdered extracts for use as an ingredient or in tablets and capsules. Some *Haematococcus* products have already been granted GRAS status, and there are more than 80 suppliers worldwide – but the market is still expanding because of the rising popularity of AX supplements (Almendarez, 2016; GRAS Notice Inventory | FDA Available online, 2022).

The list of health claims for AX as a supplement is still growing, as well as its favorable comparison with synthetic similars (Talbott et al., 2019; Capelli et al., 2021). Therefore, *Haematococcus*-derived products will most likely remain centered on human use, and the types of formulations will have little change. However, accompanying the market expansion for concentrates, there may also be an increasing offer of lower-grade, but still AX-rich leftover biomass that can become more attractive for inclusion in pet and poultry feeds and farmed crustaceans and fish. The availability of water-dispersible formulations will also, likely, increase.

There are several laboratory media suitable for the cultivation of *H. pluvialis* strains, such as 3N BBM + V, MES-Volvox Medium (UTEX, 2022; SAG, 2022), and many references to alternative media, frequently based on agro-industrial residues (de Carvalho et al., 2022). However, the added value for *Haematococcus* biomass is such that laboratory-similar synthetic media can be used industrially, provided that the salts used are food grade, not analytical grade, as is typical in laboratory settings. Many researchers compare culture media using cell count, which is adequate for vegetative motile cells (Fábregas et al., 2000; Silva et al., 2022) but complicates the comparison when cells undergo morphological changes and enhanced carotenogenesis. Comparing five culture media (Marinho et al., 2021) obtained similar specific growth rates of 0.3/d with a modified Provasoli medium and BBM, but a higher biomass concentration of 4.9 g/L on the first and 3.4 g/L on the second. However, the irradiation was continuous at 60 μmol photons/m^2/s, and Provasoli is a yeast extract-based medium, while BBM is mineral and comparably cheaper. One of the rare accounts of industrial production of *Haematococcus* used two-phase cultivation with a modified BBM, natural light, and naturally oscillating temperatures, reaching a lower specific growth rate of 0.15/d, a 0.2 to 0.36 g/L concentration, but in a 25-m^3 reactor system with a real productivity of 4 to 19 g/m^2/d (Olaizola, 2000).

Like many other microalgae, the final biomass concentrations in H. pluvialis production bioreactors are relatively low. However, the carotenoid-rich cells are larger and denser than other common algae (Baroni et al., 2019); therefore, a preliminary sedimentation step can be used for volume reduction, followed by centrifugation or filtration. It is common to mechanically disrupt cells to enhance astaxanthin bioavailability, which can be done using a high-pressure homogenizer or bead mill. The disrupted biomass can then be dried and packaged as a final AX-rich biomass or processed for extraction of the pigment using solvents. Although many solvents can be used, the high carotenoid concentration in the biomass allows SFE-CO_2, supercritical fluid extraction with CO_2, to be an effective downstream method. After SFE, the carotenoid extract is flashed to give an AX oleoresin that is usually standardized for 2–10% AX, with the addition of vegetable oils and an algae meal with 30–42% protein content, and 0.05–0.73% residual AX (Ju et al., 2012; Ma et al., 2019) (the actual concentrations depending on the solvent used and the extraction efficiency). In SFE, the CO_2 can be recovered, recompressed, and reused and leaves no residues; liquid solvents can be recovered by extract evaporation and meal drying, but tend to leave residual solvent, similarly to lipid extraction (de Carvalho et al., 2022).

Since virtually all *Haematocuccus* biomass is transformed in the three common fractions, the only relevant waste stream is spent medium. However, there are opportunities for further fractionation of the *Haematococcus* biomass, which contains about 15–40% lipids (Ma et al., 2019; Ghiggi Sorgatto et al., 2021) into a purer, low-fat astaxanthin extract, and *Haematococcus* lipids. The residual biomass meal can be further derived into protein fractions. However, these are niche applications. Another coproduct in *H. pluvialis* production is not biomass-derived but uses the spent medium for irrigation, as indicated by its effectiveness in plant tissue culture (Gollo et al., 2016). The integration into the bioprocessing of agro-industrial residues (Ghiggi Sorgatto et al., 2021; Nishshanka et al., 2021) has been proved feasible for this microalga, reducing water usage. Recycling spent medium back to the microalgae culture may not be economical (Issarapayup et al., 2011) and requires further investigation. Finally, the presence of extracellular polymeric substances in *H. pluvialis* cultures can lead to the isolation and development of new bioactive products (Li et al., 2011), as is being investigated for other microalgae (Laroche, 2022).

15.4 Spirulina and phycocyanin

Spirulina is a blue-green, unicellular, prokaryotic, photosynthesizing cyanobacterium that forms spiral, multicellular filaments. The photosynthetic assimilation of carbon, nitrogen, and phosphorus produces biomass with high values of carbohydrates, lipids, proteins, and yet phycocyanin, a blue pigment produced at about 92 mg/g dry biomass. It also has been reported to have antioxidant and anticancer properties and high content of iron, fatty acids, and vitamins, generating great interest in the food and nutraceutical industries. The most common species of this microalgae are *Spirulina platensis* and *Spirulina maxima*, usually found in different environments, such as soils, swamps, brackish water, lakes, and rivers (Costa et al., 2019; Rosero-Chasoy et al., 2022).

The consumption of *Spirulina* as a human dietary supplement has increased in recent decades, mainly in Europe and Asia. The main product on the market is powdered or granulated biomass as tablets and capsules sold at around \$1–370/kg. Spirulina has 50% of global production in feed, compared to other microalgae, and has shown positive effects on the performance of some animals. It is also used as a source of natural pigments in some fish farms. Phycocyanin is usually sold as a powder at about \$1660/kg in Japan, but methods for large-scale extraction and purification are being further explored (Costa et al., 2019) and can lead to cost reduction.

Biopolymers of microalgae origin have been considered promising for possible commercial use as an alternative to petrochemical polymers. Studies on the production of biopolymers by *Spirulina* have resulted in a 30% (w/w) concentration of PHB (polyhydroxybutyrate) under specific cultivation conditions. The production of biofuels by *Spirulina* is not common since its lipid content is lower than in other microalgae. However, studies are being conducted with *S. platensis* as a feedstock for third-generation bioethanol production because it can reach between 50 and 60% of carbohydrates when grown in specific media (Costa et al., 2019).

One of the most common medium for cultivating *Spirulina* in the laboratory and even in large-scale production is Zarrouk (1966). Still, the medium is considered costly and saturated because it contains too many salts in high concentrations. Other media options for *Spirulina* production, such as Rao's medium, CFTRI, OFERR, seawater, waste, effluent, and other organic carbon sources, are still less used than traditional Zarrouk. However, modifications of Zarrouk with adapted concentrations or changes of some reagents, or in its diluted form, have shown positive results in the production of *Spirulina* biomass on an industrial scale. Al Mahrouqi et al. (Al Mahrouqi et al., 2022), achieved productivity of about 0.16 g/L/d of *Spirulina platensis* (Table 15.3), using a 50% diluted modified Zarrouk medium, where sodium nitrate was replaced by urea (0.22 g/L).

Table 15.3: *Spirulina platensis* productivity on different media (Al Mahrouqi et al., 2022).

Microalgae	Medium	Productivity (g/L/d)
Spirulina platensis	Standard Zarrouk	0.14
Spirulina platensis	50% Zarrouk	0.13
Spirulina platensis	25% Zarrouk	0.12
Spirulina platensis	Modified Zarrouk	0.13
Spirulina platensis	50% Modified Zarrouk	0.16

Spirulina biomass has high nutritional value: obtaining dry powder with about 60–70% protein is possible, and the aminoacid profile shows a nutritional quality compared to other protein sources. The concentration of lipids in *Spirulina* is much lower, reaching 6–11% lipids in dry weight, of which the most important are palmitic and linoleic acid derivatives. It also contains about 13.6% carbohydrates, including glucose, mannose, xylose, and galactose. In addition, small concentrations of minerals (6.88%) and vitamins, such as β-carotene, B_1, B_2, B_3, B_6, and B_{12}, the last containing 0.16–0.175 mg in 100 g of biomass(Ragaza et al., 2020).

On an industrial scale, the production of *Spirulina* sp. biomass is typically carried out in raceways (open tanks) or closed photobioreactors, with constant agitation and without temperature and lighting control. For the recovery stage, filtration is used, with an average recovery of 98%, and can be in filter presses or vacuum filtration. The most industrially used forms of drying and recovery are the spray dryer and drum dryer (Costa et al., 2019).

Uebel et al. (2019) evaluated an industrial production of *Spirulina* sp. in a raceway of up to 26,000 L arranged in greenhouses without further temperature or lighting control. The inoculum, with an initial concentration of 0.3 g/L, was batch cultured in a working volume of 9,600 L. Then it was grown in semicontinuous mode at an initial concentration of 0.65 g/L and working volume of 17,500 L, with constant stirring. From Zarrouk medium diluted at 50%, it has been possible to obtain a maximum concentration of 1.64 g/L in 37 days of cultivation, reaching maximum productivity of

14.9 $g/m^2/d$ in 9 days. The biomass separation was done by pumping into a filter press, where the filtered medium returned to the culture, and the retained biomass was frozen at −80 °C and lyophilized.

Methods such as ultrasonic disruption, freeze-thawing, and autolysis are commonly used for cell lysis and phycocyanin extraction. Ammonium sulfate precipitation, ultrafiltration, and dialysis methods are used for pigment separation and purification. Shao et al. (Patent No. CN104292326A, 2013) obtained 90 kg of phycocyanin powder with a purity of 2.08 from 1,000 kg of *Spirulina* sp. powder, resuspended in distilled water at 1:10 (w/v) ratio, using three cycles of freezing (−20 °C) and thawing, after centrifugation and vacuum filtration, and further precipitation over ammonium sulfate at 30% saturation, ultrafiltration with 50 kDa membrane, and spray dryer.

The cultivation of *Spirulina* and the production of phycocyanin have received much visibility, both for the high production of biomass and its various uses and for the generation of a natural and biodegradable pigment that can be used by different industries. The use of residues, effluents, and coproducts of other refineries as a cultivation medium has been the target of studies due to the use of components and nutrients from the waste, and the use of CO_2 sources provided by burning industrial gases are sustainable options that can benefit microalgae cultivation, reduce production costs, and favor the environment (Costa et al., 2019; Brasil et al., 2017).

Despite still being a product of high production cost due to the nutritional needs, cultivation and refining equipment, and the entire process to reach the quality levels required by the market, products from Spirulina have shown promise in different areas and are innovation alternatives for several sectors (Al Mahrouqi et al., 2022; Brasil et al., 2017). In addition to nutraceuticals and animal feed products, the dried biomass of *Spirulina* has also been used directly for nutritional enrichment in some foods, reinforcing the possibility of microalgae being part of the human diet. Other products, such as biofuels and biopolymers produced from Spirulina biomass, have also been researching themes, showing results of interest for possible industrial production. The commercialization of phycocyanin pigment in powder and tablet form has increased in the market, and even attempts at encapsulated forms have been analyzed.

15.5 Dunaliella and beta-carotene

Dunaliella unicellular-flagellated green algae were described for the first time in 1838 but received their name only in 1905, in honor of Dunal, by Teodoresco. There are approximately 28 species of this Chlorophyta, the most well-known being *D. tertiolecta, Dunaliella salina, D. primolecta, D. bioculata, D. viridis,* and *D. bardawil.* The size of this eukaryote can vary according to culture conditions from 5–25 µm in length and from 3–13 µm in width, and the cell can be spherical, egg-shaped, and spindly or elliptical (Pourkarimi et al., 2020).

Dunaliella carotenoids have pro-vitamin A activity and have been approved by the USFDA for use in dietary supplements (human and animal) as a food coloring agent. Additionally, it is possible to use them in the cosmetic and pharmaceutical industries due to their antioxidant properties. *Dunaliella salina* can accumulate up to 14% carotenoids (dry biomass weight), and the most important obtained industrially is trans- β-carotene because it is the most abundant in the cell (50%) and is produced in a very short time. Other carotenoids include α-carotene, phytoene, phytofluene, lutein, and zeaxanthin (Ambati et al., 2019; Dufossé et al., 2005).

Similarly to astaxanthin, β-carotene can be produced synthetically at a relatively low cost. However, the product cannot be claimed to be natural; therefore, natural sources such as plants, fungi, and algae – perceived as high quality – can be competitive. *Dunaliella salina* is produced industrially, with market size of about $1,428 million in 2019. Important biomass producers are in Israel (Nature Beta Technologies; NBT; ~25 t biomass per year), Spain (Monzon Biotech S.R.L.; MB; ~400 kg (AFDW)), and Austria (Harvey and Ben-Amotz, 2020; Monte et al., 2018).

Besides carotenoids, *Dunaliella* can accumulate other metabolites depending on the culture conditions, e.g., lipidic components (10–25% dry weight), glycerol (>50% dry mass under high salinity), proteins (10–40% total dry weight), carbohydrates (up to 30% of the total dry weight), essential amino acids (up to 8%), and hydrocarbons. Therefore, it is possible to recover and purify different products from *Dunaliella salina* (Monte et al., 2020). Table 15.4 shows the composition of biomass for three species of *Dunaliella.*

Table 15.4: Dunaliella biomass composition based on dry weight (g/100 g dry weight) (Pourkarimi et al., 2020).

Species	Protein	Carbohydrate	Lipid
D. bioculata	49	4	8
D. salina	57	32	6
D. tertiolecta	29	14	11

β-carotene production from algal biomass occurs in three stages: cultivation, harvesting, and extraction. In the cultivation stage, it is necessary to consider strain and culture conditions. *Dunaliella* is a halophilic microalga that can grow in extreme conditions, requiring adapted culture media such as Modified Johnson, F2, Conway, and Ramaraj using different concentrations of NaCl. Culture medium costs vary widely: F2, Conway, and Jonhson range from $1.17–49.62 per cubic meter; therefore, the contribution of these costs to the final price of biomass ranged from $4.64–301.61 per kg on a dry basis (Colusse et al., 2020).

Synthetic culture media is expensive for industrial production, and for that reason, the use of seawater is a viable alternative. Reverse osmosis removes substances (anions/cations) that are soluble in water and produces purified water and a concentrated stream with high salt content, known as brine. The advantage of using brine from seawater reverse osmosis is that the salinity is equivalent to the concentration required for β-carotene production by *Dunaliella salina* (Pourkarimi et al., 2020; Yildirim et al., 2022), making it possible to couple production to other processes where desalination is used. Table 15.5 describes a few examples of culture media used to produce *Dunaliella salina.*

Table 15.5: Culture medium and productivity of *Dunaliella.*

Culture medium	Productivity	References
Seawater of 12.5% salinity	1.65 g/m^2/d	(García-González et al., 2003)
Modified artificial seawater medium	1.88 g/m^2/d	(Zhu et al., 2018)
SDC* medium	1.97 g/m^2/d	(Zhu et al., 2018)
Jonson's medium with $NaHCO_3$ and KNO_3 modified	40.8 mg/m^2/d	(Wu et al., 2017)

The production of biomass from *Dunaliella* can be done in (I) photobioreactors or (II) open ponds, which are the cheapest and most feasible on a commercial scale – a raceway in Australia has a capacity of 1,000 m^3 and China, India, Chile, USA report ponds with the capacity of 3,000 m^3 (Pourkarimi et al., 2020). Nevertheless, to produce β-carotene, it is necessary to include stress conditions, and the operation can be done in two phases. The first phase is optimized to produce biomass, and the second one accumulates carotenes. The biomass productivity in a pond culture is 5–10 g/m^2/d; β-carotene production is 10–20 mg/L, and the maximum efficiency of β-carotene production is 750 mg/m^2/d, but using continuous bioreactors in two stages, the β-carotene can be increased up to 2.45 g/m^3/d (Monte et al., 2020,Guedes et al., 2011).

The following steps include harvesting, downstream. and final purification. The harvesting depends on the volume produced; and centrifugation by continuous flow, automatic discharge centrifuges is one of the most efficient methods to recover *D. salina* despite its high cost. For the drying stage, spray drying is most recommended for β-carotene. The extraction of carotenoids from dry biomass is done with organic solvents. Finally, the separation and purification stage at the industrial level use deactivated alumina chromatographic columns followed by a crystallization stage obtaining compositions with 50–75% of 9-cis β-carotene and up to 40% of all-trans-β-carotene (Pourkarimi et al., 2020; Monte et al., 2020).

Extraction is also attempted in other ways: In a continuous production-extraction process, *D. salina* cultures are produced, and extraction occurs simultaneously (closed photobioreactors). At the end of the growth phase, carotenoid production is observed, and the medium is then enriched with the organic phase. This process can proceed for up to six weeks, though cell growth is significantly reduced. In this case, the β-carotene production efficiency can be 2.45 mg/m^2/d with a return flow of 200 ml/min in a two-phase bioreactor.

Despite the growing market for natural β-carotene, other products (glycerol, biomass, and lipids) could be economically produced on an industrial scale, if production and processing costs are reduced. Due to the high added value of some fractions and the GRAS status, *Dunaliella* can be a profitable product integrated into processes with high salinity residual streams; conversely, an increase In *Dunaliella* production could provide more biomass for valuable coproducts, especially lipids and protein fractions. However, extraction and purification processes must be improved to reduce the use of solvents.

15.5.1 Lipid-rich biomass

Many photoautotrophic microalgae produce high amounts of structural lipids that usually have long chains of fatty acids, being considered potential candidates for commercial mass cultivation to produce valuable LC-PUFAs (long-chain polyunsaturated fatty acids) (Tababa et al., 2012). These LC-PUFAs include eicosapentaenoic acid (EPA), docosapentaenoic acid (DPA), docosahexaenoic acid (DHA), and arachidonic acid (AA), which have potential way beyond biofuel production: many PUFAs are bioactive and can aid in therapies for Parkinson's, Alzheimer's, and atherosclerosis (Alishah Aratboni et al., 2019; Yates et al., 2014). The ingestion of such LC-PUFAs can affect fetal development and vision in infants, induce anti-inflammatory activities, and raise HDL cholesterol levels in the blood, decreasing heart disease risks (Mallick et al., 2019; Orozco Colonia et al., 2020).

LC-PUFA oils can be used for human and animal nutrition, in fortified foods and beverages, infant formulas, dietary supplements, clinical nutrition and medical foods, pharmaceuticals, and pet foods (Orozco Colonia et al., 2020; Schmid et al., 2013). They are also used as supplements of omega-3 oils in animal feed to increase levels of these fatty acids and their nutritional value for fish, poultry, and cattle (Orozco Colonia et al., 2020). Besides their uses in food and energy industries, these lipids may play important roles in the oleochemistry industry in the next few years by adding to or eventually substituting other oil sources used for cosmetics and chemicals (de Carvalho et al., 2022).

Lipid-rich microalgae include *Schizochytrium limacinum, Ettlia oleoabundans, Parachlorella kessleri,* and *Botryococcus braunii,* which have cell lipid contents higher than 50% (de Carvalho et al., 2022). Many other species can also accumulate PUFAs and TAGs (triacylglycerols) but in lower levels (Alishah Aratboni et al., 2019), especially after an N- or P-depleted growth phase. Table 15.6 presents some examples of lipid-rich microalgae, the culture media used, and the biomass productivity attained.

The cultivation media for lipid production by microalgae plays an essential role in lipid accumulation by nutrient stress. Nitrogen, phosphorus, and sulfur can be eliminated from the growth medium to affect lipid accumulation. Nitrogen is considered the most critical nutrient affecting lipids metabolism: TAG accumulation occurs in response to nitrogen deficiency (Tababa et al., 2012). Conversely, with C-source depletion, lipid degradation is observed (de Carvalho et al., 2022). This strategy of nutrient depletion may be effective for increasing lipid content in microalgae but overall

Table 15.6: Important oleaginous algae. *S. limacinum* and *P. kessleri* are currently industrially produced.

	Microalga species				
	Schizochytrium limacinum	**Ettlia oleoabundans**	**Parachlorella kessleri**	**Botryococcus braunii**	**Porphyridium cruentum**
Lipid content (%)	48.0	50	66.0	47.6	19.3
Principal products	DHA	Oleic acid and linoleic acid	Palmitic and oleic acids	Palmitic and palmitelaidic acids, botryococcenes, and squalene	DHA
Cultivation media	M1	BBM with glucose supplementation	UP	Effluent	F/2
Final biomass concentration, dry weight	68.0 g/L	20.9 g/L	5.8 g/L	8.6 g/L	2.1 g/L
References	(Talbierz et al., 2022)	(Morales-Sánchez et al., 2014)	(de Carvalho et al., 2022; Takeshita et al., 2018)	(de Carvalho et al., 2022; Ramaraj et al., 2016)	(Oh et al., 2009)

has a negative impact on biomass productivity; therefore other methods, which will be further discussed, are being studied to increase lipid levels in cells.

The downstream processing to obtain high-value lipids from the cultivated microalgae starts with a solid-liquid separation operation to harvest the cells. Direct centrifugation or filtration of the media is adequate for broths with high biomass concentration. Still, operations like microfiltration and flocculation may be necessary to concentrate broths with lower cell concentrations before the dewatering process (de Carvalho et al., 2022). Afterward, the concentrated broth can be stabilized by temperature to avoid lipid degradation, especially if large volumes are being treated. The broth can be chilled, washed, dewatered, or even heated for enzyme inactivation. The next step for most microalgae species is drying it to reach a water content of 3–5%. This can be done in spray-drying, tray, or fluid bed-drying (de Carvalho et al., 2022). The drying stabilizes the product for later processing, preparing it for extraction. It is important to notice that not all microalgae need to be dried before extraction. It is the case of *Schizochytrium sp.*, which can have its oils extracted from by enzymes instead of solvents. This strategy reduces the capital cost and improves the production safety by dispensing solvents (Ji and Ledesma-Amaro, 2020).

The next step can be mechanical disruption of cells if the biomass will serve as feed (because the cell wall can difficult digestibility) or if it is the case of wet processing, as mentioned before. The mechanical disruption can be done before drying, in high-pressure homogenizers or bead mills, or after drying, in ball mills or jet mills (de Carvalho et al., 2022). After that, lipid extraction takes place in agitated tanks. It can be done using preferably nonpolar solvents like hexane, but acetone, chloroform, or alcohols can also be used (de Carvalho et al., 2022). After solvent extraction, the lipids are then separated from biomass by centrifugation or decantation. The wet biomass is then dried to recover the residual solvent and generates a coproduct rich in proteins, while the miscella goes through an evaporation process to solvent recovery. The crude oil then follows to RBD steps (refining, blanching, and deodorizing).

The economic feasibility of microalgae mass cultivation to produce valuable LC-PUFAs depends on the high initial capital investment, nutritional requirements, and downstream processing (de Carvalho et al., 2022). Optimizations on each step in PUFAs production is necessary, beginning in the chosen microorganism. Genetic and biochemical engineering can enhance biomass and lipid productivity (Alishah Aratboni et al., 2019).

Cultivation conditions and photobioreactor operational parameters are extremely important when finding the balance between high biomass concentration and high productivity of PUFAs. For example, by applying a two-stage oxygen supply control strategy based on the analysis of the oxygen transfer coefficient, DHA production by *Schizochytrium* sp. was enhanced by 64% (Qu et al., 2011). Wavelength and light intensity also can cause drastic changes on microalgae growth and lipid accumulation, as well as variations in CO_2 supply rates. Besides, as the biochemical pathways related to the synthesis and accumulation of lipids are controlled by enzymes with a high sensitivity to thermal variations, temperature stress can also affect lipid induction. Heavy metal and saline stress are other strategies. As mentioned before, nutrient concentrations are factors easily manipulated to induce lipid accumulation, and are also important from the economics point of view (Tababa et al., 2012). The use of nutrient-rich wastewaters or inexpensive agricultural fertilizers is an interesting option, but depending on the destination of the oils (for example, for human consumption), those nutrient sources may not be appropriate for microalgae cultivation. Also, carbon sources cheaper than glucose are searched to reduce costs of DHA production by microalgae, as waste glycerol is the case(Kujawska et al., 2021).

Concerning the downstream processing of lipids, new extraction methods tend to become more popular, since solvent extraction requires large volumes and raises issues about solvent toxicity. For high-quality nutritional oils, supercritical fluids, like supercritical CO_2, may become important for oil production in the next years, since it is a green alternative technique that does not pollute the environment and the final product (de Carvalho et al., 2022).

15.6 Seizing the opportunities: Rationale for process integration in biorefineries

As can be seen from the processes presented in the previous sections, microalgae biomass has a high intrinsic value that can, by itself, justify its production. However, fractionation of the biomass can potentialize the applications (for example, raw biomass cannot always be used as color additive or as a concentrated PUFA source), and with the expansion of the demand and market, it is likely that biomass fractionation will gain traction. Another trend in the area is the use or residues and reuse of spent culture media. To integrate algae processing into existing agro-industrial and agroenergetic processes and vice versa – integrating new processes into algae production – it is necessary to develop economic models: analyze existing processes and markets, identify waste streams, propose integration routes based on laboratory data, and evaluate scenarios through material, techno-economical, and life cycle analyses. The many variables make the approach complex, but not necessarily complicated.

There are many heuristics to be used for process development, which are more of a thinking tool than a roadmap.

15.6.1 Resource heuristics

Are there resources to be exploited – e.g., wastewaters, flue gases or other material streams that can be converted in microalgae? In that case, the volume and composition of the residues can show which microalgae processes can be used: extremophile algae such as *Arthrospira*, *Galdieria* or *Dunaliella* can grow in basic, acidic, or hypersaline residues, for instance, while *Haematococcus* has been proven to grow in eutrophicating cassava processing wastewater or vinasses. Many microalgae have been investigated for culturing in flue gases. The biomasses from these processes can be derived into nutraceutrical and feed products, for pure cultures from agroindustrial or agroenergetic residues, while polycultures or biomass from potentially contaminated residues can be transformed into materials such as PHAs, or digested to produce biogas or biohydrogen. As this approach uses existing resources, it is the first line of process integration and the focus of intensive research today.

15.6.2 Integration heuristics

A microalgae product can be developed to potentialize an existing process. For example, agroenergetic industries can benefit from the production of biostimulant or biofertilizers from selected microalgae, to be applied again in the field, in the bioenergy crop; similarly, secondary effluents from dairy, pig, cattle, or fish farms can be used to produce nutritional and carotenoid-rich biomass for enhanced feeds. This recycles part of the nutrients and fixates extra carbon dioxide from the atmosphere. Although the focus here is developing a product that can be useful in an existing process, it can benefit from the recognition and use of process waste streams, such as N- and P-rich effluents, and CO_2-laden flue gases and biogas. The risk in recycling pathogens and heavy metals can be minimized through inclusion of filtration or sterilization steps, and quality control-informed residue purge, respectively.

15.6.3 Composition heuristics

Existing microalgae processes can be diversified by biomass fractionation, and this depends on the biomass composition. Microalgae commercial products started with biomasses, then carotenoid and phycocyanin extracts, then PUFA extracts, and more recently, exopolysaccharides and biostimulants. Other products that can be developed include proteic and peptidic extracts and cell wall polysaccharides. Ideally, all fractions of a biomass would be converted in the most valuable products. As seen in the previous sections describing commercial microalgae products, the processes usually extract a desired component and leave the rest as a residual biomass meal; the first step in diversifying products in algal biorefinery would be to look at the biomass (and residues) composition and ponder on what else could be extracted from the process. Opportunities must be evaluated for each biomass: can Spirulina or Haematococcus debris be partially digested into a protein extract and a fiber-rich fraction? Can a Haematococcus extract be profitably fractionated into pure Astaxanthin and dietary lipids?

15.6.4 Energy input and process modeling

Nutrient recycling, stream integration, and fractionation of biomass can be resource-intensive, which puts a constraint on processes. It is necessary to develop in silico process models that account first for the integration feasibility in terms of material balances, based on technical coefficients taken from the literature and from experiments in laboratory and pilot processes. After refining these material balances, efforts must be put in to evaluate the energy input in the process, preferentially using a modeling tool such as Aspen® or SuperPro® because these will facilitate tweaking the process. Utility streams must be included in this evaluation, and responsible attribution of costs and selling prices must be done. If such a process proves economically feasible in terms of operational costs, then the analysis can proceed to capital cost evaluation. Substitution of effluent treatment costs, environmental analysis, and zero-waste practices can less tangible advantages of circular, integrated processes.

15.6.5 Starting simple: biomass, biodigestion, and nutrient recovery

Agroindustrial processes where microalgae are to be integrated require large production areas for photobioreactors, and specialized processing for biomass concentration, dehydration, and eventual fractionation. Typical photobioreactor productivities are in the range of 10–30 g/m^2.day, with final biomass concentrations below 1 g/L, thus requiring large areas and volumes. Processes using more sophisticated photobioreactors reach higher concentrations, but still below the 10 g/L mark, and the bioreactor cost is higher. Even for well-known biomass products, there is a learning curve of practices and process monitoring. Therefore, the route for integration of novel microalgae processes into existing agroindustrial processes is to start with the necessary minimum – the biomass production – and slowly proceed to fractionation. Lower quality biomass – polycultures, unialgal cultures contaminated by bacteria – can be initially flocculated and biodigested; after the biomass production step is fully developed, biomass fractionation can be developed.

15.7 Conclusions and future prospects

Theoretically, algal biomass can be converted into many platform and specialty chemicals, from simple molecules such as carotenoids and lipids, to complex molecules such as phycocyanin and macromolecules such as exopolysaccharides and polyhydroxyalkanoates. However, there is a growing market for microalgae biomasses from selected strains and its raw extracts and a tangible opportunity in integrating microalgae production into agro-industrial processes, using nutrient-rich residual streams.

One of the best approaches to identify and develop these integrated processes is to build on the knowledge of successful, classical (although only decades-old) microalgae production processes such as those for *Spirulina*, *Chlorella*, *Dunaliella*, and *Haematococcus*. These biomasses can be either the primary goal for integration, or serve as templates for a next generation of microalgae strains. This process integration is complex, in the sense that it connects many diverse variables, and should start with the careful study of the existing resources, proceeding to process and scenario simulations, and then to stepwise implantation.

References

ACC. (2021). A.C.C. Guide to the Business of Chemistry Available online: https://www.americanchemistry.com/chemistry-in-america/data-industry-statistics/resources/2021-guide-to-the-business-of-chemistry (accessed on 16 October 2022).

Ahmad, M. T., Shariff, M., Md, Y. F., Goh, Y. M., & Banerjee, S. (2020). Applications of microalga chlorella vulgaris in aquaculture. Reviews in Aquaculture, *12*, 328–346. doi:10.1111/raq.12320

Al Mahrouqi, H., Vega, J., Dobretsov, S., & Abdala Díaz, R. T. (2022). The effect of medium concentration and nitrogen source on the productivity and biochemical composition of Arthrospira platensis. *Biology Bulletin*, *49*, 75–84. doi:10.1134/S1062359022020108

Alishah Aratboni, H., Rafiei, N., Garcia-Granados, R., Alemzadeh, A., & Morones-Ramírez, J. R. (2019). Biomass and lipid induction strategies in microalgae for biofuel production and other applications. *Microbial Cell Factories*, *18*, 1–17.

Almendarez, S. (October 2016). Natural products insider, pp. 9–15, web page accesible through https://www-naturalproductsinsider-com.webpkgcache.com/doc/-/s/www.naturalproductsinsider.com/sites/naturalproductsinsider.com/files/07-16INS-Astaxanthin-Redux.pdf, last time accessed in July 19, 2023.

Ambati, R. R., Gogisetty, D., Aswathanarayana, R. G., Ravi, S., Bikkina, P. N., Bo, L., & Yuepeng, S. (2019). Industrial potential of carotenoid pigments from microalgae: Current trends and future prospects. *Critical Reviews in Food Science and Nutrition*, *59*, 1880–1902. doi:10.1080/10408398.2018.1432561

Baroni, É. G., Yap, K. Y., Webley, P. A., Scales, P. J., & Martin, G. J. O. (2019). The effect of nitrogen depletion on the cell size, shape, density and gravitational settling of Nannochloropsis salina, Chlorella sp. (marine) and Haematococcus pluvialis. *Algal Research*, *39*, 101454. doi:10.1016/J.ALGAL.2019.101454

Brasil, B. S. A. F., Silva, F. C. P., & Siqueira, F. G. (2017). Microalgae biorefineries: The Brazilian scenario in perspective. New Biotechnology, *39*, 90–98. doi:10.1016/j.nbt.2016.04.007

Capelli, B., Talbott, S., Ding, L., & Capelli, F. (2021). Efficacy of Astaxanthin from different sources: Reports on the suitability for human health and nutrition. *Global Perspectives on Astaxanthin: From Industrial Production to Food, Health, and Pharmaceutical Applications*, 391–409. doi:10.1016/B978-0-12-823304-7.00027-1

Chen, J., Chen, C., Hasunuma, T., Kondo, A., Chang, C.-H., Ng, I.-S., & Chang, J.-S. (2019). Bioresource technology enhancing lutein production with mixotrophic cultivation of Chlorella sorokiniana MB-1-M12 using different bioprocess operation strategies. *Bioresource Technology*, *278*, 17–25. doi:10.1016/j.biortech.2019.01.041

Chen, J.-H., Kato, Y., Matsuda, M., Chen, C.-Y., Nagarajan, D., Hasunuma, T., Kondo, A., Dong, C.-D., Lee, D.-J., & Chang, J.-S. (2019). A novel process for the mixotrophic production of lutein with Chlorella sorokiniana MB-1-M12 using aquaculture wastewater. *Bioresource Technology*, *290*, 121786. doi:10.1016/j.biortech.2019.121786

Colusse, G. A., Mendes, C. R. B., Duarte, M. E. R., de Carvalho, J. C., & Noseda, M. D. (2020). Effects of different culture media on physiological features and laboratory scale production cost of Dunaliella salina. *Biotechnology Reports*, *27*. doi:10.1016/j.btre.2020.e00508

Coronado-Reyes, J. A., Salazar-Torres, J. A., Juárez-Campos, B., & González-Hernández, J. C. (2022). Chlorella vulgaris, a microalgae important to be used in Biotechnology: A review. *Food Science and Technology (Brazil)*, *42*, 1–11. doi:10.1590/fst.37320

Costa, J. A. V., Freitas, B. C. B., Rosa, G. M., Moraes, L., Morais, M. G., & Mitchell, B. G. (2019). Operational and economic aspects of Spirulina-based biorefinery. *Bioresource Technology*, *292*, 121946. doi:10.1016/j.biortech.2019.121946

Costa, S. S., Miranda, A. L., de Morais, M. G., Costa, J. A. V., & Druzian, J. I. (2019). Microalgae as source of polyhydroxyalkanoates (PHAs) – A review. *International Journal of Biological Macromolecules*, *131*, 536–547. doi:10.1016/j.ijbiomac.2019.03.099

de Carvalho, J. C., Cardoso, L. C., Ghiggi, V., & Woiciechowsk, A. L. (2014). Microbial Pigments. In: Brar, S. K., Dhillon, G. S., & Soccol, C. R. (Eds.), Biotransformation of Waste Biomass into High Value Biochemicals. Springer New York, New York, NY, Vol. 9781461480, ISBN 978-1-4614-8004-4.

de Carvalho, J. C., Goyzueta-Mamani, L. D., Karp, S. G., Aulestia, D. T. M., Sydney, E. B., Fanka, L. S., Pandey, A., & Soccol, C. R. (2022). Downstream processing and formulation of microbial lipids. *Biomass, Biofuels, Biochemicals*, *2022*, 261–287. doi:10.1016/B978-0-323-90631-9.00007-7 de Carvalho, J. C., Goyzueta-Mamani, L. D., Karp, S. G., Aulestia, D. T. M., Sydney, E. B., Fanka, L. S., Pandey, A., & Soccol, C. R. (2022). Downstream processing and formulation of microbial lipids. *Biomass, Biofuels, Biochemicals, 1*, 261–287. Elsevier.

de Carvalho, J. C., Goyzueta-Mamani, L. D., Molina-Aulestia, D. T., Júnior, A. I. M., Iwamoto, H., Ambati, R., Ravishankar, G. A., & Soccol, C. R. (2022). Microbial Astaxanthin production from agro-industrial wastes – Raw materials, processes, and quality. *Fermentation, 8*, 484. doi:10.3390/FERMENTATION8100484

de Carvalho, J. C., Magalhães, A. I., de Melo Pereira, G. V., Medeiros, A. B. P., Sydney, E. B., Rodrigues, C., Aulestia, D. T. M., De Souza Vandenberghe, L. P., Soccol, V. T., & Soccol, C. R. (2020). Microalgal biomass pretreatment for integrated processing into biofuels, food, and feed. *Bioresource Technology, 300*, 122719. doi:10.1016/J.BIORTECH.2019.122719

De Luca, M., Pappalardo, I., Limongi, A. R., Viviano, E., Radice, R. P., Todisco, S., Martelli, G., Infantino, V., & Vassallo, A. (2021). Lipids from microalgae for cosmetic applications. *Cosmetics, 8*. doi:10.3390/cosmetics8020052

Delrue, F., Alaux, E., Moudjaoui, L., Gaignard, C., Fleury, G., Perilhou, A., Richaud, P., Petitjean, M., & Sassi, J.-F. (2017). Optimization of Arthrospira platensis (Spirulina) growth: From laboratory scale to pilot scale. *Fermentation, 3*. doi:10.3390/fermentation3040059

Dufossé, L., Galaup, P., Yaron, A., Arad, S. M., Blanc, P., Murthy, K. N. C., & Ravishankar, G. A. (2005). Microorganisms and microalgae as sources of pigments for food use: A scientific oddity or an industrial reality? *Trends in Food Science and Technology, 16*, 389–406. doi:10.1016/j.tifs.2005.02.006

Fábregas, J., Domínguez, A., Regueiro, M., Maseda, A., & Otero, A. (2000). Optimization of culture medium for the continuous cultivation of the microalga Haematococcus pluvialis. *Applied Microbiology and Biotechnology, 53*, 530–535. doi:10.1007/S002530051652

Ganesan, R., Manigandan, S., Samuel, M. S., Shanmuganathan, R., Brindhadevi, K., Lan Chi, N. T., Duc, P. A., & Pugazhendhi, A. (2020). A review on prospective production of biofuel from microalgae. *Biotechnology Reports, 27*, e00509. doi:10.1016/j.btre.2020.e00509

García-González, M., Moreno, J., Cañavate, J. P., Anguis, V., Prieto, A., Manzano, C., Florencio, F. J., & Guerrero, M. G. (2003). Conditions for open-air outdoor culture of Dunaliella salina in southern Spain. *Journal Applied Phycology, 15*, 177–184. doi:10.1023/A:1023892520443

Ghiggi Sorgatto, V., Ricardo Soccol, C., Tatiana Molina-Aulestia, D., Aurélio de Carvalho, M., Vinícius de Melo Pereira, G., Cesar de Carvalho, J., & Mixotrophic, J. (2021). Mixotrophic cultivation of microalgae in cassava processing wastewater for simultaneous treatment and production of lipid-rich biomass. *Fuels, 2*, 521–532. doi:10.3390/FUELS2040030

Gollo, A. L., Da Silva, A. L. L., De Lima, K. K. D., Costa, J. L., Camara, M. C., Biasi, L. A., Rodrigues, C., Vandenberghe, L. P. S., Soccol, V. T., & Soccol, C. R. (2016). Developing a plant culture medium composed of vinasse originating from Haematococcus pluvialis culture. *Pakistan Journal of Botany, 48*, 295–303.

Gong, M., & Bassi, A. (2016). Carotenoids from microalgae: A review of recent developments. *Biotechnology Advance, 34*, 1396–1412. doi:10.1016/j.biotechadv.2016.10.005

GRAS Notice Inventory | FDA Available online. (2022). https://www.fda.gov/food/generally-recognized-safe-gras/gras-notice-inventory (accessed on 16 September 2022).

Guedes, A. C., Amaro, H. M., & Malcata, F. X. (2011). Microalgae as sources of carotenoids. *Marine Drugs, 9*, 625–644. doi:10.3390/md9040625

Harvey, P. J., & Ben-Amotz, A. (2020). Towards a sustainable Dunaliella salina microalgal biorefinery for 9-cis β-carotene production. *Algal Research, 50*, 102002. doi:10.1016/j.algal.2020.102002

Issarapayup, K., Powtongsook, S., & Pavasant, P. (2011). Economical review of Haematococcus pluvialis culture in flat-panel airlift photobioreactors. *Aquacultural Engineering, 44*, 65–71. doi:10.1016/J.AQUAENG.2011.03.002

Ji, X. J., & Ledesma-Amaro, R. (2020). Microbial lipid biotechnology to produce polyunsaturated fatty acids. *Trends in Biotechnology, 38*, 832–834.

Ju, Z. Y., Deng, D. F., & Dominy, W. (2012). A defatted microalgae (Haematococcus pluvialis) meal as a protein ingredient to partially replace fishmeal in diets of Pacific white shrimp (Litopenaeus Vannamei, Boone, 1931). *Aquaculture, 354–355*, 50–55. doi:10.1016/J.AQUACULTURE.2012.04.028

Khan, M. I., Shin, J. H., & Kim, J. D. (2018). The promising future of microalgae: Current status, challenges, and optimization of a sustainable and renewable industry for biofuels, feed, and other products. *Microbial Cell Factories, 17.*

Kujawska, N., Talbierz, S., Dębowski, M., Kazimierowicz, J., & Zieliński, M. (2021). Cultivation method effect on Schizochytrium sp. biomass growth and docosahexaenoic acid (dha) production with the use of waste glycerol as a source of organic carbon. *Energies (Basel), 14.* doi:10.3390/en14102952

Kumar, S., & Deshmukh, R. (2021). *Allied Market Research – Microalgae Market by Type, Application: Global Opportunity Anaysis and Industry Forecast 2021–2028.*

Laroche, C. (2022). Exopolysaccharides from microalgae and cyanobacteria: Diversity of strains, production strategies, and applications. *Marine Drugs, 20.* doi:10.3390/MD20050336

Li, H., Li, Z., Xiong, S., Zhang, H., Li, N., Zhou, S., Liu, Y., & Huang, Z. (2011). Pilot-scale isolation of bioactive extracellular polymeric substances from cell-free media of mass microalgal cultures using tangential-flow ultrafiltration. *Process Biochemical, 46*, 1104–1109. doi:10.1016/J.PROCBIO.2011.01.028

Lin, J. H., Lee, D. J., & Chang, J. S. (2015). Lutein production from biomass: Marigold flowers versus microalgae. *Bioresource Technology, 184*, 421–428. doi:10.1016/J.BIORTECH.2014.09.099

Ma, N., Long, X., Liu, J., Chang, G., Deng, D., Cheng, Y., & Wu, X. (2019). Defatted Haematococcus pluvialis meal can enhance the coloration of adult Chinese mitten crab Eriocheir sinensis. *Aquaculture, 510*, 371–379. doi:10.1016/J.AQUACULTURE.2019.05.063

Mallick, R., Basak, S., & Duttaroy, A. K. (2019). Docosahexaenoic acid,22:6n-3: Its roles in the structure and function of the brain. *International Journal of Developmental Neuroscience, 79*, 21–31.

Marinho, Y. F., Malafaia, C. B., de Araújo, K. S., da Silva, T. D., dos Santos, A. P. F., de Moraes, L. B., & Gálvez, A. O. (2021). Evaluation of the influence of different culture media on growth, life cycle, biochemical composition, and Astaxanthin production in Haematococcus pluvialis. *Aquaculture International, 29*, 757–778. doi:10.1007/S10499-021-00655-Z

Mcclure, D. D., Nightingale, J. K., Luiz, A., Black, S., Zhu, J., & Kavanagh, J. M. (2019). Pilot-scale production of lutein using Chlorella vulgaris. *Algal Research, 44*, 101707. doi:10.1016/j.algal.2019.101707

Meticulous Research. (2021). *Microalgae Market by Distribution Channel (Consumer Channel, Business Channel), Type (Spirulina, Chlorella, Dunaliella Salina, Haematococcus Pluvialis), Application (Nutraceuticals, Food & Beverages, Animal Feed, Cosmetics) – Global Forecast to 2028.*

Meticulous Research. (2022). *Chlorella Market by Technology (Open Pond), by Product Type (Extract, Capsules) by Source (Chlorella Vulgaris, Chlorella Pyrenoidosa or Sorokiniana) by Application (Nutraceutical, Food and Beverages, Animal Feed), Geography – Global Forecast to 2028.*

Mobin, S. M. A., Chowdhury, H., & Alam, F. (2019). Commercially important bioproducts from microalgae and their current applications – A review. *Proceedings of the Energy Procedia, 160*, 752–760. Elsevier Ltd.

Monte, J., Ribeiro, C., Parreira, C., Costa, L., Brive, L., Casal, S., Brazinha, C., & Crespo, J. G. (2020). Biorefinery of Dunaliella salina: Sustainable recovery of carotenoids, polar lipids and glycerol. *Bioresource Technology, 297.* doi:10.1016/j.biortech.2019.122509

Monte, J., Sá, M., Galinha, C. F., Costa, L., Hoekstra, H., Brazinha, C., & Crespo, J. G. (2018). Harvesting of Dunaliella salina by membrane filtration at pilot scale. *Separation and Purification Technology, 190*, 252–260. doi:10.1016/j.seppur.2017.08.019

Nishshanka, G. K. S. H., Liyanaarachchi, V. C., Premaratne, M., Nimarshana, P. H. V., Ariyadasa, T. U., & Kornaros, M. (2021). Wastewater-based microalgal biorefineries for the production of Astaxanthin and co-products: Current status, challenges and future perspectives. *Bioresource Technology, 342.* doi:10.1016/J.BIORTECH.2021.126018

Oh, S. H., Han, J. G., Kim, Y., Ha, J. H., Kim, S. S., Jeong, M. H., Jeong, H. S., Kim, N. Y., Cho, J. S., Yoon, W. B., et al. (2009). Lipid production in Porphyridium cruentum grown under different culture conditions. *Journal of Bioscience and Bioengineering*, *108*, 429–434. doi:10.1016/j.jbiosc.2009.05.020

Olaizola, M. (2000). Commercial production of Astaxanthin from Haematococcus pluvialis using 25,000-liter outdoor photobioreactors. *Journal Applied Phycology*, *12*, 499–506. doi:10.1023/A:1008159127672

Orozco Colonia, B. S., Vinícius de Melo Pereira, G., & Soccol, C. R. (2020). Omega-3 microbial oils from marine thraustochytrids as a sustainable and technological solution: A review and patent landscape. *Trends in Food Science and Technology*, *99*, 244–256.

Peeters, D., & Huyskens, P. (1993). Endothermicity or exothermicity of water/alcohol mixtures. *Journal of Molecular Structure*, *300*, 539–550. doi:10.1016/0022-2860(93)87046-C

Persistent Market Search. (2017). *Spirulina Market Global Market Study on Spirulina: Powder Product Form Segment Anticipated to Dominate the Global Market in Terms of Both Value and Volume During 2016–2026.*

Pourkarimi, S., Hallajisani, A., Nouralishahi, A., Alizadehdakhel, A., & Golzary, A. (2020). Factors affecting production of beta-carotene from Dunaliella salina microalgae. *Biocatalysis and Agricultural Biotechnology*, *29*, 101771. doi:10.1016/j.bcab.2020.101771

Qu, L., Ji, X. J., Ren, L. J., Nie, Z. K., Feng, Y., Wu, W. J., Ouyang, P. K., & Huang, H. (2011). Enhancement of docosahexaenoic acid production by Schizochytrium sp. using a two-stage oxygen supply control strategy based on oxygen transfer coefficient. *Letters in Applied Microbiology*, *52*, 22–27. doi:10.1111/j.1472-765X.2010.02960.x

Ragaza, J. A., Hossain, M. S., Meiler, K. A., Velasquez, S. F., & Kumar, V. (2020). A review on Spirulina: Alternative media for cultivation and nutritive value as an aquafeed. Reviews in Aquaculture, *12*, 2371–2395.

Ramaraj, R., Kawaree, R., & Unpaprom, Y. (2016). Direct transesterification of microalga Botryococcus braunii biomass for biodiesel production (Article in International Journal of Parallel Emergent and Distributed Systems Bioethanol Production from Agricultural Wastes by Separate Hydrolysis and Fermentation Method View Project Biomass Technology View Project). *Emergent Life Sciences Research*, *2*(2), 1–7.

Robertson, R., Guihèneuf, F., Schmid, M., Stengel, D. B., Fitzgerald, G., Ross, P., & Stanton, C. (2013). Algae-derived polyunsaturated fatty acids: implications for human health. Polyunsaturated fatty acids: sources, antioxidant properties and health benefits, 45–99.

Rosero-Chasoy, G., Rodríguez-Jasso, R. M., Aguilar, C. N., Buitrón, G., Chairez, I., & Ruiz, H. A. (2022). Growth kinetics and quantification of carbohydrate, protein, lipids, and chlorophyll of Spirulina platensis under aqueous conditions using different carbon and nitrogen sources. *Bioresource Technology*, *346*. doi:10.1016/j.biortech.2021.126456

Safaei, M., Maleki, H., Soleimanpour, H., Norouzy, A., Zahiri, H. S., Vali, H., & Noghabi, K. A. (2019). Development of a novel method for the purification of C-phycocyanin pigment from a local cyanobacterial strain Limnothrix sp. NS01 and evaluation of its anticancer properties. *Scientific Reports*, *9*, 1–16. doi:10.1038/s41598-019-45905-6

Safi, C., Zebib, B., Merah, O., Pontalier, P.-Y., & Vaca-Garcia, C. (2014). Morphology, composition, production, processing and applications of Chlorella vulgaris: A review. *Renewable and Sustainable, Energy Reviews*, *35*, 265–278. doi:10.1016/j.rser.2014.04.007

SAG. (2022). S. von A. Culture Media – Georg-August-Universität Göttingen Available online: https://www.uni-goettingen.de/de/www.uni-goettingen.de/de/186262.html (accessed on 15 October 2022).

Sharma, K. K., Schuhmann, H., & Schenk, P. M. (2012). High lipid induction in microalgae for biodiesel production. *Energies (Basel)*, *5*, 1532–1553. doi:10.3390/en5051532

Silva, D. L. B., de Moraes, L. B. S., Oliveira, C. Y. B., da Silva Campos, C. V. F., Bezerra, R. de S., & Gálvez, A. O. (2022). Influence of culture medium on growth and protein production by Haematococcus pluvialis. *Acta Scientiarum – Technology*, *44*. doi:10.4025/ACTASCITECHNOL.V44I1.59590

Soto-Sierra, L., Stoykova, P., & Nikolov, Z. L. (2018). Extraction and fractionation of microalgae-based protein products. *Algal Research, 36*, 175–192. doi:10.1016/j.algal.2018.10.023

Su, Y. (2021). Revisiting carbon, nitrogen, and phosphorus metabolisms in microalgae for wastewater treatment. *Science of the Total Environment, 762*, 144590, ISSN 0048-9697, https://doi.org/10.1016/j.scitotenv.2020.144590

Tababa, H. G., Hirabayashi, S., & Inubushi, K. (2012). Media optimization of Parietochloris incisa for arachidonic acid accumulation in an outdoor vertical tubular photobioreactor. *Journal Applied Phycology, 24*, 887–895. doi:10.1007/s10811-011-9709-9

Takeshita, T., Ivanov, I. N., Oshima, K., Ishii, K., Kawamoto, H., Ota, S., Yamazaki, T., Hirata, A., Kazama, Y., Abe, T., et al. (2018). Comparison of lipid productivity of Parachlorella kessleri heavy-ion beam irradiation mutant PK4 in laboratory and 150-L mass bioreactor, identification and characterization of its genetic variation. *Algal Research, 35*, 416–426. doi:10.1016/j.algal.2018.09.005

Talbierz, S., Dębowski, M., Kujawska, N., Kazimierowicz, J., & Zieliński, M. (2022). Optimization of lipid production by Schizochytrium limacinum biomass modified with ethyl methane sulfonate and grown on waste glycerol. *International Journal of Environmental Research and Public Health, 19*. doi:10.3390/ijerph19053108

Talbott, S., Hantla, D., Capelli, B., Ding, L., Li, Y., & Artaria, C. (2019). Effect of Astaxanthin supplementation on psychophysiological heart-brain axis dynamics in healthy subjects. *Functional Foods in Health and Disease, 9*, 521–531. doi:10.31989/FFHD.V9I8.636

Torres-Tiji, Y., Fields, F. J., & Mayfield, S. P. (2020). Microalgae as a future food source. *Biotechnology Advance, 41*, 107536. doi:10.1016/J.BIOTECHADV.2020.107536

Uebel, L. S., Costa, J. A. V., Olson, A. C., & de Morais, M. G. (2019). Industrial plant for production of *Spirulina* sp. LEB 18. *Brazil Journal of Chemical Engineering, 36*, 51–63. doi:10.1590/0104-6632.20180361S20170284

UTEX. (2022). C.C. of A. at the U. of T.A. Algal culture media recipes | UTEX Culture Collection of Algae Available online: https://utex.org/pages/algal-culture-media (accessed on 15 October 2022).

Wu, Z., Dejtisakdi, W., Kermanee, P., Ma, C., Arirob, W., Sathasivam, R., & Juntawong, N. (2017). Outdoor cultivation of Dunaliella salina KU 11 using brine and saline lake water with raceway ponds in northeastern Thailand. *Biotechnology and Applied Biochemistry, 64*, 938–943. doi:10.1002/bab.1537

Yates, C. M., Calder, P. C., & Ed Rainger, G. (2014). Pharmacology and therapeutics of omega-3 polyunsaturated fatty acids in chronic inflammatory disease. *Pharmacology and Therapeutics, 141*, 272–282.

Yildirim, O., Tunay, D., & Ozkaya, B. (2022). Reuse of sea water reverse osmosis brine to produce Dunaliella salina based β-carotene as a valuable bioproduct: A circular bioeconomy perspective. *Journal of Environmental Management, 302*, 114024. doi:10.1016/j.jenvman.2021.114024

Zhu, C., Zhai, X., Jia, J., Wang, J., Han, D., Li, Y., Tang, Y., & Chi, Z. (2018). Seawater desalination concentrate for cultivation of Dunaliella salina with floating photobioreactor to produce β-carotene. *Algal Research, 35*, 319–324. doi:10.1016/j.algal.2018.08.035

Jingyan Hu, Weiqi Fu*

Chapter 16
Process intensification opportunities applied to the production of microalgae specialty chemicals

Abstract: As unicellular organisms, microalgae have high carbon sequestration efficiency, based on photosynthesis, mitigating climate change through carbon neutralization. Microalgae are widely regarded as excellent feedstocks for various specialty chemicals such as carotenoids, polyunsaturated fatty acids, and bioactive polysaccharides over higher plants and other microbes due to their advantages such as high growth and carbon-fixation rate even under harsh conditions. However, due to the limitations such as production efficiency and cost, the commercialization of microalgae specialty chemicals is facing challenges. To achieve large-scale production of microalgae specialty chemicals at low cost, there have been numerous studies on process intensification applied to microalgae specialty chemicals production, including microalgae strain screening, genome editing technologies, optimization of cultivation conditions, and refinement of the production process. This chapter focuses on the significant refinement of microalgae specialty chemicals production, based on strain engineering, which can be divided into the growth-coupling type and non-growth-coupling type, according to the relationship between growth and metabolism. Recent progress on the relevant technical methods and strategies are also reviewed and discussed. In the end, challenges and future directions toward specialty chemicals production in microalgae are summarized and proposed.

Keywords: microalgae, process intensification, carbon neutrality, specialty chemicals, strain engineering, biorefinery

16.1 Introduction

Microalgae are unicellular organisms that live in every aquatic environment and can fix carbon dioxide through photosynthesis. In general, microalgae can grow rapidly and fix CO_2 about 10–50 times faster than terrestrial plants (Wang et al., 2008). The efficiency of

***Corresponding author: Weiqi Fu**, Department of Marine Science, Ocean College, Zhejiang University, Zhejiang, China; Center for Systems Biology and Faculty of Industrial Engineering, Mechanical Engineering and Computer Science, School of Engineering and Natural Sciences, University of Iceland, Reykjavík, Iceland, e-mail: weiqifu@zju.edu.cn
Jingyan Hu, Department of Marine Science, Ocean College, Zhejiang University, Zhejiang, China

https://doi.org/10.1515/9783110781267-016

carbon dioxide or bicarbonate capture in open ponds is estimated to be up to 90% under optimal conditions (Sayre, 2010). Therefore, large-scale industrial microalgae farming dramatically contributes to climate change mitigation (Yap et al., 2021). In parallel, microalgae absorb nitrogenous compounds (including NO_3, NO_2, NH_3) and phosphorus nutrients (including HPO_4^{2-} or H_2PO_4) from industrial wastewater and waste gas during carbon sequestration, and convert them to value-added chemicals such as pharmaceuticals and nutraceuticals (Apandi et al., 2019; Judd et al., 2017; Maeda et al., 2018; Sathasivam et al., 2019). To date, pigments (e.g., chlorophylls, carotenoids, and phycobiliproteins) and unsaturated fatty acids (e.g., omega-3 fatty acids, eicosapentaenoic acid (EPA), and docosahexaenoic acid (DHA) and arachidonic acid (AA)) (Griffiths et al., 2016) have received considerable attention as commercialized microalgae products for their protective benefits such as immune stimulation, antioxidant property, ability to prevent chronic diseases, anti-inflammatory antidepression capability, and promotion of brain and nerve development in infants (Zhang et al., 2018).

Faced with the growing population's increasing demand, the industrialization of microalgae shows latent capacity to essentially meet the needs for sustainable development globally (Wang et al., 2021). As rich sources of carbon compounds, microalgae, accompanied by their bioactive metabolites (e.g., proteins, polyunsaturated fatty acids, and pigments), are the feedstocks for health food supplements, pharmaceuticals, and cosmetics (Levasseur et al., 2020). The global algae products market was valued at $2.2760 billion in 2020 and is projected to reach $4.2868 billion by 2031, growing at a CAGR of 4.88% from 2022 to 2031 (https://www.alliedmarketresearch.com/algae-products-market). However, there are hurdles such as high cost, high energy consumption, and low utilization rate of stocks in the current production, which have hindered the marketing of microalgae specialty chemicals (Khan et al., 2018).

To fulfill the industrialization and commercialization of microalgae specialty chemicals on a large scale, researchers have carried out investigations on process intensification (PI) in terms of strain engineering, optimization of culture conditions, and downstream processing improvement (Gao et al., 2020; Sun et al., 2021; Verma et al., 2020). The targets of PI are less energy consumption and low waste, higher productivity, and safer processes (Stankiewicz and Moulijn, 2018). However, previous studies mostly focused on separate phases of massive production or specific chemicals such as fucoxanthin (Leong et al., 2022) and astaxanthin (Ashokkumar et al., 2021). Only a few researchers emphasized strain engineering for PI, and the different growth and metabolic relationships between microalgae species and their products were mostly overlooked. In this chapter, we first classify microalgae specialty chemicals into two different groups, i.e., growth-coupling types and non-growth-coupling types. We also review the latest progress on strain engineering, the optimization of cultivation conditions, and downstream processing. Finally, we summarize and pinpoint the opportunities and challenges for the future development of microalgae specialty chemicals.

16.2 Process intensification opportunities in microalgae specialty chemicals production

16.2.1 Production of different specialty chemicals in microalgae

Microalgae photosynthesize by fixing carbon dioxide in the atmosphere for growth and metabolism, generating a wide range of complex biological macromolecules, metabolites, and other biochemical products such as lipids, proteins, polysaccharides, and pigments (Bhattacharya and Goswami, 2020). Diverse chemicals are produced in separate parts of microalgae cells. Ideally, carbon dioxide capture by microalgae activates assimilation pigments centered on light-harvesting antenna complexes in the chloroplast, and electron transfer in the photosystem converts the light energy into chemical energy, yielding ATP and NADPH, which are used for carbon fixation in the Calvin-Benson-Bassham cycle (CBB cycle) and carbon conversion in downstream metabolism (Sun et al., 2018). The glucose molecules produced by photosynthesis in microalgae cells are lysed to make pyruvate, which synthesizes fatty acids and hydrocarbons in the endoplasmic reticulum and in the pigments of the plastids, respectively (Bermúdez et al., 2014). Microalgae possess a complex metabolism and synthesize chemicals autotrophically, heterotrophically, or mixotrophically. Each metabolic process is governed by certain conditions that can be altered by adjusting the operational parameters (Venkata Mohan et al., 2016).

Here, microalgae specialty chemicals are categorized based on their relationship between growth and metabolism into growth-coupling type and non-growth-coupling type. Table 16.1 lists the categories of products and their corresponding strain species.

16.2.1.1 Growth-coupling chemicals

Growth-coupling chemicals are typically primary metabolites, which are compounds required for microalgae to grow and reproduce themselves through metabolic processes such as lipids, proteins, polysaccharides, vitamins, etc.

Microalgae lipids are of immense interest in growth-coupling products. Dry microalgae biomass from nonstress culture conditions has a 3–20% lipid content, which varies with microalgae species and the cultivation circumstances (light, CO_2 concentration, and nitrogen deficiency) (Demirbas, 2011). Unsaturated fatty acids (PUFAs) are notable microalgae lipids and are reasonable feedstocks for food and pharmaceutical processing (Chauton et al., 2015). Docosahexaenoic acid (DHA) and eicosapentaenoic acid (EPA) are typical PUFAs produced in abundance by microalgae. Albeit, microalgae have been recognized as major producers of DHA and EPA, only a few species (*Crypthecodinium cohnii, Nannochloropsis* sp., and *Phaeodactylum tricornutum*) have achieved market success (Ramesh Kumar et al., 2019; Sahin et al., 2018). *Lobosphaera inisa* may be a microalgae source of 20-carbon omega-6 LC-PUFA dihomo-γ-linolenic

Table 16.1: Different types of high-value products by microalgae species.

Types	Products	Microalgae species	References
Growth-coupling	DHA	*Crypthecodinium, Schizochytrium, Ulkenia, Thraustochytrium, Labyrinthula, Phaeodactylum tricornutum, Monodus subterraneus, Porphyridium cruentum, Chaetoceros calcitrans, Nannochloropsis* sp., *Crypthecodinium cohnii, Isochrysis galbana, Pavlova salina*	(Hamed, 2016; Mourelle et al., 2017; Sahin et al., 2018)
	EPA	*Nannochloropsis, Phaeodactylum tricornutum, Tisochrysis lutea, M. subterraneus, Schizochytrium* sp. (ATCC 20,888)	(Hamilton et al., 2014; Han et al., 2020; Rao et al., 2020; Steinrücken et al., 2018)
	Proteins	*Arthrospira, Chlorella, Dunaliella salina*	(Becker, 2007; Christaki et al., 2011; Ejike et al., 2017)
	Polysaccharides	*Tetraselmis* sp., *Isochrysis* sp., *Porphyridium cruentum, Porphyridium purpureum, Chlorella* sp., *Rhodella reticulata, Euglena gracilis*	(Brownfield et al., 2009; Mourelle et al., 2017; A. Sun et al., 2018)
	Vitamins	*Chlorella, Spirulina, Anabaena cylindrica*	(Edelmann et al., 2019; Tarento et al., 2018)
Non-growth-coupling	Chlorophylls	*Chlorella, Scenedesmus dimorphus*	(Ferreira et al., 2016; Khanra et al., 2018; Levasseur et al., 2018)
	β-Carotene	*Scenedesmus almeriensis, D. bardawil, D. tertiolecta*	(Berthon et al., 2017; Mourelle et al., 2017)
	Astaxanthin	*Haematococcus pluvalis, Chlorella zofingiensis, Chlorococcum* sp., *Scenedesmus* sp.	(Bhalamurugan et al., 2018; Odjadjare et al., 2017)
	Lutein	*Muriellopsis* sp., *Chlorella protothecoides, Chlorella zofingiensis, Chlorococcum citriforme, Neospongiococcus gelatinosum, Scenedesmus almeriensis*	(Bhalamurugan et al., 2018; Molino et al., 2018; Ojulari et al., 2020; Yaakob et al., 2014)
	Fucoxanthin	*Halamphora coffeaeformis, Phaeodactylum tricornutum, Odontella auritacan, I. galbana T-Iso*	(Gao et al., 2021; Popovich et al., 2020; Song Xia et al., 2018)
	Violaxanthin	*Chlorella ellipsoidea, D. tertiolecta*	(Molino et al., 2018; Mourelle et al., 2017)

Table 16.1 (continued)

Types	Products	Microalgae species	References
	Zeaxanthin	*Scenedesmus almeriensis, Nannochloropsis oculata*	(Bhalamurugan et al., 2018; Hamed, 2016)
	Phycobiliproteins	*Rhodophyta phylum, Porphyridium* sp., *Arthrospira* sp., *Aphanizomenon flosaquae*	(Bhalamurugan et al., 2018; Odjadjare et al., 2017)
	Polyphenols	*Haematococcus pluvialis, Phaeodactylum tricornutum, Tetraselmis* sp., *Neochloris oleoabundans, Chlorella vulgaris, Arthrospira platensis*	(Goiris et al., 2012; Matos et al., 2020)
	Phytosterols	*Pavlova lutheri, Tetraselmis* sp. M8, *Nannochloropsis* sp. BR2	(Ahmed et al., 2015)
	Endotoxin	*Chlorella* sp., *Nostoc* sp. (ATCC 53,789)	(Gupta et al., 2013; Polyzois et al., 2020)
	Dolastatins	*Symploca* sp.	(Luesch et al., 2001)
	Bio-diols	*Chlorella ohadii*	(Benisvy-Aharonovich et al., 2020)

acid (DGLA, 20:3 n-6) as it produces large amounts of DGLA and is stable under different conditions (Abu-Ghosh et al., 2021).

Microalgae proteins are also hot spots for utilization as food proteins or dietary supplements. The crude protein content of microalgae biomass is reported to range from 6% to 63%, with most species containing more than 40% crude protein on a dry weight basis (Wang et al., 2021). *Chlorella* sp. and *Arthrospira* sp. are the most commonly used strains for industrial production due to their high protein content (51–58% of dry biomass) and balanced essential amino acid profiles (Ismail et al., 2020).

Microalgae polysaccharides are growth-coupled products with considerable industrial potential, which are mainly divided into reserve polysaccharides and structural polysaccharides. The former exists as starch in most microalgae species, and the polymerization of polysaccharides varies in different algal species; it varies with different intracellular locations. For example, in red algae, starch is stored as vesicles in the cytoplasm outside the chloroplast, and in the green microalgae chlorophyll, starch is stored in plastids (Bernaerts et al., 2019, 2018). Reserve polysaccharides are a good choice for the preparation of biofuels, and they can be easily converted into biomass at a low cost (Bharti et al., 2022). Structural polysaccharides are associated with cell walls. They have certain rheological properties, and can be used as thickeners or gelling agents (Bernaerts et al., 2019; Siqueira et al., 2018).

In addition to macronutrients, the primary metabolites of microalgae also contain many essential vitamins for the growth and development of organisms such as vitamin B12 (Edelmann et al., 2019; Smith et al., 2007), vitamin K (Tarento et al., 2018), and vitamin D (Jäpelt and Jakobsen, 2013) that do not exist in higher plants. The vitamin content of microalgae is species-dependent, for example, *D. salina* highly accumulates vitamins B as well as vitamins C and E (Udayan et al., 2017). The diatom *Skeletonema marinoi* also contains high content of vitamin C (Smerilli et al., 2019, 2017).

16.2.1.2 Non-growth-coupling chemicals

Non-growth-coupling chemicals are generally secondary metabolites, which are organic compounds with complex chemical structures, produced by microalgae growth to a certain stage. They are not directly involved in the growth or reproduction of organisms, but are essential in ecological and other activities. Pigments, phenols, steroids, toxins, etc. are some examples of secondary metabolites.

Among the non-growth-coupled products, pigments are of industrial interest, including chlorophylls, carotenoids, and phycobiliproteins (Hamed, 2016). Chlorophyll is the green pigment most closely related to photosynthesis and exists naturally in microalgae as chlorophyll a and chlorophyll b, and its content varies with environmental conditions and bacterial species (Ferreira et al., 2016; Gonçalves et al., 2019; Levasseur et al., 2018). Among the most developed strains, the most well-known microalgae are from the genus *Chlorella*, which contains about 7% of its biomass in chlorophyll, five times that of *Arthrospira* (Khanra et al., 2018). Carotenoids are a class of pigments that are abundant in microalgae. The green microalgae produce a wide range of pigments: carotene (beta-carotene, lycopene, etc.) and xanthophyll (astaxanthin, violaxanthin, zeaxanthin, neoxanthin, lutein, etc.) (Levasseur et al., 2020). Other pigments such as fucoxanthin are produced by diatoms such as *Halamphora coffeaeformis*, *Phaeodactylum tricornutum*, golden yellow algae, and green microalgae *I. galbana T-Iso* (Gao et al., 2021; Popovich et al., 2020; Xia et al., 2018). Phycobiliprotein is another pigment in microalgae, which is only found in the phylum of Rhodophyta, in some cryptoalgae, and in cyanobacteria (Gouveia et al., 2010; Levasseur et al., 2020). On an industrial scale, it is mainly produced by *Porphyridium* sp., *Arthrospira* sp., and *Aphanizomenon flosaquae* (Bhalamurugan et al., 2018; Hamed, 2016; Siqueira et al., 2018; Yaakob et al., 2014).

Polyphenols are a broad group of secondary metabolites, including phenolic acids, flavonoids, isoflavones, stilbene, lignans, and phenolic polymers (Galasso et al., 2019). Like all other active biological compounds, the composition and content of microalgae polyphenols are species-dependent. *Phaeodactylum tricornutum*, *Arthrospira platensis*, and *Chlorella* sp. are good producers of microalgae polyphenols (Matos et al., 2020).

Phytosterols are components of cell membranes and affect their stability, permeability, and fluidity by controlling the movement of fatty acid chains within them (Ahmed et al., 2015; Galasso et al., 2019). In many microalgae, 24-ethylcholesterol appears to be the main phytosterol (Galasso et al., 2019). Various microalgae can biosynthesize phytosterols, with *Pavlova lutheri* being a good producer of phytosterols with high phytosterol productivity (5,186 mg/100 g) (Ahmed et al., 2015).

In addition to the above common non-growth-coupled products of microalgae, there are also some unconventional microalgae secondary metabolites, such as toxins and alcohol. Dolastatins are a class of peptides with medicinal properties, isolated from the cyanobacteria *Symploca hydnoides* and *Lyngbya majuscule* (Mondal et al., 2020). Bio-alcohols, such as diols, have been widely used as solvents in cosmetics and pharmaceuticals, and the green microalga *Chlorella ohadii* has been reported to have the potential to produce 1,4-pentanediol (Benisvy-Aharonovich et al., 2020).

16.2.2 Strain-engineering strategies

Depending on various production scenarios, appropriate methods are chosen to improve the strains. Specifically, the growth rate or the tendency to accumulate bioactive chemicals of the strains can be improved to obtain more growth-coupling or non-growth-coupling products. It is possible to augment the production of chemicals by improving the algal strains, mainly through synthetic biology and genome editing, adaptive laboratory evolution, and such other approaches, which are generally illustrated in Figure 16.1.

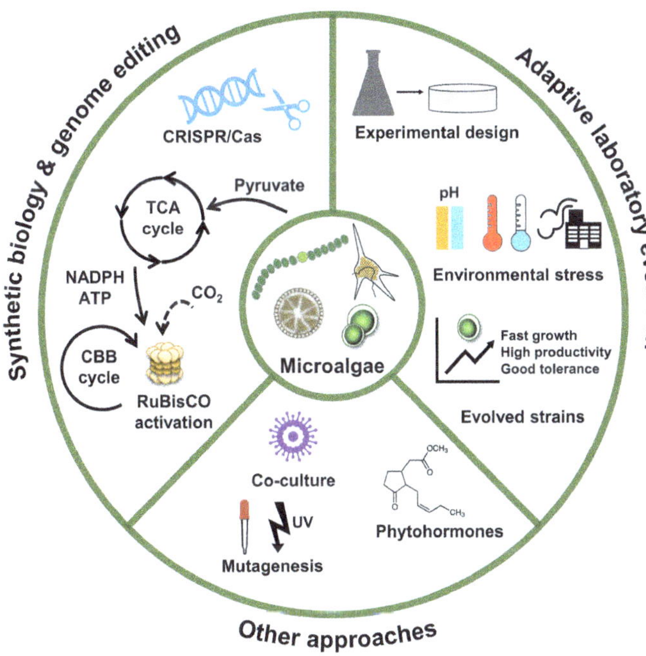

Figure 16.1: Strategies of strain engineering for enhanced performance.

16.2.2.1 Synthetic biology and genome editing technology

Current approaches to enhance the productivity of strains with synthetic biology mainly consider the improvement of CO_2 assimilation efficiency, which can be achieved by (1) improving CO_2 fixation efficiency, (2) reducing photorespiration losses and improving light energy utilization efficiency, and (3) adding additional CO_2 fixation pathways (Barati et al., 2021; Naduthodi et al., 2021).

The CBB cycle is the most important pathway for carbon dioxide fixation by microalgae, and Rubisco is the key enzyme of this cycle. However, the activity of Rubisco can only reach 25%. In recent years, Rubisco activation (Wei et al., 2017) and engineering of Rubisco mutants (Liang and Lindblad, 2017) have been explored to improve the efficiency of Rubisco action. A synthetic malonyl-CoA-glycerate (MCG) pathway was designed to enhance the CBB cycle, and implementation of this pathway in the cyanobacterium *Synechococcus elongates* PCC7942 was shown to enhance carbonate assimilation efficiency by approximately two folds (Yu et al., 2018).

In the CBB cycle, the catalytic activity of Rubisco produces the toxic compound 2-phosphoglycolate (2-PG), which leads to photorespiration, reducing the production of organic matter by microalgae. A malic acid cycle can lead to complete decarboxylation of 2-PG and increase the carbon dioxide content in chloroplasts, which has been verified in tobacco cultivation to increase biomass and its photosynthetic efficiency (South et al., 2019). To improve the efficiency of light utilization, intracellular spectral recomposition (ISR) of light in enhanced green fluorescent protein (eGFP) was conducted, with the expression of eGFP in *Phaeodactylum tricornutum*, raising the efficiency of photosynthesis by 28% (Fu et al., 2017). Optimizing the size of light-capturing antennae could significantly enhance photosynthetic performance, with an increase of 40% in biomass yield (Friedland et al., 2019). Furthermore, simulation experiments demonstrate that the conversion of light energy to biochemicals can be accelerated by inactivating flavodiiron proteins (Flv1/Flv3), to remove the naturally competing electrons (Assil-Companioni et al., 2020).

Outperforming Rubisco, the 6-phosphogluconate dehydrogenase (6PGDH) enzyme was found to be an efficient alternative to catalyze the CBB cycle, and its catalytic product could be recycled into the CBB cycle (Bar-Even, 2018). Formate dehydrogenase (FDH) converts carbon dioxide to formyl-tetrahydrofolate (THF) to produce serine with the aid of ATP, a pathway that requires fewer enzymes and utilizes residual energy from central metabolism with high efficiency (Bar-Even, 2016). Hitherto, only minor studies have been conducted to improve the photosynthetic efficiency of microalgae by designing new CO_2 fixation pathways.

Genome editing is the most radical method that may enhance photosynthetic efficiency in microalgae through nucleases, such as zinc-finger nucleases (ZFNs), transcriptional activator-like effector nucleases (TALENs), and CRISPR/Cas system (Naduthodi et al., 2021). At present, CRISPR/Cas technology is rapidly developing and is an advanced method for gene editing in microalgae. CRISPR/Cas9 has good applications in some microalgae, such as the successful knock-in of the Δ12-fatty acid desaturase (FAD12) gene in the T1 locus of *Nannochloropsis*, which resulted in a 4-fold higher yield of linoleic acid and a 1.5-fold higher yield of eicosapentaenoic acid (EPA) (Ryu et al., 2021). The fusion of Cas9 protein and the transcriptional activator, VP64, effectively increases the expression level of methylated nucleases in *Nannochloropsis* IMET1, resulting in improved photosynthetic efficiency (Wei et al., 2022). Treatment of *Porphyridium* sp. with Cas9 produces a chlorophyll synthase loss-of-function mutant (Δchs1) that increases phycoerythrin production (Jeon et al., 2021). Cas9 technology has the potential to introduce bioluminescent genes into microalgae, which is expected to solve the problem of low efficiency of light energy utilization in photobioreactors due to the high concentration of microalgae.

16.2.2.2 Adaptive laboratory evolution

Adaptive laboratory evolution (ALE) was initially designed to enhance the microorganism's adaptation to external stress conditions, and its short experimental production cycle, simple setup, and rapid screening to obtain evolved species resistant to the target stress compensate for the improved environment-specific tolerance phenotypes of microalgae that cannot be achieved by gene-editing techniques. It is now widely used to strengthen microalgae strains as well (LaPanse et al., 2021; Sandberg et al., 2019). Beginning with designing the experiment, the culture propagation mechanism of ALE includes batch cultures, chemical cultures, microfluidics, emulsion cultures, mutation accumulation, and microbial cultures (McDonald, 2019). At present, batch cultures or semicontinuous cultures in the bioreactor are still commonly used.

By setting different environmental conditions, the strains, after multigeneration culture, attain the ability to withstand extreme environmental conditions, and grow faster or accumulate more chemicals. *Chlorella* sp. Cv, obtained under simulated flue gas conditions (10% CO_2, 200 ppm NO_x and 100 ppm SO_x), tolerate flue gas with a CO_2 fixation rate of 1.2 g L^{-1} d^{-1} and carbohydrate content of 68.4% (Cheng et al., 2019). Applying a synergistic two-factor ALE strategy of low temperature and high salinity to treat *Schizochytrium* sp., the yield of DHA in the endpoint strain ALE-TF30 was 57.52% higher than the parental strain (X.-M. Sun et al., 2018). Stressful light conditions induce increased pigment accumulation in microalgae, as shown by the increased accumulation of carotenoids and lutein in ALE-iterative cultures of *Dunaliella salina* with 75% red light and 25% blue light, and increased fucoxanthin accumulation in *Phaeodactylum tricornutum* under similar conditions (Fu et al., 2013; Yi et al., 2015). Studies have shown that 60% CO_2 concentration is the optimum concentration for domesticated strains with high CO_2 tolerance. Higher concentrations of CO_2 do not necessarily promote microalgae's CO_2 tolerance due to the genotoxicity of chronic high CO_2 stress (Z. Wang et al., 2022). Therefore, setting appropriate ambient pressure conditions is one of the bottlenecks of ALE.

16.2.2.3 Other approaches for strain screening and improvement

Traditionally, radiation mutagenesis and chemical mutagenesis have been applied in mutation screening for improving the microalgae species. The co-culture of bacteria and microalgae, and chemical agents such as phytohormones, are relatively economic and eco-friendly approaches for strain screening; yet further research is needed.

Radiation mutagenesis, for example, the starch-deficient *Chlorella vulgaris* mutants (cvm5 and cvm6), produced by UV-C mutagenesis, yield 1.47-fold and 1.04-fold cell concentrations out of all colonies, and the total lipid content of the mutant cvm5 was 1.3-fold increased (Carino and Vital, 2022). Ethyl methanesulfonate (EMS) is a commonly used chemical mutagen. *Botryosphaerella* sp. and *Chlorococcus* sp. Mutants, exposed to EMS, exhibit 1.7- and 1.9-fold increases in biomass and lipid productivity over the wild type (Nojima et al., 2017). The yield of total carotenoids and astaxanthin in G1-C1 mutants of *Coelastrum* sp. after a high throughput screening with EMS and glufosinate was two times higher than that of the wild type (Tharek et al., 2021). Efficient strain screening can be achieved by fluorescence-activated cell sorting after random mutagenesis (Doppler et al., 2021; Nayak et al., 2022).

The co-culture of bacteria and microalgae was proved to augment the production efficiency of the species, and could improve the fatty acid content and lipid yield (Shokravi et al., 2022). Algal-bacteria co-culture with wastewater can produce bioenergetic products such as biodiesel (61.94%), biohydrogen (7.05%), bioelectricity (13.72%), and bioethanol (9.68%) (Zhang et al., 2021). Autotrophic microalgae form a symbiotic relationship by exchanging nutrients and metabolites with them when they are co-cultured with bacteria, yeast, fungi, or heterotrophic microalgae, which may increase biomass productivity (Alam et al., 2022).

Chemical agents such as phytohormones 3-indoleacetic acid (IAA) were claimed to be conducive to improving astaxanthin and lipid co-production in *Chromochloris zofingiensis* (Chen et al., 2020), and abscisic acid (ABA) has similar effects on fatty acid synthesis in *Scenedesmus* sp. (Kozlova et al., 2017). Methyl jasmonate can enhance the ability of *Stauroneis* sp. to produce pigments and lipids by enhancing the dark reaction of photosynthesis and increasing the rate of growth and metabolism (Gee et al., 2021). Yet, no exact cost of using biostimulants to boost chemical production in microalgae is revealed. Besides, chelated iron and iron nanoparticles boosted cell growth, and iron nanoparticles possess the potential to improve growth and lipid composition (Rana and Prajapati, 2021). It also shows that glucose with low dose supplementation can induce a 10% increase in CO_2 capture efficiency of photosynthetic autotrophy in *Chlorella vulgaris* (Fu et al., 2019).

16.2.3 Cultivation optimization and scale-up

Large-scale cultivation of microalgae is not only the core link of the commercialization process of microalgae specialty products but also the technical bottleneck in this field. According to its specific form, the microalgae culture system can be divided into an open system (outdoor raceway ponds) and a closed system (photobioreactors). For the culture process, see Figure 16.2.

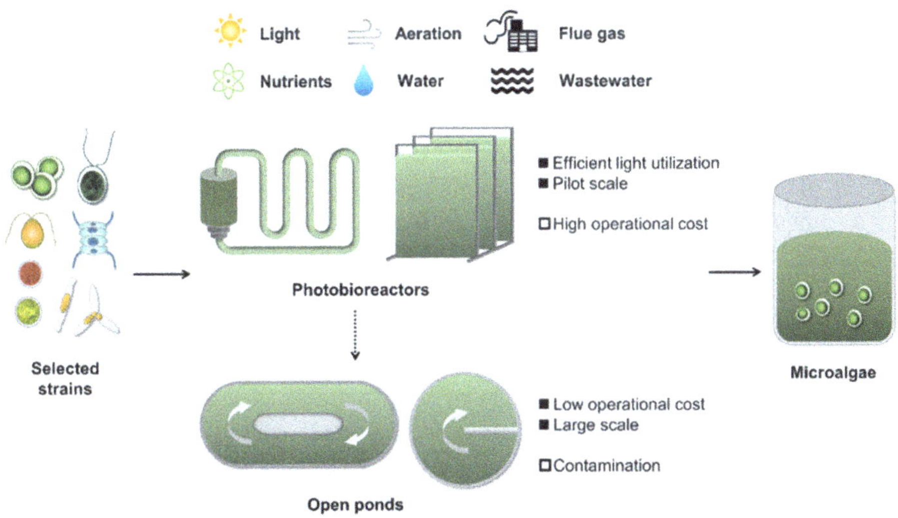

Figure 16.2: Microalgae cultivation system.

Photobioreactors are recognized for high light energy utilization, high carbon dioxide utilization, and high productivity (Shekh et al., 2022). Continuous control of the light intensity, temperature, and pH is available in PBR for its isolation from the environment, and will not be contaminated by pathogens and bacteria from the external environment (Patil et al., 2020). To expand the scale, studies on the optimization of large-scale PBR culture have been conducted, such as the diatom large-scale cultures with vertical column gas-lift photobioreactors, with maximum yields of 1.28 cm^3/L biomass (0.09–0.31 g DW/d) with complete amino acid and fatty acid contents (Eilertsen et al., 2022). The current PBR optimization is mostly based on calculations and simulations, for instance, increased fucoxanthin production using constant volume mass transfer coefficient (kLa) in 500 mL to 5 L bubble column bioreactors incubating *Tisochrysis lutea* (Mohamadnia et al., 2022). Simulations with COMSOL and experimental results showed that the efficiency of cellulase extraction of glucose from mixed microalgae could be improved by using a baffled reactor with adequate mixing (Shokrkar and Keighobadi, 2022). Additionally, a data-driven agent-based model learns the kinetic mechanisms in microalgae culture to obtain the optimal configuration of PBR using a hybrid optimization algorithm

(Otálora et al., 2021). However, the application of PBR for commercial production of microalgae products is mainly limited by expensive input and maintenance costs.

Considering the maintenance costs, open ponds are widely applied for the industrial production of microalgae products for their low investment and operating costs. Currently, the predominant type of open pond is the raceway pond, which is simple and easy to clean, but is also susceptible to contamination, making it unsuitable for the production of value-added chemicals (Shekh et al., 2022). Adding light improves the productivity of microalgae cultured in raceway ponds. Lighting with LEDs can promote sugar, protein, and chlorophyll production activities in *Nannochloropsis oceanica* in outdoor raceway ponds, only for high-value products, considering the cost (Carneiro et al., 2022). The internet of things is innovative technologies applied to the production enhancement of raceway ponds (Wang et al., 2022). A dynamic model, based on a neural network, can predict the pH in the runway, based on measured data, which may be applied to dissolved oxygen or biomass concentrations (Otálora et al., 2021).

Combining open and closed culture systems, a two-stage culture system has been developed for the production of high-value chemicals using PBR for the rapid growth of microalgae in the first stage, and exposing the microalgae cells to stress conditions in the second stage to promote the accumulation of target products such as lipids (Aziz et al., 2020; Xuehua Xiao et al., 2022). Wastewater has been studied as a nutrient source for the culture system, which can eliminate the environment of eutrophic and toxic substances while reducing costs. *Desmodesmus maximus* CN06 culture in municipal wastewater can yield high lipid content (majorly composed of hexadecanoic (C16:0) and oleic (C18:1) acids) (Purba et al., 2022). The microalga *Parachlorella kessleri* had the highest specific growth rate (0.58 d^{-1}), with 44.2% sugar accumulation, at 2.5% (v/v) of pig manure, combined with 5% CO_2 and a light/dark cycle of 20 h/4 h (Beigbeder and Lavoie, 2022).

Besides, culturing the microalgae *Auxenochlorella protothecoides* and *Chlorella sorokiniana* with volatile fatty acids and acetate instead of the conventional glucose obtained the highest biomass and lipid content of *A. protothecoides* (10.66 g/L, 33.93%) and *C. sorokiniana* (7.98 g/L, 39.80%), which is a low-cost carbon source suitable for biofuels production (Patel et al., 2022). To reduce the cost of electricity and the environmental impact of microalgae cultivation, solar-wind hybrid renewable electricity systems have a wide potential application in Africa, Australasia, and the Middle East (Dias et al., 2022b). This novel energy system brings multiple benefits, both environmentally and economically, but is greatly dependent on the geographic location (Dias et al., 2022a).

16.2.4 Downstream processing and biorefinery

The biomass of microalgae, obtained after cultivation, needs to undergo a series of important downstream treatments before becoming the final product; the main steps are shown in Figure 16.3, with biorefining focusing on the two steps of harvesting and extraction.

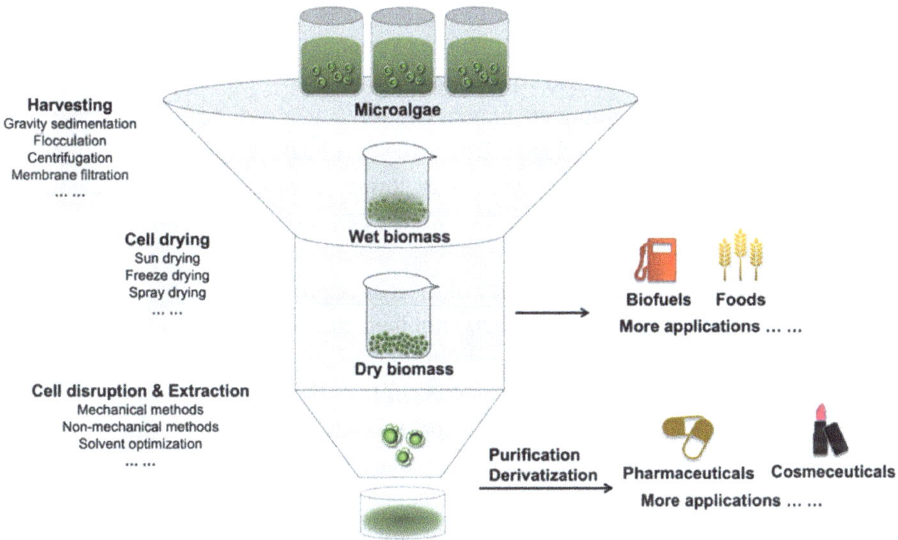

Figure 16.3: Downstream processing of microalgae products.

Harvesting is the most important and costly step in downstream treatment. The main methods applied for harvesting are gravity sedimentation, flocculation, centrifugation, and membrane filtration (Nitsos et al., 2020). Gravity sedimentation, which is collection by the gravity of the microalgae alone, is the least expensive but a time-consuming method. Centrifugal method is much more efficient than gravity settling but requires considerable power input (Khan et al., 2022). Flocculation is a relatively efficient harvesting method and includes chemical, bioflocculation, and physical flocculation. The chemical reagent, poly (diallyldimethylammonium chloride) (PDADMAC), can be used to flocculate *Chlorella vulgaris*, with an efficiency of more than 90% at a PDADMAC dose of 5 mg/L (Gerchman et al., 2017). Two newly synthesized positively charged cationic polymers, poly[2-(acryloyloxy)ethyl]trimethylammonium chloride (PAETAC) and poly(3-acrylamidopropyl)trimethylammonium chloride (PAmPTAC), have excellent flocculation effect on *Chlorella vulgaris* and *Porphyridium purpureum* (Nguyen et al., 2022). Bioflocculation, such as the use of some fungi and bacteria co-cultured with microalgae, can improve the flocculation efficiency. For example, *Citrobacter* W4 promoted flocculation of *Chlorella pyrenoidosa*, with a harvesting efficiency of 87.37% (He et al., 2022). Electroflocculation is also a cost-effective harvesting method (Krishnamoorthy et al., 2021), and in combination with flotation technology, the harvesting efficiency of *Tribonema* sp., treated with electro-flotation, reached 96.3%, with an energy consumption of 0.19 kWh/kg biomass, which is much lower compared to other harvesting technologies (Qi et al., 2022). In addition, the application of iron nanoparticles allows for a significant increase in the harvesting efficiency without affecting the downstream extraction of high-value bioproducts (Dineshkumar et al., 2017; Li et al., 2021). Membrane filtration is a promising harvesting method but is susceptible to biological contamination (Leam et al., 2020). The use of pretreated and modified membrane surfaces can control contamination during filtration (Malaguti et al., 2022).

After harvesting, the obtained biomass is wet and needs to be dried by Sun, freeze-, and spray-drying technologies. Compared to centrifuge + freeze dryer, the cost of ultrafiltration + spray dryer is lower (7.03% lower biomass cost) (Vázquez-Romero et al., 2022). After this step, some crude biomass can be obtained for biodiesel and feed production, etc.

Some high-value products from microalgae cells require cell disruption and extraction, mainly by mechanical and non-mechanical methods. Bead milling is a common mechanical method, combined with dynamic filtration treatment of *Parachlorella kessleri*, to obtain the best-fractionated harvesting (23% of total lipids, 9% of sugars, and 8% of initial biomass protein) (S. Liu et al., 2022). For the extraction of astaxanthin from *Haematococcus pluvialis*, bead milling and high-pressure homogenization showed excellent extraction results (Kim et al., 2022). Ultrasonic methods have been applied as an efficient mechanical method for cell disruption, which can be combined with other disruption methods to achieve lower costs and more efficient extraction results (Liu et al., 2022).

Non-mechanical methods include chemical and enzymatic treatments (Rahman et al., 2022). Conventional chemical treatments consist mainly of adding different types of acids or bases. Acid treatments (diluted sulfuric acid, nitric acid, hydrochloric acid, etc.) can penetrate plant cell walls more efficiently than alkalis (Phwan et al., 2019). The liquid biphasic flotation (LBF) system is a novel and environmentally friendly method; it recovered 74.44% of lipids from *Chlorella sorokiniana* CY-1 in 20 min using ethanol and ammonium sulfate 100% (v/v) (Mat Aron et al., 2022). Liquid biphasic flotation is another promising technique for the extraction of biomolecules (Aron et al., 2021), and the recovery of astaxanthin from *Haematococcus pluvialis*, combined with ultrasound, reached 83.73 ± 0.70% (Khoo et al., 2020). Lipid extraction with enzymatic hydrolysis of microalgae was carried out with a higher efficiency of hydrolysis by a combination of multiple enzymes, and 100% of total lipids were obtained by a combined enzymatic digestion of pectinase, cellulase, and xylanase (Zhang et al., 2022). Some bioactive substances are encapsulated by nanoencapsulation technology to improve their stability in downstream processing (Cai et al., 2021).

16.3 Future perspectives and recommendations

Microalgae, for the production of specialty chemicals, have proven to have great market potential. However, the complexity of microalgae metabolism and production processes limits the large-scale commercial production of microalgae-derived chemicals. Upstream improvements in the production of microalgae chemicals include strain engineering and large-scale culture optimization. Genome-scale metabolic models, as one of the strain engineering approaches, assist researchers in determining the optimal process from substrates to products, simplifying the synthesis process, or achieving more yields through genome editing or enzymatic metabolic engineering. To date, only a few genome-scale metabolic models of microalgae (i.e. model species) have been developed. Two-stage culture mode for large-scale culture optimization holds promise, with a focus on lipid production, but such a strategy is still pending for the development of other value-added chemicals. Downstream processing with new synthetic reagents or materials allows for higher harvesting and extraction efficiencies; yet the cost and environmental impact still need to be evaluated. At present, the application of computer-aided control systems has facilitated the development of the microalgae industry, mainly in large-scale cultivations, albeit the entire production process is not fully designed and realized with an auto control system (Kasani et al., 2022; K. Wang et al., 2022; Yew et al., 2020).

For future microalgae process intensification, directions should include the following: (1) comprehensively explore the whole genome-scale metabolic pathways of more microalgae that have industrial potential for commercial chemical production; (2) employ artificial intelligence, based on data analysis, such as image recognition of microalgae phenotype data, to screen for dominant species, machine learning on transcriptomics data, and reconstruction of metabolic networks; (3) develop a database, covering strain genes, genome-scale metabolic models, large-scale cultivation, downstream processing, and biorefinery; (4) provide a holistic assessment of the economic and environmental costs of each production step to support decision-making in industrial production for developing a sustainable bio-economy.

References

Abu-Ghosh, S., Dubinsky, Z., Verdelho, V., & Iluz, D. (2021). Unconventional high-value products from microalgae: A review. *Bioresource Technology*, *329*, 124895–124895.

Ahmed, F., Zhou, W., & Schenk, P. M. (2015). *Pavlova lutheri* is a high-level producer of phytosterols. *Algal Research*, *10*, 210–217.

Alam, M. A., Wan, C., Tran, D. T., Mofijur, M., Ahmed, S. F., Mehmood, M. A., Shaik, F., Vo, D.-V. N., & Xu, J. (2022). Microalgae binary culture for higher biomass production, nutrients recycling, and efficient harvesting: A review. *Environmental Chemistry Letters*, *20*, 1153–1168.

Apandi, N., Mohamed, R. M. S. R., Al-Gheethi, A., Gani, P., Ibrahim, A., & Kassim, A. H. M. (2019). *Scenedesmus* biomass productivity and nutrient removal from wet market wastewater, a bio-kinetic study. *Waste Biomass Valorization*, *10*, 2783–2800.

Ashokkumar, V., Chen, W.-H., Kumar, G., Satjarak, A., Chanthapatchot, W., & Ngamcharussrivichai, C. (2021). A biorefinery approach for high value-added bioproduct (astaxanthin) from alga *Haematococcus* sp. and residue pyrolysis for biochar synthesis and metallic iron production from hematite (Fe_2O_3). *Fuel*, *304*, 121150.

Assil-Companioni, L., Büchsenschütz, H. C., Solymosi, D., Dyczmons-Nowaczyk, N. G., Bauer, K. K. F., Wallner, S., Macheroux, P., Allahverdiyeva, Y., Nowaczyk, M. M., & Kourist, R. (2020). Engineering of NADPH supply boosts photosynthesis-driven biotransformations. *ACS Catalysis*, *10*, 11864–11877.

Aziz, M. M. A., Kassim, K. A., Shokravi, Z., Jakarni, F. M., Liu, H. Y., Zaini, N., Tan, L. S., Islam, A. B. M. S., & Shokravi, H. (2020). Two-stage cultivation strategy for simultaneous increases in growth rate and lipid content of microalgae: A review. Renewable and Sustainable Energy Reviews, *119*, 109621.

Barati, B., Zeng, K., Baeyens, J., Wang, S., Addy, M., Gan, S.-Y., & El-Fatah Abomohra, A. (2021). Recent progress in genetically modified microalgae for enhanced carbon dioxide sequestration. *Biomass Bioenergy*, *145*, 105927.

Bar-Even, A. (2018). Daring metabolic designs for enhanced plant carbon fixation. *Plant Science, Synthetic Biology Meets Plant Metabolism*, *273*, 71–83.

Bar-Even, A. (2016). Formate assimilation: The metabolic architecture of natural and synthetic pathways. *Biochemistry*, *55*, 3851–3863.

Becker, E. W. (2007). Micro-algae as a source of protein. *Biotechnology Advances*, *25*, 207–210.

Beigbeder, J.-B., & Lavoie, J.-M. (2022). Effect of photoperiods and CO_2 concentrations on the cultivation of carbohydrate-rich P. kessleri microalgae for the sustainable production of bioethanol. *Journal of CO_2 Utilization*, *58*, 101934.

Benisvy-Aharonovich, E., Zandany, A., Saady, A., Kinel-Tahan, Y., Yehoshua, Y., & Gedanken, A. (2020). An efficient method to produce 1,4-pentanediol from the biomass of the algae *Chlorella ohadi* with levulinic acid as intermediate. *Bioresource Technology Reports*, *11*, 100514.

Bermúdez, S., Romero Ogawa, M., Rittmannb, B., & Parra, R. (2014). Algae biofuels production processes, carbon dioxide fixation and biorefinery concept. *Journal of Petroleum and Environmental Biotechnology*, *5*, 2157–7463.

Bernaerts, T. M. M., Gheysen, L., Foubert, I., Hendrickx, M. E., & Van Loey, A. M. (2019). The potential of microalgae and their biopolymers as structuring ingredients in food: A review. *Biotechnology Advances*, *37*, 107419.

Bernaerts, T. M. M., Gheysen, L., Kyomugasho, C., Jamsazzadeh Kermani, Z., Vandionant, S., Foubert, I., Hendrickx, M. E., & Van Loey, A. M. (2018). Comparison of microalgae biomasses as functional food ingredients: Focus on the composition of cell wall related polysaccharides. *Algal Research*, *32*, 150–161.

Berthon, J.-Y., Nachat-Kappes, R., Bey, M., Cadoret, J.-P., Renimel, I., & Filaire, E. (2017). Marine algae as attractive source to skin care. *Free Radical Research*, *51*, 555–567.

Bhalamurugan, G. L., Valerie, O., & Mark, L. (2018). Valuable bioproducts obtained from microalgae biomass and their commercial applications: A review. *Environmental Engineering Research*, *23*, 229–241.

Bharti, R. K., Singh, A., Dhar, D. W., & Kaushik, A. (2022). Biological carbon dioxide sequestration by microalgae for biofuel and biomaterials production.

Bhattacharya, M., & Goswami, S. (2020). Microalgae – A green multi-product biorefinery for future industrial prospects. *Biocatalysis and Agricultural Biotechnology*, *25*, 101580.

Brownfield, L., Doblin, M., Fincher, G. B., & Bacic, A. (2009). Chapter 3.3.4 – Biochemical and Molecular Properties of Biosynthetic Enzymes for (1,3)-β-Glucans in Embryophytes, Chlorophytes and Rhodophytes. In: Bacic, A., Fincher, G. B. & Stone, B. A. (Eds.), Chemistry, Biochemistry, and Biology of 1–3 Beta Glucans and Related Polysaccharides. Academic Press, San Diego, pp. 283–326.

Cai, Y., Lim, H. R., Khoo, K. S., Ng, H.-S., Wang, J., Tak-Yee Chan, A., & Show, P. L. (2021). An integration study of microalgae bioactive retention: From microalgae biomass to microalgae bioactives nanoparticle. *Food and Chemical Toxicology*, *158*, 112607.

Carino, J. D., & Vital, P. G. (2023). Characterization of isolated UV-C-irradiated mutants of microalga *Chlorella vulgaris* for future biofuel application. *Environment, Development and Sustainability*, 25(2), 1258–1275.

Carneiro, M., Maia, I. B., Cunha, P., Guerra, I., Magina, T., Santos, T., Schulze, P. S. C., Pereira, H., Malcata, F. X., Navalho, J., Silva, J., Otero, A., & Varela, J. (2022). Effects of LED lighting on *Nannochloropsis oceanica* grown in outdoor raceway ponds. *Algal Research*, *64*, 102685.

Chauton, M. S., Reitan, K. I., Norsker, N. H., Tveterås, R., & Kleivdal, H. T. (2015). A techno-economic analysis of industrial production of marine microalgae as a source of EPA and DHA-rich raw material for aquafeed: Research challenges and possibilities. *Aquaculture*, *436*, 95–103.

Chen, J.-H., Wei, D., & Lim, P.-E. (2020). Enhanced coproduction of astaxanthin and lipids by the green microalga *Chromochloris zofingiensis*: Selected phytohormones as positive stimulators. *Bioresource Technology*, *295*, 122242.

Cheng, D., Li, X., Yuan, Y., Yang, C., Tang, T., Zhao, Q., & Sun, Y. (2019). Adaptive evolution and carbon dioxide fixation of *Chlorella* sp. in simulated flue gas. *Science of the Total Environment*, *650*, 2931–2938.

Christaki, E., Florou-Paneri, P., & Bonos, E. (2011). Microalgae: A novel ingredient in nutrition. *International Journal of Food Sciences and Nutrition*, *62*, 794–799.

Demirbas, A. (2011). Biodiesel from oilgae, biofixation of carbon dioxide by microalgae: A solution to pollution problems. *Applied Energy, Special Issue of Energy from Algae: Current Status and Future Trends*, *88*, 3541–3547.

Dias, R. R., Deprá, M. C., Severo, I. A., Zepka, L. Q., & Jacob-Lopes, E. (2022a). Smart override of the energy matrix in commercial microalgae facilities: A transition path to a low-carbon bioeconomy. *Sustainable Energy Technologies and Assessments*, *52*, 102073.

Dias, R. R., Deprá, M. C., Zepka, L. Q., & Jacob-Lopes, E. (2022b). *In silico* study of hybrid renewable energy in microalgae facilities: A path towards net-zero emissions. *Algal Research*, *63*, 102661.

Dineshkumar, R., Paul, A., Gangopadhyay, M., Singh, N. D. P., & Sen, R. (2017). Smart and reusable biopolymer nanocomposite for simultaneous microalgae biomass harvesting and disruption: Integrated downstream processing for a sustainable biorefinery. *ACS Sustainable Chemistry and Engineering*, *5*, 852–861.

Doppler, P., Kriechbaum, R., Singer, B., & Spadiut, O. (2021). Make microalgae cultures axenic again – A fast and simple workflow utilizing fluorescence-activated cell sorting. *Journal of Microbiological Methods*, *186*, 106256.

Edelmann, M., Aalto, S., Chamlagain, B., Kariluoto, S., & Piironen, V. (2019). Riboflavin, niacin, folate and vitamin B12 in commercial microalgae powders. *Journal of Food Composition and Analysis*, *82*, 103226.

Eilertsen, H. C., Eriksen, G. K., Bergum, J.-S., Strømholt, J., Elvevoll, E., Eilertsen, K.-E., Heimstad, E. S., Giæver, I. H., Israelsen, L., Svenning, J. B., Dalheim, L., Osvik, R., Hansen, E., Ingebrigtsen, R. A., Aspen, T., & Wintervoll, G.-H. (2022). Mass cultivation of microalgae: I. experiences with vertical column airlift photobioreactors, diatoms and CO_2 sequestration. *Applied Sciences*, *12*, 3082.

Ejike, C. E. C. C., Collins, S. A., Balasuriya, N., Swanson, A. K., Mason, B., & Udenigwe, C. C. (2017). Prospects of microalgae proteins in producing peptide-based functional foods for promoting cardiovascular health. *Trends in Food Science and Technology*, *59*, 30–36.

Ferreira, V. S., Pinto, R. F., & Sant'Anna, C. (2016). Low light intensity and nitrogen starvation modulate the chlorophyll content of *Scenedesmus dimorphus*. *Journal of Applied Microbiology*, *120*, 661–670.

Friedland, N., Negi, S., Vinogradova-Shah, T., Wu, G., Ma, L., Flynn, S., Kumssa, T., Lee, C.-H., & Sayre, R. T. (2019). Fine-tuning the photosynthetic light harvesting apparatus for improved photosynthetic efficiency and biomass yield. *Scientific Reports*, *9*, 1–12.

Fu, W., Chaiboonchoe, A., Khraiwesh, B., Sultana, M., Jaiswal, A., Jijakli, K., Nelson, D. R., Al-Hrout, A., Baig, B., Amin, A., & Salehi-Ashtiani, K. (2017). Intracellular spectral recompositioning of light enhances algal photosynthetic efficiency. *Science Advances*, *3*, e1603096.

Fu, W., Guðmundsson, Ó., Paglia, G., Herjólfsson, G., Andrésson, Ó. S., Palsson, B. Ø., & Brynjólfsson, S. (2013). Enhancement of carotenoid biosynthesis in the green microalga *Dunaliella salina* with light-emitting diodes and adaptive laboratory evolution. *Applied Microbiology and Biotechnology*, *97*, 2395–2403.

Fu, W., Gudmundsson, S., Wichuk, K., Palsson, S., Palsson, B. O., Salehi-Ashtiani, K., & Brynjólfsson, S. (2019). Sugar-stimulated CO_2 sequestration by the green microalga *Chlorella vulgaris. Science of the Total Environment*, *654*, 275–283.

Galasso, C., Gentile, A., Orefice, I., Ianora, A., Bruno, A., Noonan, D. M., Sansone, C., Albini, A., & Brunet, C. (2019). Microalgae derivatives as potential nutraceutical and food supplements for human health: A focus on cancer prevention and interception. *Nutrients*, *11*, E1226.

Gao, F., Sá, M., Cabanelas, I. T. D., Wijffels, R. H., & Barbosa, M. J. (2021). Improved fucoxanthin and docosahexaenoic acid productivities of a sorted self-settling *Tisochrysis lutea* phenotype at pilot scale. Bioresource Technology, *325*, 124725.

Gao, F., Teles, I., Wijffels, R. H., & Barbosa, M. J. (2020). Process optimization of fucoxanthin production with *Tisochrysis lutea. Bioresource Technology*, *315*, 123894.

Gee, D. M., Archer, L., Parkes, R., Fleming, G. T. A., Santos, H. M., & Touzet, N. (2021). The role of methyl jasmonate in enhancing biomass yields and bioactive metabolites in *Stauroneis* sp. (Bacillariophyceae) revealed by proteome and biochemical profiling. *Journal of Proteomics*, *249*, 104381.

Gerchman, Y., Vasker, B., Tavasi, M., Mishael, Y., Kinel-Tahan, Y., & Yehoshua, Y. (2017). Effective harvesting of microalgae: Comparison of different polymeric flocculants. *Bioresource Technology*, *228*, 141–146.

Goiris, K., Muylaert, K., Fraeye, I., Foubert, I., De Brabanter, J., & De Cooman, L. (2012). Antioxidant potential of microalgae in relation to their phenolic and carotenoid content. *Journal of Applied Phycology*, *24*, 1477–1486.

Gonçalves, V. D., Fagundes-Klen, M. R., Goes Trigueros, D. E., Kroumov, A. D., & Módenes, A. N. (2019). Statistical and optimization strategies to carotenoids production by *Tetradesmus acuminatus* (LC192133.1) cultivated in photobioreactors. *Biochemical Engineering Journal*, *152*, 107351.

Gouveia, L., Marques, A., Sousa, J., Moura, P., & Bandarra, N. (2010). Microalgae – Source of natural bioactive molecules as functional ingredients. *The Food Science and Technology Bulletin*, *7*, 21–37.

Griffiths, M., Harrison, S. T. L., Smit, M., & Maharajh, D. (2016). Major Commercial Products from Micro- and Macroalgae. In: Bux, F. & Chisti, Y. (Eds.), Algae Biotechnology: Products and Processes, Green Energy and Technology. Springer International Publishing, Cham, pp. 269–300.

Gupta, V., Ratha, S. K., Sood, A., Chaudhary, V., & Prasanna, R. (2013). New insights into the biodiversity and applications of cyanobacteria (blue-green algae) – Prospects and challenges. *Algal Research*, *2*, 79–97.

Hamed, I. (2016). The evolution and versatility of microalgae biotechnology: A review. *omprehensive Reviews in Food Science and Food Safety*, *15*, 1104–1123.

Hamilton, M. L., Haslam, R. P., Napier, J. A., & Sayanova, O. (2014). Metabolic engineering of *Phaeodactylum tricornutum* for the enhanced accumulation of omega-3 long chain polyunsaturated fatty acids. *Metabolic Engineering*, *22*, 3–9.

Han, X., Zhao, Z., Wen, Y., & Chen, Z. (2020). Enhancement of docosahexaenoic acid production by overexpression of ATP-citrate lyase and acetyl-CoA carboxylase in *Schizochytrium* sp. *Biotechnology Biofuels*, *13*, 131.

He, J., Ding, W., Han, W., Chen, Y., Jin, W., & Zhou, X. (2022). A bacterial strain *Citrobacter* W4 facilitates the bio-flocculation of wastewater cultured microalgae Chlorella pyrenoidosa. *Science of the Total Environment*, *806*, 151336.

Ismail, I., Hwang, Y.-H., & Joo, S.-T. (2020). Meat analog as future food: A review. *Journal of Animal Science and Technology*, *62*, 111–120.

Jäpelt, R. B., & Jakobsen, J. (2013). Vitamin D in plants: A review of occurrence, analysis, and biosynthesis. *Frontiers Plant Science*, *4*, 136.

Jeon, M. S., Han, S.-I., Jeon, M., & Choi, Y.-E. (2021). Enhancement of phycoerythrin productivity in *Porphyridium purpureum* using the clustered regularly interspaced short palindromic repeats/CRISPR-associated protein 9 ribonucleoprotein system. *Bioresource Technology*, *330*, 124974.

Judd, S. J., Al Momani, F. A. O., Znad, H., & Al Ketife, A. M. D. (2017). The cost benefit of algal technology for combined CO_2 mitigation and nutrient abatement. *Renewable and Sustainable Energy Reviews*, *71*, 379–387.

Kasani, A. A., Esmaeili, A., & Golzary, A. (2022). Software tools for microalgae biorefineries: Cultivation, separation, conversion process integration, modeling, and optimization. *Algal Research*, *61*, 102597.

Khan, M. I., Shin, J. H., & Kim, J. D. (2018). The promising future of microalgae: Current status, challenges, and optimization of a sustainable and renewable industry for biofuels, feed, and other products. *Microbial Cell Factories*, *17*, 36.

Khan, S., Naushad, M., Iqbal, J., Bathula, C., & Sharma, G. (2022). Production and harvesting of microalgae and an efficient operational approach to biofuel production for a sustainable environment. *Fuel*, *311*, 122543.

Khanra, S., Mondal, M., Halder, G., Tiwari, O. N., Gayen, K., & Bhowmick, T. K. (2018). Downstream processing of microalgae for pigments, protein and carbohydrate in industrial application: A review. *Food Bioprod Process*, *110*, 60–84.

Khoo, K. S., Chew, K. W., Yew, G. Y., Manickam, S., Ooi, C. W., & Show, P. L. (2020). Integrated ultrasound-assisted liquid biphasic flotation for efficient extraction of astaxanthin from *Haematococcus pluvialis*. *Ultrasonics Sonochemistry*, *67*, 105052.

Kim, B., Youn Lee, S., Lakshmi Narasimhan, A., Kim, S., & Oh, Y.-K. (2022). Cell disruption and astaxanthin extraction from *Haematococcus pluvialis*: Recent advances. *Bioresource Technology*, *343*, 126124.

Kozlova, T. A., Hardy, B. P., Krishna, P., & Levin, D. B. (2017). Effect of phytohormones on growth and accumulation of pigments and fatty acids in the microalgae *Scenedesmus quadricauda*. *Algal Research*, *27*, 325–334.

Krishnamoorthy, N., Unpaprom, Y., Ramaraj, R., Maniam, G. P., Govindan, N., Arunachalam, T., & Paramasivan, B. (2021). Recent advances and future prospects of electrochemical processes for microalgae harvesting. *Journal of Environmental Chemical Engineering*, *9*, 105875.

LaPanse, A. J., Krishnan, A., & Posewitz, M. C. (2021). Adaptive laboratory evolution for algal strain improvement: Methodologies and applications. *Algal Research*, *53*, 102122.

Leam, J. J., Bilad, M. R., Wibisono, Y., Hakim Wirzal, M. D., & Ahmed, I. (2020). Chapter 7 – Membrane Technology for Microalgae Harvesting. In: Yousuf, A. (Ed.), Microalgae Cultivation for Biofuels Production. Academic Press, Cambridge, pp. 97–110.

Leong, Y. K., Chen, C.-Y., Varjani, S., & Chang, J.-S. (2022). Producing fucoxanthin from algae – Recent advances in cultivation strategies and downstream processing. *Bioresource Technology*, *344*, 126170.

Levasseur, W., Perré, P., & Pozzobon, V. (2020). A review of high value-added molecules production by microalgae in light of the classification. *Biotechnology Advances*, *41*, 107545.

Levasseur, W., Taidi, B., Lacombe, R., Perré, P., & Pozzobon, V. (2018). Impact of seconds to minutes photoperiods on *Chlorella vulgaris* growth rate and chlorophyll a and b content. *Algal Research*, *36*, 10–16.

Li, X., Liu, B., Lao, Y., Wan, P., Mao, X., & Chen, F. (2021). Efficient magnetic harvesting of microalgae enabled by surface-initiated formation of iron nanoparticles. *Chemical Engineering Journal*, *408*, 127252.

Liang, F., & Lindblad, P. (2017). *Synechocystis* PCC 6803 overexpressing RuBisCO grow faster with increased photosynthesis. *Metabolic Engineering Communications*, *4*, 29–36.

Liu, S., Rouquié, C., Lavenant, L., Frappart, M., & Couallier, E. (2022). Coupling bead-milling and microfiltration for the recovery of lipids and proteins from *Parachlorella kessleri*: Impact of the cell disruption conditions on the separation performances. *Separation and Purification Technology*, *287*, 120570.

Liu, Y., Liu, X., Cui, Y., & Yuan, W. (2022). Ultrasound for microalgae cell disruption and product extraction: A review. *Ultrasonics Sonochemistry*, *87*, 106054.

Luesch, H., Moore, R. E., Paul, V. J., Mooberry, S. L., & Corbett, T. H. (2001). Isolation of dolastatin 10 from the marine cyanobacterium *Symploca* species VP642 and total stereochemistry and biological evaluation of its analogue symplostatin 1. *Journal of Natural Products*, *64*, 907–910.

Maeda, Y., Yoshino, T., Matsunaga, T., Matsumoto, M., & Tanaka, T. (2018). Marine microalgae for production of biofuels and chemicals. *Current Opinion in Biotechnology*, *50*, 111–120.

Malaguti, M., Novoa, A. F., Ricceri, F., Giagnorio, M., Vrouwenvelder, J. S., Tiraferri, A., & Fortunato, L. (2022). Control strategies against algal fouling in membrane processes applied for microalgae biomass harvesting. *Journal of Water Process Engineering*, *47*, 102787.

Mat Aron, N. S., Chew, K. W., Ang, W. L., Ratchahat, S., Rinklebe, J., & Show, P. L. (2022). Recovery of microalgae biodiesel using liquid biphasic flotation system. *Fuel*, *317*, 123368.

Mat Aron, N. S., Khoo, K. S., Chew, K. W., Veeramuthu, A., Chang, J.-S., & Show, P. L. (2021). Microalgae cultivation in wastewater and potential processing strategies using solvent and membrane separation technologies. *Journal of Water Process Engineering*, *39*, 101701.

Matos, J., Cardoso, C. L., Falé, P., Afonso, C. M., & Bandarra, N. M. (2020). Investigation of nutraceutical potential of the microalgae *Chlorella vulgaris* and *Arthrospira platensis*. *International Journal of Food Science and Technology*, *55*, 303–312.

McDonald, M. J. (2019). Microbial Experimental Evolution – A proving ground for evolutionary theory and a tool for discovery. *EMBO Report*, *20*, e46992.

Mohamadnia, S., Tavakoli, O., & Faramarzi, M. A. (2022). Production of fucoxanthin from the microalga *Tisochrysis lutea* in the bubble column photobioreactor applying mass transfer coefficient. *Journal of Biotechnology*, *348*, 47–54.

Molino, A., Rimauro, J., Casella, P., Cerbone, A., Larocca, V., Chianese, S., Karatza, D., Mehariya, S., Ferraro, A., Hristoforou, E., & Musmarra, D. (2018). Extraction of astaxanthin from microalga *Haematococcus pluvialis* in red phase by using generally recognized as safe solvents and accelerated extraction. *Journal of Biotechnology*, *283*, 51–61.

Mondal, A., Bose, S., Banerjee, S., Patra, J. K., Malik, J., Mandal, S. K., Kilpatrick, K. L., Das, G., Kerry, R. G., Fimognari, C., Bishayee, A., Bishayee, A., & Bishayee, A. (2020). Marine cyanobacteria and microalgae metabolites-A rich source of potential anticancer drugs. *Marine Drugs*, *18*, 476.

Mourelle, M. L., Gómez, C. P., & Legido, J. L. (2017). The potential use of marine microalgae and cyanobacteria in cosmetics and thalassotherapy. *Cosmetics*, *4*, 46.

Naduthodi, M. I. S., Claassens, N. J., D'Adamo, S., van der Oost, J., & Barbosa, M. J. (2021). Synthetic biology approaches to enhance microalgae productivity. *Trends in Biotechnology*, *39*, 1019–1036.

Nayak, M., Suh, W. I., Oh, Y. T., Ryu, A. J., Jeong, K. J., Kim, M., Mohapatra, R. K., Lee, B., & Chang, Y. K. (2022). Directed evolution of *Chlorella* sp. HS2 towards enhanced lipid accumulation by ethyl methanesulfonate mutagenesis in conjunction with fluorescence-activated cell sorting based screening. *Fuel*, *316*, 123410.

Nguyen, L. N., Vu, H. P., Fu, Q., Abu Hasan Johir, M., Ibrahim, I., Mofijur, M., Labeeuw, L., Pernice, M., Ralph, P. J., & Nghiem, L. D. (2022). Synthesis and evaluation of cationic polyacrylamide and polyacrylate flocculants for harvesting freshwater and marine microalgae. *Chemical Engineering Journal*, *433*, 133623.

Nitsos, C., Filali, R., Taidi, B., & Lemaire, J. (2020). Current and novel approaches to downstream processing of microalgae: A review. *Biotechnology Advances*, *45*, 107650.

Nojima, D., Ishizuka, Y., Muto, M., Ujiro, A., Kodama, F., Yoshino, T., Maeda, Y., Matsunaga, T., & Tanaka, T. (2017). Enhancement of biomass and lipid productivities of water surface-floating microalgae by chemical mutagenesis. *Marine Drugs*, *15*, 151.

Odjadjare, E. C., Mutanda, T., & Olaniran, A. O. (2017). Potential biotechnological application of microalgae: A critical review. *Critical Reviews in Biotechnology*, *37*, 37–52.

Ojulari, O. V., Lee, S. G., & Nam, J.-O. (2020). Therapeutic effect of seaweed derived xanthophyl carotenoid on obesity management; overview of the last decade. *International Journal of Molecular Sciences*, *21*, E2502.

Otálora, P., Guzmán, J. L., Berenguel, M., & Acién, F. G. (2021). Dynamic model for the pH in a raceway reactor using deep learning techniques. In: CONTROLO 2020. Presented at the APCA International Conference on Automatic Control and Soft Computing. Springer, Cham, pp. 190–199.

Patel, A., Krikigianni, E., Rova, U., Christakopoulos, P., & Matsakas, L. (2022). Bioprocessing of volatile fatty acids by oleaginous freshwater microalgae and their potential for biofuel and protein production. *Chemical Engineering Journal*, *438*, 135529.

Patil, R. A., Kausley, S. B., Joshi, S. M., & Pandit, A. B. (2020). Chapter 27 – Process Intensification Applied to Microalgae-based Processes and Products. In: Jacob-Lopes, E., Maroneze, M. M., Queiroz, M. I. & Zepka, L. Q. (Eds.), Handbook of Microalgae-Based Processes and Products. Academic Press, Cambridge, pp. 737–769.

Phwan, C. K., Chew, K. W., Sebayang, A. H., Ong, H. C., Ling, T. C., Malek, M. A., Ho, Y.-C., & Show, P. L. (2019). Effects of acids pre-treatment on the microbial fermentation process for bioethanol production from microalgae. *Biotechnology Biofuels*, *12*, 191.

Polyzois, A., Kirilovsky, D., Dufat, T.-H., & Michel, S. (2020). Effects of modification of light parameters on the production of cryptophycin, cyanotoxin with potent anticancer activity, in nostoc sp. *Toxins*, *12*, 809.

Popovich, C. A., Faraoni, M. B., Sequeira, A., Daglio, Y., Martin, L. A., Martínez, A. M., Damiani, M. C., Matulewicz, M. C., & Leonardi, P. I. (2020). Potential of the marine diatom *Halamphora coffeaeformis* to simultaneously produce omega-3 fatty acids, chrysolaminarin and fucoxanthin in a raceway pond. *Algal Biomass: From Bioproducts to Biofuels*, *51*, 102030.

Purba, L. D. A., Othman, F. S., Yuzir, A., Mohamad, S. E., Iwamoto, K., Abdullah, N., Shimizu, K., & Hermana, J. (2022). Enhanced cultivation and lipid production of isolated microalgae strains using municipal wastewater. *Environmental Technology and Innovation*, *27*, 102444.

Qi, S., Chen, J., Hu, Y., Hu, Z., Zhan, X., & Stengel, D. B. (2022). Low energy harvesting of hydrophobic microalgae (*Tribonema* sp.) by electro-flotation without coagulation. *Science of the Total Environment*, *838*, 155866.

Rahman, M. M., Hosano, N., & Hosano, H. (2022). Recovering microalgae bioresources: A review of cell disruption methods and extraction technologies. *Molecules*, *27*, 2786.

Ramesh Kumar, B., Deviram, G., Mathimani, T., Duc, P. A., & Pugazhendhi, A. (2019). Microalgae as rich source of polyunsaturated fatty acids. *Biocatalysis and Agricultural Biotechnology*, *17*, 583–588.

Rana, M. S., & Prajapati, S. K. (2021). Resolving the dilemma of iron bioavailability to microalgae for commercial sustenance. *Algal Research*, *59*, 102458.

Rao, A., Briskey, D., Nalley, J. O., & Ganuza, E. (2020). Omega-3 Eicosapentaenoic Acid (EPA) rich extract from the microalga nannochloropsis decreases cholesterol in healthy individuals: A double-blind, randomized, placebo-controlled, three-month supplementation study. *Nutrients*, *12*, 1869.

Ryu, A. J., Jeong, B., Kang, N. K., Jeon, S., Sohn, M. G., Yun, H. J., Lim, J. M., Jeong, S. W., Park, Y.-I., Jeong, W. J., Park, S., Chang, Y. K., & Jeong, K. J. (2021). Safe-Harboring based novel genetic toolkit for *Nannochloropsis salina* CCMP1776: Efficient overexpression of transgene via CRISPR/Cas9-Mediated Knock-in at the transcriptional hotspot. *Bioresource Technology*, *340*, 125676.

Sahin, D., Tas, E., & Altindag, U. H. (2018). Enhancement of docosahexaenoic acid (DHA) production from *Schizochytrium* sp. S31 using different growth medium conditions. *AMB Express*, *8*, 7.

Sandberg, T. E., Salazar, M. J., Weng, L. L., Palsson, B. O., & Feist, A. M. (2019). The emergence of adaptive laboratory evolution as an efficient tool for biological discovery and industrial biotechnology. *Metabolic Engineering*, *56*, 1–16.

Sathasivam, R., Radhakrishnan, R., Hashem, A., & Abd_Allah, E. F. (2019). Microalgae metabolites: A rich source for food and medicine. *Saudi Journal of Biological Sciences*, *26*, 709–722.

Sayre, R. (2010). Microalgae: The potential for carbon capture. *BioScience*, *60*, 722–727.

Shekh, A., Sharma, A., Schenk, P. M., Kumar, G., & Mudliar, S. (2022). Microalgae cultivation: Photobioreactors, CO_2 utilization, and value-added products of industrial importance. *Journal of Chemical Technology and Biotechnology*, *97*, 1064–1085.

Shokravi, Z., Shokravi, H., Atabani, A. E., Lau, W. J., Chyuan, O. H., & Ismail, A. F. (2022). Impacts of the harvesting process on microalgae fatty acid profiles and lipid yields: Implications for biodiesel production. *Renewable and Sustainable Energy Reviews*, *161*, 112410.

Shokrkar, H., & Keighobadi, A. (2022). Effect of fluid hydrodynamic situations on enzymatic hydrolysis of mixed microalgae: Experimental study and simulation. *Energy*, *241*, 122804.

Siqueira, S. F., Queiroz, M. I., Zepka, L. Q., Jacob-Lopes, E., 2018. Introductory chapter: microalgae biotechnology –a brief introduction, in: Jacob-Lopes, E., Zepka, L. Q., Queiroz, M. I. (Eds.), *Microalgal Biotechnol*. IntechOpen, London, pp. 1–8.

Smerilli, A., Balzano, S., Maselli, M., Blasio, M., Orefice, I., Galasso, C., Sansone, C., & Brunet, C. (2019). Antioxidant and photoprotection networking in the coastal diatom *Skeletonema marinoi*. *Antioxidants*, *8*, 154.

Smerilli, A., Orefice, I., Corato, F., Gavalás Olea, A., Ruban, A. V., & Brunet, C. (2017). Photoprotective and antioxidant responses to light spectrum and intensity variations in the coastal diatom *Skeletonema marinoi*. *Environmental Microbiology*, *19*, 611–627.

Smith, A. G., Croft, M. T., Moulin, M., & Webb, M. E. (2007). Plants need their vitamins too. *Current Opinion in Plant Biology, Physiology and Metabolism*, *10*, 266–275.

Xia, S., Gao, B., Fu, J.-Q., Xiong, J.-H., & Zhang, C. (2018). Production of fucoxanthin, chrysolaminarin, and eicosapentaenoic acid by Odontella aurita under different nitrogen supply regimes. *Journal of Bioscience and Bioengineering*, *126*, 723–729.

South, P. F., Cavanagh, A. P., Liu, H. W., & Ort, D. R. (2019). Synthetic glycolate metabolism pathways stimulate crop growth and productivity in the field. *Science*, *363*, eaat9077.

Stankiewicz, A., & Moulijn, J. A. (2018). Re-Engineering the Chemical Processing Plant: Process Intensification. CRC Press, Boca Raton.

Steinrücken, P., Prestegard, S. K., De Vree, J. H., Storesund, J. E., Pree, B., Mjøs, S. A., & Erga, S. R. (2018). Comparing EPA production and fatty acid profiles of three *Phaeodactylum tricornutum* strains under western Norwegian climate conditions. *Algal Research*, *30*, 11–22.

Sun, A., Hasan, M. T., Hobba, G., Nevalainen, H., & Te'o, J. (2018). Comparative assessment of the *Euglena gracilis* var. *saccharophila* variant strain as a producer of the β-1,3-glucan paramylon under varying light conditions. *Journal of Phycology*, *54*, 529–538.

Sun, H., Wu, T., Chen, S. H. Y., Ren, Y., Yang, S., Huang, J., Mou, H., & Chen, F. (2021). Powerful tools for productivity improvements in microalgae production. *Renewable and Sustainable Energy Reviews*, *152*, 111609.

Sun, H., Zhao, W., Mao, X., Li, Y., Wu, T., & Chen, F. (2018). High-value biomass from microalgae production platforms: Strategies and progress based on carbon metabolism and energy conversion. *Biotechnology Biofuels*, *11*, 227.

Sun, X.-M., Ren, L.-J., Bi, Z.-Q., Ji, X.-J., Zhao, Q.-Y., Jiang, L., & Huang, H. (2018). Development of a cooperative two-factor adaptive-evolution method to enhance lipid production and prevent lipid peroxidation in Schizochytrium sp. *Biotechnology Biofuels*, *11*, 65.

Tarento, T. D. C., McClure, D. D., Vasiljevski, E., Schindeler, A., Dehghani, F., & Kavanagh, J. M. (2018). Microalgae as a source of vitamin K1. *Algal Research*, *36*, 77–87.

Tharek, A., Yahya, A., Salleh, M. M., Jamaluddin, H., Yoshizaki, S., Hara, H., Iwamoto, K., Suzuki, I., & Mohamad, S. E. (2021). Improvement and screening of astaxanthin producing mutants of newly isolated *Coelastrum* sp. using ethyl methane sulfonate induced mutagenesis technique. *Biotechnology Reports*, *32*, e00673.

Udayan, A., Arumugam, M., & Pandey, A. (2017). Chapter 4 – Nutraceuticals from Algae and Cyanobacteria. In: Rastogi, R. P., Madamwar, D. & Pandey, A. (Eds.), Algal Green Chemistry. Elsevier, Amsterdam, pp. 65–89.

Vázquez-Romero, B., Perales, J. A., Pereira, H., Barbosa, M., & Ruiz, J. (2022). Techno-economic assessment of microalgae production, harvesting and drying for food, feed, cosmetics, and agriculture. *Science of the Total Environment, 837*, 155742.

Venkata Mohan, S., Modestra, J. A., Amulya, K., Butti, S. K., & Velvizhi, G. (2016). A circular bioeconomy with biobased products from CO_2 sequestration. *Trends in Biotechnology, 34*, 506–519.

Verma, R., Kumari, K. V. L. K., Srivastava, A., & Kumar, A. (2020). Photoautotrophic, mixotrophic, and heterotrophic culture media optimization for enhanced microalgae production. *Journal of Environmental Chemical Engineering, 8*, 104149.

Wang, B., Li, Y., Wu, N., & Lan, C. Q. (2008). CO_2 bio-mitigation using microalgae. *Applied Microbiology and Biotechnology, 79*, 707–718.

Wang, K., Khoo, K. S., Leong, H. Y., Nagarajan, D., Chew, K. W., Ting, H. Y., Selvarajoo, A., Chang, J.-S., & Show, P. L. (2022). How does the Internet of Things (IoT) help in microalgae biorefinery?. *Biotechnology Advances, 54*, 107819.

Wang, Y., Tibbetts, S. M., & McGinn, P. J. (2021). Microalgae as sources of high-quality protein for human food and protein supplements. *Foods, 10*, 3002.

Wang, Z., Cheng, J., Sun, Y., You, X., Chu, F., & Yang, W. (2022). Mutation adaptation and genotoxicity of microalgae induced by long-term high CO_2 stress. *Chemical Engineering Journal, 445*, 136745.

Wei, L., Jiang, Z., & Liu, B. (2022). A CRISPR/dCas9-based transcription activated system developed in marine microalga *Nannochloropsis oceanica. Aquaculture, 546*, 737064.

Wei, L., Wang, Q., Xin, Y., Lu, Y., & Xu, J. (2017). Enhancing photosynthetic biomass productivity of industrial oleaginous microalgae by overexpression of RuBisCO activase. *Algal Research, 27*, 366–375.

Xiao, X., Zhou, Y., Liang, Z., Lin, R., Zheng, M., Chen, B., & He, Y. (2022). A novel two-stage heterotrophic cultivation for starch-to-protein switch to efficiently enhance protein content of *Chlorella* sp. MBFJNU-17. *Bioresource Technology, 344*, 126187.

Yaakob, Z., Ali, E., Zainal, A., Mohamad, M., & Takriff, M. S. (2014). An overview: Biomolecules from microalgae for animal feed and aquaculture. *Journal of Biological Research-Thessaloniki, 21*, 6.

Yap, J. K., Sankaran, R., Chew, K. W., Halimatul Munawaroh, H. S., Ho, S.-H., Rajesh Banu, J., & Show, P. L. (2021). Advancement of green technologies: A comprehensive review on the potential application of microalgae biomass. *Chemosphere, 281*, 130886.

Yew, G. Y., Puah, B. K., Chew, K. W., Teng, S. Y., & Show, P. L. (2020). *Chlorella vulgaris* FSP-E cultivation in waste molasses: Photo-to-property estimation by artificial intelligence. *Chemical Engineering Journal, 402*, 126230.

Yi, Z., Xu, M., Magnusdottir, M., Zhang, Y., Brynjolfsson, S., & Fu, W. (2015). Photo-oxidative stress-driven mutagenesis and adaptive evolution on the marine diatom *Phaeodactylum tricornutum* for enhanced carotenoid accumulation. *Marine Drugs, 13*, 6138–6151.

Yu, H., Li, X., Duchoud, F., Chuang, D. S., & Liao, J. C. (2018). Augmenting the Calvin-Benson-Bassham cycle by a synthetic malyl-CoA-glycerate carbon fixation pathway. *Nature Communications, 9*, 1–10.

Zhang, C., Li, S., & Ho, S.-H. (2021). Converting nitrogen and phosphorus wastewater into bioenergy using microalgae-bacteria consortia: A critical review. *Bioresource Technology, 342*, 126056.

Zhang, Y., Kang, X., Zhen, F., Wang, Z., Kong, X., & Sun, Y. (2022). Assessment of enzyme addition strategies on the enhancement of lipid yield from microalgae. *Biochemical Engineering Journal, 177*, 108198.

Zhang, Y., Wang, X., Xie, D., Zou, S., Jin, Q., & Wang, X. (2018). Synthesis and concentration of 2-monoacylglycerols rich in polyunsaturated fatty acids. *Food Chemistry, 250*, 60–66.

Jonathan S. Harris, Anh N. Phan*

Chapter 17
Process integration opportunities applied to microalgae biofuels production

Abstract: Third-generation biofuels derived from microalgae are considered a promising candidate to reduce greenhouse gas emissions, especially applicable for the transportation sector, as the high lipid and carbohydrate content of microalgae results in high yields of liquid fuels compatible with current engine technologies. Microalgae have a number of advantages over conventional biomass sources including rapid growth (full cycles within days compared to weeks), use of CO_2 as a carbon source, and low nutrient requirements (can be from contaminated wastes, wastewater, or seawater). However, biofuels derived from microalgae have a few challenges due to the very high water content of the feedstock, thereby increasing drying and separation costs, and high contents of nitrogen and oxygen compounds, which result in lower quality and acidic fuels, requiring upgrading processes. To address these challenges, process intensification and integration should be focused towards maximizing nutrient recovery, minimizing the needs for bio-oil upgrading and drying of the microalgae after harvesting. This chapter will discuss current state-of-the-art technologies for biofuel production in the context of microalgae biofuel production, current challenges hindering commercialization, and how process intensification technologies can be used and integrated to make microalgae-derived biofuels more commercially viable.

Keywords: process integration, process intensification, microalgae, biofuels, sustainability, techno-economic analysis

17.1 Introduction

Continued usage of fossil fuels and the climate change associated with them remains a key issue for the world today, necessitating the development of sustainable alternatives including biofuels (gas and liquid forms) and electric vehicles (IEA, 2021a). Electric vehicles are becoming more popular but are not sustainable unless the electricity used to charge them is sustainable; they have reduced travel ranges compared to liquid fuel powered cars, which also makes them less well-suited to long distance travel

***Corresponding author: Anh N. Phan**, School of Engineering, Chemical Engineering, Newcastle University, Newcastle Upon Tyne, NE1 7RU, United Kingdom, e-mail: anh.phan@newcastle.ac.uk
Jonathan S. Harris, School of Engineering, Chemical Engineering, Newcastle University, Newcastle Upon Tyne, NE1 7RU, United Kingdom

https://doi.org/10.1515/9783110781267-017

and heavy goods vehicles (HGVs). Although many methods for producing energy sustainably are widely adopted around the world, such as solar, wind, and tidal power, the production of sustainable fuels remains low, accounting for only 3.7% of global fuel use (Ritchie, 2020), due to the higher costs of biofuel manufacture compared to fossil fuels.

Use of biofuels in the transport sector is still small (around 3%) compared to petroleum-based fuels (97%), which account for over 25% of the total greenhouse gas emissions in the U.K. (Kargbo et al., 2021). With widely adopted net-zero emission targets, it is expected that sustainable biofuel production and demand will continue to increase by 28% by 2026 (IEA, 2021b) at an estimated rate of 7% per year (WBA, 2019) with increasing use of electric and hybrid cars. Liquid biofuels (biodiesel and bioethanol) are the most common substitutes for liquid fossil fuels as they are nontoxic, biodegradable, and sulfur-free. Hydrogen fuel cells and lithium ion batteries for fueling transportation are expected to increase by 67% each year until 2026, but market share remains low, currently (Manoharan et al., 2019).

Biofuels can be produced using any form of biomass, from refined carbohydrates to plant- or animal-derived oils to lignocellulosic materials (Dahman et al., 2019). Depending upon sources of feedstock, biofuels are grouped into four generations: first-generation biofuels derived from food-based crops, typically oil rich plants, sugar and starch, second-generation biofuels, typically from agricultural and woodland residues such as switchgrass or sawdust, third-generation biofuels produced from algae, and currently developed fourth-generation biofuel; from CO_2 and water. Biodiesel and bioethanol are the most common first-generation biofuels and still currently widely used in transportation. Due to the concerns over land for food and land for fuels, second-generation biofuels have been produced from lignocellulosic materials (biomass or residues), either using the raw biomass (thermochemical processes) or using cellulose/hemicellulose extracted from residues/biomass (biological and/or chemical processes). Second-generation biofuels are also limited by the amount of waste biomass available, which leads to the prospect of algal biofuels.

Microalgae, defined as single-celled aquatic plants, can grow 15–30 times faster than terrestrial plants and require little nutrients and would be an excellent fit for biofuel production. Microalgae can be converted to biofuels via thermal, chemical, or biological approaches; the optimal method depends upon the lipid/ protein/ carbohydrate composition of the microalgae. Lipids and carbohydrates can be separated from the microalgae and converted to biofuel separately, which produces high quality fuels at the cost of extraction processing (Vasistha et al., 2021; Jablonský et al., 2018). Microalgae have a cellulose-rich cell wall over which biological and chemical methods cannot be directly applied. To address this, extraction can be performed using chemical or biochemical methods to disrupt/hydrolyze the cellulose in the microalgae cell wall, producing sugars and releasing the cell contents (Hernández et al., 2014). Thermochemical methods do not require disruption of the cell wall, but can produce cytotoxic by-products, which makes valorizing carbohydrates and proteins more challenging (Zhang and Ogden, 2017). Figure 17.1 shows conversion routes with target products from microalgae.

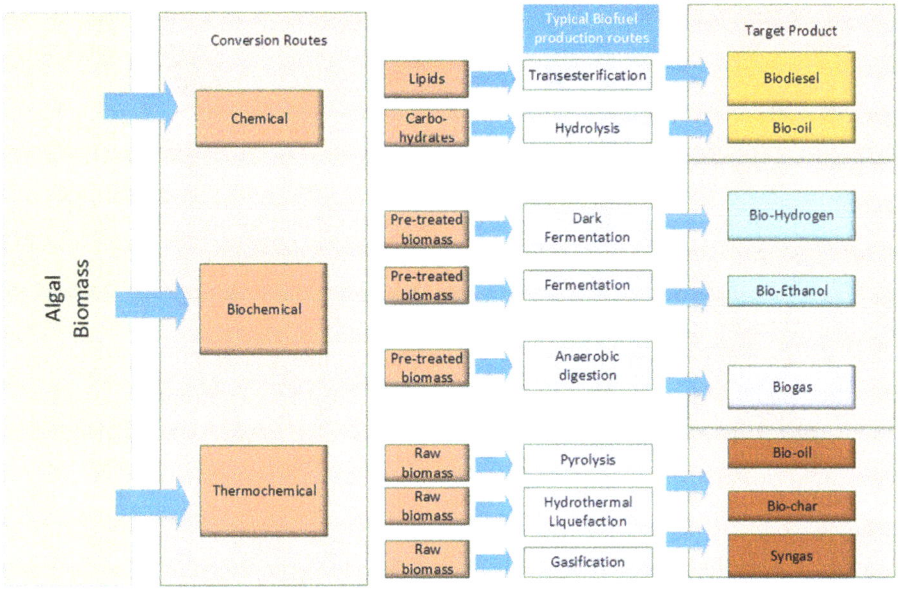

Figure 17.1: Biofuel production routes from an algae biomass feedstock.

The most common biofuel production method from microalgae is transesterification of liquid (triglycerides) to fatty acid methyl esters and by-product glycerol, even with the generally preferred alkaline catalysts being vulnerable to water and acids in the feedstock. Transesterification also uses significant amounts of methanol (>6:1 methanol: biomass ratio) to maximize conversion rates (Im et al., 2014).

Hydrothermal liquefaction (HTL) is the thermal decomposition of organic compounds in superheated water/organic solvent, generally operated at a temperature range of 160–380 °C (Chunyan et al., 2017). It is suitable for high moisture content feedstock such as microalgae, minimizing the drying energy costs. However, it can potentially generate toxic nitrogenous compounds such as imidazole (Usami et al., 2020), resulting in extensive downstream processing requirements and additional complications. Therefore, the conditions of HTL are important to recycle nutrients (e.g., nitrogen, phosphorous contents), as an upstream process would be beneficial (Xu and Savage, 2015).

Fermentation uses yeast or bacteria to convert sugars and other carbohydrates to ethanol or hydrogen (dark fermentation), which can then be separated and used as fuel. The process requires the algal biomass cells to be disrupted with a variety of pre-treatment methods such as alkali treatment or ultrasonication. Fermentation produces high yields of ethanol fuel with mild operating conditions and hence low operating costs, but has a slow reaction rate (generally takes 3–7 days to complete), mandating much larger reactors to achieve production rates comparable to other biofuel production techniques (H. Noorman et al., 2018). Similar to fermentation, anaerobic digestion is also low-cost to operate and produces high yields of fuel with a slow reaction rate and requires pretreatment to open the cell walls to produce biogas. The biogas can be either used on its own right after removing impurities and CO_2 or further converted into synthetic gas (via dry reforming) for Fischer-Tropsch synthesis to produce liquid fuels, which adds additional complexity and costs (Ward et al., 2014; Sun et al., 2020).

The most problematic challenge in producing third-generation biofuels is the separation of microalgae (which is less than 1 wt%) from the cultivation medium (Fasaei et al., 2018). After the dewatering/separation, the water content in microalgae is still high (5–10 wt%), much higher than the threshold for common approaches such as transesterification (0.05 wt%) (Gnansounou and Kenthorai Raman, 2016), thermochemical processes (i.e., pyrolysis and gasification, < 20 wt%) (Im et al., 2014). However, there is limited analysis available on how these challenges impact the environmental and economic costs of producing algae biofuel and the impact of each proposed solution to these challenges (Mahmud et al., 2021). To compare the impact of each challenge in the context of an industrial biofuel production plant and determine methods that address them, metrics that quantify the environmental impact and economic costs of the overall process are required. In addition, a holistic "cradle-to-fuels" approach needs to be investigated to identify the optimum conditions across steps to maximize the outputs.

Process integration and intensification are crucial to overcome challenges in producing biofuels from microalgae. A holistic approach must be applied, so that best interactions between unit operations/steps are explored in conjunction with intensification of unit operations to achieve the optimum output. To perform process integration holistically from cultivation to fuel use requires metrics by which process optimization can be measured. This chapter will discuss and analyze Life Cycle Analysis (LCA) for the environmental effects and Techno-Economic Analysis (TEA) for the economic implications. This will also entail a focus on how different unit operations and processes could potentially synergize by comparing required outflows and inflows of materials and energy, followed by detailed study of these combined and integrated processes.

17.2 Microalgae: current challenges and potential solutions

Algae have a different composition and growth pattern to terrestrial plant biomass (which contains high levels of cellulose and/or lignin (generally > 30 wt%) (Santos et al., 2018)).The composition of the microalgae primarily affects the methods that are suitable for valorization. Fermentation and anaerobic digestion can convert all components except cellulose if it is not pretreated (Ward et al., 2014; Borines et al., 2013), whereas transesterification is only effective on lipids. Thermochemical methods can convert all biomass components to biofuel (Chen et al., 2014), but nitrogenous components e.g., proteins produce toxic by-products in the product bio-oil (which requires upgrading) and prevents recycling of proteins as growth nutrients (Yu et al., 2011). As a result, high lipid feedstocks with relatively thin cellulose cell walls are best suited to transesterification; low cellulose feedstocks (<45 wt% cellulose) suit fermentation and anaerobic digestion; and low protein and high cellulose feedstocks are the best fit for thermochemical methods such as HTL (Roberts et al., 2013). The composition of different algae strains can be seen in Table 17.1. Microalgae contain proteins, cellulose, carbohydrates, lipids, and various high-value organic components such as omega-3 oils, and have much higher water contents (approximately 90 wt% water) after cultivation than most types of terrestrial biomass (Suutari et al., 2015). Algal proteins cannot be easily separated from biomass slurries due to their insolubility in water except at extreme pH, while the high water content interferes with most lipid extraction methods, which makes using HTL and transesterification much more complex (Patil et al., 2018). Protein content can vary between 10 wt% and 57 wt%, while carbohydrates range from 25–60 wt%. This makes selecting the optimal algae strain of key importance for producing a viable process.

Table 17.1: Typical composition of commonly used algae strains (Blifernez-Klassen et al., 2018; Tibbetts et al., 2015; Jatmiko et al., 2019; Salosso, 2019).

Strain	Carbohydrates (wt%)	Lipids (wt%)	Protein (wt%)	Other (wt%)
Botryococcus	25	29	26	20
Scenedesmus	43	16	28	14
Ulva	58	0.2	14	27
Sargassum	60	1.7	10	28
Chlorella	28	20	41	13
Nannochloropsis	27	30	41	2
Spirulina	28	8	57	7

Botryococcus braunii is lipid rich (28–34 wt%) (Tibbetts et al., 2015) and is rare in its ability to produce and excrete terpenes. The high lipid content matches well with transesterification and the terpene production results in higher hydrocarbon yields when processed into biofuel (15 wt% without decomposition vs. trace levels with other strains) (Ruangsomboon, 2012). The high lipid content also allows much higher oil yields when performing lipid extraction techniques such as supercritical CO_2 and ionic liquid extraction (Boni et al., 2018). However, *Botryococcus braunii* has a slow growth rate (0.16/d vs. 0.39/d for Scenedesmus under the same conditions), which lowers productivity (Goswami and Kalita, 2011).

Scenedesmus strains are also lipid rich (>15 wt%) (Tang et al., 2019) and hence are primarily studied for biodiesel production. *Scenedesmus* also has a much faster growth rate than *Botryococcus* (Biomass amount doubles in 8 h vs. 34). However, *Scenedesmus* requires warm water and high light intensity to grow (20–30 °C and 6,000 lux to achieve maximum growth rate) (Blifernez-Klassen et al., 2018). This makes them useful for tropical locations such as the Caribbean, Brazil, South Asia or Africa but they are less effective in colder climates.

Chlorella is commonly used to test biofuel production methods as they have extremely consistent yields (approximately 110–160 mg/L/d) and tolerance for extreme growing conditions (Chen et al., 2020). *Chlorella* can also produce lipids, but only when under stress, that is high lipid yields are only obtained when growth conditions are unfavorable, such as when there is a nitrogen deficiency (Ratomski and Hawrot-Paw, 2021).

Nannochloropsis represents a wide family of strains, but the most commonly used of these are saltwater-adapted, with high polyunsaturated fatty acid content (4–18 wt%) (Dourou et al., 2018) and rapid growth rates (0.59/day) (Manisali et al., 2019). As a result, *Nannochloropsis* strains are one of the most common choices for microalgae cultivation, biofuel production, and extracting organic chemicals (Chua and Schenk, 2017).

Spirulina strains are notably tolerant of extreme conditions with high protein content (up to 57 wt%) (Tibbetts et al., 2015). The high protein content requires upgrading bio-oils to be usable as fuels and can result in some toxic by-products. *Spirulina* strains generate low yields of bio-oils compared to other strains (36% yield at 260 °C) (Tang et al., 2016). Various nitrogenous organics such as pyrrolidine derivatives, pyrazine, and amides/amines that are valuable by-products for drug production and pesticides can be produced from thermochemical processes once separated out. These strains are therefore more suited for multistage processes to extract protein and biofuel production.

Dunaliella tertiolecta is commonly used as a source of β-carotene (>10 wt% dry basis). β-carotene is unstable at most temperatures or under high light conditions, which pairs poorly with most biofuel production methods. HTL can be used on *Dunaliella tertiolecta* producing moderate yields of bio-oils (up to 55% in 380 °C, 225 bar) (Shahi et al., 2020), which could potentially be profitably applied after β-carotene extraction to maximize value-added products.

While a range of different microalgae strains can be used for biofuel and chemical/nutrient application, there are still major problems with producing third-generation biofuels due to high energy costs of dewatering the cultivated algae broth into dried algae suitable for biofuel production, as the required extent of drying depends on the intended fuel production method. For example, transesterification catalysts are very vulnerable to water content and the required lipid extraction solvents are hydrophobic, so a feedstock with less than 0.5 wt% water is generally required (Daroch et al., 2013). By contrast, HTL uses water as the reaction medium, which permits water contents of up to 40 wt% (Gollakota et al., 2018). However, microalgae broths are less than 1 wt% algae, so significant dewatering is required, regardless of the fuel production method (Vasistha et al., 2021). The energy cost of dewatering algae to a moderate 5 wt% water accounts for one-third of the total capital costs and one-fourth of the operating costs, which is very large relative to other cost sources. Microalgae are also easily ruptured by bubble formation during extraction, which results in loss of valuable products into the extracted water stream (De

Boer et al., 2012). Conventionally, dewatering is done using evaporation or centrifugation depending on water content, algae cell size, and other factors.

Process intensification (PI) can be used to reduce the cost of microalgae dewatering, primarily by making downstream processes more tolerant of water content. An example of this would be in situ transesterification, which combines the lipid extraction step with transesterification of the extracted lipids (Kim et al., 2017). This would allow the extraction step to tolerate higher water content, particularly when the process is acid-catalyzed (Park et al., 2015). Another application for PI techniques would be in increasing the reaction rate of the slower biofuel production routes. Fermentation requires significant time to produce bioethanol and requires algae cell lysis to allow fermentable components to reach the fermenting bio-organism (Pratto et al., 2020). PI-inspired reactor designs such as airlift loop reactors can reduce losses of the fermenting organism while permitting continuous operation, which can increase throughput-to-volume ratios by a factor of 10 or more (H. J. Noorman et al., 2018). PI-inspired designs are also being looked at for algae cultivation, to allow faster growth with less light or lower nutrient supplies or easier harvesting. The potential to recycle algal biofuel refinery waste aqueous streams as a nutrient source that is part of a full process approach is also under investigation to reduce costs and environmental impact (Hu et al., 2017).

PI can also be used for dewatering microalgae in a more energy-efficient manner. A PI dewatering method that shows significant promise is the use of foam columns to separate the algae from a dilute solution (Matter et al., 2019). This technique adds a surfactant to the algae broth and then introduces air bubbles from below. Microalgae cells are entrained on the bubbles and can be collected from the foam at the top, which concentrates them significantly, achieving > 90% algae recovery while removing all water around the algae cells (Ma et al., 2020). The microalgae remains too wet to be used for transesterification (algae is naturally about 90 wt% water), but the energy cost of drying is significantly reduced (94% reduction in energy cost vs. filtration) (Fasaei et al., 2018).

PI can also be applied to downstream processing, with the most established method being reactive extraction (Mahmud et al., 2021). Reactive extraction is inducing formation of the product and separating it from solution in one reactor, often through the use of an extracting solvent. This is widely used for transesterification, where it is known as in situ transesterification (Ma et al., 2015), but it can also be used for many mixed product streams. Performing the reaction and extraction simultaneously reduces the number of processing steps required and the energy cost of extraction (Jeevan Kumar et al., 2017). This principle can also be applied to lipid extraction via hot compressed water, as this allows lipid extraction and mild HTL to occur at the same time. This allows the easier combination of HTL and transesterification while extracting amino acids as a useful by-product (Kim et al., 2017).

Optimized conditions or designs for a standalone unit or intensification of a standalone unit may no longer be optimal when they are integrated into a larger process. Therefore, more research is required into full process integration. However, there is less focus on process integration of biofuel from microalgae. Most process integration studies either attempt to combine algal biofuels with other industrial processes or focus on the separation of the algae components into multiple streams. The latter then aims to produce fuels or value-added chemicals from each component stream (Wang et al., 2022; Kumar et al., 2019). Most of these studies consider each of these streams separately, which is not ideal from a process integration perspective. Figure 17.2 below illustrates an example of partial process integration that combines fermentation, lipid extraction, and, potentially, transesterification (Dong et al., 2016), which will be discussed further in Section 17.4.

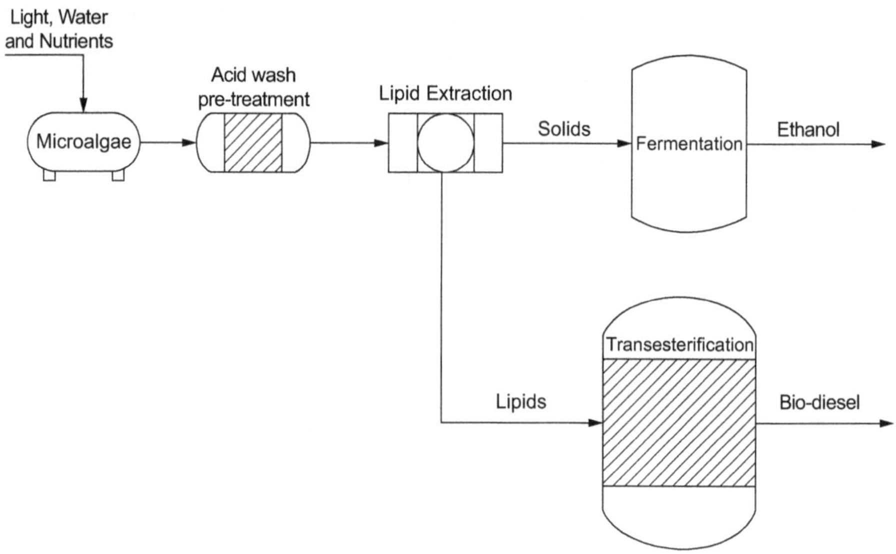

Figure 17.2: Example of partial process integration of fermentation and lipid extraction. Adapted from Dong et al. (2016), Shanmugam et al. (2021).

17.3 State-of-the-art technologies for fuel production

(HTL, fermentation, and anaerobic digestion (AD) are more favorable for algae with high moisture content, as there is little requirement for drying. However, in some high-protein (nitrogen) microalgae, AD could be an issue as this produces ammonia as a by-product. Fermentation can also tolerate water but high water content can dilute the biofuel and require additional separation. In addition, pretreatment of algae is required for AD and fermentation to disrupt cells. In situ transesterification can be potentially performed in wet conditions (up to 20% water without requiring additional methanol) (Sathish et al., 2014) with particular catalysts and process conditions.

HTL is the most suited to wet feedstocks as water acts as the working fluid. Under elevated temperatures (160–380 °C), the hot compressed water molecules interact through hydrogen bonding less than ambient water, contain more oxonium ions, and can dissolve many organics not otherwise soluble in water. This hot compressed water has a catalytic effect on decomposition processes in addition to being a reactant, which allows rapid thermal decomposition at lower temperatures than other thermochemical methods (Jindal and Jha, 2016). During HTL, hemicellulose, cellulose, and starch hydrolyze into various monosaccharides at temperatures above 150 °C, which can potentially be hydrolyzed further into levoglucosans, furfurals, and carboxylic acids (Yoon et al., 2018). However, further hydrolysis of monosaccharides is slow at temperatures below 250 °C, which results in high aqueous phase concentrations of saccharides (e.g., glucose, galactose, amylose), which can be concentrated and converted to biofuel either thermally or biologically (Ramli et al., 2020; Karemore and Sen, 2016). Acidic pH, hot compressed water during HTL greatly increases the monosaccharide hydrolysis rate, resulting in levoglucosans, furans, acetic acid, lactic acid, and formic acid(Lavarack et al., 2000; Girisuta et al., 2013; Shao et al., 2020). Proteins hydrolyze into their component amino acid during HTL at temperatures above 200 °C, though more slowly than carbohydrates due to the relative strength of peptide and glycosidic bonds (He et al., 2020).The amino acids can also hydrolyze at temperatures above 250 °C to various toxic nitrogenous compounds, CO_2, NO_x, ammonia, and water (Chua and Schenk, 2017). Minimizing the yields of these generally unwanted by-products can be performed by selecting an algae strain with low protein content or extracting the protein before HTL (Sereewatthanawut et al., 2008; Lupatini et al., 2017). Any lipids present generally hydrolyze to fatty acids and glycerol. HTL results in a polar and a nonpolar liquid phase and some solid residue, primarily undecomposed cellulose and amorphous carbons. The nonpolar phase is primarily unreacted glycerides and fatty acids with some hydrocarbons from hydrolysis reactions, while the polar phase contains the saccharides, amino acids, and glycerol (Xu and Savage, 2015; Tang et al., 2019, 2016).

Transesterification is a set of three reactions of lipids (triglycerides) and an alcohol, i.e., methanol to produce fatty acid methyl esters (FAME), releasing glycerol as a by-product (Dong et al., 2013). Transesterification is the most widely used biofuel production method from algae, but it is very intolerant of water content (requires < 0.05 wt% water) (Lin and Ma, 2020) when basic catalysts are used and the reaction rate is very low with acidic catalysts (De Boer et al., 2012; Kim et al., 2017; Thangaraj et al., 2018). The intolerance of transesterification to water content can be partially overcome by performing lipid extraction, but most lipid extraction solvents are generally nonpolar and would be kept from the algal cells by aqueous layers if water is present, complicating lipid extraction from wet biomass. Combining the extraction and transesterification steps, known as in situ transesterification reduces the number of processing steps and reduces capital and operating costs through reduced utility requirements (Ehimen et al., 2010; Li et al., 2011). This process can tolerate some water content when nonpolar cosolvents or acidic catalysts are used, as the methanol used as a reactant during transesterification can act as a cosolvent for extracting lipids. For example, in situ transesterification performed in a 36.4% H_2SO_4, 73.6% methanol solution at 300 °C under vacuum distillation conditions resulted in 85 wt% FAME yield over a 2-h reaction time (Torres et al., 2017). The use of a cosolvent also reduces the required amount of methanol to obtain complete transesterification by around 25% without compromising the reaction rates (Park et al., 2017). The largest downside of in situ transesterification is that significant volumes of methanol (up to 300:1 depending on conditions) are required, which are generally produced from fossil oils (Ehimen et al., 2010; Velasquez et al., 2012). Other alcohols such as isopropanol can be used but are more expensive due to requirements for low water content to prevent catalyst poisoning. In situ transesterification can also be combined with other PI techniques such as microwaves and ultrasound to further increase yields or improve lipid extraction. For example, FAME yields of > 50% can be obtained without an extraction solvent by using ultrasonication to break up algal cells and release lipids (Patle et al., 2021; Rokicka et al., 2020; Luo et al., 2014).

Anaerobic digestion (AD) uses microorganisms such as bacteria to decompose biomass to biogas and solid residues. The biogas is predominantly methane (50–70%vol) and CO_2 (30–40%vol), with low levels of hydrogen sulfide (<3%vol) and hydrogen (<2% vol) (Sialve et al., 2009; Passos et al., 2014). AD is a very good fit for microalgae processing as it operates in a water-rich medium, produces biogas in a single reaction stage, and is tolerant of variations in local conditions. Some dewatering is still required to maintain a certain reaction rate e.g., 5 wt% algae solutions are sufficiently dewatered for AD (Ward et al., 2014). Another advantage of AD is that the waste products can be recycled as nutrients for algae cultivation. However, the reaction rate is extremely slow if the algal cells are not disrupted. Certain algae species also emit hydrogen sulfide, which can poison the microorganisms and reduce reaction rates drastically (Chen et al., 2008). Cellulase and/or lipase can enhance digestion rates by breaking up the algae cell walls, which also increases biogas yields by up to 20 wt% (Nitsos et al., 2020) though this has only been used at lab-scale, to date. The presence of lipids during AD can also slow reaction rates as many lipids intermediate decomposition products inhibit bacterial activity. Lipids produce less biogas and are converted more slowly than carbohydrates, making AD well-suited to carbohydrate-rich and lipid-poor algae strains (Ward et al., 2014). Biogas needs to be separated and purified to remove CO_2 and other impurities before being used as fuels. Alternatively, biogas can be converted to methanol and other liquid fuels via reforming and hydrogenation, but yields are currently low (around 8 wt%) (Giuliano et al., 2020).

Fermentation is most commonly used for producing liquid biofuel, e.g., ethanol from carbohydrate components. Fermentation can be performed in aerobic or anaerobic conditions, though aerobic conditions generally result in higher ethanol yields (Zhu et al., 2013). It operates by utilizing ethanol producing microorganisms to convert sugars to ethanol. The fermentation rate depends on temperature, pH, microorganism growth rates, and tolerance to ethanol and potential inhibitors that may be present in the algae cells. The microorganisms are also incapable of fermenting cellulose or other glucose polymers, so saccharification processes are generally performed as a pretreatment step (Pratto et al., 2020). This is conventionally performed chemically using hydrochloric or sulfuric acid or thermally or biologically using cellulase enzymes. Enzymatic saccharification is compatible with fermentation, which could reduce processing steps and improve fermentation rates if these processing steps could be combined (Chohan et al., 2020). A key factor in fermentation-derived biofuels is that fermentation only produces ethanol up to around 12–15% vol as the microorganisms cannot survive beyond this point. Therefore, distillation needs to be done, resulting in additional energy costs that typically account for 60% or more of the total energy costs (Chen et al., 2018).

17.4 Process intensification/integration

There is a wide variety of different methods to produce biofuels. However, these processes are often considered as standalone systems without integrating with others. A well-known example of process intensification/integration is in situ transesterification where lipid extraction and transesterification are combined. It reduces the number of reactors required for transesterification but still requires algae with moisture content ≤ 20% (Sathish et al., 2014). The main issue with in situ transesterification is high amounts of solvent requirement (e.g., up to 300:1 methanol-to-algae ratio) to achieve complete transesterification within a 4-h residence time (Velasquez et al., 2012) compared to 1–2 h reaction time at a ratio of 9:1 for conventional transesterification. This results in an increased requirement for solvent recovery and larger reactors and separation systems. This can potentially be resolved using ultrasonication to extract lipids without a nonpolar solvent, another example of PI, which reduces the methanol requirement and separation costs and hence becomes less expensive than conventional transesterification for wet feedstocks. Intensified reactor designs can also be used to reduce reactor sizes and extent of downstream processing and can be used in the integration system. Intensified designs have been tested for transesterification, achieving 99% conversion within 5–20 min instead of 1–2 h in a standard reactor under the same conditions (Patle et al., 2021). Microwaves and ultrasonication can be used to extract lipids more effectively during in situ transesterification, which can achieve 35–60 wt% biodiesel yields without a cosolvent (Luo et al., 2014).

Two-stage HTL is the most widely proposed intensification technique for high-protein content algae as it drastically reduces toxic by-product formation and hence reduces downstream processing requirements. For two-stage HTL, the pretreatment stage operates at 140–180 °C to induce hydrolysis to sugars and amino acids for extraction and further processing and breaking down the cellulose rich cell wall to sugars, without producing toxic amino acid decomposition products (Sereewatthanawut et al., 2008). The second HTL stage at > 200 °C produced bio-oil yields from 25 to 28 wt% and deoxygenates the bio-oil from 30.5% to 19.6% oxygen, which minimizes the upgrading steps, therefore reducing the operational and capital costs (Usami et al., 2020; Prapaiwatcharapan et al., 2015). The process layout of two-stage HTL is illustrated below in Figure 17.3, showing how dewatered microalgae conversion to lipids and bio-oil would be laid out and illustrating the nutrient recycling step.

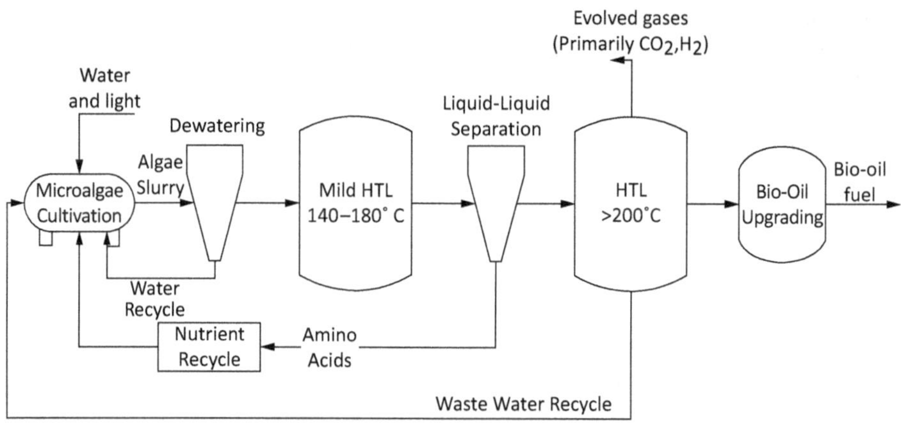

Figure 17.3: Process layout of two-stage hydrothermal liquefaction.

Biochemical processes are generally limited by mass transfer of nutrients or oxygen to the yeast/bacteria (H. Noorman et al., 2018), while microorganisms are often destroyed by many improved mixing techniques. To enhance the mass transfer limitation, biochemical processes can be operated continuously to reduce reactor size 10–100 fold, and intensified mixing techniques that do not cause cavitation or bubbles can increase fuel production rate by up to 260% (H. Noorman et al., 2018). As fermentation generally uses a saccharification pretreatment step, combining saccharification and fermentation with PI techniques can improve ethanol yield by 20–25% from improved mixing and increased saccharification efficiency as a result of in situ sugar consumption through fermentation (Jiang et al., 2020).

Process integration can be applied to combinations of the abovementioned intensified processes to greatly reduce waste, increase production of fuels and/or coproducts and decrease energy requirements. One example is the combination of transesterification and HTL, which are most effective for conversion of lipids and carbohydrates, respectively (Gollakota et al., 2018; Moazeni et al., 2019). Lipid extraction can be performed under mild HTL conditions. This has the advantage that the lipids can be easily separated from the bulk aqueous phase by cooling and depressurizing, causing the lipids to accumulate into a separate layer. The lipids can then be transesterified to fatty acid esters (biodiesel), while the de-lipidated biomass can be converted to bio-oil using HTL, minimizing solid waste and producing additional fuel (Shahi et al., 2020). Mild HTL can also decompose proteins to amino acids and extract them, where they can be used as fertilizers or nutrient for algae cultivation (Jazrawi et al., 2015).

Process integration of algal biofuel production with other industrial processes that produce wastewater streams has also been considered, especially those that are rich in nitrogenous compounds and low in toxins (Zhou et al., 2014). These wastewater streams can be used as a source of nutrients for algae cultivation. The use of aqueous discharge from water treatment plants is of particular interest in reducing pollution discharged to local waterways (Rahman et al., 2020).

Another example of process integration is the combination of algae cultivation with carbon capture by utilizing algae to convert CO_2 from the air or concentrated CO_2 to solids that can be pyrolyzed to solid carbon for storage. This can greatly reduce the energy cost of carbon capture through avoiding the need to deoxygenate carbon dioxide or filter it from the air (Mona et al., 2021).

Economically viable production of algae-derived biofuels could greatly reduce usage of fossil fuels and slow down climate change. However, many challenges remain, primarily around the capability of producing the fuel cheaply enough to inspire investment. Process integration and intensification are crucial to overcome challenges in producing biofuels from microalgae. A holistic approach must be applied, so that best interactions between unit operations/steps are explored in conjunction with intensification of unit operations to achieve the optimum output. To perform process integration holistically from cultivation to fuel usage requires metrics by which process optimization can be measured. These can be measured via environmental impact sources from across the whole life cycle, from feedstock cultivation or collection to disposal of final products (life cycle analysis (LCA)) and economic viability (techno-economic analysis, TEA). These also generally include sensitivity analyses to determine where small process changes have great impacts on financial costs, which can help identify the best areas for cost savings and any simulation limitations.

Table 17.2 compares "cradle-to-fuel" TEA studies of third-generation biofuel production methods, alongside the required fuel cost for the biofuel production method to break even over predicted plant lifespans. These TEA studies include the costs for the whole production process, from transportation of required nutrients to the algae cultivation vessels, electricity costs to run the plant and all other expected operating costs to transportation of the biofuel to the end-user. Most of these studies are examples of PI, with some examples of partial process integration that focusses on separating the algae by composition and then converting each component type separately, with nutrients recycled to the algae cultivation stage. This means there is still potential to optimize any of these methods further, which will be discussed below.

The cost for biofuel was compared against that for petrol from crude oil, which was around £1.53/L (January 2022). Biofuel from HTL had the lowest price at large scales, which is due to the minimized dewatering requirements (water < 95–98 wt%) (Gollakota et al., 2018). Lower dewatering requirements generally translate into cheaper

Table 17.2: Comparison of "cradle-to-fuel" TEA studies for Intensified and non-intensified biofuel production routes.

Technique	Plant size (MT/ year)	Feedstock	Product	Production cost (£/L)	References
Transesterification	100	Microalgae	Biodiesel	2.45	(Taylor et al., 2013; Sano Coelho et al., 2021)
HTL	2000	Microalgae	Biodiesel	0.54	(Ou et al., 2015)
Lipid extraction + HTL	3000	Microalgae	Biodiesel	2.05	(Davis, Aden and Pienkos, 2011)
Solvent extraction	2000	Microalgae	Biodiesel	2.64	(Sun et al., 2011)
Anaerobic digestion(productivity: 30 g /m^2/day dry algae)	73	Microalgae	Liquefied biogas	0.73	(Zamalloa et al., 2011)
Intensified / Integrated processes					
In situ transesterification	105	Microalgae	Biodiesel	0.86	(Nagarajan et al., 2013)
Integrated MOTU with co-products	0.187	Microalgae	Bio-oil	0.25	(Wiatrowski et al., 2022)
Integrated MA with co-products	0.187	Microalgae	Biodiesel/ Bioethanol	0.16	(Wiatrowski et al., 2022)
Catalytic HTL/hydrogenation	0.072	Microalgae	Bio-oil	1.75	(Masoumi and Dalai, 2021)
CAP	–	Microalgae	Biodiesel/ Bioethanol	1.63	(Katiyar, Banerjee and Arora, 2021)
Fermentation	–	Microalgae	Bioethanol	0.45	(DeRose et al., 2019)
Integrated transesterification and AD with PHB and Astaxanthin coproducts	0.043	Microalgae	Biodiesel + Biogas	1.08	(Prieto et al., 2017)

fuel, as dewatering accounts for 33% of the capital costs (Batan et al., 2016) and 20–30% of the operating costs (Fasaei et al., 2018) for a 5% water feedstock, with this proportion rising if drier feedstocks are required.

Assuming that algae growth rates can be maintained at 30 g/m^2/day, AD produces fuel at comparable cost to HTL, also due to a high degree of water tolerance in the feedstock (Zamalloa et al., 2011). However, AD produces biogas as the main product instead of liquid biofuels, which is largely incompatible with current engines (Yu et al., 2020). Conversion of current engines to use biogas is possible but the gaseous fuel needs to be pressurized to obtain a decent vehicle range and designs for these modified engines show limited uptake (Yu et al., 2020). The biogas can also be converted to bio-oil by Fischer-Tropsch synthesis, but this adds significant extra costs.

Basic conventional transesterification catalysts, such as KOH, are well-known to be converted to soaps by water and acids, while acidic catalysts are around 1/1,000th the activity of a basic catalyst. As a result, algal biodiesels are 3–5× more costly than other biodiesels due to the requirement for low water content feedstock (£2.5/L vs. £0.5–0.9/ L for other biodiesels).

By contrast, in situ transesterification biodiesels are significantly cheaper to produce (£0.86/L), as the extraction solvent is acidic and dewatering requirements are minimal due to the acidic catalysts being unaffected by water (Kim et al., 2017; Velasquez et al., 2012). As a result, the in situ transesterification biodiesel price remains comparable to thermochemical and biological methods, despite the requirement for large amounts of methanol and a significantly slower reaction rate (Davis et al., 2011).

Integrating AD into a transesterification process has been performed, where the AD-derived biogas is used as a power source for the transesterification of the extracted lipids. This has been found to reduce the cost of fuel production by 35%, while also reducing process wastage (Harun et al., 2011).

In addition to simple methods that use feedstock to produce a small range of useful fuel products, methods have been developed that are more thoroughly integrated, which includes recycling nitrogen and/or inorganics in the biomass as algae growth nutrients (Liu et al., 2019), producing valuable by-products such as polyhydroxybutrate monomers(Jablonský et al., 2018) and the use of process intensification techniques. While this may introduce additional capital and operational costs due to the increased complexity, the energy costs are likely to be reduced due to recycling, and the by-products add an additional source of income (Kumar et al., 2020).

In situ transesterification produces biofuel at low cost of £0.86/L through the use of ultrasonication to extract lipids from the algae cells during transesterification. This allows for avoiding cosolvent or acidic catalyst use at the cost of still requiring extensive dewatering to < 0.5 wt% water(Rokicka et al., 2020). To offset the energy cost, the defatted algal residue is collected and used to generate biogas to power the process via AD (Ward et al., 2014). Nutrient recycling is also performed on the remains after AD, reducing input costs. A TEA study for integrated conventional transesterification with nutrient recycling is also available, which after dewatering of the algae performs lipid extraction with hexane and then, transesterification to biodiesel with methanol. Any unreacted methanol is recovered, any polyhydroxybutrate compatible chemicals are fermented to polyhydroxybutrate, and the glycerol is sold as a coproduct. Any unreacted algae biomass is either processed to astaxanthin or converted to electricity via AD followed by a combined heat and power system (Prieto et al., 2017). Overall, the fuel cost is slightly above that of in situ transesterification, likely due to the additional input of hexane to extract lipids and processes to separate it back out from the lipids (Qu et al., 2018) and the increased dewatering requirements of conventional transesterification.

Integrated MOTU and MA also implements extensive process integration, based around the hydrogenation/upgrading and fermentation processes, respectively (Masoumi and Dalai, 2021). MOTU uses solvent extraction to remove the lipids from the algae, which are then saponified and phosphates recycled as algae nutrients. The defatted algae were hydrolyzed enzymatically while all liquids and enzyme products are slightly oxidized and then upgraded using acid catalysts. This produces biofuels rich in hydrocarbons, with the nutrients remaining in the aqueous phase being recycled. The soaps formed from the lipids are then cold-pressed, with the saturated fats being upgraded and the unsaturated fats converted to polyurethane plastics as a by-product. This process produces minimal waste, which keeps costs down, but the real reason for the low fuel price is the high value of the by-product. The fuel price assumes a polyurethane sell price of £3.7/kg (taken 2021), which is within current price variations as of July 2022 (Usman et al., 2012). The simplified process layout for MOTU is shown below in Figure 17.4.

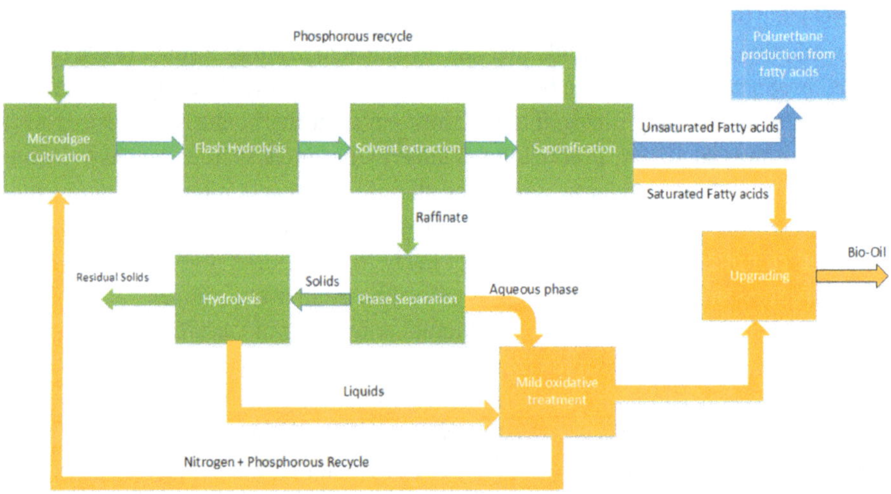

Figure 17.4: Flowchart for integrated mild oxidative treatment and upgrading (Wiatrowski et al., 2022; Davis et al., 2020).

MA also uses solvent extraction and lipid saponification, followed by enzymatic hydrolysis of the defatted algae. The liquids and sugars derived from this are fermented to MA and separated. Saturated fats from cold-pressing are also converted to alcohols using lipases and unsaturated fats to polyurethane. MA produces a lower-cost fuel than MOTU per kg of fuel with the same assumptions (DeRose et al., 2019), but the alcohols have a lower calorific value than the hydrocarbons produced with MOTU (Kumar and Strezov, 2021), that is, the cost-to-energy ratio is similar.

Catalytic HTL and hydrogenation use a methanol/water HTL fluid to allow operation at less severe reaction conditions and utilizes char produced during the process as catalyst supports, which reduces catalyst costs (Fu et al., 2013). The methanol present also esterifies any acids present in the HTL oils, helping to stabilize the products. The HTL-derived oil is the hydrotreated to increase its calorific content and improve the characteristics of the fuel (Verdier et al., 2021). The result is a high-quality fuel that is more expensive than most intensification processes listed here. However, the extent of process integration in this method is low, requiring hydrogen as a process input and venting evolved gases and water to waste. The cost of operating this process would likely decrease if water was used as the hydrogen donor via hydrodeoxygenation (HDO) and the waste streams recycled for algae cultivation or as fuels (Duan et al., 2018).

Combined algal processing (CAP), which is based around fermentation of acid-treated algae, is an integrated process that also produces oils as a by-product by distilling the fermentation broth. The general process layout can be seen in Figure 17.5.

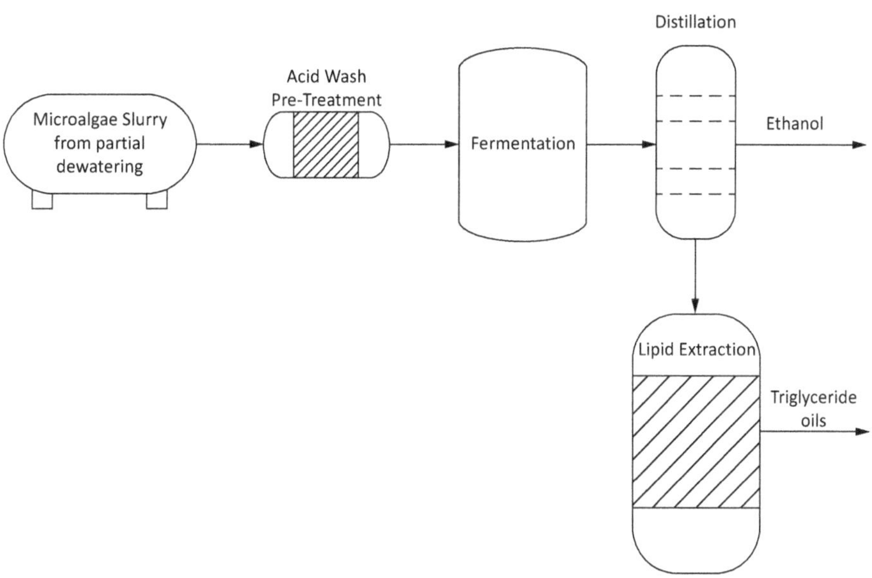

Figure 17.5: Process layout of combined algal processing (CAP).

As fermentation requires minimal energy inputs compared to other techniques, the fuel price is mostly due to the use of two distillation steps, which are energy-intensive. The process also produces 15–25% solid residues, which can still contain unfermented sugars, and more efficiently valorizing this would also assist with reducing fuel prices (Pratto et al., 2020). Fermentation, when set up without extensive distillation, can reduce fuel costs significantly. This source also uses fermentation of acid-hydrolyzed algae, but further converts any nonalcoholic components to bio-oil using HTL, which minimizes waste and produces another fuel stream (Shahi et al., 2020). Production of struvite, a useful fertilizer, as a by-product also helps the process economics, while the remaining biochar from HTL can be sold as solid fuel or as catalyst supports. This more extensive valorization of waste has more potential to reduce costs than a more brute-force approach.

Microalgae could also be used to photolyze water to bio-hydrogen, but this is considered fourth-generation biofuel and not the focus of this chapter (Khetkorn et al., 2017). This process has minimal waste due to the simplistic feedstocks but currently is more expensive than fossil fuels in terms of cost per GGE (Goswami et al., 2021). Further work on bio-hydrogen would be needed to reduce the costs sufficiently to make their adoption economically feasible.

In addition to the economic outlook, the environmental impact of biofuels must also be quantified at all stages of the process from initial feedstock to final fuel use. All the LCA studies below in Table 17.3 consider all stages of the biofuel production process, including:
- the end use of the fuel
- distributing the fuels and collecting the feedstock (e.g., pumping costs)
- fuel production from the feedstock

As well as these stages, all by-product emissions must also be included (Singh and Olsen, 2011), such as:
- losses of produced fuels to the environment
- venting or flaring of gaseous by-products
- CO_2 content loss in soils from land use changes

Carbon capture from algae growth is also included for all cases, which may also include energy-related emissions from collecting the CO_2 from the atmosphere if pressurized CO_2 or hydrogen carbonate is used (Somers and Quinn, 2019).

The values available in LCA studies vary greatly due to differences in methodology and variables such as feedstock logistics emissions and recycling waste products. HTL generally produces emissions lower than gasification or fossil fuels but often slightly higher than AD or in situ transesterification (Bennion et al., 2015). Significant amounts of these emissions are associated with electricity used in producing the fuel. If the electricity is sourced from environmentally friendly sources, all the above techniques produce up to 40% lower emissions, especially HTL (Fasaei et al., 2018). Transesterification is shown to have a low environmental impact, primarily emissions from electricity usage. In situ transesterification has a higher environmental impact, due to a significantly increased requirement for methanol, which is primarily sourced from natural gas. Fermentation also produces low environmental impact values despite hydrolysis pretreatment, while AD LCA studies are not available on algae due to exten-

Table 17.3: Third-generation biofuels production methods and accompanying LCA emissions on a "cradle-to-fuel" basis.

Technique	Feedstock	Product	Emissions by product yield ($kgCO_{2eq}$/kg fuel)	References
Fossil fuel For comparison	Crude Oil	Diesel	2.66	
Gasification	Microalgae	Syngas	1.05–4.29	(Azadi et al., 2015; Bošnjaković and Sinaga, 2020)
HTL	Microalgae	Bio-oil	0.36–2.02	(Frank et al., 2013; Fortier et al., 2014)
Transesterification	Microalgae	Biodiesel	0.14–0.57	(Frank et al., 2013; Sander and Murthy, 2010; Adesanya et al., 2014)
In situ transesterification	Microalgae	Biodiesel	0.81	(Brentner et al., 2011)
Fermentation	Seaweed	Bioethanol	0.35–0.96	(Alvarado-Morales et al., 2013; Luo et al., 2010)
Integrated/Intensified studies				
Catalytic HTL/Hydrogenation	Microalgae	Bio-oil	1.60	(Masoumi and Dalai, 2021)
Combined transesterification and AD	Microalgae	Biodiesel	0.26	(Gnansounou and Kenthorai Raman, 2016)
Combined transesterification, fermentation, and AD	Microalgae	Biodiesel + succinic acid	0.42	(Gnansounou and Kenthorai Raman, 2016)
Transesterification with hot compressed water (HCW) lipid extraction	Microalgae	Biodiesel	0.10	(DeRose et al., 2019; Ponnusamy et al., 2014)

sive pretreatment requirements and no useful liquid fuel products without extensive downstream processing (Solé-Bundó et al., 2019).

By comparison, integrated processes based around lipid extraction to transesterification with fermentation and/or AD to convert the residual carbohydrates to an additional biofuel stream result in significant reductions in CO_2 emissions relative to a process that uses only one technique. This is most likely due to the extraction of lipids into a nonpolar solvent, that is, water content during transesterification is minimal without additional drying, while the cell residues are valorized by water-tolerant processes. This would allow relatively little drying to be performed. HTL could also be used instead of fermentation/AD, which would produce a more readily usable biofuel at the cost of making many of the recyclable nutrients (e.g., proteins) present in the algae into toxic chemicals, complicating recycling and increasing costs. This can be seen in Table 17.3, as the use of integrated HTL remains relatively high, though still well below fossil fuels. This issue could be addressed by extracting the proteins during lipid extraction as shown in Table 17.2, which would make the HTL outflows nontoxic, but this has not been shown in an LCA study.

Hot compressed water can also be used for lipid extraction before transesterification, which results in very low emissions (Table 17.2) (DeRose et al., 2019; Ponnusamy et al., 2014). Transesterification can easily be combined with HTL by extracting lipids and then performing transesterification, while performing HTL in the extraction vessel on the now lipid-free algae.

17.5 Conclusions and recommendation

While utilization of microalgae remains challenging regardless of the production route chosen, algal biofuels are becoming more feasible due to fossil fuel cost increases and more optimized biofuel production methods. Microalgae biorefineries are already used for extracting valuable organics, but biofuels remain economically unviable currently (Harvey and Ben-Amotz, 2020; Marzorati et al., 2020; Hannon et al., 2010), though relative costs are coming close.

Currently, HTL, AD, transesterification, and fermentation, all show promise in producing biofuels from microalgae but have not been able to become financially competitive with other biofuels and the energy and nutrient/chemical requirements reduce the sustainability of each process. HTL efficiently produces fuel from all organic components in microalgae with a moderate energy cost but complicates process integration in the form of nutrient recycling.

Process intensification (PI) and integration are beginning to be employed to optimize the complete algal biofuel production process. The whole processes, from cultivation to fuel use that produce the cheapest fuel with the lowest emissions utilize waste minimization, wherever possible. The dominant method for valorizing microalgae according to current TEA and LCA studies is the combination of HTL, fermentation, transesterification, and AD listed above with process intensification and integration wherever possible. One particularly effective route would be using lipid extraction to remove the lipids and proteins, performing transesterification on the lipids and collecting the protein as a by-product or as nutrients, and then performing HTL/fermentation/AD on the extracted cells. This adds additional steps, but lipid extraction would mean the feedstock for transesterification is almost water-free lipids without additional drying, and the HTL/AD feedstock contains little to no proteins that would produce toxic by-products. This allows the largest downsides of each process to be bypassed, reducing resulting fuel cost and emissions by 53% and 74%, respectively, compared to HTL alone (Gnansounou and Kenthorai

Raman, 2016; Wiatrowski et al., 2022; Masoumi and Dalai, 2021). The removal of proteins during de-lipidating also allows high value organics to be extracted alongside biofuel production, introducing a large source of additional income to be tapped while producing biofuels. This also integrates well with upstream optimizations such as flocculation, especially when lipid extraction methods that tolerate water are used.

The issues with combinations of techniques are the requirement for separating out certain cell components (generally lipids and proteins) and the added processing steps introduced by using multiple techniques. Process intensification is well suited to removing or mitigating these issues, with several methods for optimizing these integrated processes already studied. The added costs of solvents (e.g., hexane) to perform lipid extraction can be removed by using ultrasonication or HCW to extract lipids in an aqueous or otherwise polar working fluid, which would also form a second layer once the vibrations or pressure is removed. Using HCW is also potentially helpful for combining processing steps, as the reactor used for HCW extraction can be made completely compatible with HTL, making lipid extraction and HTL of the de-lipidated biomass possible with just one reactor stage. Dewatering costs can be reduced using more energy-efficient methods such as flocculation or membrane separation, while the extent of dewatering required can be reduced by using more water tolerant processes.

Full processes, which are based around HTL or fermentation in isolation, are also viable, especially when proteins are extracted or algae cells are broken open as preprocessing, respectively. These single-method routes produce a single fuel type with lowered complexity compared to combined processes, though the resulting minimum fuel costs are higher when revenue from by-products are not included.

With more integrated process plans being studied and current rises in fuel costs, the largest barrier to the implementation of integrated methods is the relative lack of complete "cradle-to-fuel" TEA and LCA studies and pilot-scale studies for integrated processes, which are required as proof of concept to minimize project risks and increase commercial and industrial interest.

References

Adesanya, V. O., et al. (2014). Life cycle assessment on microalgal biodiesel production using a hybrid cultivation system. *Bioresource Technology, 163*, 343–355.

Alvarado-Morales, M., et al. (2013). Life cycle assessment of biofuel production from brown seaweed in Nordic conditions. *Bioresource Technology, 129*, 92–99.

Azadi, P., et al. (2015). Simulation and life cycle assessment of algae gasification process in dual fluidized bed gasifiers. *Green Chemistry, 17*(3), 1793–1801.

Batan, L. Y., Graff, G. D., & Bradley, T. H. (2016). Techno-economic and Monte Carlo probabilistic analysis of microalgae biofuel production system. *Bioresource Technology, 219*, 45–52.

Bennion, E. P., et al. (2015). Lifecycle assessment of microalgae to biofuel: Comparison of thermochemical processing pathways. *Applied Energy, 154*, 1062–1071.

Blifernez-Klassen, O., et al. (2018). Metabolic survey of Botryococcus braunii: Impact of the physiological state on product formation. *PloS One, 13*(6), e0198976–e0198976.

Boni, J., Aida, S., & Kalsum, L. (2018). Lipid Extraction method from microalgae botryococcus braunii as raw material to make biodiesel with soxhlet extraction. *Journal of Physics: Conference Series, 1095*, 012004.

Borines, M. G., De Leon, R. L., & Cuello, J. L. (2013). Bioethanol production from the macroalgae Sargassum spp. *Bioresource Technology, 138*, 22–29.

Bošnjaković, M., & Sinaga, N. (2020). The perspective of large-scale production of algae biodiesel. *Applied Sciences, 10*(22), 8181.

Brentner, L. B., Eckelman, M. J., & Zimmerman, J. B. (2011). Combinatorial Life Cycle Assessment to Inform Process Design of Industrial Production of Algal Biodiesel. *Environmental Science & Technology, 45*(16), 7060–7067.

Chen, H., et al. (2018). Novel distillation process for effective and stable separation of high-concentration acetone–butanol–ethanol mixture from fermentation–pervaporation integration process. *Biotechnology for Biofuels, 11*(1), 286.

Chen, P. H., et al. (2020). Nutrient recycle from algae hydrothermal liquefaction aqueous phase through a novel selective remediation approach. *Algal Research, 46*, 101776.

Chen, W.-T., et al. (2014). Hydrothermal liquefaction of mixed-culture algal biomass from wastewater treatment system into bio-crude oil. *Bioresource Technology, 152*, 130–139.

Chen, Y., Cheng, J. J., & Creamer, K. S. (2008). Inhibition of anaerobic digestion process: A review. *Bioresource Technology, 99*(10), 4044–4064.

Chohan, N. A., et al. (2020). Valorisation of potato peel wastes for bioethanol production using simultaneous saccharification and fermentation: Process optimization and kinetic assessment. *Renewable Energy, 146*, 1031–1040.

Chua, E. T., & Schenk, P. M. (2017). A biorefinery for Nannochloropsis: Induction, harvesting, and extraction of EPA-rich oil and high-value protein. *Bioresource Technology, 244*, 1416–1424.

Chunyan, T., Liu, Z., & Zhang, Y. (2017). Hydrothermal Liquefaction (HTL): A promising pathway for biorefinery of algae, *Algal Biofuels: Recent advances and future prospects Recent advances and future prospects*, 361–391.

Dahman, Y., et al. (2019). Biofuels: Their Characteristics and Analysis. In: Biomass, Biopolymer-based Materials, and Bioenergy. Elsevier, Woodhead Publishing series in composites, science and engineering, pp. 277–325.

Daroch, M., Geng, S., & Wang, G. (2013). Recent advances in liquid biofuel production from algal feedstocks. *Applied Energy, 102*, 1371–1381.

Davis, R., Aden, A., & Pienkos, P. T. (2011). Techno-economic analysis of autotrophic microalgae for fuel production. *Applied Energy, 88*(10), 3524–3531.

Davis, R., et al. (2020). Conceptual Basis and Techno-Economic Modeling for Integrated Algal Biorefinery Conversion of Microalgae to Fuels and Products (2019 NREL TEA Update: Highlighting Paths to Future Cost Goals via a New Pathway for Combined Algal Processing). National Renewable Energy Lab. (NREL), Golden, CO (United States); DWH ....

de Boer, K., et al. (2012). Extraction and conversion pathways for microalgae to biodiesel: A review focused on energy consumption. *Journal of Applied Phycology, 24*(6), 1681–1698.

DeRose, K., et al. (2019). Integrated techno economic and life cycle assessment of the conversion of high productivity, low lipid algae to renewable fuels. *Algal Research, 38*, 101412.

Dong, T., et al. (2013). Two-step in situ biodiesel production from microalgae with high free fatty acid content. *Bioresource Technology, 136*, 8–15.

Dong, T., et al. (2016). Combined algal processing: A novel integrated biorefinery process to produce algal biofuels and bioproducts. *Algal Research, 19*, 316–323.

Dourou, M., et al. (2018). Fish farm effluents are suitable growth media for Nannochloropsis gaditana, a polyunsaturated fatty acid producing microalga. *Engineering in Life Sciences, 18*(11), 851–860.
Duan, P., et al. (2018). Integration of hydrothermal liquefaction and supercritical water gasification for improvement of energy recovery from algal biomass. *Energy, 155*, 734–745.
Ehimen, E. A., Sun, Z. F., & Carrington, C. G. (2010). Variables affecting the in situ transesterification of microalgae lipids. *Fuel, 89*(3), 677–684.
Fasaei, F., et al. (2018). Techno-economic evaluation of microalgae harvesting and dewatering systems. *Algal Research, 31*, 347–362.
Fortier, M.-O. P., et al. (2014). Life cycle assessment of bio-jet fuel from hydrothermal liquefaction of microalgae. *Applied Energy, 122*, 73–82.
Frank, E. D., et al. (2013). Life cycle comparison of hydrothermal liquefaction and lipid extraction pathways to renewable diesel from algae. *Mitigation and Adaptation Strategies for Global Change, 18*(1), 137–158.
Fu, X., et al. (2013). A microalgae residue based carbon solid acid catalyst for biodiesel production. *Bioresource Technology, 146*, 767–770
Girisuta, B., et al. (2013). A kinetic study of acid catalysed hydrolysis of sugar cane bagasse to levulinic acid. *Chemical Engineering Journal, 217*, 61–70.
Giuliano, A., et al. (2020). Valorization of OFMSW digestate-derived syngas toward methanol, hydrogen, or electricity: Process simulation and carbon footprint calculation. *Processes, 8*(5), 526.
Gnansounou, E., & Kenthorai Raman, J. (2016). Life cycle assessment of algae biodiesel and its co-products. *Applied Energy, 161*, 300–308.
Gollakota, A. R. K., Kishore, N., & Gu, S. (2018). A review on hydrothermal liquefaction of biomass. *Renewable and Sustainable Energy Reviews, 81*, 1378–1392.
Goswami, R. D., & Kalita, M. (2011). Scenedesmus dimorphus and Scenedesmus quadricauda: Two potent indigenous microalgae strains for biomass production and CO2 mitigation – A study on their growth behavior and lipid productivity under different concentration of urea as nitrogen source. *Journal of Algal Biomass Utilization, 2*(4), 2–4.
Goswami, R. K., et al. (2021). Advanced microalgae-based renewable biohydrogen production systems: A review. *Bioresource Technology, 320*, 124301.
Hannon, M., et al. (2010). Biofuels from algae: Challenges and potential. *Biofuels, 1*(5), 763–784.
Harun, R., et al. (2011). Technoeconomic analysis of an integrated microalgae photobioreactor, biodiesel and biogas production facility. *Biomass and Bioenergy, 35*(1), 741–747.
Harvey, P. J., & Ben-Amotz, A. (2020). Towards a sustainable Dunaliella salina microalgal biorefinery for 9-cis β-carotene production. *Algal Research, 50*, 102002.
He, S., et al. (2020). Hydrothermal liquefaction of low-lipid algae Nannochloropsis sp. and Sargassum sp.: Effect of feedstock composition and temperature. *Science of the Total Environment, 712*, 135677.
Hernández, D., et al. (2014). Biofuels from microalgae: Lipid extraction and methane production from the residual biomass in a biorefinery approach. *Bioresource Technology, 170*, 370–378.
Hu, Y., et al. (2017). Investigation of aqueous phase recycling for improving bio-crude oil yield in hydrothermal liquefaction of algae. *Bioresource Technology, 239*, 151–159.
IEA. (2021). Net Zero by 2050. Paris.
IEA. 2021b. *Renewables 2021*, Paris.
Im, H., et al. (2014). Concurrent extraction and reaction for the production of biodiesel from wet microalgae. *Bioresource Technology, 152*, 534–537
Jablonský, M., et al. (2018). Extraction of value-added components from food industry based and agro-forest biowastes by deep eutectic solvents. *Journal of Biotechnology, 282*, 46–66.
Jatmiko, T., et al. (2019). Nutritional Evaluation of Ulva sp. from Sepanjang Coast, Gunungkidul, Indonesia. In: IOP Conference Series: Earth and Environmental Science. IOP Publishing.
Jazrawi, C., et al. (2015). Two-stage hydrothermal liquefaction of a high-protein microalga. *Algal Research, 8*, 15–22.

Jeevan Kumar, S. P., et al. (2017). Sustainable green solvents and techniques for lipid extraction from microalgae: A review. *Algal Research*, *21*, 138–147.

Jiang, F., et al. (2020). Effect of hydraulic retention time on anaerobic baffled reactor operation: Enhanced biohydrogen production and enrichment of hydrogen-producing acetogens. *Processes*, *8*(3), 339.

Jindal, M. K., & Jha, M. K. (2016). Effect of process parameters on hydrothermal liquefaction of waste furniture sawdust for bio-oil production. *RSC Advances*, *6*(48), 41772–41780.

Karemore, A., & Sen, R. (2016). Downstream processing of microalgal feedstock for lipid and carbohydrate in a biorefinery concept: A holistic approach for biofuel applications. *RSC Advances*, *6*, 29486–29496.

Kargbo, H., Harris, J. S., & Phan, A. N. (2021). "Drop-in" fuel production from biomass: Critical review on techno-economic feasibility and sustainability. *Renewable and Sustainable Energy Reviews*, *135*, 110168.

Katiyar, R., Banerjee, S., & Arora, A. (2021). Recent advances in the integrated biorefinery concept for the valorization of algal biomass through sustainable routes. *Biofuels, Bioproducts and Biorefining*, *15*(3), 879–898.

Khetkorn, W., et al. (2017). Microalgal hydrogen production – A review. *Bioresource Technology*, *243*, 1194–1206.

Kim, B., Chang, Y. K., & Lee, J. W. (2017). Efficient solvothermal wet in situ transesterification of Nannochloropsis gaditana for biodiesel production. *Bioprocess and Biosystems Engineering*, *40*(5), 723–730.

Kim, B., et al. (2017). Catalyst-free production of alkyl esters from microalgae via combined wet in situ transesterification and hydrothermal liquefaction (iTHL). *Bioresource Technology*, *244*(Pt 1), 423–432.

Kumar, G., et al. (2019). A comprehensive review on thermochemical, biological, biochemical and hybrid conversion methods of bio-derived lignocellulosic molecules into renewable fuels. *Fuel*, *251*, 352–367.

Kumar, R., & Strezov, V. (2021). Thermochemical production of bio-oil: A review of downstream processing technologies for bio-oil upgrading, production of hydrogen and high value-added products. *Renewable and Sustainable Energy Reviews*, *135*, 110152.

Kumar, R., Ghosh, A. K., & Pal, P. (2020). Synergy of biofuel production with waste remediation along with value-added co-products recovery through microalgae cultivation: A review of membrane-integrated green approach. *Science of the Total Environment*, *698*, 134169.

Lavarack, B. P., Griffin, G. J., & Rodman, D. (2000). Measured kinetics of the acid-catalysed hydrolysis of sugar cane bagasse to produce xylose. *Catalysis Today*, *63*(2), 257–265.

Li, Y., et al. (2011). One-step production of biodiesel from Nannochloropsis sp. on solid base Mg-Zr catalyst. *Applied Energy*, *88*(10), 3313–3317.

Lin, C.-Y., & Ma, L. (2020). Influences of water content in feedstock oil on burning characteristics of fatty acid methyl esters. *Processes*, *8*(9), 1130.

Liu, F., et al. (2019). Development of a closed-loop process for fusel alcohol production and nutrient recycling from microalgae biomass. *Bioresource Technology*, *283*, 350–357.

Luo, D., et al. (2010). Life cycle energy and greenhouse gas emissions for an ethanol production process based on blue-green algae. *Environmental Science and Technology*, *44*(22), 8670–8677.

Luo, J., Fang, Z., & Smith, R. L. (2014). Ultrasound-enhanced conversion of biomass to biofuels. *Progress in Energy and Combustion Science*, *41*, 56–93.

Lupatini, A. L., et al. (2017). Potential application of microalga Spirulina platensis as a protein source. *Journal of the Science of Food and Agriculture*, *97*(3), 724–732.

Ma, G., et al. (2015). In situ heterogeneous transesterification of microalgae using combined ultrasound and microwave irradiation. *Energy Conversion and Management*, *90*, 41–46.

Ma, J., et al. (2020). Magnetic flocculation of algae-laden raw water and removal of extracellular organic matter by using composite flocculant of Fe3O4/cationic polyacrylamide. *Journal of Cleaner Production*, *248*, 119276.

Mahmud, R., et al. (2021). Integration of techno-economic analysis and life cycle assessment for sustainable process design – A review. *Journal of Cleaner Production*, *317*, 128247.

Manisali, A. Y., Sunol, A. K., & Philippidis, G. P. (2019). Effect of macronutrients on phospholipid production by the microalga Nannochloropsis oculata in a photobioreactor. *Algal Research, 41*, 101514.

Manoharan, Y., et al. (2019). Hydrogen fuel cell vehicles; current status and future prospect. *Applied Sciences, 9*(11), 2296.

Marzorati, S., et al. (2020). Carotenoids, chlorophylls and phycocyanin from Spirulina: Supercritical CO2 and water extraction methods for added value products cascade. *Green Chemistry, 22*(1), 187–196

Masoumi, S., & Dalai, A. K. (2021). Techno-economic and life cycle analysis of biofuel production via hydrothermal liquefaction of microalgae in a methanol-water system and catalytic hydrotreatment using hydrochar as a catalyst support. *Biomass and Bioenergy, 151*, 106168.

Matter, I. A., et al. (2019). Flocculation harvesting techniques for microalgae: A review. *Applied Sciences, 9*(15), 3069.

Moazeni, F., Chen, Y.-C., & Zhang, G. (2019). Enzymatic transesterification for biodiesel production from used cooking oil, a review. *Journal of Cleaner Production, 216*, 117–128.

Mona, S., et al. (2021). Towards sustainable agriculture with carbon sequestration, and greenhouse gas mitigation using algal biochar. *Chemosphere, 275*, 129856.

Nagarajan, S., et al. (2013). An updated comprehensive techno-economic analysis of algae biodiesel. *Bioresource Technology, 145*, 150–156.

Nitsos, C., et al. (2020). Current and novel approaches to downstream processing of microalgae: A review. *Biotechnology Advances, 45*, 107650.

Noorman, H. J., et al. (2018). Chapter 1, Intensified Fermentation Processes and Equipment. In: Intensification of Biobased Processes. The Royal Society of Chemistry, pp. 1–41, Green Chemistry Series, published18th June 2018, DOI:10:1039/9781788010320.

Ou, L., et al. (2015). Techno-economic analysis of transportation fuels from defatted microalgae via hydrothermal liquefaction and hydroprocessing. *Biomass and Bioenergy, 72*, 45–54.

Park, J. Y., et al. (2015). Advances in direct transesterification of algal oils from wet biomass. *Bioresource Technology, 184*, 267–275.

Park, J., et al. (2017). Wet in situ transesterification of microalgae using ethyl acetate as a co-solvent and reactant. *Bioresource Technology, 230*, 8–14.

Passos, F., et al. (2014). Pretreatment of microalgae to improve biogas production: A review. *Bioresource Technology, 172*, 403–412.

Patil, P. D., et al. (2018). Extraction of bio-oils from algae with supercritical carbon dioxide and co-solvents. *The Journal of Supercritical Fluids, 135*, 60–68.

Patle, D. S., et al. (2021). Ultrasound-intensified biodiesel production from algal biomass: A review. *Environmental Chemistry Letters, 19*(1), 209–229.

Ponnusamy, S., et al. (2014). Life cycle assessment of biodiesel production from algal bio-crude oils extracted under subcritical water conditions. *Bioresource Technology, 170*, 454–461.

Prapaiwatcharapan, K., et al. (2015). Single- and two-step hydrothermal liquefaction of microalgae in a semi-continuous reactor: Effect of the operating parameters. *Bioresource Technology, 191*, 426–432.

Pratto, B., et al. (2020). Experimental optimization and techno-economic analysis of bioethanol production by simultaneous saccharification and fermentation process using sugarcane straw. *Bioresource Technology, 297*, 122494.

Prieto, C. V. G., et al. (2017). Optimization of an integrated algae-based biorefinery for the production of biodiesel, astaxanthin and PHB. *Energy, 139*, 1159–1172.

Qu, Z., et al. (2018). Hydrothermal cell disruption of Nannochloropsis sp. and its influence on lipid extraction. *Algal Research, 35*, 407–415.

Rahman, A., et al. (2020). A review of algae-based produced water treatment for biomass and biofuel production. *Water, 12*(9), 2351.

Ramli, R. N., Lee, C. K., & Kassim, M. A.(2020). Extraction and characterization of starch from microalgae and comparison with commercial corn starch. *IOP Conference Series: Materials Science and Engineering, 716*, 012012.

Ratomski, P., & Hawrot-Paw, M. (2021). Influence of nutrient-stress conditions on Chlorella vulgaris biomass production and lipid content. *Catalysts, 11*(5), 573.

Ritchie, H., Roser, M., & Rosado, P. (2020). Energy. *Our World in Data*, https://ourworldindata.org/energy.

Roberts, G. W., et al. (2013). Promising pathway for algal biofuels through wastewater cultivation and hydrothermal conversion. *Energy and Fuels, 27*(2), 857–867.

Rokicka, M., et al. (2020). Effects of ultrasonic and microwave pretreatment on lipid extraction of microalgae and methane production from the residual extracted biomass. *BioEnergy Research, 14*, 752–760.

Ruangsomboon, S. (2012). Effect of light, nutrient, cultivation time and salinity on lipid production of newly isolated strain of the green microalga, Botryococcus braunii KMITL 2. *Bioresource Technology, 109*, 261–265.

Salosso, Y. (2019). Nutrient and alginate content of macroalgae Sargassum sp. from Kupang Bay waters, East Nusa Tenggara, Indonesia. *Aquaculture, Aquarium, Conservation and Legislation, 12*(6), 2130–2136.

Sander, K., & Murthy, G. S. (2010). Life cycle analysis of algae biodiesel. *The International Journal of Life Cycle Assessment, 15*(7), 704–714.

Sano Coelho, R., et al. (2021). Techno-economic assessment of heterotrophic microalgae biodiesel production integrated with a sugarcane bio-refinery. *Biofuels, Bioproducts and Biorefining, 15*(2), 416–429.

Santos, C. I., et al. (2018). Integrated 1st and 2nd generation sugarcane bio-refinery for jet fuel production in Brazil: Techno-economic and greenhouse gas emissions assessment. *Renewable Energy, 129*, 733–747.

Sathish, A., Smith, B. R., & Sims, R. C. (2014). Effect of moisture on in situ transesterification of microalgae for biodiesel production. *Journal of Chemical Technology and Biotechnology, 89*(1), 137–142.

Sereewatthanawut, I., et al. (2008). Extraction of protein and amino acids from deoiled rice bran by subcritical water hydrolysis. *Bioresource Technology, 99*(3), 555–561.

Shahi, T., et al. (2020). Bio-oil production from residual biomass of microalgae after lipid extraction: The case of Dunaliella Sp. *Biocatalysis and Agricultural Biotechnology, 23*, 101494.

Shanmugam, S., et al. (2021). Recent developments and strategies in genome engineering and integrated fermentation approaches for biobutanol production from microalgae. *Fuel, 285*, 119052.

Shao, Y., et al. (2020). Acidic seawater improved 5-hydroxymethylfurfural yield from sugarcane bagasse under microwave hydrothermal liquefaction. *Environmental Research, 184*, 109340.

Sialve, B., Bernet, N., & Bernard, O. (2009). Anaerobic digestion of microalgae as a necessary step to make microalgal biodiesel sustainable. *Biotechnology Advances, 27*(4), 409–416.

Singh, A., & Olsen, S. I. (2011). A critical review of biochemical conversion, sustainability and life cycle assessment of algal biofuels. *Applied Energy, 88*(10), 3548–3555.

Solé-Bundó, M., et al. (2019). Co-digestion strategies to enhance microalgae anaerobic digestion: A review. *Renewable and Sustainable Energy Reviews, 112*, 471–482.

Somers, M. D., & Quinn, J. C. (2019). Sustainability of carbon delivery to an algal biorefinery: A techno-economic and life-cycle assessment. *Journal of CO2 Utilization, 30*, 193–204.

Sun, A., et al. (2011). Comparative cost analysis of algal oil production for biofuels. *Energy, 36*(8), 5169–5179.

Sun, L., et al. (2020). Methane activation and utilization: Current status and future challenges. *Energy Technology, 8*(8), 1900826.

Suutari, M., et al. (2015). Macroalgae in biofuel production. *Phycological Research, 63*(1), 1–18.

Tang, S., et al. (2019). Hydrotreatment of biocrudes derived from hydrothermal liquefaction and lipid extraction of the high-lipid Scenedesmus. *Green Chemistry, 21*(12), 3413–3423.

Tang, X., et al. (2016). Element and chemical compounds transfer in bio-crude from hydrothermal liquefaction of microalgae. *Bioresource Technol*, *202*, 8–14

Taylor, B., et al. (2013). Techno-economic assessment of carbon-negative algal biodiesel for transport solutions. *Applied Energy*, *106*, 262–274.

Thangaraj, B., et al. (2018). Catalysis in biodiesel production – A review. *Clean Energy*, *3*(1), 2–23.

Tibbetts, S. M., Milley, J. E., & Lall, S. P. (2015). Chemical composition and nutritional properties of freshwater and marine microalgal biomass cultured in photobioreactors. *Journal of Applied Phycology*, *27*(3), 1109–1119.

Torres, S., et al. (2017). Direct transesterification of microalgae biomass and biodiesel refining with vacuum distillation. *Algal Research*, *28*, 30–38.

Usami, R., Fujii, K., & Fushimi, C. (2020). Improvement of bio-oil and nitrogen recovery from microalgae using two-stage hydrothermal liquefaction with solid carbon and HCl acid catalysis. *ACS Omega*, *5*(12), 6684–6696.

Usman, M., Adeosun, S., & Osifeso, G. (2012). Optimum calcium carbonate filler concentration for flexible polyurethane foam composite. *Journal of Minerals and Materials Characterization and Engineering*, *11*, 311–320.

Vasistha, S., et al. (2021). Current advances in microalgae harvesting and lipid extraction processes for improved biodiesel production: A review. *Renewable and Sustainable Energy Reviews*, *137*, 110498.

Velasquez, S., Lee, J., & Harvey, A. (2012). Alkaline in situ transesterification of Chlorella vulgaris. *Fuel*, *94*, 544–550.

Verdier, S., et al. (2021). Pilot-scale hydrotreating of catalytic fast pyrolysis biocrudes: Process performance and product analysis. *Sustainable Energy and Fuels*, *5*(18), 4668–4679.

Wang, S., et al. (2022). Integrated microalgal biorefinery – Routes, energy, economic and environmental perspectives. *Journal of Cleaner Production*, *348*, 131245.

Ward, A. J., Lewis, D. M., & Green, F. B. (2014). Anaerobic digestion of algae biomass: A review. *Algal Research*, *5*, 204–214.

WBA. (2019). *Global Bioenergy Statistics 2019*.

Wiatrowski, M., et al. (2022). Techno-economic assessment for the production of algal fuels and value-added products: Opportunities for high-protein microalgae conversion. *Biotechnology for Biofuels and Bioproducts*, *15*(1), 8.

Xu, D., & Savage, P. E. (2015). Effect of reaction time and algae loading on water-soluble and insoluble biocrude fractions from hydrothermal liquefaction of algae. *Algal Research*, *12*, 60–67.

Yoon, J., et al. (2018). Kinetics of the hydrolysis of xylan based on ether bond cleavage in subcritical water. *The Journal of Supercritical Fluids*, *135*, 145–151.

Yu, G., et al. (2011). Distributions of carbon and nitrogen in the products from hydrothermal liquefaction of low-lipid microalgae. *Energy and Environmental Science*, *4*(11), 4587–4595.

Yu, X., et al. (2020). Suitability of energy sources for automotive application – A review. *Applied Energy*, *271*, 115169.

Zamalloa, C., et al. (2011). The techno-economic potential of renewable energy through the anaerobic digestion of microalgae. *Bioresource Technology*, *102*(2), 1149–1158.

Zhang, B., & Ogden, K. (2017). Recycled wastewater from anaerobic digestion of lipid extracted algae as a source of nutrients. *Fuel*, *210*705–712.

Zhou, W., et al. (2014). Environment-enhancing algal biofuel production using wastewaters. *Renewable and Sustainable Energy Reviews*, *36*, 256–269.

Zhu, G., et al. (2013). Reducing sugars production from sugarcane bagasse wastes by hydrolysis in sub-critical water. *Clean Technologies and Environmental Policy*, *15*(1), 55–61.

Aparna Gautam, Sushil Kumar, Dipesh S. Patle*

Chapter 18
Process intensification opportunities in the production of microalgal biofuels

Abstract: A significant population boom has resulted in a massive increase in energy consumption, necessitating the search for alternative energy sources that may meet the current demand, both quantitatively and qualitatively. One of the renewable feedstocks is microalgae, which contains essential components, including carbohydrates, lipids, and proteins that can be utilized to produce a variety of sustainable biofuels. Microalgae can grow fast on nonarable land over the years without compromising with the food supply chain and also help to prevent global warming by capturing CO_2 from the environment. The major hurdle in biofuels commercialization is the availability of raw material that can be overcome by using microalgae as feedstock. Also, low-culture cell density, inefficient harvesting, and downstream processes, are currently impeding large-scale production of algal biofuels. Process intensification encompasses utilizing scientific and engineering principles to create unique equipment and processing methods that enhance energy efficiency, economic profitability, environmental impact, safety, and overall process footprint. At the industrial scale, the type of microalgae species, growth systems, pH, temperature, rate of nutrient transfer, nutrient quality, aeration, transesterification conditions, catalysts, etc. are crucial factors in biofuel production. Therefore, intensification of many of the above aspects can yield favorable outcomes. This chapter mainly focuses on the various process intensification techniques that may be employed at different steps in biofuel production. Also, the effects of using these process intensification techniques on biofuel production are thoroughly discussed in this chapter.

Keywords: microalgae, process intensification, bioethanol, biodiesel, biogas, bio-oil

18.1 Introduction

The massive growth in population has resulted in a huge increase in energy consumption, necessitating the search for alternative fuels that are relatively inexpensive, renewable, and meet the current demand, both quantitatively and qualitatively. Also,

***Corresponding author: Dipesh S. Patle**, Department of Chemical Engineering, Motilal Nehru National Institute of Technology Allahabad, Prayagraj 211004, Uttar Pradesh, India,
e-mail: dipesh-patle@mnnit.ac.in
Aparna Gautam, Sushil Kumar, Department of Chemical Engineering, Motilal Nehru National Institute of Technology Allahabad, Prayagraj 211004, Uttar Pradesh, India

https://doi.org/10.1515/9783110781267-018

excessive usage of fossil fuels has caused an unexpected price increase and greenhouse gas emissions (Saravanan et al., 2021; Kumar et al., 2020). So, bioenergy production from biomass has recently received considerable attention because of the abundance of biomass, finite supply of fossil fuels, escalating levels of carbon dioxide (CO_2), air pollutants, and other greenhouse gases in the atmosphere (Kumar et al., 2018; Thakur et al., 2018). Microalgae, which belong to the category of third-generation feedstock, possess considerable amount of carbohydrates, lipids, and proteins, and utilizes sunlight and carbon dioxide to increase the biomass (Panahi et al., 2019). Various kinds of biofuels, like bioethanol, biodiesel, bio-butanol, biochar, biogas, and bio-oil, are produced from algal biomass.

The major bottleneck is the high cost of production of biofuels from microalgae, which can be mitigated by assessing biorefinery strategies that generates multiple products. Process intensification (PI) can lead to significant savings by reducing the amount of materials used and by lowering the operating temperature (Joshi and Gogate, 2018). While employing different kinds of process intensification approaches in order to produce competitive products, there are some crucial aspects in both the upstream and downstream side that must be considered during large-scale manufacturing. PI's ultimate purpose is to create processes that are relatively greener, safer, simpler, and more energy-efficient (Patil et al., 2020).

The current chapter focuses on the various process intensification techniques employed at the different steps of biofuels production. The benefits of using these process intensification techniques in biofuel production in terms of enhancement in performance are thoroughly discussed in this chapter.

18.1.1 Need for biofuels

Biofuels refers to energy-rich compounds that can be solid, liquid, or gas, produced by biochemical processes or acquired from biological biomass like microalgae, plants, and bacteria. These are employed for a number of reasons, such as energy security, environmental issues, foreign exchange savings, and regional socioeconomic challenges (Demirbas, 2008). One of the primary reasons for the use of biofuel is that it can be utilized in modern engines, variety of vehicles, and infrastructure without requiring major modifications. Biofuel is similar to petroleum diesel fuel in terms of storage, combustion, and pumping. It is also safe to use in blended or pure form (Festel, 2008). Disruption of fossil fuels supplies, rising energy prices, and inadequate fuel sources are just a few of the numerous threats to energy security (Ayoo, 2020). Also, investment in biofuels will lead to an increase in economic growth. This implies that the sector will provide more employment and new revenue streams for workers and farmers. Developing countries profit from the increase in global energy demand due to economic expansion (Subramaniam and Masron, 2021). If biofuel is produced and use sufficiently, it will reduce a considerable quantity of greenhouse gas emission from the transportation sector, which is responsible for more than a fifth of the total greenhouse gas emissions. Any country's economy will come to a halt without a consistent supply of affordable energy to run power plants, transportation, and heat homes (Bhat et al., 2009). This opens up the possibility of addressing some of the most imperative issues we face today, such as fuel quality and emissions (Moschini et al., 2012).

18.1.2 Microalgae as feedstock for biofuel production

Microalgae have lately sparked a lot of attention throughout the world because of their extensive applications in sustainable power, biopharmaceutical, and nutraceutical industries. These are single or multicellular photosynthetically driven living organisms that are found in a humid environment and float in a variety of water, e.g., ground water, salt water, and wastewater that is exposed to sunlight or artificial light (Hossain et al., 2019). They may grow alone or in symbiosis with other species and can endure a warmer climate, range of salt concentration, different pH values, varying light intensities, and in environments such as reservoirs or deserts (Evangelista et al., 2008). The primary advantages of algae in comparison with plants are rapid growth rate, large CO_2 absorption capacity, and low nutrient requirements, making them a promising technology for the future. Algae can be categorized as *Rhodophyta* (red algae), *Phaeophyta* (brown algae), and *Chlorophyta* (green algae) (Khan et al., 2018).

Some microalgae species along with their hydrocarbon content (Lipid, protein, and carbohydrates) and class are shown in Table 18.1. Microalgae, with significant amount of protein, carbohydrate, and fat content, are used as feed for producing different kinds of biofuels. Due to their 40–70% protein contents, *Spirulina*, *Scenedesmus*, and *Chlorella* sp. are known as single-cell protein sources. Carbohydrates in the microalgae biomass are composed of sugars, cellulose, starch, and some other polysaccharides. Lipids consist of triglycerides, diglycerides, monoglycerides, glycerol, and fatty acids that can be esterified/transesterified to alkyl esters (i.e., biodiesel). Of all the lipids, triglycerides are abundant in cyanobacteria, accounting up to 80% of the total lipid content (Roy and Pal, 2015).

Table 18.1: Protein, carbohydrate, and lipid contents of few microalgae species, with their class and naturally found color (González-Delgado and Kafarov, 2011; Saifullah et al., 2014; Roy and Pal, 2015; Patle et al., 2020).

Microalgae	Class and color	Lipid (%)	Protein (%)	Carbohydrates (%)
Anabaena sp.	Cyanophyceae (Blue-green)	4–7	48	25–30
Chlorella sp.	Trebouxiophyceae (Green)	14–22	51–58	12–17

Table 18.1 (continued)

Microalgae	Class and color	Lipid (%)	Protein (%)	Carbohydrates (%)
Chlamydomonas sp.	Chlorophyceae (Green)	14–22	43–56	2.9–17
Dunaliella sp.	Chlorophyceae (Green)	6–8	49–57	4–32
Euglena sp.	Euglenoidea (Green/Red)	14–20	39–61	14–18
Enteromorpha intestinalis	Ulvophyceae (Green)	7.13	6.15	30.58
Spirulina platensis	Cyanophyceae (Blue-green)	4–9	50–65	8–14
Porphyridium sp.	Porphyridiophyceae (Red)	–	28–39	50–57
Spirulina maxima	Cyanophyceae (Blue-green)	6–7	60–71	13–16
Lola capillaris	Cyanophyceae (Green)	4.05	40.87	22.32
Nannochloropsis sp.	Eustigmatophyceae (Yellow- green)	27	52.0	16
Tetraselmis sp.	Chlorodendrophyceae (Yellow-green)	16–45	52	15
Scenedesmus sp.	Chlorophyceae (Green)	12–14	50–56	10–52
Prymnesium parvum	Prymnesiophyceae (Golden)	22–38	28–45	25–33
Spirogyra sp.	Zygnematophyceae (Green)	11–21	6–20	33–64
Chlorococum sp.	Chlorophyceae (Green)	19.3	–	–
Schizocytrium sp.	Labyrinthulomycetes (White-red-orange)	50–77	–	–
Botrycoccus braunii	Trebouxiophyceae (Red-green)	25–75	–	–

18.1.2.1 Identification, isolation, and selection of microalgae strain

Biofuel production from microalgae is being examined on a large scale as a possible green energy source, from the economic and environmental aspect. Millions of microalgae species reside on the Earth, and according to an estimate, over 40,000 species have been discovered (Wang, 2010). Despite the fact that numerous microalgae species have been isolated and extracted for carbohydrates, proteins, and lipid production, there is still no knowledge of the species that produces the most as different species are supposed to thrive in different aquatic, geographic, and climatic conditions (Duong et al., 2012). Choosing an appropriate microalgae strain with high biomass productivity and lipid content is a crucial prerequisite for microalgae-based biofuel (Peng et al., 2020). Microalgae strains that have high potential of producing the desired component will also result in a high yield of the product. The respective strain should 1) be efficient at photosynthesizing, 2) consume CO_2 at a high rate, 3) have the ability to grow in a variety of waters at a fast growth rate, and 4) produce maximum lipids. Each crucial parameter must be carefully selected and optimized (Patil et al., 2020). Isolation is also a mandatory part in the selection of microalgae. Cell dilution, single-cell isolation, and automatic cell isolation techniques are mostly used for the isolation of microalgae species.

Microalgae biomass comprises proteins, carbohydrates, and lipids (oil) that are converted into a wide range of bioproducts, livestock feed, food supplements, biofertilizers, and biofuels. In a nutshell, the selection of a proper strain becomes absolutely crucial from an economic prospective and the type of end products (Chowdhury and Loganathan, 2019).

18.1.2.2 Cultivation

Microalgae cultivation is an essential part of biofuel production. After the selection of the most suitable strain, the cultivation environment, such as the microalgae's habitat (freshwater, saltwater, and wastewater), must be considered. Because specific varieties of water containing specific nutrients that may affect the growth of cell, the growth characteristics of microalgae and their elemental composition significantly depend on the cultivation conditions (Chew et al., 2018). There are four major types of cultivation conditions for microalgae: photoautotrophic, heterotrophic, mixotrophic, and photoheterotrophic (Chen et al., 2011). Microalgae grown in a photoautotrophic environment use sunlight as an energy source and inorganic carbon (e.g. CO_2) as a carbon source to produce chemical energy through photosynthesis (Huang et al., 2010). Under heterotrophic cultivation, microalgae species require organic carbon (glucose, acetate, glycerol, sucrose, galactose, lactose, etc.) as both the energy and carbon source under dark conditions (Liang et al., 2009). The mixotrophic cultivation of microalgae utilizes photosynthetic organic and inorganic (CO_2) compounds as a carbon source for growth. As a result, microalgae can live in both phototrophic and heterotrophic environments

(Brennan and Owende, 2010). Under photoheterotrophic cultivation, microalgae need light energy and they use sole organic compounds as a source of carbon (Cruz et al., 2018).

There are two algae cultivation systems deployed, based on the open pond (lakes, lagoons, and artificial ponds: raceways) and closed photobioreactor technologies. When compared to closed photobioreactors, open pond frameworks are less efficient in terms of biomass productivity due to losses in evaporation, fluctuation in growth media temperature, inefficient mixing, and CO_2 deficiency (Chen et al., 2011). Closed systems include tubular, flat plate, and column photobioreactors which overcome the limitations of open pond systems, such as pollution and contamination risk and are also more appropriate for sensitive strains due to proper control of contamination, resulting in a higher cell mass productivity (Chew et al., 2018).

In photobioreactors, carbon supply decoupling and mixing helps in the appropriate supply of CO_2 and removal of oxygen where process intensification strategies can be applied. Basically, the process intensification of a photobioreactor focuses on optimizing the level of light within the reactor, thus improving the gas exchange and ensuring uniformity of the level of nutrients throughout the cultivation medium to improve the biomass production. Carvalho et al. (2006) reported the use of hollow fiber membrane to address the problem of inefficient transfer of nutrients to an extent. Abbott et al. (2015) developed a photobioreactor by using oscillatory a baffled reactor, which increases the gas transfer, thus improving the overall economics of microalgae production.

The primary focus of process intensification in open type systems is to optimize the configuration of the ponds in order to lower the energy consumption throughout the agitation and mixing. Sompech et al. (2012) (re)designed a raceway pond by placing three flow detector baffles at each end, thereby preventing the dead zone. The configuration resulted in reduced energy consumption (up to 20%) compared to the standard configuration (from 3,464 to 2,852 W). Hybrid cultivation system, also known as two-stage cultivation, is a relatively new approach to overcome the drawbacks of open and closed cultivation systems. This two-stage cultivation usually produces enormous biomass in closed photobioreactors under a controlled environment, followed by a transfer of this biomass to a nutrient-depleted open raceway. The cultivation of microalgae can also be made economically sustainable using low-cost feedstock like agro-industrial waste water and flue gases as nutrients (Patil et al., 2020).

18.1.2.3 Harvesting

Harvesting is the process of separating microbial biomass from the cultivated froth and then transforming it into the desired product using the steps of biomass recovery, dewatering, and drying. Figure 18.1 shows the harvesting techniques for microalgae. Physical harvesting methods for microalgae dewatering can be categorized as: sedimentation, filtration, centrifugation, and floatation. Chemical methods (coagulation/flocculation) seem to offer the best chances for larger-scale microalgae harvesting. In these methods, the harvested biomass sinks to the bottom of the cultivating apparatus due to its high density (Milledge et al., 2013). Electrolytic methods are also used to avoid the use of expensive and hazardous chemicals (Uduman et al., 2010).

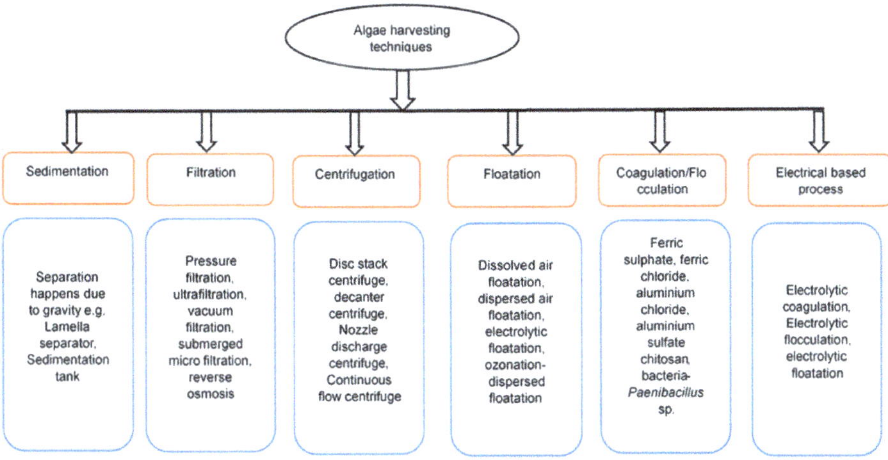

Figure 18.1: Microalgae harvesting techniques (Khan et al., 2021).

The harvesting process must be effective for high biomass concentration, considering aspects such as low operational cost, less energy requirement, and lowest maintenance cost. Upon selecting a suitable harvesting method, cell size, cell density, and the total quantity of the product to be separated must be considered (Joshi and Gogate, 2018). After the cell harvesting step, the obtained thick algal slurry is further processed in the drying step, which is used to prolong the effectiveness of the biomass and of the desired product. Sun drying, drum drying, spray drying, rotary drying, freeze drying, fluidized bed drying, and refractance window technology are methods that have been used for biomass drying (Brennan and Owende, 2010).

Advanced harvesting techniques and application of process intensification include sedimentation assisted by polymers (Zheng et al., 2015), flocculation assisted by magnetic microparticles (Vergini et al., 2016), electro-coagulation-filtration (ECF) (Gao et al., 2010), magnetic membrane filtration (Bilad et al., 2013), and electrochemical harvesting (ECH) (Misra et al., 2015). Low-frequency ultrasound may also be used on growing microalgae cells to reduce their buoyancy and promote sedimentation, resulting in a harvesting efficiency of 90–92% (Kim et al., 2013). Depending on the type of the intended product or cell wall structure, cell disruption may occur after the drying process.

18.1.2.4 Cell disruption

Microalgae are unicellular organisms with a complex cell wall that protects the plasma membrane. The chemical components, intramolecular and intermolecular connections, and overall structure of the algal cell walls vary considerably. Lipid, cellulose, protein, glycoprotein, and polysaccharide are the most prevalent components of microalgae cell walls (Lee et al., 2017).

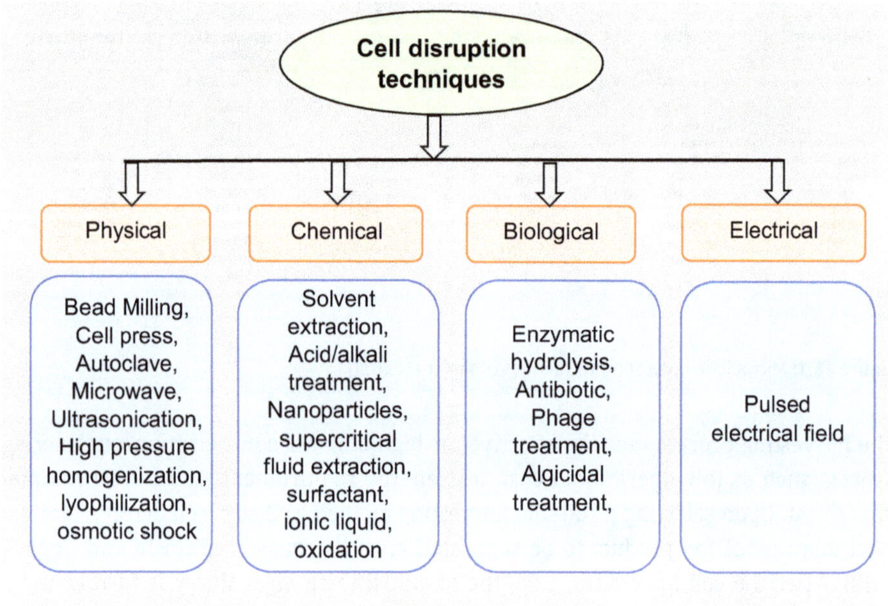

Figure 18.2: Microalgae cell disruption techniques.

The key biochemical products utilized in biofuel production are usually situated inside the cellular structures of microalgae. So, cell disruption becomes a crucial downstream processing step as it releases the stated intracellular products from the cellular matrix of microalgae aiming to a high yield and lower product contamination (Halim et al., 2012). There are several cell disruption techniques presented in Figure 18.2. Physical treatment methods rely on the energy, and cause cell wall disruption. These methods are less dependent on microalgae species, and they show low product contamination. However, when compared to chemical and biological pretreatments, they require more advanced equipment and larger energy inputs for processing, and the resulting heat can degrade the final products (D'Hondt et al., 2017).

Zou et al. (2021) utilized a combined form of bead milling and solvent extraction, aiding ultrasonication, for the purpose of lipid extraction. They found that in comparison with the sole ultrasound-assisted approach, the disruption efficiency was considerably better with relatively smaller particle size. In the microwave (frequency of 2,450 MHz) cell disruption treatment, local heating occurs because of inter and intramolecular movement, causing an efficient cell lysis and release of the cellular content (Dvoretsky et al., 2017). Sandani et al. (2022) adopted an advanced Electro-Fenton process (EFP) as a promising cell disruption technique to extract lipids from the wet biomass of microalgae *Chlorella homosphaera*. They were able to get 18.29% lipid yield in 40 min time from biomass having a concentration of 4.38 g/L. For the recovery of oil and protein from wet microalgae, Halim et al. (2022) used a novel cell disruption technique combination: autolytic incubation and hypotonic osmotic shock, combined with high pressure homogenization (HPH) (Mechanical method) or pH 12 (Chemical method). HPH combinations only cause physical disruption and could only release soluble proteins, whereas pH 12 combinations allowed chemical links to be hydrolyzed, and could release both soluble and structural proteins. Safi et al. (2014) used a variety of cell disruption techniques to release soluble proteins, including manual grinding, ultrasonication, alkaline treatment, and high-pressure treatments. Recent studies (Yao et al., 2018; Zinkoné et al., 2018; Zhang et al., 2019; González-González et al., 2019; Tavanandi et al., 2020) have focused on the new technology advancements, parameters optimization, pretreatments in conjunction with the disruption method, environmental impact assessment, and analysis of costs for recovering cellular components from microalgae.

18.1.2.5 Conversion

The essential components of microalgae biomass are lipids, proteins, and carbohydrates, which determine their effectiveness. They also contain organic (C, O, N) and inorganic

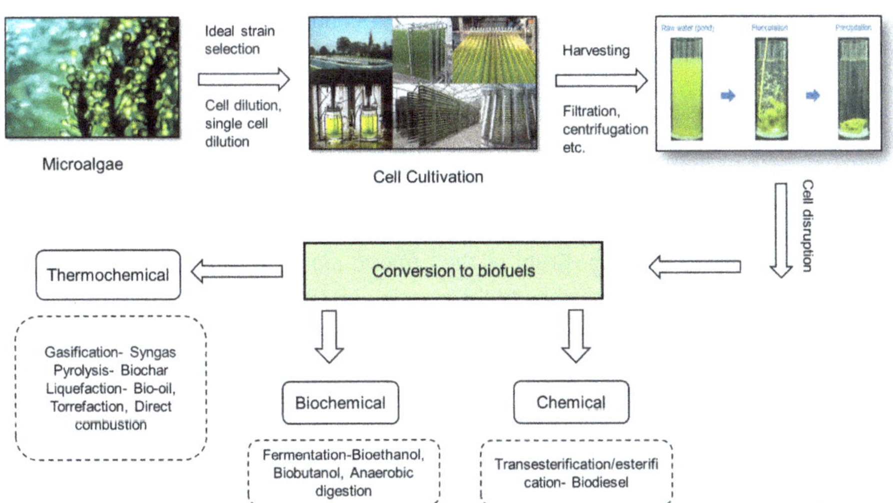

Figure 18.3: Conversion of microbial biomass.

elements, known as ultimate composition. The chemical composition of each species differs from one another, and depends on the environment in which they are cultivated.

Microalgae grown in waste water contains huge amount of inorganic elements like nitrogen, phosphorous, potassium, sodium, magnesium, calcium, and silicon as major elements and zinc, molybdenum, aluminum, copper, iron and manganese as minor elements, as compared to those species that grow in fresh water (Hossain, 2019). The conversion method must be selected in such a way that the production of biofuel must be economically feasible and environmental friendly. The conversion process is affected by a variety of parameters, including composition of biomass, targeted product, process time, operational conditions and economics. Microalgae conversion methods are categorized as biochemical, chemical, and thermochemical as shown in Figure 18.3. Among the thermochemical conversion techniques, direct combustion, torrefaction, hydrothermal liquefaction, gasification, and pyrolysis are propitious due to their higher yield and also the value-added by-product formation using suitable catalyst and high temperature to produce char, oil, and gas. Thermal-based processes, such as torrefaction and hydrothermal carbonization, are classified as low-temperature processes ($T < 300$ °C), whereas gasification, combustion, and pyrolysis are classified as high-temperature processes ($T > 300$ °C) (Quereshi et al. 2021). Direct combustion of biomass for biofuel production usually occurs in the temperature range from 1,000 to 2,000 °C in the presence of air (Hossain et al., 2019).

Gasification is the conversion of carbonaceous materials into clean fuel gases or synthetic gases. This can be accomplished by processing the biomass in the temperature range (800–1,000 °C) without combustion and with controlled amount of air. Initially, the breakdown of biomass occurs thermochemically; later, it produces synthesis gas (H_2: 6–55%, CO: 8–53%, and CH_4: 2–26%). There are three stages involved in microalgae biomass gasification: drying, devolatilization, and combustion. There are numerous chemical reactions involved in the gasification as described below:

$$C + H_2O \rightarrow CO + H_2 \tag{18.1}$$

$$C + CO_2 \rightarrow 2CO \tag{18.2}$$

$$CO + H_2O \rightarrow H_2 + CO_2 \tag{18.3}$$

Also, by the thermal decomposition of the organic materials, methane is also produced during gasification, according to the reaction (4) (Raheem et al., 2015):

$$CO_2 + 4H_2 \rightarrow CH_4 + 2H_2O \tag{18.4}$$

Pyrolysis is one of the most extensively employed thermochemical process for degrading components like cellulose, hemicellulose, and lignin, which results in the formation of biofuel via endothermic reaction in the absence of oxygen (Osman et al., 2021). Microalgae pyrolysis is further categorized, according to use of the carrier gas, catalyst, and heating route: slow, fast, catalytic, microwave, and hydropyrolysis (Azizi et al., 2018). Slow pyrolysis is carried out by burning the biomass slowly under a lower heating rate

5–50 °C/ min and with residence time more than 10 s in the absence of air. In fast pyrolysis, the material is quickly heated (heating rate > 103 °C/min) and with a shortened contact time (up to 3 s). Catalytic pyrolysis process requires the presence of catalysts like zeolite, Cu/Al_2O_3, Co/Mo/Z, Fe_2O_3, and Ni-CaO-C to enhance the selectivity toward the desired products. Microwave pyrolysis converts biomass into chemical energy by using microwave as the input energy (Azizi et al., 2018; Osman et al., 2021). Hydrothermal liquefaction processes occur in the presence of solvents such as sub or supercritical water, organic solvents and a combination of water & organic solvents, and operates at mild operating parameters such as temperature range of 250–500 °C, 5–35 MPa, and the contact time of 5–60 min (Yang and Yang, 2019; Akalin et al., 2017). Another thermochemical process is torrefaction, which is an endothermic pathway requiring temperature in the range from 200–300 °C and heating rate less than 50 °C/min in the absence of air. Volatilization, polymerization, and carbonization are three fundamental reactions that occur during the torrefaction of biomass, and the torrefied product is known as bio-coal or biochar (Chen et al., 2021).

Yeast microalgae biomass can be converted via biochemical ways such as fermentation using microbes, anaerobic digestion, and biomass degradation photobiologically (Pittman et al., 2011). These types of conversion techniques require specific microorganisms and enzymes to produce particular biofuels from biomass. However, despite being ecologically friendly, these conversion methods are usually not used in industries because of the tedious reaction steps, lower conversion efficiency, and high operating costs (Suali and Sarbatly, 2012). Microalgae biomass generates biohydrogen, methane, methanol, and syngas by processes such as anaerobic digestion and fermentation. The fermentation process converts organic substrates like sucrose, starch, cellulose, and bagasse into ethanol through microbial activities. Those microalgae species that contain a high amount of carbohydates are suitable to produce bioethanol. Sugar components such as agar, alginate, sulphated polysaccharide, and mannitol are broken down into ethanol and carbon dioxide by yeast as per the reaction given below (Choo et al., 2020):

$$C_6H_{12}O_6 \rightarrow 2C_2H_5OH + 2CO_2 \tag{18.5}$$

Biobutanol is produced by the fermentation (ABE: Acetone-Butanol-Ethanol) of biomass where bacterium (e.g., C. acetobutylicum) is introduced for anaerobic digestion (Yeong et al., 2018). ABE fermentation consists of two major processes, acidogenesis, which happens during the logarithmic phase of bacterial growth, and solventogenesis, which starts after the logarithmic phase (Abomohra et al., 2019). Anaerobic digestion (AD) is the process of converting organic matter into biogas, primarily CH_4 and CO_2, in the absence of oxygen or with the help of hydrogen. Hydrolysis, fermentation, and methanogenesis are three successive steps in the anaerobic digestion process. The first stage involves the hydrolysis of polysaccharides into simple sugars, and then acidogenesis and acetogenesis (production of acetic acid and carbon dioxide) occur. Then, the methanogenesis step produces CH_4 and CO_2 up to 70% (v/v) and 30%, respectively; it also produces some by-product gases like NH_3 and H_2S (Antoni et al., 2007; Choo et al., 2020).

The characteristics of the feedstocks are one of the primary issues with biochemical conversion processes. So, improvements in feedstocks that require pretreatment operations need significant capital cost, which raise the manufacturing cost. Furthermore, the biochemical conversion of biomass generally shows lower efficiency. Also, unlike the thermochemical pathways that might generate different kinds of end products from a single process, the biochemical pathway produces only a single end product (Ahorsu et al., 2018).

Depending on the feedstock composition, the chemical conversion of microbial biomass can be accomplished through transesterification or esterification reactions. The transesterification reaction converts lipids (mainly triglycerides, 90–98%) present in the biomass into glycerine and fatty acid alkyl ester (biodiesel), catalyzed by acid/base or enzymatic catalysts. Due to the reversible nature of transesterification reaction, it requires an excess amount of alcohol to shift the reaction equilibrium toward the product formation. Direct transesterification is a fast and simple method that simultaneously carries out lipid extraction and transesterification in a single stage. By combining the aforementioned steps, the extraction efficiency improves along with the FAME yield, compared to the conventional two-step technique (Pragya et al., 2013; Gautam et al., 2021).

The next section explains the numerous intensification techniques, including their underlying concepts and mechanisms, and how they might be useful in biofuel production.

18.2 Process intensification techniques

The major challenges associated with the conventional biodiesel manufacturing processes, such as prolonged reaction time, high manufacturing costs, high energy demand, and low efficiency in downstream processing, may be addressed using a variety of novel process intensification technologies (Maddikeri et al., 2012). Process intensification (PI) technologies target process miniaturization, energy efficiency, environmental safety, product quality improvement, and capital cost reduction. The goal of process intensification technologies is to formulate and design multipurpose devices for heat and mass transfer operation, separation of the desired products, and extraction and mixing operations (Kumar and Nigam, 2012). Figure 18.4 lists several important process intensifications techniques that can be used at different stages of biofuel production.

Microwave energy, which possesses wavelength in the range of 1 nm to 1 m, directly transfers energy to the reactants by two mechanism, namely, ionic conduction and dipolar polarization that help to accelerate the reaction. In the dipolar polarization method, when a substance with a dipole moment is exposed to EMR (Electromagnetic radiation), it rotates to align with the electromagnetic field; the resulting rotation in molecules lead to the generation of heat. Ionic conduction is another dissipation mechanism in which the dissolved ions migrate due to an oscillating electric field, causing frictional losses and generating heat. Microwave-assisted technologies

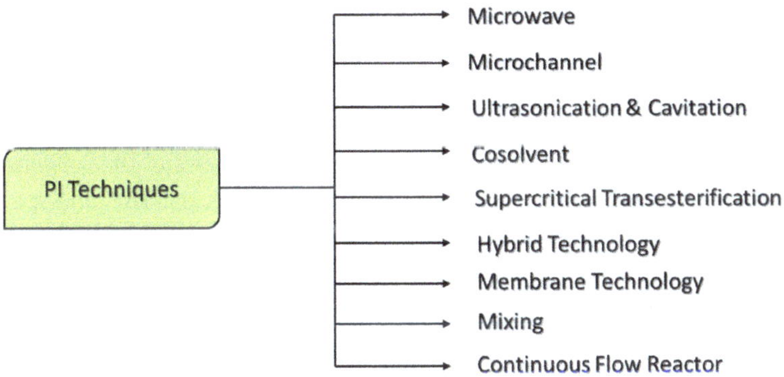

Figure 18.4: Some process intensification techniques.

improve the reaction selectivity, yield, and reduce the reaction time (Qiu et al., 2010; Kumar and Nigam, 2012; Muley and Bolder, 2018).

Microchannels are now gaining importance in chemical processes due to their considerable high transport rates, short residence time, and compact design that require small amount of reagents and catalyst and provides a sense of safety for intensely exothermic and endothermic reactions. Microchannels reactors have channels with diameter in the range of 10–500 μm and a larger surface area-to-volume ratio (10,000–50,000 m^2/m^3) that provide excellent heat and mass transfer (Pohar and Plazl, 2009; Natarajan et al., 2019). Cavitational reactors works by two mechanisms: ultrasonication (utilize acoustic energy) and hydrodynamic cavitation (utilise flow energy). Cavities emerge due to the decline in the static pressure of the liquid below the liquid's vapor pressure when subjected to high pressure. These small cavities (bubbles or voids filled with vapor) collapse and generate huge magnitude of energy. In ultrasonication, sonication is carried out in the reaction vessel and an ultrasonication probe is responsible for creating the ultrasound waves. Hydrodynamic cavitation involves the formation of cavities when liquid passes through constricted devices such as a valve, an orifice plate, venture, etc. Physical phenomena such as liquid streaming, associated with intense turbulence, as well as chemical effects, such as the production of hot spots and reactive free radicals, lead to the intensification of different physical and chemical processing applications (Gogate et al., 2006; Gogate, 2008). Solvents play a crucial role in the extraction of carbohydrates and lipids from microalgae and in their transformation. In the transesterification reaction, alcohol shows less solubility in the lipid/oil reactant, causing mass transfer limitation and poor biodiesel yield. To enhance the reaction efficiency and to overcome the mass transfer limitation, different cosolvents such as tetrahydrofuran, cyclohexane, diethyl ether, ether, toluene, and diethyl ether are used. By incorporating the solvent into the transesterification reaction step, the reaction mixture transforms from a single phase to a two-phase system because of the increase in the mutual solubility of alcohol and oil at low reaction temperatures, based on the concept of "like to dissolve like" (Pena et al., 2008; Jiang and Tan, 2012). Supercritical fluid extractions (SFEs) have emerged as viable alternatives to traditional organic solvent-based extraction techniques with regard to health and safety concerns and environmental consequences. Supercritical fluid is a highly compressible fluid, with characteristics of both gases and liquids at a temperature and pressure above its critical point. The reactor pressure and temperature are adjusted in a supercritical fluid process to impact the thermophysical characteristics of the solvents (ethanol, water, methanol, etc.), such as viscosity, dielectric constant, polarity, and specific gravity. SCFs (CO_2, ethane, methanol, and propane) also allow rapid mass transfer and adjustable solvent power (Harris et al., 2018).

The employment of a combined or a sequential approach of multiple intensification techniques, such as microwave, ultrasound, or hydrodynamic cavitation, may sometimes considerably speed up chemical processes and other processing applications. The remarkable acceleration effect might be attributed to a combination of microwave-induced forced heat transfer and ultrasonic or hydrodynamic cavitation-induced intense mass transfer at phase boundaries (Onumaegbu et al., 2018; Patle et al., 2020). The development of a membrane reactor might be a feasible option for resolving the constraints of conventional biodiesel production processes. In the membrane reactor, reaction and separation steps proceed in a single chamber, ensuring that the reversible reaction shifts in the forward direction with an effective separation of the desired products from the reaction mixture, resulting in an increased yield. The simultaneous withdrawal of a product from a reaction mixture improves the reaction rates, and the permeate produced is in pure form, which requires less processing downstream (Cao et al., 2007; Dube et al. 2007).

Mixing is one of the fundamental unit operations for enhancing the reaction yield and separation efficiency. Proper mixing is critical for green synthesis since it intensifies the chemical reactions and the synthesis processes. Proper mixing reduces the concentration and temperature gradients while also preventing secondary and runaway reactions. Numerous mixers are used for biofuel production such as impinging jet mixer, vortex mixer, static mixer, oscillating baffled reactor, and coiled flow inverter (Kumar and Nigam, 2012). The continuous microwave reactor (CMR) and ultrasonic reactor (CUR) were designed by Payakkawan et al. (2014) for biomass carbonization and by Cintas et al. (2010) for synthesis of biodiesel. The application of process intensification technologies at various steps of the biofuel production processes, such as bioethanol/biobutanol, biodiesel, and biogas/biohydrogen, are discussed in the following section.

18.3 Process intensification techniques applied to microalgae fuel

18.3.1 Bioethanol/ biobutanol

Bioethanol is gaining popularity as a way to reduce the use of gasoline. It can be made from renewable biomass sources that contain sugar and starch components and it offers a number of advantages, including the fact that it is biodegradable. Bioethanol production from microalgae does not have the same food-versus-fuel issue as the first-generation bioethanol. Also, it can be cost-effective to produce bioethanol due to the simplified pretreatment and hydrolysis processes. The development of suitable hydrolysis processes is crucial in bioethanol production for a cost effective, simple, and higher yield of reducing sugar (Patil et al. 2020).

Table 18.2: Process intensification techniques applied at the different stages of bioethanol/ biobutanol production.

Microalgae	PI technique	Step	Condition	Performance improvement	References
Chlorella vulgaris	Ultrasonication	Carbohydrates extraction	320 W, 20 min	Increase in carbohydrate conversion efficiency up to 99.5%	Nasirpour et al., 2022
Chlorella vulgaris	Microwave	Hydrolysis	170 °C, 10 min	Increase in reducing sugar content from 7.31 to 98.11 g/L	Yu et al., 2020
Chlorococum sp.	SCF	Lipid extraction	60 °C, 400 mL/ min CO_2	After lipid extraction, microalgae shows ethanol concentration of 3.83 g/L	Harun et al., 2010
Chlorella sorokiniana	Chemo-enzymatic hydrolysis	Pretreatment	HMF: 0.106 ± 0.042	Increment in reducing sugar yield by about one-third	Constantino et al., 2021
Chlorococcum sp.	Bead-beating	Cell disruption	4 min., 1:3 beads to culture	Average disruption (about 17.5%) of the initial intact cells	Halim et al., 2012
Chlorella sp.	Chemical	Hydrolysis (HCl and $MgCl_2$)	30 °C, 40 h	sugar concentration is 12%, of sugar recovery (about 83%)	Zhou et al., 2011
Dunaliella sp.	Autoclave and chemical	Pretreatment (H_2SO_4)	1% H_2SO_4, 121 °C, 15 min	10.7 times increment in bioethanol yield, to 7.26 g/L	Karatay et al., 2016

Table 18.2 (continued)

Microalgae	PI technique	Step	Condition	Performance improvement	References
Schizocytrium sp.	Hydrothermal and enzymatic	Hydrothermal fractionation	Hydrothermal: 115.5 °C, 46.7 min, enzymatic: 350(GAU)/g glucoamylase	Maximum sugar yield of about 19.4 wt.%	Kim et al., 2012
Scenedesmus sp. WZKMT	Liquid hot water (LHW) and enzymatic	Pretreatment	LHW: 147 °C, 40 min and enzymatic: pH 4.5, 37 °C, 48 h	five-fold higher recovery of sugar than without LHW	Yuan et al., 2016
Chlorococum humicola	Ultrasonication and enzymatic	Cell disruption and hydrolysis	40 kHz, 130 W, for 25 min and 0.02 enzyme/g substrate	Glucose yield: 64.2% (w/w)	Harun and Danquah, 2011
Arthrospira platensis	Thermochemical	Pretreatment	0.1 mM H_2SO_4, 100 °C	positively influenced ethanol and butanol production	Efremenko et al., 2012
Laminaria digitata	Enzymatic hydrolysis	Hydrolysis	10% v/w, 24 h	Higher butanol: ABE molar ratio and high butanol yield	Hou et al., 2017
Chlamydomonas mexicana	Sonication	Pretreatment	15 min at 50 °C, 40 kHz	Residual sugar increases in comparison to no treatment	El-Dalatony et al., 2016
Chlamydomonas mexicana	Sonication with enzyme hydrolysis	Pretreatment	Cellulose > 1 U per mg solid	445 mg/g release of total reducing sugar	El-Dalatony et al., 2016
Chlamydomonas mexicana	Acidic hydrolysis and PBR	Pretreatment and cultivation	1 M H_2SO_4, 180 µmol m^{-2} s^{-1} Light intensity	high biomass productivity: 0.28 ± 0.001 g $L^{-1}d^{-1}$	Onay, 2018

The findings on process intensification-assisted bioethanol/biobutanol synthesis as well as its performance enhancement during biofuel production are included in Table 18.2. Nasirpour et al. (2022) investigated the pretreatment and carbohydrate hydrolysis of microalgae with the aid of ultrasonication. They found that ultrasonication-assisted acidic method gives a higher conversion efficiency (98.3%). Due to the absence of lignin in microalgae, it becomes easily hydrolysable. Generally, cells are disrupted in acidic conditions as done by Zhou et al. (2011). The authors reported a 83% yield for reducing sugars, containing xylose, glucose, and arabinose, and the cell disruption was carried out by the addition of 2% hydrochloric acid along with 2.5% magnesium chloride. Yu et al. (2020) utilized microalgae hydrolysate under diluted acidic pretreatment. Wet torrefaction was then done using the microwave technique to produce bioethanol at an yield of 0.0761 g ethanol/ (g microalgae). Harun et al. (2010) reported a supercritical extraction of lipid from microalgae cell as an efficient method; it was carried out at a temperature of 60 °C and 400 mL min^{-1} supply of CO_2 gas. The lipid extracted microalgae displayed a significantly higher ethanol concentration of 3.83 g/L. A chemo-enzymatic hydrolysis strategy for microalgae biomass was developed by Constantino et al. (2021). They performed acidic pretreatment at a high pressure and temperature, followed by incubation with enzymes (Amyloglucosidase and α-Amylase). Halim et al. (2012) studied high-pressure homogenization, ultrasonication, bead beating, and sulfuric acid treatment, among others, at laboratory scale. They have found that the high-pressure homogenization technique disrupted approximately 73.8% cells, followed by sulfuric acid treatment, which showed an average disruption of about 33.2% of the initial intact cells, whereas bead beating showed an average disruption of approximately 17.5% of the initial intact cells. Karatay et al. (2016) discovered that dilute acidic hydrolysis with 1% sulfuric acid in an autoclave was immensely effective in saccharifying algal biomass, yielding 7.26 g/L of bioethanol, which was 10.7-fold greater than the value obtained using microalgae biomass that had not been pretreated. Yuan et al. (2016) investigated the effects of liquid hot water (LHW) pretreatment on *Scenedesmus sp. WZKMT*. The enzymatic efficiency by LHW pretreatment was considerably improved, making it an attractive pretreatment approach for glucose recovery from microalgae. Genetic engineering in microalgae cellular structure also plays a significant role in bioethanol production from microalgae, and this can be achieved by genetic modifications. By enhancing the activity of enzymes (like alcohol dehydrogenase and pyruvate dehydrogenase), the stored carbon may be converted to ethanol. High carbohydrate accumulation has been accomplished in microalgae through process intensification techniques, but further study is required to convert it straight to ethanol. (De Farias Silva and Bertucco, 2016).

Biobutanol is one such impactful biofuel with properties similar to gasoline, allowing for easier public distribution using existing oil and gas infrastructure. Owing to a couple of advantages, biobutanol is viewed as a viable alternative to conventional fuels. It is also more effective as a biofuel than biomethanol or bioethanol due to its higher energy density and structural resemblance to gasoline (Yeong et al., 2018). Various carbohydrates such as monosaccharides, disaccharides, polysaccharides, or oligosaccharides are processed through ABE (acetone-biobutanol-ethanol) fermentation to produce biobutanol. ABE fermentation is carried out by *Clostridia species* in two stages: acidogenesis and solventogenesis. During acidogenesis, the carbohydrate-rich biomass is degraded into organic acids (such as acetic and butyric acid) and subsequently, the accumulated acids are transformed into the final ABE products.

In the fermentation of lignocellulosic or cellulosic microalgae biomass, saccharification is the primary step. A number of methods such as acid hydrolysis, alkaline hydrolysis, thermolysis and physical saccharification techniques (high-pressure homogenization, sonication, microwaving and heat) can be used to saccharify raw feedstocks using diluted or concentrated enzyme digestion. Enzymatic hydrolysis of the brown seaweed *Laminaria digitata* containing high sugar compounds was performed by Hou et al. (2017). They were able to get the maximum butanol yield (0.42 g/g-consumed-substrates) by adding Cellulases mixture (NS81016; Novozymes A/S), which contains a mixture of different -glucanases and -glucosidases. El-Dalatony et al. (2016) investigated the use of immobilized yeast for fermentation and sonication in conjunction with the enzymatic hydrolysis step. It was discovered that combining sonication with hydrolysis resulted in a higher yield of total reducing sugars of 445 mg/mg. It was also observed that SSF (Simultaneous saccharification fermentation) produced more ethanol than SHF (Separate hydrolysis fermentation), and that the energy recovery of the process was increased by the use of immobilized yeast cells. Efremenko et al. (2012) utilized immobilized *Clostridium acetobutylicum* cells, immobilized into poly (vinyl alcohol) cryogel, for ABE fermentation, and thermal decomposition technique was used to pretreat *Nannochloropsis sp.* immobilized biomass. Onay (2018) reported the use of waste water in flat photobioreactor to cultivate the *chlorella sp. DEE006*. They obtained both a high biobutanol content (6.23 ± 0.19 g/L) and a high bioethanol yield, 0.16 ± 0.005 g/g sugar.

18.3.2 Biodiesel

Biodiesel production through the transesterification reaction is one such method that can be performed using conventional methods like heating plates, oil or sand bath, and water heated jacketed reactor, coupled with mechanical mixing, which requires a longer reaction time. Novel techniques such as ultrasonication, microwave, and hydrodynamic cavitation can reduce the reaction time considerably under optimum conditions and enhance the biodiesel yield simultaneously, according to some studies that are summarized in Table 18.3.

A process intensification concept applied to biodiesel production processes must ensure a shorter reaction time, a low molar ratio of alcohol to oil, lowest catalyst concentration, high conversion efficiency, and most importantly a lower the operating costs and energy consumption for downstream processing of biodiesel and glycerol, catalyst separation, and enable excess alcohol recovery.

Microalgae have a thick cell wall that results in expensive cell disruption. To overcome the limitation associated with cell disruption, Zheng et al. (2011) intensified the process by using the bead-milling technique that involves the bead-beater chamber being filled with glass beads (0.4–0.6 mm size range) and rotating it at 1500 rpm. This

Table 18.3: Process ntensification used at different stages of biodiesel production.

Microalgae	PI technique	Step	Condition	Performance improvement	References
Chlorella Vulgaris	Bead- Milling	Cell disruption	Time = 20 min, 1500 rpm	Up to 10% lipid concentration greater than manual grinding	Zheng et al., 2011
Chlorella Vulgaris	Ultrasonication	Transesterification	Time = 20 min, 40% amplitude	85% biodiesel yield	Cercado et al., 2018
Chlorella pyrenoidosa	Hydrodynamic cavitation	Cell disruption	Time = 180 min, 5 bar	40% reduction treatment time than ultrasonication	Waghmare et al., 2019
Scenedesmus sp.	Microwave	Cell disruption	Time = 40 min, T = 95 °C	20.8% lipid yield from dry algae powder	Mamo and Mekonnen, 2020
Nannochloropsis sp.	Microwave + Ultrasound	Direct transesterification, Cultivation	T = 60 °C, 5 min.	Better biodiesel yield in single stage TE with MW and US than two-stage TE	Koberg et al., 2011
Nannochloropsis (CCMP1776)	SCF	In situ transesterification	P = 1200 psi, 250 °C, 25 min	Reduced energy consumption, from 107.3 MJ to 42.3 MJ/kg biodiesel	Patil et al., 2012
Chlorella pyrenoidosa	Organic solvents (2-methyltetrahydrofuran)	In situ transesterification	150 min,	Higher purity, 95–99%, with high % of PUFAs	de Jesus et al., 2020
Nannochloropsis sp.	Ionic liquid [EMIM][MeSO$_4$]	Lipid extraction and transesterification	25 min, MeOH:IL 1:0.5	42.22% biodiesel yield per dry biomass	Wahidin et al., 2018
Chlorella sp.	Continuous reactor (Packed bed)	Direct transesterification	100 °C, 30 min	Less residence time required, 30 min as compared to batch process (12 h)	Jazie et al., 2020
Chlorella sp.	HPH	Cell disruption	1,070 bar	50% cells ruptured per pass	Spiden et al., 2013

technique was found to be more efficient than manual grinding with quartz sand. Ultrasound-assisted transesterification reaction was carried out by Cercado et al. (2018). They prepared a heterogeneous catalyst called K-pumice, and the reaction completed in only 10 min to obtain 85% FAME yield; the time is significantly shorter than the reaction carried out without the use of ultrasonication. Waghmare et al. (2019) used hydrodynamic cavitation (HC) and ultrasonication (US) for microalgae cell disruption and found that HC was 6.3 times more energy efficient than US and it reduced the time by approximately 40%. A microwave digester–assisted solvent extraction method was used by Mamo and Mekonnen (2020) to recover algal oil from dried microalgae biomass. As a result, they were able to get approximately 20.8% algal oil per gram. Koberg et al. (2011) carried out direct conversion of microalgae into biodiesel using microwave and ultrasonication techniques. The authors used flue gas from a coal-burning power station to cultivate microalgae. These techniques were able to significantly reduce the costs related to algae production and its conversion. Furthermore, direct transesterification of the as-harvested *Nannochloropsis* biomass (a one-step process) yielded a higher biodiesel yield (32.8%) than a two-step process (18.9%). Patil et al. (2012) used the supercritical methanol method to perform the direct conversion of algal biomass into biodiesel. They observed that the SCM method produced more purified products, consumed less energy in the separation and purification step, and required minimal chemicals in the overall biodiesel production process. de Jesus et al. (2020) suggested the direct synthesis of biodiesel from *Chlorella pyrenoidosa* using cyclopentyl methyl ether as a green co-solvent to extract the lipids. This method resulted in higher biodiesel purity (95–99%) with a larger percentage of polyunsaturated fatty acid and higher biodiesel yield (67–92%). Direct transesterification of wet algal biomass was performed by Wahidin et al. (2018) under microwave radiation and Ionic liquid. The synergetic effect by combining IL-methanol solvent with microwave heating, under optimal reaction conditions, resulted in a high biodiesel yield of 42.22% per dry biomass. Jazie et al. (2020) investigated biodiesel production from *Chlorella sp.* oil using a dodecylbenzenesulfonic acid catalyst (DBSA) in a packed bed reactor with cylindrical glass raschig rings. Authors discovered that using the DBSA catalyst in a continuous packed bed reactor for only 30 min would make the process more cost effective than using sulfuric acid catalyst in a batch process that takes more than 12 h. For an efficient rupture of the cell wall of different microalgae species, Spiden et al. (2013) adopted a high-pressure homogenization technique. The rupturability of *Chlorella sp.*, *T. suecica* and *Nannochloropsis sp.* were compared to that of *S. cerevisiae.* It was found that *T. suecica* was most susceptible to rupture by high pressure homogenization, followed by *Chlorella sp.* and *S. cerevisiae.* The disruption of *Nannochloropsis sp.* was more difficult.

18.3.3 Biogas/ Biohydrogen

Biogas production is an anaerobic process in which specialized organisms help to decompose organic materials in order to produce a gas. It is primarily composed of CH_4 (60–70%) and CO_2 (20–40%), with minor amounts of H_2, N_2, water vapor, and H_2S. Complex compounds of the cell wall of microalgae like cellulose, hemicellulose, and pectin, cause bacterial attack difficult; so biomass pretreatment is required to make microalgae anaerobic digestion more feasible. Hence, research efforts are focused on pretreatment of substrates to increase the anaerobic biodegradability (Kendir and Ugurlu, 2018).

Table 18.4 shows that the intensification in physical, chemical, and biological processes improved microbial biomass breakdown and anaerobic biodegradability at the cell disruption and conversion step of biogas/biohydrogen production. Biological treatment of microalgae biomass was done by Passos et al. (2016). Cellulase, glucohydrolase, and xylanase were reported to be the most efficient enzyme combination for improving microalgae biomass solubilization and methane production. The enzyme combination increased biomass solubilization by 126% and methane yield by 15%, whereas cellulase increased biomass solubilization by 110% and methane yield by 8%. Varol and Ugurlu (2016) reported the use of *Spirulina platensis*, a blue–green algae and studied the continuous biogas production. Biogas production and volatile solids (VS) reductions were improved when waste sewage sludge was co-digested with *S. platensis*. With 2,880 mL/day total biogas production (640 mL biogas/g VS day), 62.5% VS reductions were achieved throughout this operation. Xiao et al. (2020) reported the use of hydrothermal pretreatment that improved the biogas production from microalgae biomass via anaerobic digestion. It was also reported that biogas production from microalgae with solar-driven HTP (hydrothermal pretreatment) obtained the lowest LCOE at \$0.17 per m^3 (levelized cost of energy), with biogas yield of 166.13 g $CO_{2\text{-eq}}$/(kWh biogas). To enhance the methane productivity, Alzate et al. (2014) performed ultrasonic pretreatment using the lipid extraction technique and the biomass was then thermally hydrolyzed to produce biogas. A novel technique, ozonation, in the pretreatment of biomass was investigated by Cardeña et al. (2017). They were able to get the highest methane production (432.7 mL CH_4/g VS_{algal}, equaling to 259.6 mL CH_4/g COD_{in}) using the highest dosage of ozone (382 mg O_3/g VS_{algal}) during retreatment. Another effective approach to enhance the methane productivity by anaerobic digestion is hydrothermal pretreatment (Xiao et al. 2019a). To improve the energy efficiency in hydrothermal pretreatment, a solar-driven hydrothermal retreatment system was proposed by the authors. Methane generated from microalgae biomass prepared by the solar-driven hydrothermal pretreatment system increased by 57% through anaerobic digestion. Later, Xiao et al. (2019b) performed an exergy analyses of biogas production from microalgae biomass via anaerobic digestion.

Table 18.4: Process intensification used at different stages of biogas/ biohydrogen production.

Microalgae	PI technique	Step	Condition	Performance improvement	References
Microalgae biomass	Enzymatic	Pretreatment	1% cellulase and other, 6 h	110% increment in biomass solubilization by cellulose and methane yield by 8%.	Passos et al., 2016
Spirulina platensis	Anaerobic digestion	Co-digestion (two-stage digestion)	36 °C, 200 rpm	Improved biogas production (2,880 mL/ day); 62.5% VS reduction	Varol and Ugurlu, 2016
Chlorella sp.	Anaerobic digestion	Solar driven hydrothermal pretreatment	160 °C, HRT 28 days	Less cost of energy than hydrothermal pretreatment	Xiao et al., 2020
Nannochloropsis gaditana	Ultrasonic	Pretreatment	10,000 kJ/kg TS	Enhanced methane productivity	Alzate et al., 2014
Scenedesmus sp.	Ozonation	Pretreatment	94 mgO_3/g VS	Methane yield increased by various extents (6–66%).	Cardeña et al., 2017
Chlorella pyrenoidosa	Hydrothermal	Pretreatment	160 °C, 2 Mpa	57% CH_4 production achieved, compared to without microalgae pretreatment	Xiao et al., 2019a
Eucheuma spinosum	Membrane bioreactor	Conversion	21.58 ± 1.59 L/L-d H_2, 3 h	High hydrogen production rate than previously reported	Sim et al., 2021
C. reinhardtii 137c	Torus shaped photobioreactor	Cultivation	110 µmol/ m^2/s, pH 7.5	Maximum H_2 production of about 2.5 mL l^{-1} h^{-1}	Fouchard et al., 2008
Chlorella sacchrarophila FACHB-4	Acid hydrolysis (1% H_2SO_4)	Dark fermentation and waste utilization	50 µmol/ m/s. 12/12 h light/dark, 7.0	Cultivation and wastewater treatment cost were reduced	Mu et al., 2020

Microalgae is a potential viable clean energy source to produce biohydrogen, which produces just water vapor as a by-product and does not contaminate the environment like fossil fuels do when burned. The two-stage method is used to produce hydrogen from microalgae: stage 1 for carbon fixation and stage 2 for anaerobic digestion and hydrogen production. Dark fermentation is one of the methods for biohydrogen production from microalgae. Mu et al. (2020) used duckweed as feedstock, which contains a high amount of starch to carry out dark fermentation, and fermentative waste was used to produce microalgae lipids. The diluted acid was used to pretreat the biomass.

The costs of microalgae cultivation and wastewater treatment were reduced as a consequence of the simultaneous generation of biohydrogen and waste utilization, resulting in a cost-effective and environmentally friendly technique for producing biofuels. Fouchard et al. (2008) used a torus-shaped photobioreactor to cultivate the microalgae under fully controlled condition. Recently, for a continuous biohydrogen production, Sim et al. (2021) used a dynamic membrane bioreactor fed by pretreated *Echeuma spinosum* with the highest average hydrogen production rate (HPR) of 21.58 ± 1.59 L/L-d at HRT 3 h, which was higher than previous reports for continuous H_2 production from biomass feedstock.

18.3.4 Bio-oil

Bio-oils produced from microalgae pyrolysis mostly comprise aliphatic and aromatic hydrocarbons, apart from oxygen- and nitrogen-containing compounds. Bio-oils derived from lignocellulosic sources are unstable, viscous, acidic, and contain oxygenates and solids in large quantities. Bio-oils made from highly proteinaceous microalgae are more stable, have a higher calorific value, and contain less oxygen. The existing approaches of converting microalgae to biofuels are oil extraction using organic solvents, hydrothermal liquefaction, and pyrolysis. However, these methods have a number of drawbacks, including a low conversion rate, a long conversion time, and high energy consumption. So, microwave-assisted pyrolysis (MAP) is a new conversion route that has recently gained a lot of attention. Table 18.5 presents the usage of process intensification techniques in bio-oil production from microalgae.

Xie et al. (2015) used a microwave-based system as a process intensification technique for catalytic co-pyrolysis of microalgae and scum to produce bio-oil. A HZSM-5 catalyst was used at an optimal temperature of 550 °C and microalgae to scum ratio of 1:2. Another process intensification technique, sub- and supercritical water liquefaction, was studied by Zou et al. (2010). The highest oil yield obtained was 36.9%, with a reaction temperature of 360 °C, holding time of 30 min, and feedstock ratio of materials to water of 1: 10. Bio-oil from the microalgae, *Dunaliella tertiolecta*, has been proposed as a potentially lucrative and environmentally sustainable feedstock candidate for biofuels and chemicals. A fast pyrolysis was adopted by Miao and Wu (2004) to produce bio-oil from heterotrophic *Chlorella protothecoides* microalgae species. They manipulated the metabolic pathway of microalgae and obtained 3.4 times higher bio-oil yield than that from autotrophic cells by fast pyrolysis. Heterotrophic growth was achieved in conventional microbial bioreactors, resulting in higher biomass yield and lower microalgae biomass production costs.

Table 18.5: Process intensification used in bio-oil production.

Microalgae	PI technique	Step	Condition	Performance improvement	References
Nannochloropsis sp.	Microwave	Co-pyrolysis	750 W, 2,450 MHz, 550 °C,	Max. 22 wt.% bio-oil yield	Xie et al., 2015
Dunaliella tertiolecta	Sub/ Super-critical water	Thermochemical liquefaction	360 °C, 30 min	Min. material input, energy efficient and 37% bio-oil yield	Zou et al., 2010
Chlorella protothecoides	Fast pyrolysis	Conversion	450 °C	3.4 times higher bio-oil (58%) yield than from autotrophic cells	Miao and Wu, 2004

Overall, this chapter reviews the process intensification aspects applied to biofuels production from microalgae. The first section summarizes the different varieties of microalgae and their identification, isolation and cultivation procedures, followed by a discussion of the various methods for harvesting and cell disruption, and then conversion to biofuels from microalgae. This chapter shows that process intensification techniques can be applied successfully to produce biofuels like bioethanol, biodiesel, biogas, biohydrogen, and bio-oil. A review of the literature is presented to provide guidelines for various intensification approaches and their operating parameters. The benefits that may be obtained are also highlighted. Overall, this chapter is expected to provide critical information about various intensification techniques and their projected advantages, in addition to detailing the information about the processes that may be useful to improve reactor designs for appropriately treating or utilizing the biomass.

18.4 Conclusion and recommendations

Biofuels derived from microalgae can be considered as a viable alternative to fossil fuels. To commercialize the biofuel production economically, currently, limited technologies are available. Therefore, improvements in current technological scenario are required to address issues related to the production and separation to obtain desired products that will result in a cost-effective process. The advantages of process intensification technology over commercial biofuel processes include feedstock flexibility, reaction homogenization, high conversion and product yield, and environmental friendliness. As shown in this chapter, applying process intensification approaches at various stages of processing can lead to an efficient process in terms of time and energy along with commercial viability. Biofuel production, facilitated by process intensification technologies like microwave and ultrasonication, were observed to be most energy- and time-efficient means. Ultrasound- and microwave-assisted intensification have a number of advantages in bioprocessing, such as better profitability and ease of use, when compared to their counterparts. The aforementioned process intensification techniques decrease the reaction time by 10–12 folds.

It is worth noting that the activities intended to increase the efficiency of bioprocesses are not isolated tasks as these involve a wide range of expertise. As a consequence, it is indeed critical to emphasize the formation of multidisciplinary team in order to improve biofuel production using intensified technologies, such as experts from process engineering, genetic engineering, biotechnology and chemical engineering. Further, model-based approaches can be utilized to produce biofuel in a systematic way with proper design, control, and optimization. Hybrid technologies for biofuel production may be a better option for sustainable development. Promising results have been obtained at a pilot scale and more focus is required for scale up of process intensification technologies to commercial level.

References

Abbott, M. S. R., Brain, C. M., Harvey, A. P., Morrison, M. I., & Valente, G. (2015). Liquid culture of microalgae in a photobioreactor (PBR) based on oscillatory baffled reactor (OBR) technology— A feasibility study. *Chemical Engineering Science*, *138*, 315–323.

Abomohra, A.E.F., Elshobary, M. (2019). Biodiesel, Bioethanol, and Biobutanol Production from Microalgae. In: Alam, M., Wang, Z. (Eds) *Microalgae Biotechnology for Development of Biofuel and Wastewater Treatment*. Springer, Singapore, 293–321.

Ahorsu, R., Medina, F., & Constantí, M. (2018). Significance and challenges of biomass as a suitable feedstock for bioenergy and biochemical production: A review. *Energies*, *11*(12), 3366.

Akalın, M. K., Tekin, K., & Karagöz, S. (2017). Supercritical fluid extraction of biofuels from biomass. *Environmental Chemistry Letter*, *15*(1), 29–41.

Alzate, M. E., Muñoz, R., Rogalla, F., Fdz-Polanco, F., & Pérez-Elvira, S. I. (2014). Biochemical methane potential of microalgae biomass after lipid extraction. *Chemical Engineering Journal*, *243*, 405–410.

Antoni, D., Zverlov, V. V., & Schwarz, W. H. (2007). Biofuels from microbes. *Applied Microbiology and Biotechnology*, *77*(1), 23–35.

Ayoo, C. (2020). Towards Energy Security for the Twenty-first Century. Energy Policy. IntechOpen, London.

Azizi, K., Moraveji, M. K., & Najafabadi, H. A. (2018). A review on bio-fuel production from microalgae biomass by using pyrolysis method. *Renewable and Sustainable Energy Reviews*, *82*, 3046–3059.

Bhat, I. K., & Prakash, R. (2009). LCA of renewable energy for electricity generation systems—A review. *Renewable and Sustainable Energy Reviews*, *13*(5), 1067–1073.

Bilad, M. R., Discart, V., Vandamme, D., Foubert, I., Muylaert, K., & Vankelecom, I. F. J. (2013). Harvesting microalgae biomass using a magnetically induced membrane vibration (MMV) system: Filtration performance and energy consumption. *Bioresource Technology*, *138*, 329–338.

Brennan, L., & Owende, P. (2010). Biofuels from microalgae—A review of technologies for production, processing, and extractions of biofuels and co-products. *Renewable and Sustainable Energy Reviews*, *14*(2), 557–577.

Cao, P., Tremblay, A. Y., Dubé, M. A., & Morse, K. (2007). Effect of membrane pore size on the performance of a membrane reactor for biodiesel production. *Industrial and Engineering Chemistry Research*, *46*, 52–58.

Carvalho, A.P., Meireles, L.A. and Malcata, F.X. (2006). Microalgal reactors: a review of enclosed system designs and performances. *Biotechnology progress*, *22*, 1490–1506.

Cardeña, R., Moreno, G., Bakonyi, P., & Buitrón, G. (2017). Enhancement of methane production from various microalgae cultures via novel ozonation pretreatment. *Chemical Engineering Journal*, *307*, 948–954.

Cercado, A. P., Ballesteros Jr, F., & Capareda, S. (2018). Ultrasound assisted transesterification of microalgae using synthesized novel catalyst. *Sustainable Environment Research*, *28*(5), 234–239.

Chen, C. Y., Yeh, K. L., Aisyah, R., Lee, D. J., & Chang, J. S. (2011). Cultivation, photobioreactor design and harvesting of microalgae for biodiesel production: A critical review. *Bioresource Technology*, *102*(1), 71–81.

Chen, W. H., Lin, B. J., Lin, Y. Y., Chu, Y. S., Ubando, A. T., Show, P. L., Ong, H. C., Chang, J. S., Ho, S. H., Culaba, A. B., & Pétrissans, A. (2021). Progress in biomass torrefaction: Principles, applications and challenges. *Progress in Energy and Combustion Science*, *82*, 100887.

Chew, K. W., Chia, S. R., Show, P. L., Yap, Y. J., Ling, T. C., & Chang, J. S. (2018). Effects of water culture medium, cultivation systems and growth modes for microalgae cultivation: A review. *Journal of the Taiwan Institute of Chemical Engineers*, *91*, 332–344.

Choo, M.-Y.; Oi, L.E.; Ling, T.C.; Ng, E.-P.; Lee, H.V.; Juan, J.C. (2020). Conversion of microalgae biomass to biofuels. In: Abu Yousuf (Eds) *Microalgae Cultivation for Biofuels Production*; Elsevier: Amsterdam, Netherlands, 149–161.

Chowdhury, H., & Loganathan, B. (2019). Third-generation biofuels from microalgae: A review. *Current Opinion in Green and Sustainable Chemistry*, *20*, 39–44.

Cintas, P., Mantegna, S., Gaudino, E. C., & Cravotto, G. (2010). A new pilot flow reactor for high-intensity ultrasound irradiation. Application to the synthesis of biodiesel. *Ultrasonics Sonochemistry*, *17*(6), 985–989.

Constantino, A., Rodrigues, B., Leon, R., Barros, R., & Raposo, S. (2021). Alternative chemo-enzymatic hydrolysis strategy applied to different microalgae species for bioethanol production. *Algal Research*, *56*, 102329.

Cruz, Y.R., Aranda, D., Seidl, P.R., Diaz, G.C., Carliz, R.G., Fortes, M.M., et al. (2018). Cultivation systems of microalgae for the production of biofuels. In: Biernat K, (Eds) *Biofuels-State of Development*, London: IntechOpen; 199–218.

D'Hondt, E., Martín-Juárez, J., Bolado, S., Kasperoviciene, J., Koreiviene, J., Sulcius, S., Elst, K., & Bastiaens, L. (2017). Cell disruption technologies. In: Cristina G-F., Raul M., (Eds) *Microalgae-Based Biofuels and Bioproducts*, Woodhead Publishing, UK, 133–154.

de Farias Silva, C. E., & Bertucco, A. (2016). Bioethanol from microalgae and cyanobacteria: A review and technological outlook. *Process Biochemical*, *51*(11), 1833–1842.

de Jesus, S. S., Ferreira, G. F., Moreira, L. S., & Maciel Filho, R. (2020). Biodiesel production from microalgae by direct transesterification using green solvents. *Renew Energy*, *160*, 1283–1294.

Demirbas, A. (2008). Biofuels sources, biofuel policy, biofuel economy and global biofuel projections. *Energy Conversion and Management*, *49*(8), 2106–2116.

Dube, M. A., Tremblay, A. Y., & Liu, J. (2007). Biodiesel production using a membrane reactor. *Bioresource Technology*, *98*, 639–647.

Duong, V. T., Li, Y., Nowak, E., & Schenk, P. M. (2012). Microalgae isolation and selection for prospective biodiesel production. *Energies*, *5*(6), 1835–1849.

Dvoretsky, D., Dvoretsky, S., Temnov, M., Akulinin, E., & Zuorro, A. (2017). The effect of the complex processing of microalgae Chlorella vulgaris on the intensification of the lipid extraction process. *Chemical Engineering Transactions*, *57*, 721–726.

Efremenko, E. N., Nikolskaya, A. B., Lyagin, I. V., Senko, O. V., Makhlis, T. A., Stepanov, N. A., Maslova, O. V., Mamedova, F., & Varfolomeev, S. D. (2012). Production of biofuels from pretreated microalgae biomass by anaerobic fermentation with immobilized Clostridium acetobutylicum cells. *Bioresource Technology*, *114*, 342–348.

El-Dalatony, M. M., Kurade, M. B., Abou-Shanab, R. A., Kim, H., Salama, E. S., & Jeon, B. H. (2016). Long-term production of bioethanol in repeated-batch fermentation of microalgae biomass using immobilized Saccharomyces cerevisiae. *Bioresource Technology*, *219*, 98–105.

Evangelista, V., Barsanti, L., Frassanito, A. M., Passarelli, V., & Gualtieri, P. (eds.) (2008). Algal Toxins: Nature, Occurrence, Effect and Detection. Springer Science & Business Media, Springer Dordrecht.

Festel, G. W. (2008). Biofuels-economic aspects. *Chemical Engineering and Technology*, *31*(5), 715–720.

Fouchard, S., Pruvost, J., Degrenne, B., & Legrand, J. (2008). Investigation of H2 production using the green microalga Chlamydomonas reinhardtii in a fully controlled photobioreactor fitted with on-line gas analysis. *International Journal of Hydrogen Energy*, *33*(13), 3302–3310.

Gao, S., Yang, J., Tian, J., Ma, F., Tu, G., & Du, M. (2010). Electro-coagulation-flotation process for algae removal. *Journal of Hazardous Material*, *177*, 336–343.

Gautam, A., Bhagat, P. R., Kumar, S., & Patle, D. S. (2021). Dry route process and wet route process for algal biodiesel production: A review of techno-economical aspects. *Chemical Engineering Research and Design*, *174*, 365–385.

Gogate, P. R. (2008). Cavitational reactors for process intensification of chemical processing applications: A critical review. *Chemical Engineering and Processing: Process Intensification*, *47*(4), 515–527.

Gogate, P. R., Tayal, R. K., & Pandit, A. B. (2006). Cavitation: A technology on the horizon. *Current Science*, *91*, 35–46.

González-Delgado, Á. D., & Kafarov, V. (2011). Microalgae based biorefinery: Issues to consider. *CT&F - Ciencia, Tecnología y Futuro*, *4*(4), 5–22.

González-González, L. M., Astals, S., Pratt, S., Jensen, P. D., & Schenk, P. M. (2019). Impact of osmotic shock pre-treatment on microalgae lipid extraction and subsequent methane production. *Bioresource Technology*, *7*, 100214.

Halim, R., Harun, R., Danquah, M. K., & Webley, P. A. (2012). Microalgae cell disruption for biofuel development. *Applied Energy*, *91*(1), 116–121.

Halim, R., Papachristou, I., Chen, G. Q., Deng, H., Frey, W., Posten, C., & Silve, A. (2022). The effect of cell disruption on the extraction of oil and protein from concentrated microalgae slurries. *Bioresource Technology*, *346*, 126597.

Harris, J., Viner, K., Champagne, P., & Jessop, P. G. (2018). Advances in microalgae lipid extraction for biofuel production: A review. *Biofuels, Bioproducts and Biorefining*, *12*(6), 1118–1135.

Harun, R., & Danquah, M. K. (2011). Enzymatic hydrolysis of microalgae biomass for bioethanol production. *Chemical Engineering Journal*, *168*(3), 1079–1084.

Harun, R., Danquah, M. K., & Forde, G. M. (2010). Microalgae biomass as a fermentation feedstock for bioethanol production. *Journal of Chemical Technology and Biotechnology*, *85*(2), 199–203.

Hossain, N., Mahlia, T. M. I., & Saidur, R. (2019). Latest development in microalgae-biofuel production with nano-additives. *Biotechnology Biofuels*, *12*(1), 1–16.

Hossain, S. Z. (2019). Biochemical conversion of microalgae biomass into biofuel. *Chemical Engineering and Technology*, *42*(12), 2594–2607.

Hou, X., From, N., Angelidaki, I., Huijgen, W. J., & Bjerre, A. B. (2017). Butanol fermentation of the brown seaweed Laminaria digitata by Clostridium beijerinckii DSM-6422. *Bioresource Technology*, *238*, 16–21.

Huang, G., Chen, F., Wei, D., Zhang, X., & Chen, G. (2010). Biodiesel production by microalgae biotechnology. *Applied Energy*, *87*(1), 38–46.

Jazie, A. A., Abed, S. A., Nuhma, M. J., & Mutar, M. A. (2020). Continuous biodiesel production in a packed bed reactor from microalgae Chlorella sp. using DBSA catalyst. *Engineering Science and Technology, an International Journal*, *23*(3), 642–649.

Jiang, -J.-J., & Tan, C.-S. (2012). Biodiesel production from coconut oil in supercritical methanol in the presence of cosolvent. *Journal of the Taiwan Institute of Chemical Engineers*, *2012*(43), 102–107.

Joshi, S., & Gogate, P. (2018). Process Intensification of Biofuel Production from Microalgae. In Energy from Microalgae. Springer, Cham.

Karatay, S. E., Erdoğan, M., Dönmez, S., & Dönmez, G. (2016). Experimental investigations on bioethanol production from halophilic microalgae biomass. *Ecological Engineering*, *95*, 266–270.

Kendir, E., & Ugurlu, A. (2018). A comprehensive review on pretreatment of microalgae for biogas production. *International Journal of Energy Research*, *42*(12), 3711–3731.

Khan, M. I., Shin, J. H., & Kim, J. D. (2018). The promising future of microalgae: Current status, challenges, and optimization of a sustainable and renewable industry for biofuels, feed, and other products. *Microbial Cell Factories*, *17*(1), 1–21.

Khan, S., Naushad, M., Iqbal, J., Bathula, C., & Sharma, G. (2021). Production and harvesting of microalgae and an efficient operational approach to biofuel production for a sustainable environment. *Fuel*, *311*, 122543.

Kim, J., Yoo, G., Lee, H., Lim, J., Kim, K., Kim, C. W., et al. (2013). Methods of downstream processing for the production of biodiesel from microalgae. *Biotechnology Advances*, *31*, 862–876.

Kim, J. K., Um, B. H., & Kim, T. H. (2012). Bioethanol production from micro-algae, Schizocytrium sp., using hydrothermal treatment and biological conversion. *Korean Journal of Chemical Engineering*, *29*(2), 209–214.

Koberg, M., Cohen, M., Ben-Amotz, A., & Gedanken, A. (2011). Bio-diesel production directly from the microalgae biomass of Nannochloropsis by microwave and ultrasound radiation. *Bioresource Technology*, *102*(5), 4265–4269.

Kumar, M., Sun, Y., Rathour, R., Pandey, A., Thakur, I. S., & Tsang, D. C. (2020). Algae as potential feedstock for the production of biofuels and value-added products: Opportunities and challenges. *Science of the Total Environment*, *716*, 137116.

Kumar, M., Sundaram, S., Gnansounou, E., Larroche, C., & Thakur, I. S. (2018). Carbon dioxide capture, storage and production of biofuel and biomaterials by bacteria: A review. *Bioresource Technology*, *247*, 1059–1068.

Kumar, V., & Nigam, K. D. P. (2012). Process intensification in green synthesis. pp.79–107.

Lee, S. Y., Cho, J. M., Chang, Y. K., & Oh, Y. K. (2017). Cell disruption and lipid extraction for microalgae biorefineries: A review. *Bioresource Technology*, *244*, 1317–1328.

Liang, Y., Sarkany, N., & Cui, Y. (2009). Biomass and lipid productivities of Chlorella vulgaris under autotrophic, heterotrophic and mixotrophic growth conditions. *Biotechnology Letters*, *31*(7), 1043–1049.

Maddikeri, G. L., Pandit, A. B., & Gogate, P. R. (2012). Intensification approaches for biodiesel synthesis from waste cooking oil: A review. *Industrial and Engineering Chemistry Research*, *51*(45), 14610–14628.

Mamo, T. T., & Mekonnen, Y. S. (2020). Microwave-assisted biodiesel production from microalgae, scenedesmus species, using goat bone–made nano-catalyst. *Biotechnology and Applied Biochemistry*, *190*(4), 1147–1162.

Miao, X., & Wu, Q. (2004). High yield bio-oil production from fast pyrolysis by metabolic controlling of Chlorella protothecoides. *Journal of Biotechnology*, *110*(1), 85–93.

Milledge, J. J., & Heaven, S. (2013). A review of the harvesting of micro-algae for biofuel production. *Reviews in Environmental Science BioTechnology*, *12*, 165–178.

Misra, R., Guldhe, A., Singh, P., Rawat, I., Stenstrom, T. A., & Bux, F. (2015). Evaluation of operating conditions for sustainable harvesting of microalgae biomass applying electrochemical method using non sacrificial electrodes. *Bioresource Technology*, *176*, 1–7.

Moschini, G., Cui, J., & Lapan, H. (2012). Economics of biofuels: An overview of policies, impacts and prospects. *Bio-based and Applied Economics*, *1*(3), 269–296.

Mu, D., Liu, H., Lin, W., Shukla, P., & Luo, J. (2020). Simultaneous biohydrogen production from dark fermentation of duckweed and waste utilization for microalgae lipid production. *Bioresource Technology*, *302*, 122879.

Muley, P., & Boldor, D. (2018). Process intensification and parametric optimization in biodiesel synthesis using microwave reactors. *Green Chemistry for Sustainable Biofuel Production*, *614*, 167–204.

Nasirpour, N., Ravanshad, O., & Mousavi, S. M. (2022). Ultrasonic-assisted acid and ionic liquid hydrolysis of microalgae for bioethanol production. *Biomass Conversion and Biorefinery*, 1–14.

Natarajan, Y., Nabera, A., Salike, S., Tamilkkuricil, V. D., Pandian, S., Karuppan, M., & Appusamy, A. (2019). An overview on the process intensification of microchannel reactors for biodiesel production. *Chemical Engineering and Processing: Process Intensification*, *136*, 163–176.

Onay, M. (2018). Investigation of biobutanol efficiency of Chlorella sp. cultivated in municipal wastewater. *Journal of Geosciences and Environment*, *6*(10), 40–50.

Onumaegbu, C., Mooney, J., Alaswad, A., & Olabi, A. G. (2018). Pre-treatment methods for production of biofuel from microalgae biomass. *Renewable and Sustainable Energy Reviews*, *93*, 16–26.

Osman, A. I., Mehta, N., Elgarahy, A. M., Al-Hinai, A., Al-Muhtaseb, A. A. H., & Rooney, D. W. (2021). Conversion of biomass to biofuels and life cycle assessment: A review. *Environmental Chemistry Letters*, *19*(6), 4075–4118.

Panahi, Y., Khosroushahi, A. Y., Sahebkar, A., & Heidari, H. R. (2019). Impact of cultivation condition and media content on Chlorella vulgaris composition. *Advanced Pharmaceutical Bulletin*, *9*(2), 182.

Passos, F., Hom-Diaz, A., Blanquez, P., Vicent, T., & Ferrer, I. (2016). Improving biogas production from microalgae by enzymatic pretreatment. *Bioresource Technology*, *199*, 347–351.

Patil, P. D., Gude, V. G., Mannarswamy, A., Cooke, P., Nirmalakhandan, N., Lammers, P., & Deng, S. (2012). Comparison of direct transesterification of algal biomass under supercritical methanol and microwave irradiation conditions. *Fuel*, *97*, 822–831.

Patil, R. A., Kausley, S. B., Joshi, S. M., & Pandit, A. B. (2020). Process Intensification Applied to Microalgae-based Processes and Products. In Handbook of Microalgae-Based Processes and Products. Academic Press, UK.

Patle, D. S., Shrikhande, S., & Rangaiah, G. P. (2020). Process development, design and analysis of microalgae biodiesel production aided by microwave and ultrasonication. In: Adrian B-P, GP Rangaiah, editor. *Process Systems Engineering for Biofuels Development*, Wiley, UK, 259–284.

Payakkawan, P., Areejit, S., & Sooraksa, P. (2014). Design, fabrication and operation of continuous microwave biomass carbonization system. *Renewable energy*, *66*, 49–55.

Peña, R., Romero, R., Martínez, S. L., Ramos, M. J., Martínez, A., & Natividad, R. (2008). Transesterification of castor oil: Effect of catalyst and co-solvent. *Industrial and Engineering Chemistry Research*, *48*, 1186–1189.

Peng, L., Fu, D., Chu, H., Wang, Z., & Qi, H. (2020). Biofuel production from microalgae: A review. *Environmental Chemistry Letter*, *18*(2), 285–297.

Pittman, J. K., Dean, A. P., & Osundeko, O. (2011). The potential of sustainable algal biofuel production using wastewater resources. *Bioresource Technology*, *102*(1), 17–25.

Pohar, A., & Plazl, I. (2009). Process intensification through microreactor application. *Chemical and Biochemical Engineering Quarterly*, *23*(4), 537–544.

Pragya, N., Pandey, K. K., & Sahoo, P. K. (2013). A review on harvesting, oil extraction and biofuels production technologies from microalgae. *Renewable and Sustainable Energy Reviews*, *24*, 159–171.

Qiu, Z., Zhao, L., & Weatherley, L. (2010). Process intensification technologies in continuous biodiesel production. *Chemical Engineering and Processing: Process Intensification*, *49*(4), 323–330.

Quereshi, S., et al. (2021, Jan 1). Overview of Sustainable Fuel and Energy Technologies. In: Dutta, S. & Mustansar Hussain, C. (Eds.), Sustainable Fuel Technologies Handbook. Academic Press, Cambridge, MA, USA, pp. 3–25.

Raheem, A., Azlina, W. W., Yap, Y. T., Danquah, M. K., & Harun, R. (2015). Thermochemical conversion of microalgae biomass for biofuel production. *Renewable and Sustainable Energy Reviews*, *49*, 990–999.

Roy, S. S., & Pal, R. (2015, June). Microalgae in aquaculture: A review with special references to nutritional value and fish dietetics. In Proceedings of the Zoological Society. Springer India, pp. 1–8.

Safi, C., Ursu, A. V., Laroche, C., Zebib, B., Merah, O., Pontalier, P. Y., & Vaca-Garcia, C. (2014). Aqueous extraction of proteins from microalgae: Effect of different cell disruption methods. *Algal Research*, *3*, 61–65.

Saifullah, A. Z. A., Karim, M. A., & Ahmad-Yazid, A. (2014). Microalgae: An alternative source of renewable energy. *American Journal of Engineering Research*, *3*(3), 330–338.

Sandani, W. P., Premaratne, M., Ariyadasa, T. U., & Premachandra, J. K. (2022). Novel strategy for microalgae cell disruption and wet lipid extraction by employing electro-Fenton process with sacrificial steel anode. *Bioresource Technology*, *343*, 126110.

Saravanan, A., Kumar, P. S., Jeevanantham, S., Karishma, S., & Vo, D. V. N. (2021). Recent advances and sustainable development of biofuels production from lignocellulosic biomass. *Bioresource Technology*, *344*, 126203.

Sim, Y. B., Jung, J. H., Baik, J. H., Park, J. H., Kumar, G., Banu, J. R., & Kim, S. H. (2021). Dynamic membrane bioreactor for high rate continuous biohydrogen production from algal biomass. *Bioresource Technology*, *340*, 125562.

Sompech, K., Chisti, Y., & Srinophakun, T. (2012). Design of raceway ponds for producing microalgae. *Biofuels*, *3*(4), 387–397.

Spiden, E. M., Yap, B. H., Hill, D. R., Kentish, S. E., Scales, P. J., & Martin, G. J. (2013). Quantitative evaluation of the ease of rupture of industrially promising microalgae by high pressure homogenization. *Bioresource Technology*, *140*, 165–171.

Suali, E., & Sarbatly, R. (2012). Conversion of microalgae to biofuel. *Renewable and Sustainable Energy Reviews*, *16*(6), 4316–4342.

Subramaniam, Y., & Masron, T. A. (2021). The impact of economic globalization on biofuel in developing countries. *Energy Conversion and Management: X*, *10*, 100064.

Tavanandi, H. A., & Raghavarao, K. S. M. S. (2020). Ultrasound-assisted enzymatic extraction of natural food colorant C-Phycocyanin from dry biomass of Arthrospira platensis. *Lebensmittel-Wissenschaft & Technologie*, *118*, 108802.

Thakur, I. S., Kumar, M., Varjani, S. J., Wu, Y., Gnansounou, E., & Ravindran, S. (2018). Sequestration and utilization of carbon dioxide by chemical and biological methods for biofuels and biomaterials by chemoautotrophs: Opportunities and challenges. *Bioresource Technology*, *256*, 478–490.

Uduman, N., Qi, Y., Danquah, M. K., Forde, G. M., & Hoadley, A. (2010). Dewatering of microalgae cultures: A major bottleneck to algae-based fuels. *Journal of Renewable and Sustainable Energy*, *2*, 012701–012715.

Varol, A., & Ugurlu, A. (2016). Biogas production from microalgae (Spirulina platensis) in a two stage anaerobic system. *Waste and Biomass Valorization*, *7*(1), 193–200.

Vergini, S., Aravantinou, A. F., & Manariotis, I. D. (2016). Harvesting of freshwater and marine microalgae by common flocculants and magnetic microparticles. *Journal Applied Phycology*, *28*, 1041–1049.

Waghmare, A., Nagula, K., Pandit, A., & Arya, S. (2019). Hydrodynamic cavitation for energy efficient and scalable process of microalgae cell disruption. *Algal Research*, *40*, 101496.

Wahidin, S., Idris, A., Yusof, N. M., Kamis, N. H. H., & Shaleh, S. R. M. (2018). Optimization of the ionic liquid-microwave assisted one-step biodiesel production process from wet microalgae biomass. *Energy Conversion and Management*, *171*, 1397–1404.

Wang, B. (2010). Microalgae for Biofuel Production and CO2 Sequestration. Nova Science Publishers, UK.

Xiao, C., Fu, Q., Liao, Q., Huang, Y., Xia, A., Chen, H., & Zhu, X. (2020). Life cycle and economic assessments of biogas production from microalgae biomass with hydrothermal pretreatment via anaerobic digestion. *Renew Energy*, *151*, 70–78.

Xiao, C., Liao, Q., Fu, Q., Huang, Y., Chen, H., Zhang, H., Xia, A., Zhu, X., Reungsang, A., & Liu, Z. (2019a). A solar-driven continuous hydrothermal pretreatment system for biomethane production from microalgae biomass. *Applied Energy*, *236*, 1011–1018.

Xiao, C., Liao, Q., Fu, Q., Huang, Y., Xia, A., Shen, W., Chen, H., & Zhu, X. (2019b). Exergy analyses of biogas production from microalgae biomass via anaerobic digestion. *Bioresource Technology*, *289*, 121709.

Xie, Q., Addy, M., Liu, S., Zhang, B., Cheng, Y., Wan, Y., Li, Y., Liu, Y., Lin, X., Chen, P., & Ruan, R. (2015). Fast microwave-assisted catalytic co-pyrolysis of microalgae and scum for bio-oil production. *Fuel*, *160*, 577–582.

Yang, J., & Yang, L. (2019). A review on hydrothermal co-liquefaction of biomass. *Applied Energy*, *250*, 926–945.

Yao, S., Mettu, S., Law, S. Q., Ashokkumar, M., & Martin, G. J. (2018). The effect of high-intensity ultrasound on cell disruption and lipid extraction from high-solids viscous slurries of Nannochloropsis sp. biomass. *Algal Research*, *35*, 341–348.

Yeong, T. K., Jiao, K., Zeng, X., Lin, L., Pan, S., & Danquah, M. K. (2018). Microalgae for biobutanol production–Technology evaluation and value proposition. *Algal Research*, *31*, 367–376.

Yu, K. L., Chen, W. H., Sheen, H. K., Chang, J. S., Lin, C. S., Ong, H. C., Show, P. L., & Ling, T. C. (2020). Bioethanol production from acid pretreated microalgae hydrolysate using microwave-assisted heating wet torrefaction. *Fuel*, *279*, 118435.

Yuan, T., Li, X., Xiao, S., Guo, Y., Zhou, W., Xu, J., & Yuan, Z. (2016). Microalgae pretreatment with liquid hot water to enhance enzymatic hydrolysis efficiency. *Bioresource Technology*, *220*, 530–536.

Zhang, R., Grimi, N., Marchal, L., Lebovka, N., & Vorobiev, E. (2019). Effect of ultrasonication, high pressure homogenization and their combination on efficiency of extraction of bio- molecules from microalgae Parachlorella kessleri. *Algal Research*, *40*, 101524.

Zheng, H., Yin, J., Gao, Z., Huang, H., Ji, X., & Dou, C. (2011). Disruption of Chlorella vulgaris cells for the release of biodiesel-producing lipids: A comparison of grinding, ultrasonication, bead milling, enzymatic lysis, and microwaves. *Biotechnology and Applied Biochemistry*, *164*(7), 1215–1224.

Zheng, Y., Roberts, M., Kelly, J., Zhang, N., & Walker, T. (2015). Harvesting microalgae using the temperature-activated phase transition of thermoresponsive polymers. *Algal Research*, *11*, 90–94.

Zhou, N., Zhang, Y., Wu, X., Gong, X., & Wang, Q. (2011). Hydrolysis of Chlorella biomass for fermentable sugars in the presence of HCl and MgCl2. *Bioresource Technology*, *102*(21), 10158–10161.

Zinkoné, T. R., Gifuni, I., Lavenant, L., Pruvost, J., & Marchal, L. (2018). Bead Milling disruption kinetics of microalgae: Process modeling, optimization and application to biomolecules recovery from Chlorella sorokiniana. *Bioresource Technology*, *267*, 458–465.

Zou, S., Wu, Y., Yang, M., Li, C., & Tong, J. (2010). Bio-oil production from sub-and supercritical water liquefaction of microalgae Dunaliella tertiolecta and related properties. *Energy and Environmental Science*, *3*(8), 1073–1078.

Zou, X., Xu, K., Chang, W., Qu, Y., & Li, Y. (2021). Rapid extraction of lipid from wet microalgae biomass by a novel buoyant beads and ultrasound assisted solvent extraction method. *Algal Research*, *58*, 102431.

Rafaela Basso Sartori, Eduardo Jacob-Lopes*

Chapter 19
Process integration approaches applied to carbon dioxide capture and use from microalgae

Abstract: The environmental impacts caused by greenhouse gas emissions are a global concern, and for this reason, the search for ways to reduce these emissions has become a critical factor in our society. Although carbon capture and storage (CCS) methods have been extensively worked on, they are not economy-heavy, and their long-term environmental safety is a constant concern. Alternatively, carbon dioxide (CO_2) biosequestration using photosynthetic microorganisms has emerged as a promising way in biomass application and production process and, consequently, at the same time, reducing carbon dioxide emissions. Although microalgae are a viable source of several valuable compounds, their industrial applications still face many challenges and are unfeasible. Therefore, it is essential to promote the capture of carbon through an integrated process capable of combining several operations and/or processes to add value to CO_2 conversion and reduce costs. In that regard, an overview of carbon capture and utilization and the ability of microalgae to assimilate CO_2 and synthesize it into a wide range of chemicals is initially addressed. Finally, the main integration processes within a microalgae biorefinery and the potential for using CO_2 in integrating these bioprocesses are presented.

Keywords: greenhouse gas, carbon capture, CO_2 mitigation, photosynthesis, biorefinery

19.1 Introduction

Global concern about climate change has grown massively in recent decades, mainly due to greenhouse gas emissions (Gayathri et al., 2021). The increase in the emission of carbon dioxide (CO_2) that leads to the increase in the world temperature is one of the biggest environmental challenges that the world currently faces (Bharti et al., 2021). The main sources of the underlying increase in greenhouse gas emissions are mainly

*Corresponding author: **Eduardo Jacob-Lopes**, Bioprocess Intensification Group, Federal University of Santa Maria, UFSM, Roraima Avenue 1000, 97105-900, Santa Maria, RS, Brazil,
e-mail: ejacoblopes@gmail.com
Rafaela Basso Sartori, Bioprocess Intensification Group, Federal University of Santa Maria, UFSM, Roraima Avenue 1000, 97105-900, Santa Maria, RS, Brazil

https://doi.org/10.1515/9783110781267-019

related to deforestation and the burning of fossil fuels by the energy and industry sectors. While much data indicates that this share will decline, these categories will still account for more than half of all CO_2 emissions by mid-century (Jesus et al., 2021).

In view of this scenario, intensive carbon capture and storage (CCS) technologies for CO_2 sequestration were largely developed based on the urgent need to reduce carbon dioxide emissions into the atmosphere. These technologies were initially addressed by the Kyoto protocol elaborated in 1997, with the objective of containing and reversing the accumulation of CO_2 in the atmosphere and with the intention of minimizing environmental impacts (Jin et al., 2020).

Although the global annual CO_2 capture capacity has improved substantially, relative energy consumption still remains the main issue. For example, inherent disadvantages of capture and storage are primarily related to storage in geological formations, leakage probability, and long-term liability issues (Styring et al., 2011). Still,

capture projects are very expensive and may require significant development to develop cheaper alternatives (Zangh and Liu, 2021).

Therefore, the mitigation of CO_2 by photosynthetic organisms has effectively become an interesting configuration for carbon capture and utilization. Fundamentally, through photosynthesis, water and CO_2 are converted into organic compounds without increasing energy consumption or causing adjacent pollution (Jesus et al., 2021). In this case, the capture mediated by processes based on microalgae is designed to obtain an effective rate of CO_2 bioconversion and generate, at the same time, several products of commercial interest (e.g., biofuels, nutraceuticals, dyes, food supplements, among many others) (Andrade et al., 2020).

Finally, for the production of microalgae to become attractive, it is necessary to develop economically competitive processes. In this sense, several process integration options have been proposed in order to improve the cost-effectiveness of microalgae-based systems (Severo et al., 2020). In this case, the integration of processes through the use of energy, mass, and water aims to optimize the total biomass chain, focused mainly on connecting operations, equipment, or process techniques, significantly improving productivity, sustainability, and more specifically, economic viability within a microalgae biorefinery (Deprá et al., 2018).

In this chapter, therefore, a comprehensive overview of CO_2 capture and use by microalgae-based systems is presented. Here, we discuss the integration of processes focused on the reuse of surplus flows in order to decrease the demand for resources and, at the same time, increase the beneficiation of a microalgae biorefinery plant. In this case, recovering and integrating mass and energy from the exhaust gases directly from the photobioreactor seems to be the most beneficial and efficient way for these processes.

19.2 CO_2 capture and storage methods

Carbon capture, whether for storage or use, is an emerging technology to reduce greenhouse gases (GHGs) since energy production still has a high need for fossil fuels (Netto et al., 2021). Technically, CCS and carbon capture and utilization (CCU) involve the same initial individual components of capture, separation, and transport, but differ in the final destination of the captured carbon. CCS stores carbon underground, while CCU converts carbon dioxide into commercial products (Figure 19.1) (Pieri et al., 2018).

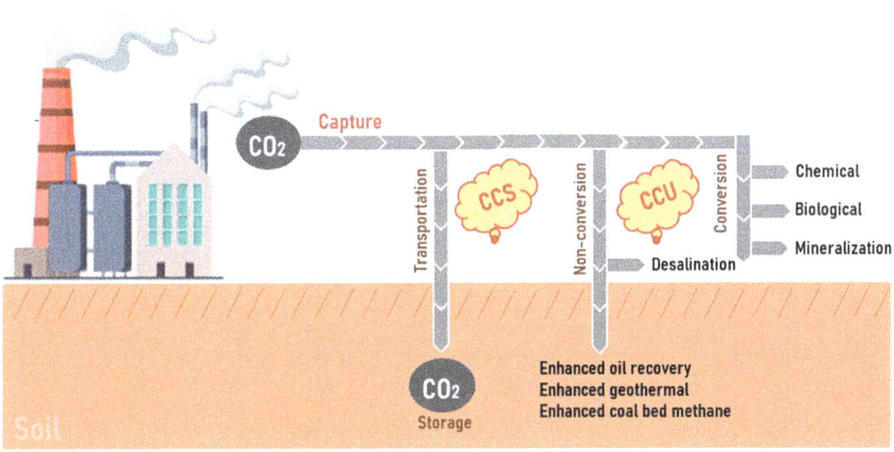

Figure 19.1: General process carbon capture and storage (CCS) and carbon capture and utilization (CCU). Adapted from Sekera et al. (2020).

Initially, the capture process provides high-concentration, high-pressure sources of CO_2 from point sources through appropriate technology. An example of this is post-conversion capture, pre-conversion, and oxy-fuel combustion (Madejski et al., 2022). Post-conversion capture refers to the separation of CO_2 from the exhaust gases from burning fossil fuels with air. The small amount of carbon dioxide in the exhaust gases (about 10% by volume) is captured by the dissolution of CO_2 and can be adapted to be included in power plants and in industries where fossil fuels are burned (Plaza and Pevita, 2019). In pre-conversion capture, a mixture of carbon monoxide (CO) and hydrogen (H_2) gases (syngas) is produced from the partial oxidation of fossil fuels or carbon-containing biomass, which are then subjected to a water gas displacement in the reaction and separation of gases. This reaction provides CO_2 in higher concentration and pressure, which favors the subsequent separation and transport processes (Raza et al., 2018). In the capture by oxy-fuel combustion, unlike the pre- and post-combustion methods, oxygen is used instead of air, for the combustion of pulverized coal under a very high temperature, in contrast to air. Although many studies report the applicability of oxy-combustion and pre-combustion technologies, post-combustion seems to be the most viable option for retrofitting existing facilities today (Gizer et al., 2022).

After capture, the CO_2 is then transported to a safe location for further storage or use. The resulting CO_2 in higher concentration and pressure is then compressed into a liquid and supercritical fluid to be transported to the location where it will be stored or used (Pieri et al., 2018). The transport of CO_2 can occur in two ways: (i) pipelines, for large flows being mainly designed for enhanced oil recovery (EOR), or (ii) railroad tanks, used for low concentrations due to lower cost. Still, the cost of freight can vary considerably depending on the economic situation of the location (Leung et al., 2014; Yáñez et al., 2020).

The CO_2 is usually stored in geological formations up to 1 km deep. Pilot scales are already demonstrated in some countries and according to data from the International Energy Agency, CCS could represent a reduction of more than 15% in carbon emissions by 2050 (Chen and Bollas, 2018). Despite these promises, several studies point out that considerable disadvantages are still associated with the CCS chain. This is faced with the possibility of gas leakage from the pipelines or storage reservoirs. In addition, CO_2 would need to be stored for at least 100 years, which leads to long-term liability issues (Styring et al., 2011).

Other alternatives suggest that the use of CO_2 in the manufacture of fuels and other chemical products represents a definitive step towards reducing the emission of carbon dioxide in the atmosphere associated with the objective of the sustainable chemical industry (Madejski et al., 2022). Enhanced oil recovery (EOR) is a process that involves injecting a fluid into a reservoir to drain residual oil from a rock formation. In this case, the possibility of injecting anthropogenic CO_2 to replace natural sources, such as natural gas, has been considered an alternative for the chemical use of CO_2 in recent years (Pieri et al., 2018).

The CO_2 can also be used as a feedstock for synthetic applications in the chemical and fuel industries through reduction and carboxylation reactions. Products such as nitrogen fertilizers, adhesives and binding agents, salicylic acid, and polycarbonate-based plastics are practical examples of these processes (Alper and Orhan, 2017). Also, conversion routes for the formation of methane, methanol, and long-chain hydrocarbons present excellent results in the use of CO_2. However, these processes are not cost-effective and efficient enough to meet industrial deployment and commercialization requirements. For this reason, future research in this area should look for improvements, new designs, and optimization in perfectly integrated processes (Cuéllar-Franca and Azapagic, 2015; Yu et al., 2018). In that case, most problems can be circumvented through biological carbon capture using organisms capable of assimilating CO_2. The advantages of biological capture reveal that in addition to CO_2 mitigation, the biomass generated by this method can be very advantageous; in fact, there are many applications for different industrial sectors (Severo et al., 2019).

19.3 CO_2 capture by microalgae

Once captured, CO_2 can be used directly or indirectly by different photosynthetic microorganisms. Microalgae, for example, are the most important microorganisms in aquatic ecosystems for the global carbon budget, playing crucial roles in CO_2 fixation (Deprá et al., 2021). Through photosynthesis, several pathways of carbon assimilation are involved in its biotransformation capable of synthesizing a wide range of chemicals depending on its metabolic pathway (Masojidek et al., 2021).

The steps of central carbon metabolism can be divided into two phases. Initially, it absorbs light energies at different wavelengths between photosystem I (700 nm) and photosystem II (680 nm), and later, CO_2 reduction occurs, powered by ATP and NADPH (Legrand et al., 2021). Microalgae carry out the process of carbon fixation through the Calvin-Benson cycle discovered in the 1950s (Bassham et al., 1950). This cycle occurs in the stroma and can be highlighted in three main phases: (i) carbon fixation, carried out by a carboxylase/oxygenase enzyme called Rubisco, responsible for capturing and fixing atmospheric CO_2; (ii) reduction of structures, in which each one of the molecules is phosphorylated by an ATP and reduced by an NADPH, forming the structure 1,3 bisphosphoglycerate and glyceraldehyde-3-phosphates (G3P), serving as a building block for other biomolecules of interest (e.g., carbohydrates, lipids, proteins); and finally characterized by the (iii) regeneration step that involves a series of reactions, taking in three ATP molecules to regenerate ribulose-phosphate (Figure 19.2) (Ferreira et al., 2020).

By knowing in depth the metabolite process of carbon fixation by microalgae, its high power of efficiency in the biomitigation of atmospheric CO_2 becomes clear. In fact, it is reported in the literature that for every 1,000 grams of microalgae dry cell culture, more than 1,800 grams of CO_2 is fixed (Benemann et al., 2003; Anguselvi et al., 2019). Furthermore, recent research indicates that fixation in biomass corresponds to approximately 5% of the CO_2 bioconverted in photobioreactors. In this case, almost 95% of the conversion is channeled to volatile organic compounds (Jacob-Lopes et al., 2010; Deprá et al., 2019). Several studies have reported that a large number of microalgae have the ability to remove atmospheric CO_2 and have great efficiency in carbon fixation rates and productivity, as shown in Table 19.1.

19.4 Biorefinery of microalgae

By observing the different biomitigation approaches, and presenting, among other advantages, high rates of carbon fixation, it is clear that microalgae are a promising alternative for CO_2 sequestration due to their unique and inherent characteristics (Ighalo et al., 2022). Above all, the simple and versatile nutritional needs of microalgae give them the ability to survive in an inhospitable environment, which puts them

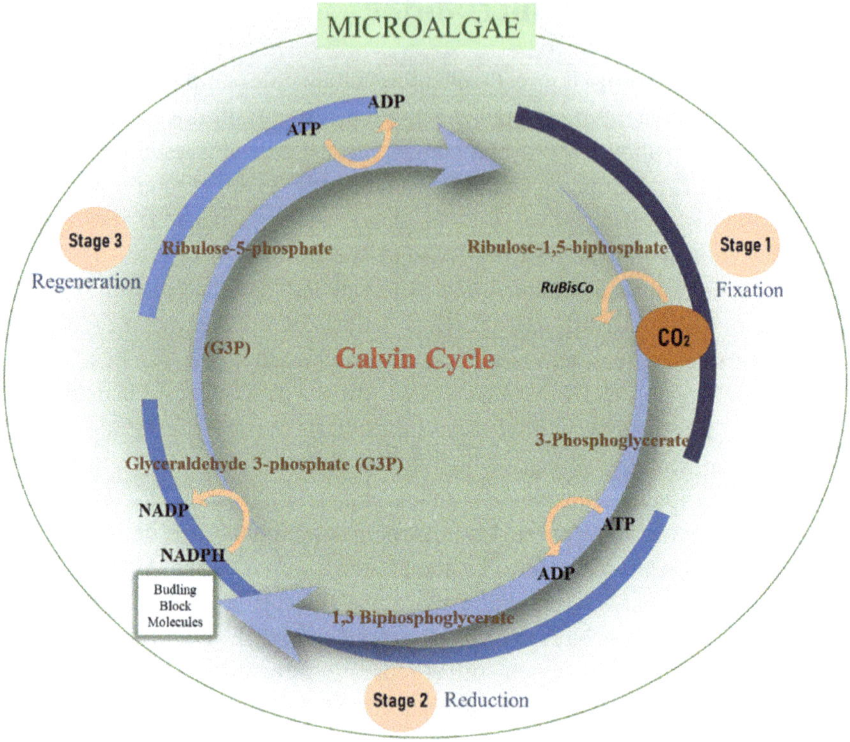

Figure 19.2: Process of carbon fixation through the Calvin-Benson cycle. Adapted from Senatore et al. (2020).

ahead of other sources. In addition, flue gases, industrial effluents, and various other residual sources still offer us the opportunity to efficiently convert into high-value products while generating less harmful forms for the environment (Onyeaka et al., 2021).

In fact, the fate of the supplied carbon will be converted into several metabolites and other biochemical compounds (including lipids, proteins, and carbohydrates) in addition to acting as a precursor of a multitude of microalgae-based bioproducts. However, despite great prominence, microalgae production for low-value bulk products (e.g., food, feed, nutraceuticals, and biofuels) has so far not reached an economically viable form. Thus, several strategies and evaluations have emerged recently in order to overcome these challenges. It seems that the best way would be to realize the potential production and fully use the biomass in an integrated biorefinery setup where every component is extracted, processed, and valued (Singh and Dhar, 2022).

To simplify, a microalgae biorefinery is best visualized in Figure 19.3. It basically involves a few processes, specifically called upstream and downstream processes.

Table 19.1: Removal rate, fixation, and productivity of carbon dioxide from various microalgae strains.

Microalgae	CO_2 removal rate (g/L/d)	CO_2 fixation rate (%v)	Productivity rate (g/L/d)	References
Aphanothece m. Nageli	2.64	15	nd	Jacob-Lopes et al. (2008)
Aphanothece m. Nageli	4.8	15	nd	Jacob-Lopes et al. (2010)
Chlorella sp.	3.72	9.5	0.12	Pourjamshidian et al. (2019)
Chlorella sp.	5.03	1.75	0.18	Pourjamshidian et al. (2019)
Chlorella vulgaris	3.89	15	0.62	Maroneze et al. (2019)
Desmodesmus sp.	0.04	10	0.02	Swarnalath et al. (2015)
Dunaliella sp.	0.03	27	0.17	Wang et al. (2008)
Dunaliella tertiolecta	0.08	5	nd	Van Den Hende et al. (2011)
Phormidium sp.	7.4	15	nd	Francisco et al. (2010)
Haematococcus pluvialis	0.15	16–34	0.08	Huntley and Redalje (2007)
Nannochlorophsis oculata	0.39	15	0.21	Chiu et al. (2009)
Scendesmus obliquus	4.60	15	nd	Francisco et al. (2010)
Scendesmus obliquus	0.03	12	0.02	Basu et al. (2014)

The choice of methods for the processes mentioned above can significantly influence economic viability, efficiency, energy requirements, and miscellaneous expenses (Severo et al., 2021). The upstream and downstream phases are composed of several regular unit operations, whose process flow usually includes the selection of strains, types of cultivation systems, harvesting, drying, extraction, and fractionation, where different conversion approaches are elaborated in order to achieve the desired target, in this case, biofuels and/or value-added products. It is worth mentioning that the main factors involved in the first stage, that is, the choice of nutrients (nitrogen and phosphorus), water, light, and CO_2, directly influence the productivity of microalgae and the result of the final products (Deprá et al., 2018).

Although many studies have reported the successful use of microalgae for the production of bioproducts within a biorefinery approach, economic viability is not realized and the microalgae biorefinery is still very expensive. In this case, to achieve optimal viability and sustainability, processes need to be streamlined and efficiently integrated (Sivaramakrishnan et al., 2022). With that in mind, we present below the main integration approaches found in a microalgae biorefinery for a more realistic implementation on a commercial scale.

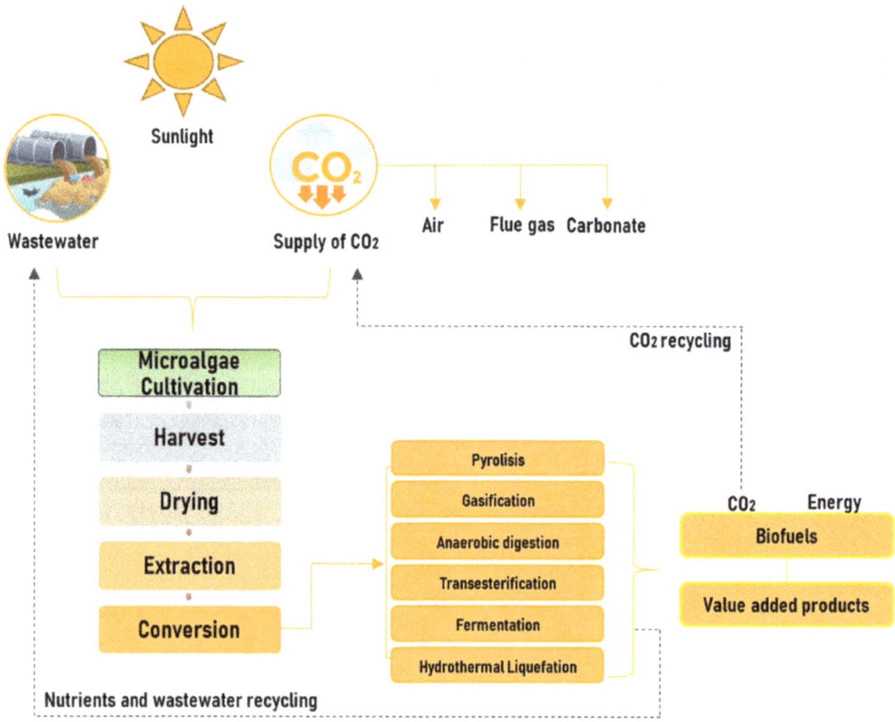

Figure 19.3: Overview of a microalgae-based biorefinery system. Adapted from Singh and Dhar (2022).

19.5 Process integration in microalgae-based systems

In the state of the art, an integrated process basically consists of combining several operations and/or processes in order to add value to the CO_2 conversion while reducing costs. In this regard, in order to be viable and environmentally sustainable, the integration of bioprocesses mediated by microalgal biorefinery systems requires the global integration of the chain (Deprá et al., 2019).

Overall, the reuse of surplus streams (i.e., energy, mass, and water) decreases resource demand and, at the same time, increases plant beneficiation. In this case, these flows are directly recovered and reused to input a new cycle, where the focus is on the production of biomass for later efficient conversion into bioproducts for different commodities (Oliveira et al., 2022).

Energy integration is the oldest method of integration. Since the mid-1950s, heat recovery by thermal energy has been applied to reduce consumption between hot and cold utilities. Over the years, mass integration, through minimizing the use of materials, has become common. Furthermore, water integration, which until then belonged to mass integration, refers today to the reduction of water consumption and the reuse and recycling of different water networks, such as wastewater (Severo et al., 2019). These three types of integration and their main characteristics are best visualized in Figure 19.4.

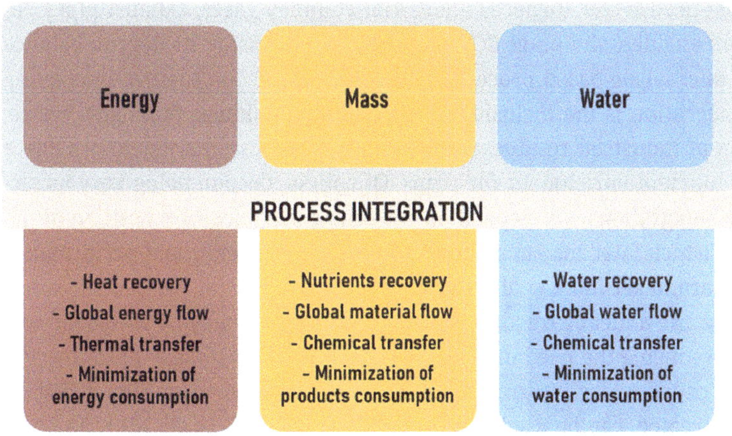

Figure 19.4: Main characteristics of the processes integration in microalgae-based systems. Adapted from Sivaramakrishnan et al. (2022).

Initially, energy integration considers all forms of energy within the process, fundamentally based on thermodynamic principles (Sinnott and Towler, 2020). In this case, microalgae installations refer mainly to heating, cooling, and other types of energy generation and/or consumption. This principle aims to minimize energy consumption and maximize internal heat recovery by identifying the main sources (equipment, unit operations, and diverse systems) in order to improve conservation and energy efficiency, and thus reduce costs within the facilities (Severo et al., 2020).

Several methodologies have been developed within microalgae biorefinery systems. For example, the use of generated biomass can be converted into energy using different thermochemical and biochemical processes. More specifically, energy integration can take place during the recovery of waste heat produced by gasification technologies, pyrolysis, liquefaction, hydrogenation, fermentation, transesterification, or direct thermal combustion, the stages of which produce synthesis gas, electricity, and biofuels, with substantial energy content (Lee et al., 2019). Furthermore, other forms of integration, whether by co-combustion (with other fossil fuels) or direct combustion of biomass (integrated gasification (combined cycle)) can be considered essential to increase the performance of the energy process (IEA, 2017).

In contrast to energy integration, bulk integration is an approach that provides an understanding of the overall flow of materials within the process. This can be reflected through the supply of essential nutrients, such as carbon, nitrogen, phosphorus, potassium, and sulfur, which are efficiently assimilated by microalgae aiming at the metabolic maintenance of their cellular structures (Sarwer et al., 2022).

As already mentioned, microalgae bioconvert free CO_2 during photosynthesis, which is not always efficient due to the low level of CO_2 in the air. Thus, the integration of direct CO_2 by highly concentrated stationary sources (e.g., the use of gas streams from coal-fired boiler furnaces, industrial chimney gases, exhaust pipes, internal combustion engines, and other CO_2 emitters) is a promising strategy to balance the economy of microalgae-based processes (Prasad et al., 2021). Another interesting form of mass integration is the inclusion of agricultural residues, municipal waste, and the insertion of industrial residues in microalgae systems, in order to provide a cheap source of nutrient enrichment for crops. Of course, despite being very attractive, any of these integrations will depend on the physicochemical composition of the compounds used, which later may or may not impair the efficiency and performance of microalgae (Sivaramakrishnan et al., 2022).

Comparatively, the aqueous fraction from industrial waste may be viable for several applications, reducing the need or demand for water in microalgae biorefineries. This process, also called water integration, aims to recycle water that, after being directly or partially treated, can be used in other processes (Severo et al., 2019). The destination is usually the cleaning of equipment, cooling or heating of the process, or even the incorporation into the cultivation system itself, always considering the quality standards established by legislation (EPA, 2012).

19.5.1 The potential utilization of CO_2 in processes integration

Under photoautotrophic conditions, atmospheric air enriched with CO_2 up to a certain limit favors cell growth. From the point of view of bioprocess engineering, this enrichment of CO_2 both in atmospheric air and in gases from various industrial emissions must be integrated into the final destination of the obtained biomass, as discussed previously. Therefore, in terms of carbon capture and use by microalgae, it is possible to recover and integrate mass and energy from the exhaust gases directly from the photobioreactor (Pruvost et al., 2022).

In photobioreactors, microalgae cultures can be monitored, enabling the control of cultivation conditions, such as cell density and concentration of nutrients, for example, CO_2. At the same time, unlike traditional utilization approaches, the trend in recent years has focused on integrated bioprocess from the efficient use of CO_2 in microalgae-based processes. Thus, the integration method to be implemented must be done taking into account the product to be obtained and its application (Shekh et al., 2021).

In fact, parametric analyses indicate a strong dependence of the CO_2 mass transfer constant on the growth rate of the microalgae culture, that is, the more efficient the addition of dissolved CO_2 in the culture, the faster it will develop. For example, in microalgae processes, the mass of CO_2 is integrated directly into a photobioreactor, which converts part of this compound into high-value products from photosynthetic metabolism (e.g., biomass enriched with pigments, proteins, lipids, carbohydrates, among others) (Rampi, 2021). At the same time, VOCs, O_2, and CO_2 that were not converted during this process have the possibility to be reused. Thus, the integration of mass and energy occurs through the recovery of exhaustion of these gases from the photobioreactor, which, in addition to containing O_2 and VOCs, also contains part of CO_2 in the bioprocess, which is so reused simultaneously as oxidant, gaseous fuels, and diluent of nitrogen in order to improve the energy efficiency of combustion systems (Deprá et al., 2019).

These two types of integrated processes, i.e., mass and energy, can be better visualized in Figure 19.5. Here, different sources of CO_2 (atmospheric air, industrial gases, and use of microalgae in photobioreactors) are compared in order to obtain a better view of the efficiency of each one.

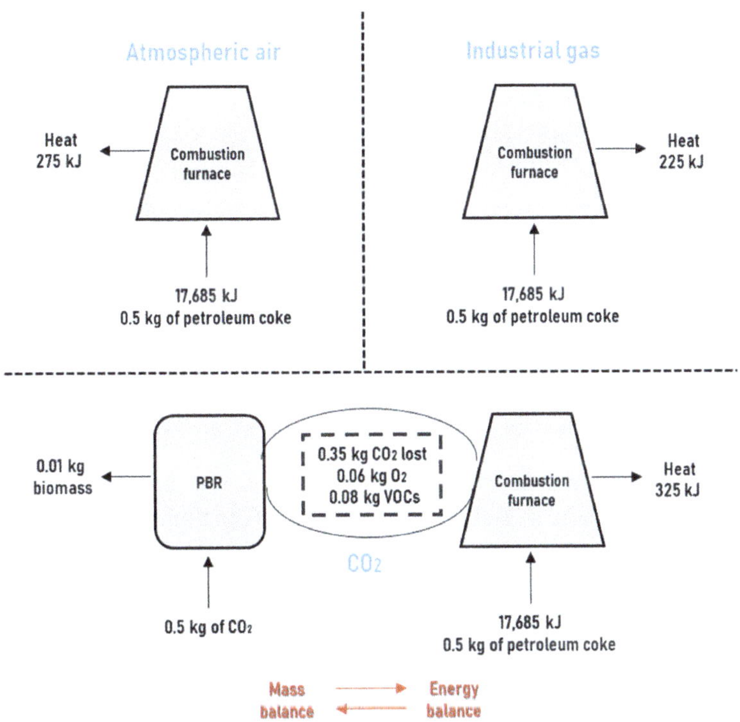

Figure 19.5: Integrated process of capturing and using carbon from different sources of CO_2. Adapted from Severo et al. (2019).

Although many studies report the success of the integration of exhaust gases with microalgal photobioreactors, some limitations must be overcome for this to happen. An example of this would be maximizing efficiency in process performance regarding the conversion of CO_2 through current projects of photobioreactors, since today only a part of them has the capacity to efficiently convert the CO_2 (~30%). Issues of implementation of the cultivation plant would also lead to process efficiency and economy. For example, the choice of location, size, and distribution will effectively influence the costs and productivity of the system. Actual estimates of economic analyses, in addition to the values obtained on a laboratory scale or on small-scale plants, would considerably facilitate the decision-making for the future for the real implementation of these systems from the conversion and integration of CO_2 by microalgae (Jacob-Lopes et al., 2018).

19.6 Concluding remarks

The integration of biologically-based processes with established industrial processes is an appropriate and innovative method to meet the requirements of green technology, through the recovery and reuse of surplus energy, mass, and water from industrial processes. However, there are still many bottlenecks related to these approaches for it to become an industrial reality. In this case, one of the biggest challenges is the complete understanding of process flows and their value in biological systems.

As more aspects of integration are considered, the complexity of the system design problem increases, but so do the overall benefits such as reducing capital and operating expenses, maximizing the quality, quantity, and value of bioproducts, as well as minimizing environmental impacts. The most important point is to synergistically consider environmental and economic assessments, as well as the integrated biorefinery approach, in an attempt to find an optimal scenario for large-scale implementation.

The capture and use of carbon dioxide in microalgae photobioreactors thus emerge as a new context in the attempt to reduce the concentration of atmospheric pollutants. At the same time, microalgae have several characteristics that make them promising for obtaining several bioproducts of commercial interest. Also, in photobioreactors, microalgae cultures can be monitored, enabling the control of cultivation conditions, such as cell density and CO_2 concentration. Thus, the integration of processes from the conversion and integration of CO_2 by microalgae directly from photobioreactors will result in safer and more sustainable technologies and should make microalgae biotechnology competitive and commercially more attractive in the future.

References

Alper, E., & Orhan, O. Y. (2017). CO_2 utilization: Developments in conversion processes. *Petroleum*, *3*(1), 109–126. doi:10.1016/j.petlm.2016.11.003

Andrade, D. S., Telles, T. S., Castro, G. H. L. (2020). The Brazilian microalgae production chain and alternatives for its consolidation. *Journal of Cleaner Production*, *250*, 119526. doi: 10.1016/j.jclepro.2019.119526

Anguselvi, V., Masto, R. E., Mukherjee, A., & Singh, P. K. (2019). CO_2 capture for industries by algae. *Algae Intechopen*, *1*, 11. doi:10.5772/intechopen.81800

Bassham, J. A., Benson, A. A., & Calvin, M. (1950). The path of carbon in photosynthesis. *The Journal of Biological Chemistry*, *185*(2), 781–787. PMID: 14774424.

Basu, S., Roy, A. S., Mohanty, K., & Ghoshal, A. K. (2014). CO_2 biofixation and carbonic anhydrase activity in Scenedesmus obliquus SA1 cultivated in large scale open system. *Bioresource Technology*, *164*, 323–330. doi:10.1016/j.biortech.2014.05.017

Benemann, J. R., Van Olst, J. C., Massingill, M. J., Weissman, J. C., & Brune, D. E. (2003). The Controlled Eutrophication Process: Using Microalgae for CO_2 Utilization and Agricultural Fertilizer Recycling. In: Gale, J., & Kaya, Y. (Eds.), Greenhouse Gas Control Technologies. Elsevier Sciences, Kyoto, Japan. doi:10.1016/B978-008044276-1/50227-0

Bharti, R., Grimm, D. G. (2021) Current challenges and best-practice protocols for microbiome analysis. *Briefings in Bioinformatics*, *22*(1), 178–193. doi: 10.1093/bib/bbz155

Chen, C., & Bollas, G. M. (2018). Efficiency Analysis of Chemical-looping Fixed Bed Reactors Integrated in Combined Cycle Power Plants. In: Eden, M. R., Lerapetritou, M. G., & Towler, G. P. (Eds.), 13th International Symposium on Process Systems Engineering (PSE 2018). Computer Aided Chemical Engineering, vol. 44, pp. 2389–2394. doi:10.1016/B978-0-444-64241-7.50393-1

Chiu, S. Y., Kao, C. Y., Tsai, M. T., Ong, S. C., Chen, C. H., & Lin, C. S. (2009). Lipid accumulation and CO(2) utilization of nannochloropsis oculata in response to CO(2) aeration. *Bioresource Technology*, *100*(2), 833–838. doi:10.1016/j.biortech.2008.06.061

Cuéllar-Franca, R. M., & Azapagic, A. (2015). Carbon capture, storage and utilisation technologies: A critical analysis and comparison of their life cycle environmental impacts. *Journal of CO_2 Utilization*, *9*, 82–102. doi:10.1016/j.jcou.2014.12.001

Deprá, M. C., Dos Santos, A. M., Severo, I. A., Santos, A. B., Zepka, L. Q., & Jacob-Lopes, E. (2018). Microalgal biorefineries for bioenergy production: Can we move from concept to industrial reality? *BioEnergy Researcha*. doi:10.1007/s12155-018-9934-z

Deprá, M. C., Ramírez-Merida, L. G., De Menezes, C. R., Zepka, L. Q., & Jacob-Lopes, E. (2019). A new hybrid photobioreactor design for microalgae culture. *Chemical Engineering Research and Design*, *144*, 1–10. doi:10.1016/j.cherd.2019.01.023

Deprá, M. C., Severo, I. A., Dias, R. R., Zepka, L. Q., Jacob-Lopes, E. (2021). Photobioreactor design for microalgae culture. In: Galanakis, C. M.(Ed.), *Microalgae Cultivation, Recovery of Compounds and Applications*, Elsevier, 35–61. doi: 10.1016/B978-0-12-821218-9.09995-3

EPA. Environmental Protection Agency of United States (2012). *Guidelines for Water Reuse*, AR-1530. EPA/600/R12/618.

Ferreira, R., Skrekas, C., Nielsen, J., David, F. (2020). Multiplexed CRISPR/Cas9 Genome Editing and Gene Regulation. *Biology*, *7*, 10–15. doi: 10.1021/acssynbio.7b00259

Francisco, E. C., Neves, D. B., Jacob-Lopes, E., & Franco, T. T. (2010). Algae as feedstock for biofuel production: Carbon dioxide sequestration, lipid production and biofuel quality. *Journal of Chemistry Technology Biotechnology*, *85*, 395–403. doi:10.1002/jctb.2338

Gayathri, R., Mahboob, S., Govindarajan, M., Al-Ghanim, K. A., Ahmed, Z., Vijayalakshmi, S. (2021). A review on biological carbon sequestration: A sustainable solution for a cleaner air environment, less pollution and lower health risks. *Journal of King Saud University – Science*, *33*(2), 101282. doi: 10.1016/j.jksus.2020.101282.

Gizer, S. G., Manoj, O. P., & Sahiner, N. (2022). Recent developments in CO_2 capture, utilization, related materials, and challenges. *Energy Research*, *46*, 16241–16263. doi:10.1002/er.8347

Huntley, M. E., & Redalje, D. G. (2007). CO_2 mitigation and renewable oil from photosynthetic microbes: A new appraisal. *Mitigation and Adaptation Strategies for Global Change*, *12*, 573–608. doi:10.1007/s11027-006-7304-1

IEA Bioenergy (2017). State of Technology Review – Algae Bioenergy An IEA Bioenergy Inter-Task Strategic Project. Available from: https://www.ieabioenergy.com/wp-content/uploads/2017/02/IEABioenergy-Algae-report-update-Final-template-20170131.pdf. Accessed in: October 6, 2022.

Ighalo, J. O., Dulta K., c, Kurniawan, S. B., Omoarukhe, F. O., Ewuzie, U., Eshiemogie, S. E., Ojo, A. U., Abdullah, S. R. S. (2022) Progress in Microalgae Application for CO2 Sequestration. *Cleaner Chemical Engineering*, *3*, 100044. doi: 10.1016/j.clce.2022.100044

Jacob-Lopes, E., Revah, S., Hernandez, S., Shirai, K., & Franco, T. T. (2009). Development of operational strategies to remove carbon dioxide in photobioreactors. *The Chemical Engineering Journal*, *153*, 120–126. doi:10.1016/j.cej.2009.06.025

Jacob-Lopes, E., Scoparo, C. H. G., Franco, T. T. (2008) Rates of CO2 removal by Aphanothece microscopica Nageli in tubular photobioreactors. *Chemical Engineering and Process*, *47*, 1365–73. doi:10.1016/j.cep.2007.06.004

Jacob-Lopes, E., Scoparo, C. G. H., Queiroz, M. I., & Franco, T. T. (2010). Biotransformations of carbon dioxide in photobioreactors. *Energy Conversion and Manage*, *51*, 894–900. doi:10.1016/j.enconman.2009.11.027

Jacob-Lopes, E., Zepka, L. Q., & Queiroz, M. I. (2018). Energy from Microalgae (First Edition). Springer International Publishing, Switzerland.

Jesus, G. A. C., Rocha, M. F., Chibite, E. E. A., Motoyama, M. H., & Conte, H. (2021). Biotechnology as a tool for carbon sequestration: Bacteria, microalgae and genetically modified trees. *Revista Ibero-Americana de Ciências Ambientais*, *12*(11), 246–255. doi:http://doi.org/10.6008/CBPC2179-6858.2021.011.0021

Jin, L., Chang, Y. H., Wang, M., Zheng, X. Z., Yang, J. X., Gu, J. (2020) The dynamics of CO_2 emissions, energy consumption, and economic development: Evidence from the top 28 greenhouse gas emitters. *Environmental and Science Pollution Research*, *29*, 36565–36574. doi: 10.1007/s11356-021-18069-y

Lee, S. Y., Sankaran, R., Chew, K. W., Tan, C. H. (2019) Waste to bioenergy: a review on the recent conversion technologies. *BMC Energy*, *1*, 4. doi: 10.1186/s42500-019-0004-7

Legrand, J., Artu, A., & Pruvost, J. (2021). A review on photobioreactor design and modelling for microalgae production. *Reaction Chemistry and Engineering*, *7*, 1134–1151. doi:10.1039/D0RE00450B

Leung, D. Y. C., Caramanna, G., & Maroto-Valer, M. (2014). An overview of current status of carbon dioxide capture and storage technologies. *Renewable and Sustainable, Energy Reviews*, *39*, 426–443. doi:10.1016/j.rser.2014.07.093

Madejski, P., Chmiel, K., Subramanian, N., Kus, T. (2022) Methods and Techniques for CO_2 Capture: Review of Potential Solutions and Applications in Modern Energy Technologies. *Energies*, *15*(3), 887. doi: 10.3390/en15030887

Masojídek, J., Ranglová, K., Lakatos, G. E., Benavides, A. M. S., & Torzillo, G. (2021). Variables governing photosynthesis and growth in microalgae mass cultures. *Processes*, *9*, 820. doi:10.3390/pr9050820

Maroneze, M. M., Deprá, M. C., Zepka, L. Q., & Jacob-Lopes, E. (2019). Artificial lighting strategies in photobioreactors for bioenergy production by Scenedesmus obliquus CPCC05. *SN Applied Sciences*, *1*, 1695. doi:10.1007/s42452-019-1761-0

Netto, A. L. A., Alves, V. H., Peyer, H., & Jacobi, P., dos Santos, E. M. (2021). Overview of public policies and strategies for the deployment of carbon capture and storage: Reflections for Brazil. *Journal of Environmental Management and Sustainability*, *10*(1), 1–21. 19305 doi:10.5585/geas.v10i1.19305

Oliveira, M. C., Vieira, S. M., Iten, M., Matos, H. A. (2022) Optimisation of Water-Energy Networks in Process Industry: Implementation of Non-Linear and Multi-Objective Models. *Frontiers Chemical and Engineering, 23*. doi: 10.3389/fceng.2021.750411

Onyeaka, L., Miri, T., Obileke, K. C., Hart, A., Anumudu, C., & Al-Sharify, Z. T. (2021). Minimizing carbon footprint via microalgae as a biological capture. *Carbon Capture Science and Technology, 1*, 100007. doi:10.1016/j.ccst.2021.100007

Plaza, M. G., & Pevida, C. (2019). Current Status of CO_2 Capture from Coal Facilities. In: New Trends in Coal Conversion. pp. 31–58. doi:10.1016/B978-0-08-102201-6.00002-9. International Journal of Energy Research.

Pieri, T., Nikitas, A., Castillo-Castillo, A., & Dimakis, A. A. (2018). Holistic assessment of carbon capture and utilization value chains. *Environments, 5*, 108. doi:10.3390/environments5100108

Pourjamshidian, R., Abolghasemi, H., Esmaili, M., Amrei, H. D. (2019) Carbon dioxide biofixation by chlorella sp. in a bubble column reactor at different flow rates and CO2 concentrations. *Brazilian Journal of Chemical Engineering, 36*(2), 639–645. doi: 10.1590/0104-6632.20190362s20180151

Prasad, R., Gupta, S. K., Shabnam, N., Oliveira, C. Y. B., Nema, A. K., Ansari, F. A., Bux, F. (2021). Role of Microalgae in Global CO2 Sequestration: Physiological Mechanism, Recent Development, Challenges, and Future Prospective. *Sustainability, 13*(23), 13061. doi: 10.3390/su132313061

Pruvost, P., Le Gouic, B., & Cornet, J.-F. (2022). Kinetic modeling of CO2 biofixation by microalgae and optimization of carbon supply in various photobioreactor technologies. *ACS Sustainable Chemistry and Engineering, 10*, 12826–12842. doi:10.1021/acssuschemeng.2c03927

Rampi, M. G. (2021). Modelagem E Simulação de Fotobiorreatores a Fluxo de Ar de Elevação Com Tubos Seriados Para Cultivo de Microalgas. Tese Dissertação, Universidade Federal do Paraná.

Raza, A., Gholami, R., Rezaee, R., Rasouli, V., & Rabiei, M. (2018). Significant aspects of carbon capture and storage – A review. *Petroleum*. doi:10.1016/j.petlm.2018.12.007

Sarwer, A., Hamed, S. M., & Osman, A. I. (2022). Algal biomass valorization for biofuel production and carbon sequestration: A review. *Environmental Chemistry Letter, 20*, 2797–2851. doi:10.1007/s10311-022-01458-1

Sekera, J., Lichtenberger, A. (2020) Assessing Carbon Capture: Public Policy, Science, and Societal Need. *Biophysical Economics and Sustainability, 5*(3), 14. doi:10.1007/s41247-020-00080-5

Senatore, A., Lania, I., Corrente, G. A., Basile, A. (2020) CO2 capture by bacteria and their enzymes Advances in Carbon Capture. In: Mohammad, F., Amin, M. M. (Ed.),: *Advances in Carbon Capture: Methods, Technologies and Applications*, Elsevier, 407–429. doi: 10.1016/B978-0-12-819657-1.00018-9

Severo, I. A., Depra, M. C., Dias, R. R., Jacob-Lopes, E. (2020). Process integration applied to microalgae-based systems. In: Settre, C. M. M., Connor, J. D., Wheeler, S. A. (Eds), *Handbook of Microalgae-Based Processes and Products*. Elsevier. doi: 10.1016/B978-0-12-818536-0.00026-9

Severo, I. A., Deprá, M. C., Zepka, L. Q., & Jacob-Lopes, E. (2019). Carbon dioxide capture and use by microalgae in photobioreactors. *Bioenergy with Carbon Capture and Storage*. doi:10.1016/B978-0-12-816229-3.00008-9

Severo, I. A., Dos Santos, A. M., Deprá, C. M., Barin, J. S., & Jacob-Lopes, E. (2021). Microalgae photobioreactors integrated into combustion processes: A patent-based analysis to map technological trends. *Algal Research, 60*, 102529. doi:10.1016/j.algal.2021.102529

Shekh, A., Sharma, A., Schenk, P. M., Kumar, G., & Mudliar, S. (2021). Microalgae cultivation: Photobioreactors, CO_2 utilization, and value-added products of industrial importance. *Journal of Chemical Technology and Biotechnology*. doi:10.1002/jctb.6902

Singh, J., & Dhar, D. W. (2019). Overview of carbon capture technology: Microalgal biorefinery concept and state-of-the-art. *Frontiers in Marine Science, 6*, 29. doi:10.3389/fmars.2019.00029

Sinnot, R., Towler, G. (2020). Fundamentals of Energy Balances and Energy Utilization. *Chemical Engineering Design, 6*, 75–140. doi: 10.1016/B978-0-08-102599-4.00003-5

Sivaramakrishnan, R., Suresh, S., Kanwal, S., Ramadoss, G., Ramprakash, B., & Incharoensakdi, A. (2022). Microalgal biorefinery concepts' developments for biofuel and bioproducts: Current perspective and bottlenecks. *International Journal of Molecular Science*, *23*, 2623. doi:10.3390/ijms23052623

Styring, P., Jansen, D., Coninck, D., Reith, H., & Armstrong, K. (2011). Carbon Capture and Utilisation in the Green Economy. Using CO_2 to Manufacture Fuel, Chemicals and Materials. Centre for Low Carbon Futures 501, ISBN: 978-0-9572588-1-5.

Swarnalath, G. V., Hegde, N. S., Chauhan, V. S., & Sarada, R. (2015). The effect of carbon dioxide rich environment on carbonic anhydrase activity, growth and metabolite production in indigenous freshwater microalgae. *Algal Research*, *9*, 151–159. doi:10.1016/j.algal.2015.02.014

Van Den Hende, S., Vervaeren, H., & Boon, N. (2012). Flue gas compounds and microalgae: (Bio-)chemical interactions leading to biotechnological opportunities. *Biotechnology Advances*, *30*, 1405–1424. doi:10.1016/j.biotechadv.2012.02.015

Van Den Hende, S., Vervaeren, H., Desmet, S., & Boon, N. (2011). Bioflocculation of microalgae and bacteria combined with flue gas to improve sewage treatment. *New Biotechnology*, *29*(1), 23–31. doi:10.1016/j.nbt.2011.04.009

Wang, B., Li, Y., Wu, N., & Lan, C. Q. (2008). CO2 bio-mitigation using microalgae. *Applied Microbiology Biotechnology*, *79*, 707–718. doi:10.1007/s00253-008-1518-y

Yáñez, E., Ramírez, A., Núñez-López, V., Castillo, E., & Faaij, A. (2020). Exploring the potential of carbon capture and storage-enhanced oil recovery as a mitigation strategy in the Colombian oil industry. *International Journal of Greenhouse Gas Control*, *94*, 102938. doi:10.1016/j.ijggc.2019.102938

Yu, J., Dow, A., & Pingali, S. (2013). The energy efficiency of carbon dioxide fixation by a hydrogen-oxidizing bacterium. *International Journal of Hydrogen Energy*, *38*(21), 8683–8690. doi:10.1016/j.ijhydene.2013.04.153

Yu, B. Y., Elbuken, C., Shen, C., Huissoon, J. P., Ren, C. L. (2018). An integrated microfluidic device for the sorting of yeast cells using image processing. *Scientific Reports*, *8*, 3550. doi: 10.1038/s41598-018-21833-9

Zhang, S., & Liu, Z. (2021). Advances in the biological fixation of carbon dioxide by microalgae. *Journal of Chemical Technology and Biotechnology*, *96*(6), 1475–1495. doi:10.1002/jctb.6714

Olatunde Akinbuja, Kamelia Boodhoo, Sharon Velasquez Orta*

Chapter 20
Algae-based bioelectrochemical systems

Abstract: Bioelectrochemical systems (BES) are a technology under development that utilizes microorganisms to convert chemical energy to electrical energy or vice versa. BES have been used for applications in wastewater treatment, production of electricity, and material recovery. Microbial fuel cells (MFCs), the most common type of BES, have been widely tested for producing electricity from wastewater treatment. Conventionally, the cathodic part of an MFC is abiotic but recent research has been exploring the use of microalgae because of their ability to naturally produce oxygen, which is required for the cathodic reactions in an MFC. In this chapter, we provide a review on the status of the algae-based MFC technology. We discuss the main components used in this technology as well as the results of microalgae tested systems. As it is common for anodic biofilms to improve the performance of microbial fuel cells, a case has been made for the influence of cathodic biofilms on the performance of the system. Therefore, we discuss the possibility of immobilizing microalgae in the cathodic compartment of microbial fuel cell, highlighting the electrode surface immobilization techniques such as entrapment, adsorption and affinity immobilization. Finally, the prospects of adapting other forms of microalgae immobilization used in other technologies, into microbial fuel cell systems, have been highlighted.

Keywords: microbial fuel cells, bioelectrochemical systems, microalgae as cathodes, immobilization, cathode

20.1 Introduction to bioelectrochemical systems

Bioelectrochemical systems (BES) use microorganisms to enhance the transfer of electrons between organic compounds and electrodes (Mook et al., 2013; Luo et al., 2016; Sleutels et al., 2012). The systems can convert chemical to electrical energy or vice versa. There are many types of BES, including microbial fuel cell (MFC), microbial electrolysis cells (MEC), plant microbial fuel cells (PMFC), microbial desalination cells (MDC), and microbial solar cells (MSC) (Bajracharya et al., 2016). All of them have similar but slightly different configurations, depending on their functions (Table 20.1).

***Corresponding author: Sharon Velasquez Orta**, School of Engineering, Merz Court, Newcastle University, Newcastle Upon Tyne, United Kingdom, e-mail: sharon.velasquez-orta@newcastle.ac.uk
Olatunde Akinbuja, Kamelia Boodhoo, School of Engineering, Merz Court, Newcastle University, Newcastle Upon Tyne, United Kingdom

https://doi.org/10.1515/9783110781267-020

Table 20.1: Types of BES and some of their features.

BES	Functions
Microbial fuel cell (MFC)	Electricity generation and waste treatment
Microbial electrosynthesis (MES)	Volatile fatty acids (VFAs) and alcohols production
Microbial electrolysis cell (MEC)	Electricity generation and hydrogen production
Microbial desalination cell (MDC)	Seawater desalination
Microbial solar cells (MSC)	Electricity, hydrogen, methane, ethanol, hydrogen peroxide
Plant microbial fuel cells (PMFC)	Electricity

The conventional BES (Figure 20.1) comprises an anode section and a cathode section, which are usually separated by a proton exchange membrane. The reactions on either or both sides are catalyzed by microbes, and the electrons are transported via an external circuit (Santoro et al., 2017).

Many bacteria and algae transfer the electrons to electrodes through mediators, biofilms, c-type cytochromes and highly conductive pili, also known as nanowires (Choudhury et al., 2017; Rabaey and Rozendal, 2010). BES have a number of applications, which include material recovery, biocompound production, bioelectricity production, and wastewater treatment (Sleutels et al., 2012). Some reports suggest that the energy inherent in wastewater that can be potentially harvested might be 5–10

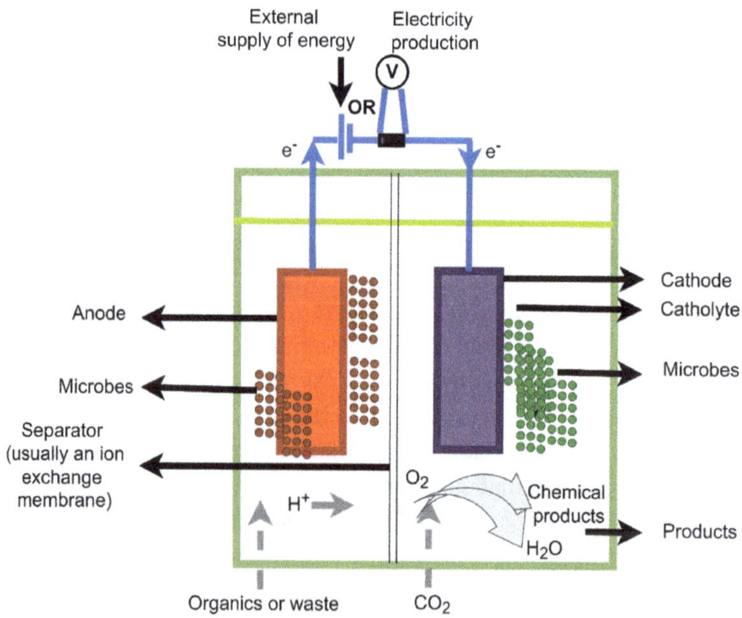

Figure 20.1: A schematic diagram of a bioelectrochemical system (BES). Microbes can be immobilized in the cathodic and/or anodic parts of the BES.

times the energy required for its treatment (Gandiglio et al., 2017). This suggests that wastewater treatment process could produce a net positive energy if an ideal technology is utilized. An estimated 1.93 kWh/m^3 energy content is assumed to be inherent in a wastewater with 500 mg/L COD (McCarty et al., 2011).

Energy can be extracted in the form of electricity, hydrogen, methane, and other chemicals (Yuan et al., 2016). Therefore, it is important to employ a technology that can simultaneously harness this energy and treat waste while maintaining environmental integrity. Bioelectrochemical systems promise to achieve or contribute to this if successfully scaled up over time.

20.2 Microbial fuel cells (MFC)

In the past three decades, microbial fuel cells have received more attention due to the possibility of changing chemical energy in organic waste to electricity, and consequently being a potential part of the solution to energy demands (Santoro et al., 2017). Microbial Fuel Cells (MFCs) are electrochemical systems that use microorganisms as biological catalysts to extract and convert energy from sustainable fuel or biomass present in nature, industrial, and domestic waste. Microbial fuel cells (MFC) are a type of BES that produces electricity. It has been successfully used to treat a wide range of waste streams, and produces less sludge than with conventional methods (Oh et al., 2010; Arbianti et al., 2018). MFC utilizes microbes as biocatalysts in electrochemical reactions (Moqsud et al., 2013; Choudhury et al., 2017; Walter et al., 2015). The conventional MFC consists of a biological catalyst that breaks down organic and/or inorganic matter by oxidizing them. Consequently, electrons are released to the anode, which then flows through an external circuit into the cathode as shown in Figure 20.2. Protons are also released in the anodic chamber, which diffuse into the cathode chamber through proton exchange membrane (PEM) in a two-chamber system. The electrons and protons combine with oxygen in the cathode chamber to form water (Logan et al., 2006). The reactions that take place in both chambers of a typical MFC, with a carbon source at the anode and oxygen as the electron acceptor at the cathode, are shown in eqs. (20.1) and (20.2) (Santoro et al., 2017; Ucar et al., 2017). Other reactions are possible, depending on the substrate and the electron acceptor utilized in the anode and cathode, respectively. Glucose and acetate are common anode substrates, while oxygen, nitrate, and permanganate have been utilized as electron acceptors at the cathode (Lu and Li, 2012; Logan et al., 2006). The energy produced by MFCs is clean and renewable. They have been applied in pollution treatment and monitoring, energy generation, and remediation technology (Chen et al., 2016; Abourached et al., 2016). Although a lot of research has been carried out on different aspects of this technology, there is still a lot of work required for industrialization to be achieved (Santoro et al., 2017).

The cathodic reaction is known as the oxygen reduction reaction (ORR) when oxygen is the electron acceptor (reaction 20.2). This reaction takes place at $E^0 = 1.229$ V vs SHE (Standard Hydrogen Electrode) at standard temperature and pressure (Erable et al., 2012; Moore et al., 2013). The ORR is one of the most important reactions in a microbial fuel cell and it has a significant effect on the performance of the system (Moore et al., 2013). This is because the kinetics of the reaction are slow and the reaction has associated losses, which affect the overall performance of an MFC (Rismani-Yazdi et al., 2008).

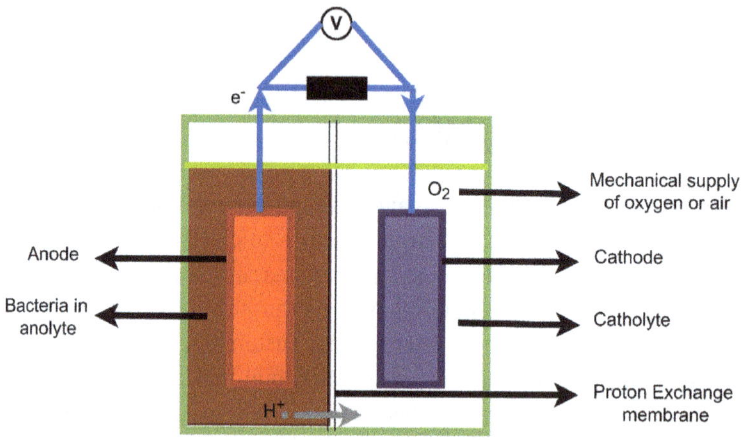

Figure 20.2: A schematic diagram of an MFC with biotic anode and abiotic cathode.

Anodic reaction

$$C + 4OH^- \rightarrow CO_2 + 2H_2O + 4e^- \tag{20.1}$$

C – Organic carbon

Cathodic reaction

$$O_2 + 4H^+ + 4e^- \rightarrow 2H_2O \tag{20.2}$$

For an MFC that uses acetate as anodic substrate, the Nernst anodic potential (E_{an}) is −0.3 V while the Nernst cathodic potential (E_{ca}) is 0.805 V (Yasri et al., 2019). The cell potential ($E_{cell} = E_{ca} - E_{an}$) is therefore 1.105 V. For a spontaneous reaction, the Gibbs free energy is less than zero and it is related to the cell potential by eq. (20.3).

$$\Delta G = -nFE_{cell} \tag{20.3}$$

Where: ΔG is Gibbs free energy, n is the number of electrons, F is the Faraday constant, and E_{cell} is the cell potential.

As the cell potential for an MFC is positive, the Gibbs free energy is negative, according to eq. (20.3). A negative Gibbs free energy connotes spontaneity (Yasri et al., 2019). Therefore, the reactions in a microbial fuel cell are spontaneous.

MFC is a more sustainable treatment process than conventional wastewater treatment processes in that it does not require input electricity (if oxygen is provided by algae), while it produces clean microelectricity (Toczylowska-Maminska, 2017). Also, Huggins et al. (2013) operated a waste treatment MFC and a traditional aeration chamber for waste treatment. They concluded that the MFC reduced sludge generation by 50%.

20.2.1 Oxygen reduction reaction (ORR)

In aqueous solutions, ORR can follow the 4-electron reduction pathway to water or the 2-electron reduction pathway to hydrogen peroxide, as detailed in Table 20.2. The more desirable pathway is the direct 4-electron pathway because the 2-electron pathway is slower and produces H_2O_2 (Table 20.2), which can act as an oxidizing agent that attacks the catalytic centers of the cathode and reduces the cathode potential (Harnisch and Schroder, 2010). The 4-electron pathway is favored by platinum electrodes while the 2-electron pathway is favored by carbon-based electrodes (Roche and Scott, 2009; Erable et al., 2012). The kinetic of oxygen reduction reaction is normally slow (Santoro et al., 2017). It is therefore necessary to catalyze the reaction to properly harness its way forward. In the last century, platinum has been the most commonly used catalyst in both the anode and cathode of fuel cell systems due to its effectiveness in adsorption and dissociation of gasses (Lamy et al., 2002; Brouzgou et al., 2012). Also, it is inert to most chemicals and has been identified to be a better catalyst for ORR than other metals such titanium, nickel, and cobalt (Norskov et al., 2004; Roche and Scott, 2009; Zhao et al., 2005; Wei et al., 2011).

Table 20.2: Reaction pathways of ORR adapted from Erable et al. (2012).

Electrolyte	ORR reactions	Electrode potential vs SHE (V)
Acidic	$O_2 + 4H^+ + 4e^- \rightarrow 2H_2O$	1.229
	$O_2 + 2H^+ + 2e^- \rightarrow H_2O_2$	0.70
	$H_2O_2 + 2H^+ + 2e^- \rightarrow 2H_2O$	1.76
Alkaline	$O_2 + 2H_2O + 4e^- \rightarrow 4OH^-$	0.401
	$O_2 + H_2O + 2e^- \rightarrow HO_2^- + OH^-$	−0.065
	$HO_2^- + H_2O + 2e^- \rightarrow 3OH^-$	0.867

However, this does not mean that platinum is the optimal catalytic material for ORR. It has some limitations for applications due to its cost, rarity, carbon monoxide poisoning, and unsustainability (Baschuk and Li, 2001; Ponrouch et al., 2009). To address these limitations, a few non-noble metals and carbon materials have been explored in recent research (Peera et al., 2021; Yuan et al., 2016). Examples of carbon-based materials utilized as electrodes for ORR are activated carbon, graphite, and glassy carbon. Activated carbon is the most used for ORR among this group of catalysts, possibly because of its high surface area that promotes reaction (Santoro et al., 2017; Wang et al., 2014). Although some of these have been commercialized as anode, there is still much work required for the development and commercialization of an inexpensive cathode, such as improving the effectiveness and speed of ORR (Norskov et al., 2004; Li et al., 2017; Santoro et al., 2014).

20.2.2 Immobilization of microbes on the cathode

Immobilization of microbes involves localizing an organism to a part of the system being investigated such that the organism is limited from free movement in such a system (Chen and Lin, 1994; Dunnill, 1988). Some of the methods of cell immobilization are entrapment, adsorption, and affinity immobilization (Lu et al., 2020). Immobilization helps to provide a better system design than suspension systems in terms of resilience (Mallick, 2002). Moreover, when cells are immobilized, the cell density and cell wall permeability are greatly improved (Sharma, 2017). Furthermore, immobilization improves products separation and reduces cost of environmental catalyst (Cortez et al., 2017). Immobilization can enable an otherwise catalytically inert material to become functionalized (Bernal et al., 2017). In bioelectrochemical systems, the formation of a biofilm has been linked to a higher system performance, but too thick biofilms may limit electron transfers (Wang et al., 2012). Baranitharan et al. (2015) observed in their work on a palm oil mill effluent treatment MFC that there was a significant power density increase as the initial biofilm was developing. Immobilization on the electrodes, therefore, helps to provide a uniform, high cell density biofilm. The physical biofilm immobilization protects cells from eventual high-shear stream disruption.

Microorganisms can also be immobilized to form a biocomposite. Biocomposites are made up of two or more distinct constituent materials; one of which is naturally derived, combining to yield a new material with a better performance when compared with the performance of the individual components. The constituent materials are the matrix and the reinforcing component (Desmet et al., 2015; Touloupakis et al., 2016; Sawa et al., 2017). An example of a matrix used is paper, while carbon fiber, glass fiber, and cellulose microfibrils are examples of reinforcing materials (Orts et al., 2005).

In preparing an algae biocomposite, incorporating the algae into the production process of a porous immobilization substrate (e.g. paper) is better than applying the microbes to the surface of the substrate as it allows microbes to be attached to all parts of a porous substrate, instead of only the surface. This, in effect, increases the functionality of the resulting biocomposite, when applied to bioelectrochemical processes (Bernal et al., 2017). Furthermore, a higher cell concentration per unit surface area is achieved when the cells are attached to the surfaces within the porous biocomposite structure (Ekins-Coward et al., 2019).

Paper has been available for a long time, and it is produced from renewable resources (Rojas and Hubbe, 2004). It is available almost everywhere in the world and the experience garnered in the production process over the years can be applied to novel uses of the material. Paper is simple and inexpensive to make (Fairchild, 2000). It is also relatively environmentally friendly when compared to platinum, which is the most commonly used catalyst in the cathode of microbial electrolysis cells as it is obtained from a renewable source (Desmet et al., 2015; Bernal et al., 2017; Sawa et al., 2017). Also, paper is porous and this may be a good advantage in its application in BES as it has been observed that BES systems with porous electrodes show a considerable increase in performance (Bajracharya et al., 2016).

The development of biocoated algal paper electrodes in bioelectrochemical systems a novel advancement in bioelectrochemical systems. Sawa et al. (2017) recently developed a thin-film paper-based biophotovoltaic cell. They digitally printed *Synechocystis* sp PCC 6803 on a carbon nanotube conducting surface. Then, they incorporated this in a photovoltaic cell, which was capable of producing current in light and dark.

20.3 A case for the use of microalgae in MFC

Research has shown that the photosynthetic efficiency of green algae is higher than that of the higher terrestrial plants (Keffer and Kleinheinz, 2002). The CO_2 sequestration efficiency of algae can be as high as 80 % to 99 %, depending on the temperature, the physiology of the species, and other factors (Sayre, 2010). This results in the production of approximately 75 % of oxygen on Earth (Liu et al., 2015). Since ORR is one of the most important reactions in an MFC, the presence of oxygen-producing algae offers the potential to contribute significantly to the improvement of performance of the system (Santoro et al., 2017). Powell et al. (2009) suggested that algae can act as electron acceptors using redox mediators. They suggested that electrons reduce the mediator and the algae absorbs the reduced mediator (Figure 20.3). The algae then extract the electron into the CO_2 to form the carbohydrate pathway, thereby oxidizing the mediator for reutilization. This process makes algae a good biocatalyst for cathodic reactions in the MFC where an electron acceptor is required. Furthermore, introducing algae into an MFC set up may enable the system to directly harness solar energy for conversion to chemical energy and electrical energy.

Utilization of algae also suggests pollution control through carbon sequestration (Melnicki et al., 2008; Rozendal et al., 2008). Algae have been known to thrive in various conditions, even when nutrients are not abundant (Saratale et al., 2017). Therefore, introducing algae into MFCs could enhance waste treatment potential because algae have been used for removal of toxic waste from wastewater and to monitor

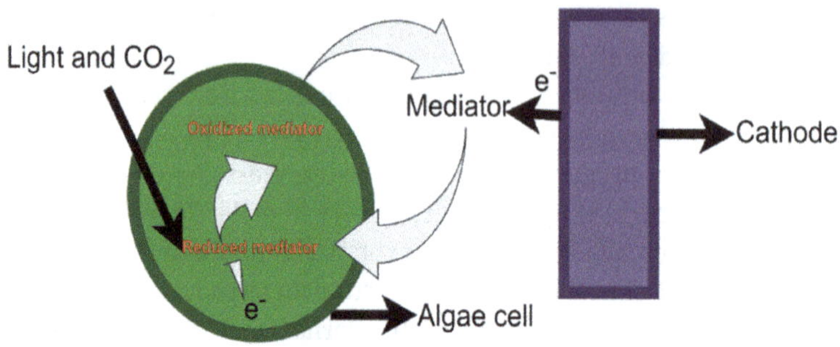

Figure 20.3: Algae acting as an electron acceptor using mediating redox electron compounds. Adapted from Powell (2009).

some metals such as zinc and copper (Labro et al., 2017; Santoro et al., 2017). Other wastes such as phosphorus can be removed from the cathodic part of an MFC by algae since it serves as a nutrient for algae (Yang et al., 2018). Oxygen production by the algae in an MFC means that the energy requirement for aeration is eliminated, thereby reducing the cost. Kokabian and Gude (2013) observed that when an algae species (*Chlorella vulgaris*) was used as a cathode in desalination cells, electric current generation and BOD removal were enhanced, in comparison with an air cathode. Similarly, Neethu and Ghangrekar (2017) observed a sediment removal efficiency of 77.6 % in their sediment microbial fuel cell (SMFC), which was 16.6 % higher than the efficiency observed for a similar system with no algae in the cathode. They also observed 3.65 times higher power density with their algae cathode SMFC than the power density in the system without algae. In other studies, algae utilization in the cathode has been shown to improve wastewater treatment efficiency and the overall performance of an MFC (Yuan et al., 2016; Reddy et al., 2019). For instance, Yang et al. (2018) achieved nitrogen and phosphorus removal efficiencies of 96.0 % and 91.5 %, respectively, using an algae biofilm-assisted MFC, while an abiotic MFC achieved 34.0 % and 10.8 % removal efficiencies for the same parameters.

Chlorella vulgaris has been well studied in bioelectrochemical systems and has a high photosynthetic productivity over a broad temperature range (Singh and Singh, 2015; Velasquez-Orta et al., 2009).

20.4 State-of-the-art applications of algae-based MFC

Studies on algae-based MFC indicate that algae provides the oxygen requirement for the cathodic reactions and eliminates the need for mechanical aeration (Xiao et al., 2015). In their work on an MFC with a photoautotrophic cathode, Kakarla and Min

(2014) used *Scenedesmus Obliquus* as the biological component of the cathode. They had wastewater in their anode chamber. They observed a maximum power of 153 mW/m^2. This was higher than the maximum power of 116 mW/m^2 when they used mechanical aeration in the cathode. They also observed that it took about 3 days for an algal biofilm to be formed after the introduction of the microorganism to the cathode. This resulted in a delay in power generation as the oxygen source was not initially close to the cathode. They concluded that proximity of the oxygen source to the cathode influences the performance of the electrode and power output. This suggests that application of a biocomposite electrode, as proposed in this project, would be a quicker way of bringing the source of oxygen close to the electrode.

Xiao et al. (2012) investigated an MFC for waste treatment and electricity production. They applied anaerobic sludge and a green algae, *Pseudokirchneriella subcapitata*, in their anode and cathode compartments, respectively. They observed more than 90% removal of pollutants and a peak power density of 2.2 W/m^3. This high percentage of pollutant removal rate is typical for MFCs (Gude, 2016).

Investigation of the effect of light intensity on the photosynthetic cathode has also been carried out. Gouveia et al. (2014) studied an MFC that had a microbial consortium in the anode compartment and *Chlorella vulgaris* (suspension) in the cathode compartment. They observed a 6-fold increase in power density when they increased the light intensity from 26 $\mu E/(m^2\ s)$ to 96 $\mu E/(m^2\ s)$. They were able to achieve a power density of 67.2 mW/m^2. They suggested the light intensity had a positive influence on the photosynthetic activities of the algae and, consequently, on the performance of the MFC. As algae require natural light in their natural environment and the supply of light is not constant, researchers have attempted to mimic the natural light fluctuation in an MFC environment and investigate how this affects the performance of the system. del Campo et al. (2013) applied a 12-hour light/dark cycle in an MFC that had activated sludge in the anode and *Chlorella vulgaris* in the cathode. They monitored the dissolved oxygen in the cathode and corresponding system power density. They observed that the cell voltage increased as the dissolved oxygen increased. They were able to achieve a maximum power density of 13.5 mW/m^2 when the system reached steady state conditions. Also, Commault et al. (2017) observed a maximum power density of 15.6 ± 9.7 mW/m^2 with an abiotic cathode and this was less than half of what was observed with a *Chlorella vulgaris* cathode. They achieved a maximum power density of 34.2 ± 10.0 mW/m^2 in the light regime which then completely reduced in the dark regime. This performance was found to be related to oxygen production by the *Chlorella vulgaris* cells in the cathodic compartment of the MFC. However, they observed that the correlation between the oxygen concentration and the power output was not very strong as most of the algal community responsible for power production had formed a biofilm on the electrode surface while the dissolved oxygen concentration was associated mostly with the algal cells in the bulk liquid. This emphasizes the importance of biofilms in the cathode compartment of an MFC. It also suggests that artificial immobilization of the algae may improve the performance of an MFC by reducing the need for diffusion of oxygen from the bulk liquid to the electrode surface. Bazdar et al. (2018) studied the effects of different light regimes on a *Chlorella vulgaris* cathode MFC. They observed that continuous illumination produced the best volumetric power density. The system achieved a volumetric power density of 126 mW/m^3 during continuous illumination that was 74.8% higher than the performance when the illumination regime was 12 hr / 12 hr light /dark cycle and 12.7% higher than the performance when the illumination regime was 16 hr / 8 hr light /dark cycle. This suggests that continuous illumination favours the power production of an algae cathode MFC. However, there are concerns about the longevity of such systems as continuous illumination is not favourable for effective algae growth (Csavina et al., 2011).

Cui et al. (2014) went a step further by feeding a bacteria consortium present in an anaerobic activated sludge sample in an anodic chamber of an MFC with dead *Scenedesmus obliquus* biomass. They had *Chlorella vulgaris* in the cathodic chamber as the biocatalyst and recycled the CO_2 generated in the anode chamber into the cathodic chamber for photosynthetic activity. They compared the system with a system where the anode chamber was fed with acetate, instead of algal biomass. They observed a maximum power density of 1926 mW/m^2 with the algae-fed system, while 1110 mW/m^2 was achieved when acetate was fed. They also observed that the algae-fed anodic chamber produced more CO_2 and consequently promoted better growth in the cathodic chamber than the acetate-fed system. This is because the CO_2 supported the growth of *Chlorella vulgaris* and, consequently, the cathodic performance. In a similar experiment carried out by Liu et al. (2015) with the same algae species (*Chlorella vulgaris*) in the cathode but without an algae biomass feed in the anode, they fed the anode with acetate, and a maximum power of 187 mW/m^2 was achieved. This power output is less than what was observed by (Cui et al., 2014), possibly because the supply of CO_2 was insufficient when compared with the algae-fed anode system.

From the aforementioned, it is evident that the introduction of algae biomass into the anode compartment and/or having an algal cathode compartment of an MFC enhanced the power production of the system. However, there is still a need for further improvement such as immobilization of the algae on cathode, which may enhance the performance, as it is a way of creating an artificial biofilm that has been shown to promote a better performance in bioelectrochemical systems (Baranitharan et al., 2015; Kakarla and Min, 2014).

20.4.1 Immobilized microalgae cathodes

Some researchers have explored the effects of immobilization of algae in the cathodic compartment of an MFC on the performance of the system. Ruiz-Marin et al., 2010 immobilized *Chlorella vulgaris* using sodium alginate and calcium chloride to form algal beads. These algal beads were utilized by Zhou et al. (2012) as cathodic biocatalyst in microbial fuel cells and compared with *Chlorella vulgaris* suspension in the cathode.

They observed that although the immobilization limited the growth rate of the algal beads, the MFC with immobilized algal beads produced 464 mV of voltage, while the suspended *Chlorella vulgaris* cells produced a lower value of 443 mV. Similarly, the stability of the voltage produced by the immobilized algae system was 20 hours longer than for the suspension system. Their research suggested that immobilizing the *Chlorella vulgaris* in the cathode chamber enhanced the overall performance of the system by possibly reducing the oxygen mass transport limitation between the bulk liquid and the cathode. This suggests that immobilization of the cells on the cathode may be a better way of reducing mass transport limitation, further improving the performance.

Logrono et al. (2017) immersed carbon fiber cathodes in a culture of *Chlorella vulgaris* for 12 days to form an immobilized algae layer on the cathode. This was applied in a microbial fuel cell. They had an abiotic cathode as the blank. The cathodes were exposed directly to air. They recorded an open circuit potential (OCP) of 182 mV for the immobilized algae cathode, and this increased to 387 mV after 5 days. On the other hand, the blank had an OCP of 130 mV, which increased to 250 mV. The initial and final OCPs of the immobilized algae cathode were more than the recorded values for the blank. Other performance indicators, such as wastewater color and COD removals, were also examined. A maximum color removal efficiency of 93 % was observed for the immobilized algae cathode, while the blank had recorded 79 %. Similarly, a maximum COD removal of 98 % was observed for the immobilized algae cathode while the blank had recorded 95 %. The research suggests that immobilized *Chlorella vulgaris* enhanced the production of electricity and the overall performance of the microbial fuel cell system. However, this study did not compare the performance of the immobilized algae with algae in suspension in the cathodic compartment of the MFC. Our laboratory is currently researching the impact of a biocomposite paper electrode being used as an immobilised cathodic structure. Early results indicate much promise for this technology in BES.

20.5 Conclusions and recommendations

Within the different bioelectrochemical systems under study, microalga has mainly been applied in microbial fuel cells. In MFCs, the literature has shown the feasibility of using microalgae to facilitate completion of a redox reaction. It is inferred that microalgae facilitates the production of oxygen in the system and its transfer to the cathode electrode. More work can be undertaken to determine the mechanism by which microalgae facilitate this redox transfer and whether electron transfer is facilitated via the use of mediators, the production of oxygen from autotrophic growth, or via direct electron transfer. Research should focus on this area in order to facilitate understanding of the role of microalgae that can lead to optimization of the process conditions. Furthermore, immobilization techniques applied in other systems can be adapted for microalgae cathodes in microbial fuel cells for improved performance and ease of scale up.

References

Abourached, C., English, M. J., & Liu, H. (2016). Wastewater treatment by Microbial Fuel Cell (MFC) prior irrigation water reuse. *Journal of Cleaner Production, 137*, 144–149.

Arbianti, R., Utami, T. S., Leondo, V., Elisabeth, Putri, S. A., & Hermansyah, H. (2018). Effect of biofilm and selective mixed culture on microbial fuel cell for the treatment of tempeh industrial wastewater. *Quality in Research: International Symposium on Materials, Metallurgy, and Chemical Engineering, 316*(1), 012073.

Bajracharya, S., Sharma, M., Mohanakrishna, G., Benneton, X. D., Strik, D. P. B. T. B., Sarma, P. M., & Pant, D. (2016). An overview on emerging bioelectrochemical systems (BESs): Technology for sustainable electricity, waste remediation, resource recovery, chemical production and beyond. *Renewable Energy, 98*, 153–170.

Baranitharan, E., Khan, M. R., Prasad, D. M. R., Teo, W. F. A., Tan, G. Y. A., & Jose, R. (2015). Effect of biofilm formation on the performance of microbial fuel cell for the treatment of palm oil mill effluent. *Bioprocess and Biosystems Engineering, 38*, 15–24.

Baschuk, J. J., & Li, X. G. (2001). Carbon monoxide poisoning of proton exchange membrane fuel cells. *International Journal of Energy Research, 25*, 695–713.

Bazdar, E., Roshandel, R., Yaghmaei, S., & Mardanpour, M. M. (2018). The effect of different light intensities and light/dark regimes on the performance of photosynthetic microalgae microbial fuel cell. *Bioresource Technology, 261*, 350–360.

Bernal, O. I., Pawlak, J. J., & Flickinger, M. C. (2017). Microbial paper: Cellulose fiber-based photo-absorber producing hydrogen gas from acetate using dry-stabilized Rhodopseudomonas palustris. *Bioresources, 12*, 4013–4030.

Brouzgou, A., Song, S. Q., & Tsiakaras, P. (2012). Low and non-platinum electrocatalysts for PEMFCs: Current status, challenges and prospects. *Applied Catalysis B-Environmental, 127*, 371–388.

Chen, K. C., & Lin, Y. F. (1994). Immobilization of microorganisms with phosphorylated polyvinyl-alcohol (PVA) gel. *Enzyme and Microbial Technology, 16*, 79–83.

Chen, Z. J., Niu, Y. Y., Zhao, S., Khan, A. A., Ling, Z. M., Chen, Y., Liu, P., & Li, X. K. (2016). A novel biosensor for p-nitrophenol based on an aerobic anode microbial fuel cell. *Biosensors and Bioelectronics, 85*, 860–868.

Choudhury, P., Uday, U. S. P., Mahata, N., Tiwari, O. N., Ray, R. N., Bandyopadhyay, T. K., & Bhunia, B. (2017). Performance improvement of microbial fuel cells for waste water treatment along with value addition: A review on past achievements and recent perspectives. *Renewable and Sustainable Energy Reviews, 79*, 372–389.

Commault, A. S., Laczka, O., Siboni, N., Tamburic, B., Crosswell, J. R., Seymour, J. R., & Ralph, P. J. (2017). Electricity and biomass production in a bacteria-Chlorella based microbial fuel cell treating wastewater. *Journal of Power Sources, 356*, 299–309.

Cortez, S., Nicolau, A., Flickinger, M. C., & Mota, M. (2017). Biocoatings: A new challenge for environmental biotechnology. *Biochemical Engineering Journal, 121*, 25–37.

Csavina, J. L., Stuart, B. J., Riefler, R. G., & Vis, M. L. (2011). Growth optimization of algae for biodiesel production. *Journal of Applied Microbiology, 111*, 312–318.

Cui, Y. F., Rashid, N., Hu, N. X., Rehman, M. S. U., & Han, J. I. (2014). Electricity generation and microalgae cultivation in microbial fuel cell using microalgae-enriched anode and bio-cathode. *Energy Conversion and Management*, *79*, 674–680.

Del Campo, A. G., Canizares, P., Rodrigo, M. A., Fernandez, F. J., & Lobato, J. (2013). Microbial fuel cell with an algae-assisted cathode: A preliminary assessment. *Journal of Power Sources*, *242*, 638–645.

Desmet, C., Marquette, C. A., Blum, L. J., & Doumeche, B. (2015). Paper electrodes for bioelectrochemistry: Biosensors and biofuel cells. *Biosensors and Bioelectronics*, *76*, 145–163.

Dunnill, P. (1988). Immobilized cells – Principles and applications – J. Tampion and MD Tampion. *Nature*, *332*, 316–317.

Ekins-Coward, T., Boodhoo, K. V. K., Velasquez-Orta, S., Caldwell, G., Wallace, A., Barton, R., & Flickinger, M. C. (2019). A microalgae biocomposite-integrated spinning disk bioreactor (SDBR): Toward a scalable engineering approach for bioprocess intensification in light-driven CO2 absorption applications. *Industrial and Engineering Chemistry Research*, *58*, 5936–5949.

Erable, B., Feron, D., & Bergel, A. (2012). Microbial catalysis of the oxygen reduction reaction for microbial fuel cells: A review. *Chemsuschem*, *5*, 975–987.

Fairchild, C. A. (2000). Arnold Grummer's complete guide to easy papermaking. *Library Journal*, *125*, 160–160.

Gandiglio, M., Lanzini, A., Soto, A., Leone, P., & Santarelli, M. (2017). Enhancing the energy efficiency of wastewater treatment plants through co-digestion and fuel cell systems. *Frontiers in Environmental Science*, *5*(70).

Gouveia, L., Neves, C., Sebastiao, D., Nobre, B. P., & Matos, C. T. (2014). Effect of light on the production of bioelectricity and added-value microalgae biomass in a photosynthetic alga microbial fuel cell. *Bioresource Technology*, *154*, 171–177.

Gude, V. G. (2016). Wastewater treatment in microbial fuel cells – An overview. *Journal of Cleaner Production*, *122*, 287–307.

Harnisch, F., & Schroder, U. (2010). From MFC to MXC: Chemical and biological cathodes and their potential for microbial bioelectrochemical systems. *Chemical Society Reviews*, *39*, 4433–4448.

Huggins, T., Fallgren, P. H., Song, J., & Ren, Z. J. (2013). Energy and performance comparison of microbial fuel cell and covectional aeration treating of wastewater. *Microbial and Biochemical Technology* S6:002.

Kakarla, R., & Min, B. (2014). Photoautotrophic microalgae Scenedesmus obliquus attached on a cathode as oxygen producers for microbial fuel cell (MFC) operation. *International Journal of Hydrogen Energy*, *39*, 10275–10283.

Keffer, J. E., & Kleinheinz, G. T. (2002). Use of Chlorella vulgaris for CO2 mitigation in a photobioreactor. *Journal of Industrial Microbiology and Biotechnology*, *29*, 275–280.

Kokabian, B., & Gude, V. G. (2013). Photosynthetic microbial desalination cells (PMDCs) for clean energy, water and biomass production. *Environmental Science-Processes and Impacts*, *15*, 2178–2185.

Labro, J., Craig, T., Wood, S. A., & Packer, M. A. (2017). Demonstration of the use of a photosynthetic microbial fuel cell as an environmental biosensor. *International Journal of Nanotechnology*, *14*, 213–225.

Lamy, C., Lima, A., Lerhun, V., Delime, F., Coutanceau, C., & Leger, J. M. (2002). Recent advances in the development of direct alcohol fuel cells (DAFC). *Journal of Power Sources*, *105*, 283–296.

Li, S., Cheng, C., & Thomas, A. (2017). Carbon-based microbial-fuel-cell electrodes: From conductive supports to active catalysts. *Advanced Materials*, *29*(8).

Liu, T., Rao, L., Yuan, Y., & Zhuang, L. (2015). Bioelectricity generation in a microbial fuel cell with a self-sustainable photocathode. *Scientific World Journal*, *2015*, 864568.

Logan, B. E., Hamelers, B., Rozendal, R. A., Schrorder, U., Keller, J., Freguia, S., Aelterman, P., Verstraete, W., & Rabaey, K. (2006). Microbial fuel cells: Methodology and technology. *Environmental Science and Technology*, *40*, 5181–5192.

Logrono, W., Perez, M., Urquizo, G., Kadier, A., Echeverria, M., Recalde, C., & Rakhely, G. (2017). Single chamber microbial fuel cell (SCMFC) with a cathodic microalgal biofilm: A preliminary assessment of the generation of bioelectricity and biodegradation of real dye textile wastewater. *Chemosphere, 176*, 378–388.

Lu, J. S., Peng, W. F., LV, Y., Jiang, Y. J., Xu, B., Zhang, W. M., Zhou, J., Dong, W. L., Xin, F. X., & Jiang, M. (2020). Application of cell immobilization technology in microbial cocultivation systems for biochemicals production. *Industrial and Engineering Chemistry Research, 59*, 17026–17034.

Lu, M., & LI, S. F. Y. (2012). Cathode reactions and applications in microbial fuel cells: A review. *Critical Reviews in Environmental Science and Technology, 42*, 2504–2525.

Luo, S., Sun, H. Y., Ping, Q. Y., Jin, R., & He, Z. (2016). A review of modeling bioelectrochemical systems: Engineering and statistical aspects. *Energies, 9*(2), 111.

Mallick, N. (2002). Biotechnological potential of immobilized algae for wastewater N, P and metal removal: A review. *Biometals, 15*, 377–390.

Mccarty, P. L., Bae, J., & Kim, J. (2011). Domestic wastewater treatment as a net energy producer-Can this be achieved?. *Environmental Science and Technology, 45*, 7100–7106.

Melnicki, M., Bianchi, L., Dephilippis, R., & Melis, A. (2008). Hydrogen production during stationary phase in purple photosynthetic bacteria. *International Journal of Hydrogen Energy, 33*, 6525–6534.

Mook, W. T., Aroua, M. K. T., Chakrabarti, M. H., Noor, I. M., Irfan, M. F., & Low, C. T. J. (2013). A review on the effect of bio-electrodes on denitrification and organic matter removal processes in bio-electrochemical systems. *Journal of Industrial and Engineering Chemistry, 19*, 1–13.

Moore, M., Putz, A., & Secanell, M. (2013). Investigation of the ORR using the double-trap intrinsic kinetic model. *Journal of the Electrochemical Society, 160*, F670–F681.

Moqsud, M. A., Omine, K., Yasufuku, N., Hyodo, M., & Nakata, Y. (2013). Microbial fuel cell (MFC) for bioelectricity generation from organic wastes. *Waste Management, 33*, 2465–2469.

Neethu, B., & Ghangrekar, M. M. (2017). Electricity generation through a photo sediment microbial fuel cell using algae at the cathode. *Water Science and Technology, 76*, 3269–3277.

Norskov, J. K., Rossmeisl, J., Logadottir, A., Lindqvist, L., Kitchin, J. R., Bligaard, T., & Jonsson, H. (2004). Origin of the overpotential for oxygen reduction at a fuel-cell cathode. *Journal of Physical Chemistry B, 108*, 17886–17892.

Oh, S. T., Kim, J. R., Premier, G. C., Lee, T. H., Kim, C., & Sloan, W. T. (2010). Sustainable wastewater treatment: How might microbial fuel cells contribute. *Biotechnology Advances, 28*, 871–881.

Orts, W. J., Shey, J., Imam, S. H., Glenn, G. M., Guttman, M. E., & Revol, J. F. (2005). Application of cellulose microfibrils in polymer nanocomposites. *Journal of Polymers and the Environment, 13*, 301–306.

Peera, S. G., Maiyalagan, T., LIU, C., Ashmath, S., Lee, T. G., Jiang, Z. Q., & Mao, S. (2021). A review on carbon and non-precious metal based cathode catalysts in microbial fuel cells. *International Journal of Hydrogen Energy, 46*, 3056–3089.

Ponrouch, A., Garbarino, S., & Guay, D. (2009). Effect of the nanostructure on the CO poisoning rate of platinum. *Electrochemistry Communications, 11*, 834–837.

Powell, E. E., Mapiour, M. L., Evitts, R. W., & Hill, G. A. (2009). Growth kinetics of Chlorella vulgaris and its use as a cathodic half cell. *Bioresource Technology, 100*, 269–274.

Rabaey, K., & Rozendal, R. A. (2010). Microbial electrosynthesis – Revisiting the electrical route for microbial production. *Nature Reviews Microbiology, 8*, 706–716.

Reddy, C. N., Nguyen, H. T. H., Noori, M. T., & Min, B. (2019). Potential applications of algae in the cathode of microbial fuel cells for enhanced electricity generation with simultaneous nutrient removal and algae biorefinery: Current status and future perspectives. *Bioresource Technology, 292*, 122010.

Rismani-Yazdi, H., Carver, S. M., Christy, A. D., & Tuovinen, I. H. (2008). Cathodic limitations in microbial fuel cells: An overview. *Journal of Power Sources, 180*, 683–694.

Roche, I., & Scott, K. (2009). Carbon-supported manganese oxide nanoparticles as electrocatalysts for oxygen reduction reaction (orr) in neutral solution. *Journal of Applied Electrochemistry, 39*, 197–204.

Rojas, O. J., & Hubbe, M. A. (2004). The dispersion science of papermaking. *Journal of Dispersion Science and Technology*, *25*, 713–732.
Rozendal, R. A., Hamelers, H. V. M., Rabaey, K., Keller, J., & Buisman, C. J. N. (2008). Towards practical implementation of bioelectrochemical wastewater treatment. *Trends in Biotechnology*, *26*, 450–459.
Ruiz-Marin, A., Mendoza-Espinosa, L. G., & Stephenson, T. (2010). Growth and nutrient removal in free and immobilized green algae in batch and semi-continuous cultures treating real wastewater. *Bioresource Technology*, *101*, 58–64.
Santoro, C., Arbizzani, C., Erable, B., & Ieropoulos, I. (2017). Microbial fuel cells: From fundamentals to applications. A review. *Journal of Power Sources*, *356*, 225–244.
Santoro, C., Artyushkova, K., Babanova, S., Atanassov, P., Ieropoulos, I., Grattieri, M., Cristiani, P., Trasatti, S., LI, B. K., & Schuler, A. J. (2014). Parameters characterization and optimization of activated carbon (AC) cathodes for microbial fuel cell application. *Bioresource Technology*, *163*, 54–63.
Saratale, R. G., Kuppam, C., Mudhoo, A., Saratale, G. D., Periyasamy, S., Zhen, G. Y., Kook, L., Bakonyi, P., Nemestothy, N., & Kumar, G. (2017). Bioelectrochemical systems using microalgae – A concise research update. *Chemosphere*, *177*, 35–43.
Sawa, M., Fantuzzi, A., Bombelli, P., Howe, C. J., Hellgardt, K., & Nixon, P. J. (2017). Electricity generation from digitally printed cyanobacteria. *Nature Communications*, *8*, 1327.
Sayre, R. (2010). Microalgae: The potential for carbon capture. *Bioscience*, *60*, 722–727.
Sharma, A., & ARYA, K. S. (2017). Hydrogen from algal biomass: A review of production process. *Biotechnology Reports*, *15*, 63–69.
Singh, S. P., & Singh, P. (2015). Effect of temperature and light on the growth of algae species: A review. *Renewable and Sustainable Energy Reviews*, *50*, 431–444.
Sleutels, T. H. J. A., Ter Heijne, A., Buisman, C. J. N., & Hamelers, H. V. M. (2012). Bioelectrochemical systems: An outlook for practical applications. *ChemSusChem*, *5*, 1012–1019.
Toczylowska-Maminska, R. (2017). Limits and perspectives of pulp and paper industry wastewater treatment – A review. *Renewable and Sustainable Energy Reviews*, *78*, 764–772.
Touloupakis, E., Rontogiannis, G., Benavides, A. M. S., Cicchi, B., Ghanotakis, D. F., & Torzillo, G. (2016). Hydrogen production by immobilized Synechocystis sp PCC 6803. *International Journal of Hydrogen Energy*, *41*, 15181–15186.
Ucar, D., Zhang, Y. F., & Angelidaki, I. (2017). An overview of electron acceptors in microbial fuel cells. *Frontiers in Microbiology*, *8*, 634.
Velasquez-Orta, S. B., Curtis, T. P., & Logan, B. E. (2009). Energy from algae using microbial fuel cells. *Biotechnology and Bioengineering*, *103*, 1068–1076.
Walter, X. A., Greenman, J., Taylor, B., & Ieropoulos, I. A. (2015). Microbial fuel cells continuously fuelled by untreated fresh algal biomass. *Algal Research-Biomass Biofuels and Bioproducts*, *11*, 103–107.
Wang, Q. F., Suraweera, N. S., Keffer, D. J., Deng, S. X., & Mays, J. (2012). Atomistic and coarse-grained molecular dynamics simulation of a cross-linked sulfonated poly(1,3-cyclohexadiene)-based proton exchange membrane. *Macromolecules*, *45*, 6669–6685.
Wang, Z. J., Cao, C. L., Zheng, Y., Chen, S. L., & Zhao, F. (2014). Abiotic oxygen reduction reaction catalysts used in microbial fuel cells. *ChemElectroChem*, *1*, 1813–1821.
Wei, J. C., Liang, P., & Huang, X. (2011). Recent progress in electrodes for microbial fuel cells. *Bioresource Technology*, *102*, 9335–9344.
Xiao, L., Young, E. B., Berges, J. A., & He, Z. (2012). Integrated photo-bioelectrochemical system for contaminants removal and bioenergy production. *Environmental Science and Technology*, *46*, 11459–11466.
Xiao, L., Young, E. B., Grothjan, J. J., Lyon, S., Zhang, H. S., & He, Z. (2015). Wastewater treatment and microbial communities in an integrated photo-bioelectrochemical system affected by different wastewater algal inocula. *Algal Research-Biomass Biofuels and Bioproducts*, *12*, 446–454.

Yadav, G., Sharma, I., Ghangrekar, M., & Sen, R. (2020). A live bio-cathode to enhance power output steered by bacteria-microalgae synergistic metabolism in microbial fuel cell. *Journal of Power Sources, 449*.

Yang, Z. G., Pei, H. Y., Hou, Q. J., Jiang, L. Q., Zhang, L. J., & Nie, C. L. (2018). Algal biofilm-assisted microbial fuel cell to enhance domestic wastewater treatment: Nutrient, organics removal and bioenergy production. *Chemical Engineering Journal, 332*, 277–285.

Yasri, N., Roberts, E. P. L., & Gunasekaran, S. (2019). The electrochemical perspective of bioelectrocatalytic activities in microbial electrolysis and microbial fuel cells. *Energy Reports, 5*, 1116–1136.

Yuan, H. Y., Hou, Y., Abu-Reesh, I. M., Chen, J. H., & He, Z. (2016). Oxygen reduction reaction catalysts used in microbial fuel cells for energy-efficient wastewater treatment: A review. *Materials Horizons, 3*, 382–401.

Zhao, F., Harnisch, F., Schroder, U., Scholz, F., Bogdanoff, P., & Herrmann, I. (2005). Application of pyrolysed iron(II) phthalocyanine and CoTMPP based oxygen reduction catalysts as cathode materials in microbial fuel cells. *Electrochemistry Communications, 7*, 1405–1410.

Zhou, M. H., He, H. H., Jin, T., & Wang, H. Y. (2012). Power generation enhancement in novel microbial carbon capture cells with immobilized Chlorella vulgaris. *Journal of Power Sources, 214*, 216–219.

Mariany Costa Deprá*, Leila Queiroz Zepka, Eduardo Jacob-Lopes

Chapter 21 Process integration and process intensification approaches as enhancers of industrial sustainability in microalgae-based systems

Abstract: Nowadays, planning and designing a more viable industrial process is one of the fundamental challenges in cleaner production progress. Microalgae-based systems have emerged in the scientific field as they are cleaner sources of bioenergy, offering bioproducts with several advantages over conventional cultures. However, criticisms attributed to these processes center on their technological practicality, and economic and environmental feasibility. In this way, processes integration and intensification strategies were implemented in the microalgae context to offer a technological leap and to advance the development of cleaner and more sustainable processes. In the light of this development, we review the theory and practice applied to the integration and intensification processes associated with microalgae-based systems in this chapter. Besides, this work presents an overview of the status of process integration and intensification and their positive relationships with sustainability aspects. To support these assumptions, recent studies have been gathered and presented, which correlated with the engineering strategies and reductions in environmental burdens.

Keywords: microalgae, engineering processes, environmental indicators, sustainability approaches

21.1 Background

Sustainability touches all facets of society. The industrial manufacturing sector, as one of the largest dependents on natural resources, plays a crucial role in decreasing the magnitude of environmental burden (Walmsley et al., 2018). A proof of this is that, today, processing industries are constantly pressured to implement methods and

*Corresponding author: Mariany Costa Deprá, Bioprocess Intensification Group, Federal University of Santa Maria, UFSM, Roraima Avenue 1000, 97105-900, Santa Maria, RS, Brazil, e-mail: marianydepra@gmail.com
Leila Queiroz Zepka, Eduardo Jacob-Lopes, Bioprocess Intensification Group, Federal University of Santa Maria, UFSM, Roraima Avenue 1000, 97105–900, Santa Maria, RS, Brazil

https://doi.org/10.1515/9783110781267-021

practical solutions to maintain economic viability while adapting to environmentally sustainable models.

To this end, engineering tools and methods based on process integration and intensification have emerged in recent years. In essence, process integration can be defined as a holistic design and operation approach that focuses on the working of the process plant. Its first application was in heat recovery. However, the similarity between materials with regard heat and mass, gave rise to new variations of integration, necessitating the emergence of the concepts of mass and water integration (BOX 1). Regardless of the specific integration, all its variations are supported by the identification of quality and quantity aspects for a given planning problem (Kong et al., 2021). In this way, it can be said that the integration process consists of a systematic way of identifying and correcting inefficiencies in a process since it analyzes the global process and its interactions between the different stages at an individual level before paying.

Box 1: Definitions of energy, mass, and water integration

Energy integration is one of the integration processes based on thermodynamic principles. In short, energy integration encompasses all forms of energy (such as heating, cooling, energy generation/consumption, pressurization, and fuels) and exergy balances (García-García et al., 2019; Sinnott and Towler, 2020).

Mass integrations are based on the identification of chemical constituents of a physical body. However, they have some obstacles related to the mass conservation law, which is dependent on the properties of material flows and not necessarily on chemical compositions (Klemes, 2013). In this way, mass integration provides an understanding of material flows using mass balances in order to track the best path of allocation, separation, recovery, and generation flow of mass species (El-Halwagi, 2017).

Water integration is a technique used to minimize freshwater usage in process industries. The most common forms of water integration in the industry is in operations such as cooling water and steam production in water treatment systems and inside the design of the heat exchanger network itself (Elena Savulescu and Alva-Argaez, 2013).

On the other hand, process intensification is an opportunity to integrate operations or functions in order to improve existing processes. Technically, intensification occurs when improvement strategies are implemented, thus substantially reducing the relationship between the equipment sizes and production, waste generation, as well as in the consumption of inputs (feedstock and energy). These changes result in environmentally and economically sustainable operating plants. Nevertheless, its use can also be considered as a process design and, therefore, potentially ensure a more cost-effective process at an early stage of development (Baharudin et al., 2021).

Given these aspects in microalgae-based systems, both integration and process intensification are promising approaches for a more sustainable manufacturing industry. This is because a combination of both can provide even better results regarding economic and ecological efficiency and allow for greater industrial application of the biotechnological processes (Meyer et al., 2021).

However, as they come from different areas, the viewpoint through which process engineering recognizes and identifies obstacles, integration and intensification are, sometimes, divergent from the way environmental engineering sees them. In practice, most strategies implemented focus on the central theme of minimizing the consumption of resources and energy through recycling, recovery, and integration (Walmsley et al., 2019). In contrast, although they present the same preservation bias, sustainable development values entire progress much greater than simply minimizing damage.

Therefore, the full sustainable picture of process integration and intensification must be subjected to analysis to ensure social, environmental, and economic performance. Thus, sustainability metrics and indicators are mathematical tools used to estimate the major environmental, economic, and social impacts attributed to the industrial chain (Jacob-Lopes et al., 2021).

In the light of this approach, the present chapter aims to review the main aspects of the theory and practice of the application of process integration and intensification techniques applied to microalgae-based systems. In addition, the approach of environmental management tools through sustainability metrics and indicators are discussed from the perspective of enhancing the sustainable profile of processes and products based on microalgae.

21.2 How are the processes of integration and intensification of microalgae-based systems in theory and practice?

Microalgae-based systems are conventional bioprocesses that typically comprise three steps: (a) upstream steps; (b) biotransformation/bioreaction steps; and (c) downstream separation and purification. In general, each of these steps offers several opportunities to intensify and/or integrate a given process.

Thus, microalgae biotechnology has used process intensification as a key parameter to establish improvement indicators such as, for example, an increase in the production of the bioproducts, in relation to cell concentrations, lower production time units, lower work volumes of the bioreactor, as well as cost reduction criteria (see Box 2 for more details about the definition and structure of intensification processes). However, conceptual misconceptions between process intensification and process optimization are often confusing and slight distinctions between these two terms must be made (López-Guajardo et al., 2021). In fact, both intensification and optimization are about performance improvements. However, intensification is responsible for a substantial increase in process improvement, which does not only correspond to increases in productivity but also encompasses technological innovations that result in more satisfactory environmental and economic indicators such as the reduction of energy consumption, carbon, as well as expenditure or operational capital indicators (Boodhoo et al., 2022). Under this concept, examples applied to batch processing or continuous processing in microalgae bioreactors can be seen as promising alternatives within the intensification approach. In addition, the use of multifunctional bioreactors is also an example of intensification measures. On the other hand, the simple optimization of processes can be understood as an incremental improvement in productivity or in the use of utilities, such as energy demand. For instance, in microalgae biotechnological processes, today, genetic engineering – a technique used to increase the productivity of specific products such as pigments and lipids – is seen as process optimization since it adjusts and directs the composition of the biomass to slightly higher yields when compared to conventional processes. However, the extraction and recovery of microalgae biocompounds with the application of ultrasonic methods – which increase the efficiency of the process by up to 90% – are considered consolidated alternatives for intensification (Gonzalez-Balderas et al., 2020). Even so, it must of course be considered that the optimization can be superimposed on the intensification step. Besides, it is appropriate to refer to optimization as the first step toward intensification (Varbanov et al., 2021).

Box 2: Definition and structure of intensification processes – Past and present

In its first definition, in the mid-1995s, process intensification was characterized as any processing project that presented a reduction in size by a factor of 100, maintaining the same target production objective (Ramshaw, 1995). Evidently, over the years, this concept has become outdated and – indeed – vague. Hence, a new concept, defined as any chemical engineering development that would lead to a substantially smaller, cleaner, and more energy-efficient technology through innovation, concomitantly with the valorization of customers through just-in-place manufacturing time (accurate production model between demand for raw materials, production chain, and inventory) evolved (Segovia-Hernández and Sánchez-Ramírez, 2022). To reinforce this new foundation, a framework was established to characterize the processes intensification in four domains of action: spatial, thermodynamic, functional, and temporal (Figure 21.1) (Rivas et al., 2020). Furthermore, this classification is supported by the four principles of process intensification:

i. maximize the effectiveness of intramolecular and intermolecular events;
ii. give each molecule the same processing experience;
iii. optimize the driving forces and maximize the specific areas to which these forces apply;
iv. maximize the synergistic effects between the partial processes.

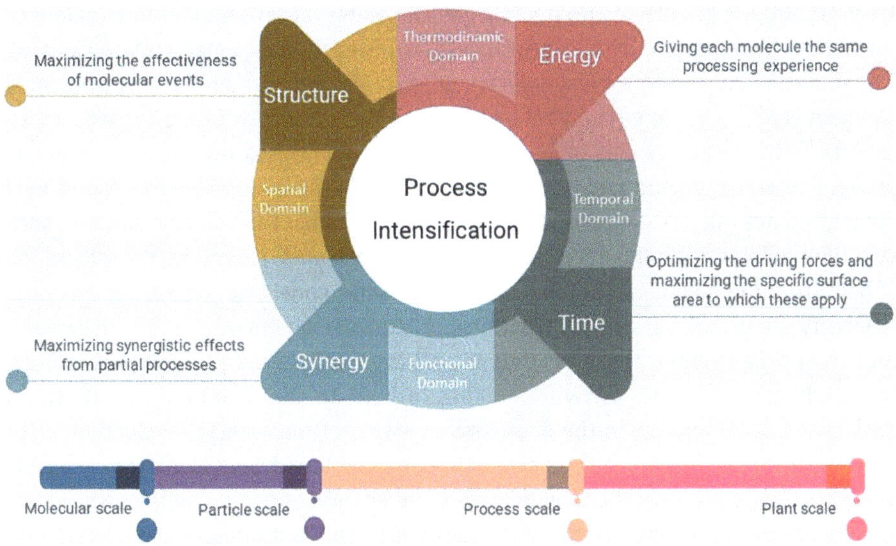

Figure 21.1: Fundamental framework for process intensification: principles, approaches, and the scales.

Additionally, process intensification can be performed from different perspectives by integrating/hybridizing under the most varied levels of abstraction:

(a) Integration of known unit operations: hybrid reaction-reaction, reaction-separation, or separation-separation systems;
(b) Integration of functions: incorporating new functionalities into a known operation based on the limitations of the bioprocess;
(c) Integration of phenomena: identify the main target phenomena to carry out biotransformation and customize the bioprocess design to put them together.

However, microalgae-based systems have only proved viable on a large scale through integrated projects. In fact, various integration approaches, including the integration of industrial platforms, inter-industry exchanges and reuse of waste and products, use of smart logistics, as well as biomass optimization are essential to achieve sustainable and economically viable process consolidation (Kasani et al., 2022).

Thus, several process integration options have been proposed for microalgae-based systems to improve cost effectiveness and environmental burdens. This is because, as is well known, microalgae biotechnological processes are intensive in natural resources, placing substantial demands on energy, water, and nutrients. Therefore, it is possible that the three main types of integration are applied: energy, water, and mass.

The starting point of energy integration in microalgae-based systems is given at the cultivation stage, that is, at the bioreactor level. Although significant efforts have been made in the progress and performance of bioreactors, there is still a specific problem associated with this process step: temperature. Studies show that microalgae productivity in cultivation systems is closely related to the ideal temperature, between

25–35 °C, for the growth of the microorganism. However, at temperatures below or above this range, most microalgae strains show lower than expected growth rates, although there may be some exceptions when adaptive pretreatments are implemented (Rehman et al., 2022). To circumvent temperature fluctuations, whether for reasons associated with closed systems or because of variations noticed during the day or because of seasonality (cold and hot seasons), strategies for the implementation of heat exchangers are often installed close to cultivation systems. Thus, in cold seasons, such as autumn/winter, the cultivation systems need to be heated. Therefore, the integration of heat from the deployment of devices such as solar panels for generating heat and electricity is a promising strategy (Dias et al., 2022a). On the other hand, in hot seasons, such as spring/summer, these systems often have overheating problems. Therefore, heat exchanger strategies allow for cooling without an additional expense (Rasheed et al., 2022). Additionally, in the downstream stages of microalgae production, steps such as drying can be energy-intensive. It is estimated that drying alone is responsible for approximately 85% of the total energy of the process due to its high moisture content (Deprá et al., 2020a). To this end, energy integration strategies, associated with ways that have energy co-productions such as bioenergy co-generation, can be reintegrated into the system to help reduce energy demand (Severo et al., 2020).

Similarly, the mass integration of microalgae-based systems predominantly presents the nutrient integration base for microalgae growth. In short, key elements for cell maintenance, such as carbon, nitrogen, phosphorus, and sulfur (in smaller amounts) are integrated into the cultivation system. In the last 15 years, studies have focused on integrating carbon in its inorganic form into microalgae cultivation systems, aiming at biosequestration and carbon capture through the photosynthetic mechanism in photobioreactors, in order to assist in aspects associated with environmental management (Deprá et al. al., 2020b; Severo et al., 2021). On the other hand, the integration of organic carbon and other nutrients such as nitrogen and phosphorus were investigated with respect to the reuse of industrial wastewater from the dairy and agro industries, under heterotrophic cultivations. In parallel, nutrients integration from effluents also enables the integration of water since the liquid effluent also serves as a culture medium for microalgae cells (dos Santos et al., 2020).

As seen before, water integration – in addition to the reuse of industrial effluents – can also be recovered mainly in three other ways: (i) water from oceans; (ii) rainwater; and (iii) recycling of water used in the system itself. In fact, using ocean waters has the advantages of saving freshwater resources and low cost, in addition to inhibiting the growth of competing and predatory algae (Gao et al., 2022). Also, when seawater was used for freshwater microalgae cultivation, the high saline concentration proved to be a positive stressor, as lipid productivities doubled (Zhang et al., 2018). Additionally, debates about the potential advantages and disadvantages of reusing/recycling water from the culture medium for further microalgae cultivation have been discussed (Farooq et al. 2015). In fact, recycling water after harvesting the biomass substantially reduces the overall water footprint of the process. However, some studies suggest that recycled cultures showed an increase in saline concentration by up to 12% due to water evaporation, increasing their ash-free dry weight. In addition, constant concerns about the quality of recycled water as well as the collection methods are topics that still need to be investigated in greater depth (Pinho and Mateus, 2021).

In view of the above, as the processes attributed to microalgae systems are integrated and intensified, it is possible that an ideal microalgae biorefinery model will be proposed to achieve the desired objectives (Siddiki et al., 2022) (Figure 21.2). In fact, these countless possibilities open paths for the development of cleaner industrial productions and, therefore, more sustainable production. However, its consolidation can only be achieved with a combination of technical, economic, and environmental principles, whose quantitative results are feasible.

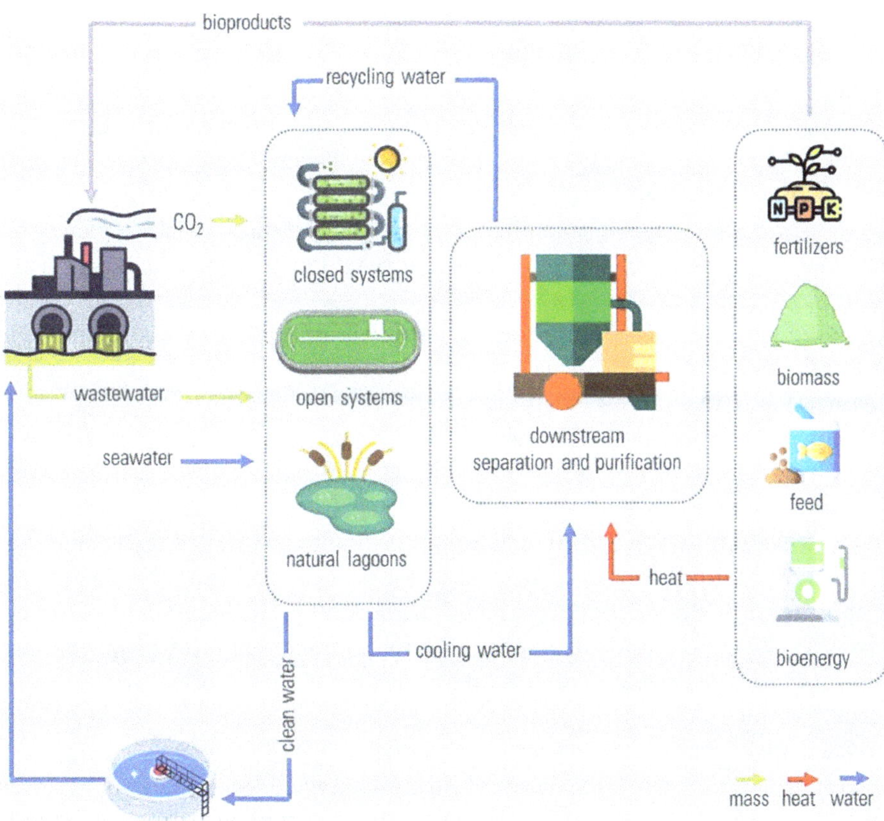

Figure 21.2: Processes integration concept for microalgae-based systems under the biorefinery approach.

21.3 Sustainability approach

In the current scenario, the practical concept of sustainability is abstract. In practice, it is difficult to translate into goals, performance measures, and indicators, something as complex as unifying the environmental, social, and economic scores of a process. Therefore, sustainability approaches are often discouraging for business decision-makers, who often only measure operational and economic results (Henao and Sarache, 2022).

However, in recent years, it has become impossible to ignore the consequences of the industry's lax activities on the damage to the ecosystem. As a result, sustainability challenges have shifted to prioritizing the preservation of the environment. Thus, the redesign of operational processes through the integration and intensification of processes was proposed as a timely alternative to reduce environmental impacts and offer a lower compensation value (Dias et al., 2022b).

Notably, the numerous possibilities for improving integration and process intensification have been extended to microalgae-based systems in order to make them sustainable. However, due to its complexities, there are still bottlenecks that need to be elucidated, and these can only be well explained by the application of environmental performance tools (Severo et al.2020).

In view of these aspects, quantifying sustainability requires accounting for material flows. In an attempt to obtain this quantitative information, methodological mathematical tools, such as life cycle assessment, were developed and standardized (ISO, 2006; Gani et al., 2022). With its propagation, several studies were carried out in order to spread the environmental benefits of integrating and intensifying the processes (Sharma et al., 2021).

Among these findings in the literature, microalgae cultures were integrated into wastewater that has pollutants such as pesticides, and studies showed that they were able to fully remove chemical compounds such as alachlor (100%), chlorfenvinphos (100%), malaoxon (100%), and fenthion oxon (100%) (García-Galán et al., 2020). These compounds constitute an ecotoxicological threat to aquatic organisms, which results in an increase in environmental burdens associated with eutrophication categories. However, when in contact with humans, high concentrations of these compounds are associated with neurotoxic problems that can directly impact categories such as human toxicity.

Similarly, studies carried out by dos Santos et al. (2020) showed that the process integration using effluents from agro industries reduced the eutrophication potential from 80×10^4 kg eq PO_4/y to 5×10^4 kg eq PO_4/y. This reduction is attributed to the high capacity of microalgae systems to recover/remove organic matter (~97%), total nitrogen (~85%), and total phosphorus (~93%). At the same time, wastewater integration not only provides benefits against mass integration but is also the primary way to minimize the water footprint. This strategy is estimated to reduce freshwater demand by up to 90% (Gupta et al., 2016).

Additionally, the predominance of the photosynthetic mechanism of microalgae allows the integration of gaseous pollutants into cultivation systems, mainly the conversion of carbon dioxide (Cheirsilp and Maneechote, 2022). Thus, through biological carbon capture and utilization, microalgae cultures were able to reduce the global warming potential by up to 41% when the culture was fed with concentrations of 15% carbon dioxide, simulating industrial exhaust gases (Deprá et al., 2019; Deprá et al., 2020b). Similarly, studies conducted by Severo et al. (2020) reduced carbon dioxide-equivalent emissions by ~80% when compared to conventional combustion, under an oxy-combustion model integrated with microalgae bioreactors. In addition, the success of carbon biocapture by microalgae is also associated with process intensification. This is because genetic engineering, which today is just a process optimization, could substantially improve photosynthetic mechanisms and biochemical pathways, in the long term, in order to increase carbon biofixation capacities, increase biomass productivity, and further reduce system carbon footprints (Barati et al., 2021).

Still, one of the most challenging strategies applied to microalgae-based systems is energy integration. As a potential strategy, the use of residual heat from combustion gases was proposed to assist in the drying of microalgae biomass. Since about 10% of this energy is lost through exhaust gases, this integration could reduce the environmental impacts associated with the demand for energy resources. However, as approximately > 60% of the world's energy demand comes from coal, the burning of the generated gaseous pollutants to obtain thermoelectric energy directly impacts categories such as global warming potential, human toxicity, ozone depletion, and photochemical oxidation potential (Wang and Song et al., 2021). Thus, the most sensible strategy would be to reduce energy inputs to the system or reintegrate potentially lost energies, as in the case of using volatile organic compounds released from cultivation systems. The energy generation rate of these compounds is estimated to reach ~86 MJ/kg; this allows them to be reported as promising heat sources (Jacob-Lopes et al., 2017, 2020). In addition, another resource with potential possibility of reducing environmental impacts is the integration of solar panels in commercial microalgae installations as a smart replacement of the energy matrix, since environmental impacts are reduced by up to 87% (Dias et al., 2022c).

In parallel, the use of process intensification strategies as an alternative to reduce environmental impacts is also being applied to microalgae-based systems. The focus is on the downstream steps of the microalgae biomass production, usually in the steps of extraction and purification of intracellular compounds, such as lipids and pigments. At the level of exemplification, lipid extraction aimed for the production of biodiesel has been an experimental focus for direct transesterification through the intensification of processes through the application of ultrasound or microwaves (Patle et al., 2021). Among the advantages, there are notable differences regarding the reduction of reaction times, solvent requirements, process yields, separation times, and reduced energy consumption (Gude and Martinez-Guerra, 2018). As a consequence, they result in significant energy savings as well as severe reductions in the impact categories inherent to this demand.

Most sustainability approaches presented here, despite showing promising results, are still at an early stage of development. However, an increasing number of technical-scientific data has been made available in order to demonstrate the sustainable feasibility of applying process integration and intensification strategies for microalgae-based systems. Furthermore, it is important to note that it is clear algae technologies currently have difficulty competing with conventional approaches. However, as progress in integration and intensification projects is strengthened, microalgae-based systems will be able to grow as an independent and sustainable industry.

21.4 Conclusion

Over the past few decades, scientists have continued to innovate strategies to make microalgae-based systems a sustainable reality. However, what in the past was clear to see, today, with emerging needs in the face of global environmental challenges, has resulted in even more questions than answers.

In order to fulfill the sustainable promise, continuous efforts to advance and develop these systems were and still are necessary to increase the biomass yield and reduce the demands on natural resources and their inherent inputs.

Fortunately, the integration and intensification of processes have shown to be promising strategies to accelerate this development, although the results obtained refer to laboratory data, and on a small scale, the theoretical effects in scale up remain favorable.

In this incessant search for the consolidation of sustainable microalgae processes, the central point is to remember that the development of process integration and intensification is iterative and must be seen as a whole. Year after year, potential discoveries are elucidated in order to fill in the gaps and add to this construction. Therefore, recent advances are an impetus for intensive research in this field.

References

Baharudin, L., Watson, M. J., & Yip, A. C. (2021). Process intensification in multifunctional reactors: A review of multi-functionality by catalytic structures, internals, operating modes, and unit integrations. *Chemical Engineering and Processing-Process Intensification*, *168*, 108561. doi:10.1016/j.cep.2021.108561

Barati, B., Zeng, K., Baeyens, J., Wang, S., Addy, M., Gan, S. Y., & Abomohra, A. E. F. (2021). Recent progress in genetically modified microalgae for enhanced carbon dioxide sequestration. *Biomass and Bioenergy*, *145*, 105927. doi:10.1016/j.biombioe.2020.105927

Boodhoo, K. V. K., Flickinger, M. C., Woodley, J. M., & Emanuelsson, E. A. C. (2022). Bioprocess intensification: A route to efficient and sustainable biocatalytic transformations for the future. *Chemical Engineering and Processing-Process Intensification*, 108793. doi:10.1016/j.cep.2022.108793

Cheirsilp, B., & Maneechote, W. (2022). Insight on zero waste approach for sustainable microalgae biorefinery: Sequential fractionation. *Conversion and Applications for High-to-low Value-added Products. Bioresource Technology Reports*, 101003. doi:10.1016/j.biteb.2022.101003

Deprá, M. C., Dias, R. R., Severo, I. A., de Menezes, C. R., Zepka, L. Q., & Jacob-Lopes, E. (2020b). Carbon dioxide capture and use in photobioreactors: The role of the carbon dioxide loads in the carbon footprint. *Bioresource Technology, 314*, 123745. doi:10.1016/j.biortech.2020.123745

Deprá, M. C., Mérida, L. G., de Menezes, C. R., Zepka, L. Q., & Jacob-Lopes, E. (2019). A new hybrid photobioreactor design for microalgae culture. *Chemical Engineering Research and Design*, *144*, 1–10. doi:10.1016/j.cherd.2019.01.023

Deprá, M. C., Severo, I. A., dos Santos, A. M., Zepka, L. Q., & Jacob-Lopes, E. (2020a). Environmental impacts on commercial microalgae-based products: Sustainability metrics and indicators. *Algal Research*, *51*, 102056. doi:10.1016/j.algal.2020.102056

Dias, R. R., Deprá, M. C., Severo, I. A., Zepka, L. Q., & Jacob-Lopes, E. (2022c). Smart override of the energy matrix in commercial microalgae facilities: A transition path to a low-carbon bioeconomy. *Sustainable Energy Technologies and Assessments*, *52*, 102073. doi:10.1016/j.seta.2022.102073

Dias, R. R., Deprá, M. C., Zepka, L. Q., & Jacob-Lopes, E. (2022a). In silico study of hybrid renewable energy in microalgae facilities: A path towards net-zero emissions. *Algal Research*, *63*, 102661. doi:10.1016/j.algal.2022.102661

Dias, R. R., Deprá, M. C., Zepka, L. Q., & Jacob-Lopes, E. (2022b). Roadmap to net-zero carbon emissions in commercial microalgae-based products: Environmental sustainability and carbon offset costs. *Journal of Applied Phycology*, 1–14. doi:10.1007/s10811-022-02725-y

dos Santos, A. M., Deprá, M. C., dos Santos, A. M., Cichoski, A. J., Zepka, L. Q., & Jacob-Lopes, E. (2020). Sustainability metrics on microalgae-based wastewater treatment system. *Desalination and Water Treatment*, *185*, 51–61. doi:10.5004/dwt.2020.25397

Elena Savulescu, L., & Alva-Argaez, A. (2013). Process integration concepts for combined energy and water integration. In: Handbook of Process Integration (PI), pp. 461–483. doi:10.1533/9780857097255.4.461

El-Halwagi, M. M. (2017). Sustainable Design through Process Integration: Fundamentals and Applications to Industrial Pollution Prevention, Resource Conservation, and Profitability Enhancement. Butterworth-Heinemann, Amsterdam, Netherlands.

Farooq, W., Suh, W. I., Park, M. S., & Yang, J. W. (2015). Water use and its recycling in microalgae cultivation for biofuel application. *Bioresource Technology*, *184*, 73–81. doi:10.1016/j.biortech.2014.10.140

Gani, A., James, A. T., Asjad, M., & Talib, F. (2022). Development of a manufacturing sustainability index for MSMEs using a structural approach. *Journal of Cleaner Production*, *353*, 131687. doi:10.1016/j.jclepro.2022.131687

Gao, F., Zhang, X. L., Zhu, C. J., Huang, K. H., & Liu, Q. (2022). High-efficiency biofuel production by mixing seawater and domestic sewage to culture freshwater microalgae. *Chemical Engineering Journal*, *443*, 136361. doi:10.1016/j.cej.2022.136361

García-Galán, M. J., Arashiro, L., Santos, L. H., Insa, S., Rodríguez-Mozaz, S., Barceló, D., . . . Garfi, M. (2020). Fate of priority pharmaceuticals and their main metabolites and transformation products in microalgae-based wastewater treatment systems. *Journal of Hazardous Materials*, *390*, 121771. doi:10.1016/j.jhazmat.2019.121771

García-García, J. C., Ponce-Rocha, J. D., Marmolejo-Correa, D., & Morales-Rodriguez, R. (2019). Exergy Analysis for Energy Integration in a Bioethanol Production Process to Determine Heat Exchanger Networks Feasibility. In: Computer Aided Chemical Engineering. Elsevier, Vol. 46, pp. 475–480. doi:10.1016/B978-0-12-818634-3.50080-1

Gonzalez-Balderas, R. M., Velasquez-Orta, S. B., & Ledesma, M. O. (2020). Biorefinery process intensification by ultrasound and ozone for phosphorus and biocompounds recovery from microalgae. *Chemical Engineering and Processing-Process Intensification*, *153*, 107951. doi:10.1016/j.cep.2020.107951

Gude, V. G., & Martinez-Guerra, E. (2018). Green chemistry with process intensification for sustainable biodiesel production. *Environmental Chemistry Letters*, *16*(2), 327–341. doi:10.1007/s10311-017-0680-9

Gupta, S., Pandey, R. A., & Pawar, S. B. (2016). Microalgae bioremediation of food-processing industrial wastewater under mixotrophic conditions: Kinetics and scale-up approach. *Frontiers of Chemical Science and Engineering*, *10*(4), 499–508. doi:10.1007/s11705-016-1602-2

Henao, R., & Sarache, W. (2022). Sustainable performance in manufacturing operations: The cumulative approach vs. trade-offs approach. *International Journal of Production Economics*, *244*, 108385. doi:10.1016/j.ijpe.2021.108385

ISO. (2006). International Standardization Organization. Environmental management – Life cycle assessment-principles and framework, ISO 14040, 2006. Available at: http://www.iso.org/iso/catalogue_detail?csnumber=37456 (Accessed in: 06. 12.2022).

Jacob-Lopes, E., Santos, A. B., Severo, I. A., Depra, M. C., Maroneze, M. M., & Zepka, L. Q. (2020). Dual production of bioenergy in heterotrophic cultures of cyanobacteria: Process performance, carbon balance, biofuel quality and sustainability metrics. *Biomass and Bioenergy*, *142*, 105756. doi:10.1016/j.biombioe.2020.105756

Jacob-Lopes, E., Severo, I. A., Bizello, R. S., Barin, J. S., Menezes, C. R. D., Cichoski, A. J., et al. (2017). Process and system for re-using carbon dioxide transformed by photosynthesis into oxygen and hydrocarbons used in an integrated manner to increase the thermal efficiency of combustion systems. Patent WO2017/112984 A1, 2017.

Jacob-Lopes, E., Zepka, L. Q., & Deprá, M. C. (2021). Sustainability Metrics and Indicators of Environmental Impact: Industrial and Agricultural Life Cycle Assessment. Elsevier, Amsterdam, Netherlands.

Kasani, A. A., Esmaeili, A., & Golzary, A. (2022). Software tools for microalgae biorefineries: Cultivation, separation, conversion process integration, modeling, and optimization. *Algal Research*, *61*, 102597. doi:10.1016/j.algal.2021.102597

Klemes, J. J. (Ed.). (2022). Handbook of process integration (PI): minimisation of energy and water use, waste and emissions. Woodhead Publishing, Elsevier, Amsterdam, Netherlands.

Kong, K. G. H., How, B. S., Teng, S. Y., Leong, W. D., Foo, D. C., Tan, R. R., & Sunarso, J. (2021). Towards data-driven process integration for renewable energy planning. *Current Opinion in Chemical Engineering*, *31*, 100665. doi:10.1016/j.coche.2020.100665

López-Guajardo, E. A., Delgado-Licona, F., Álvarez, A. J., Nigam, K. D., Montesinos-Castellanos, A., & Morales-Menendez, R. (2021). Process intensification 4.0: A new approach for attaining new, sustainable and circular processes enabled by machine learning. *Chemical Engineering and Processing-Process Intensification*, 108671. doi:10.1016/j.cep.2021.108671

Meyer, F., Johannsen, J., Liese, A., Fieg, G., Bubenheim, P., & Waluga, T. (2021). Evaluation of process integration for the intensification of a biotechnological process. *Chemical Engineering and Processing-Process Intensification*, *167*, 108506. doi:10.1016/j.cep.2021.108506

Patle, D. S., Pandey, A., Srivastava, S., Sawarkar, A. N., & Kumar, S. (2021). Ultrasound-intensified biodiesel production from algal biomass: A review. *Environmental Chemistry Letters*, *19*(1), 209–229. doi:10.1007/s10311-020-01080-z

Pinho, H. J., & Mateus, D. M. (2021). Sustainable production of reclaimed water by constructed wetlands for combined irrigation and microalgae cultivation applications. *Hydrology*, *8*(1), 30. doi:10.3390/hydrology8010030

Ramshaw, C. (Ed.) (1995). Process Intensification for the Chemical Industry (No. 18). Wiley-Blackwell, Belgium.

Rasheed, R., Thaher, M., Younes, N., Bounnit, T., Schipper, K., Nasrallah, G. K., . . . & Pruvost, J. (2022). Solar cultivation of microalgae in a desert environment for the development of techno-functional feed ingredients for aquaculture in Qatar. *Science of the Total Environment*, 155538. doi:10.1016/j.scitotenv.2022.155538

Rehman, M., Kesharvani, S., Dwivedi, G., & Suneja, K. G. (2022). Impact of cultivation conditions on microalgae biomass productivity and lipid content. *Materials Today: Proceedings*. doi:10.1016/j.matpr.2022.01.152

Rivas, D. F., Boffito, D. C., Faria-Albanese, J., Glassey, J., Afraz, N., Akse, H., . . . & Weber, R. S. (2020). Process intensification education contributes to sustainable development goals. Part 1. *Education for Chemical Engineers*, *32*, 1–14. doi:10.1016/j.ece.2020.04.003

Segovia-Hernández, J. G., & Sánchez-Ramírez, E. (2022). Current status and future trends of computer-aided process design, applied to purification of liquid biofuels, using process intensification: A short review. *Chemical Engineering and Processing-Process Intensification*, 108804. doi:10.1016/j.cep.2022.108804

Severo, I. A., Deprá, M. C., Dias, R. R., & Jacob-Lopes, E. (2020). Process Integration Applied to Microalgae-Based Systems. In: Handbook of Microalgae-Based Processes and Products. Academic Press, London, United Kingdom, pp. 709–735. doi:10.1016

Severo, I. A., Dos Santos, A. M., Deprá, M. C., Barin, J. S., & Jacob-Lopes, E. (2021). Microalgae photobioreactors integrated into combustion processes: A patent-based analysis to map technological trends. *Algal Research*, *60*, 102529. doi:10.1016/j.algal.2021.102529

Sharma, P., Srinivas, G. L. K., Varjani, S., & Kumar, S. (2021). Emerging microalgae-based technologies in biorefinery and risk assessment issues: Bioeconomy for sustainable development. *Science of the Total Environment*, 152417. doi:10.1016/j.scitotenv.2021.152417

Siddiki, S. Y. A., Mofijur, M., Kumar, P. S., Ahmed, S. F., Inayat, A., Kusumo, F., . . . Mahlia, T. M. I. (2022). Microalgae biomass as a sustainable source for biofuel, biochemical and biobased value-added products: An integrated biorefinery concept. *Fuel*, *307*, 121782. doi:10.1016/j.fuel.2021.121782

Sinnott, R., & Towler, G. P. (2020). Chemical Engineering Design: Ray Sinnott, Gavin Towler. Butterworth-Heinemann, Elsevier, London, United Kingdom.

Varbanov, P. S., Jia, X., & Lim, J. S. (2021). Process assessment, integration and optimisation: The path towards cleaner production. *Journal of Cleaner Production*, *281*, 124602. doi:10.1016/j.jclepro.2020.124602

Walmsley, T. G., Ong, B. H., Klemeš, J. J., Tan, R. R., & Varbanov, P. S. (2019). Circular integration of processes, industries, and economies. *Renewable and Sustainable, Energy Reviews*, *107*, 507–515. doi:10.1016/j.rser.2019.03.039

Walmsley, T. G., Varbanov, P. S., Su, R., Ong, B., & Lal, N. (2018). Frontiers in process development, integration and intensification for circular life cycles and reduced emissions. *Journal of Cleaner Production*, *201*, 178–191. doi:10.1016/j.jclepro.2018.08.041

Wang, Q., & Song, X. (2021). Why do China and India burn 60% of the world's coal? A decomposition analysis from a global perspective. *Energy*, *227*, 120389. doi:10.1016/j.energy.2021.120389

Zhang, L., Pei, H., Chen, S., Jiang, L., Hou, Q., Yang, Z., & Yu, Z. (2018). Salinity-induced cellular cross-talk in carbon partitioning reveals starch-to-lipid biosynthesis switching in low-starch freshwater algae. *Bioresource Technology*, *250*, 449–456. doi:10.1016/j.biortech.2017.11.067

Ahmad Farhad Talebi*, Sara Kabirnataj, Elham Soleimani

Chapter 22 Patents related to process integration and process intensification applied to microalgae-based systems

Abstract: The growing world demand for food and energy has given prominence to the potential use of microalgae in the context of green energy. Bioscientists from several countries have confirmed the great potential of algal bio-products to reduce the pressure on nonrenewable resources. On the other hand, myriad technological bottlenecks in large-scale cultivation have resulted in an increased costs and energy consumption, which has further curbed the commercialization of microalgae-based products. Hence, several researches have been carried out to reduce the final production cost through minimizing energy consumption as well as maximizing biomass productivity. Approaches such as minimizing resources usage, recycling of energy, reusing of material, decreasing emission, and also increasing the yield of products in up- and mainstream processes have been surveyed through patents in order to integrate and intensify processes related to microalgae-based systems. This chapter reviews international patents related to technical methodologies that can be employed to increase the productivity and also decrease pollution and energy waste. Information derived from online databases was meta-analyzed to throw light on the current status of intensification of microalgae-based systems. Process integration and intensification through research and innovation helps to well-organize energy management systems, making them considerably smaller, with the ability to deliver cleaner and safer processes. Resources integration by the efficient usage/recycling of energy and material will finally reduce the production cost; accordingly, concentrated economic benefits will provide more cost-effective products for the manufacturer and the consumer, simultaneously.

Keywords: Patent mining, Innovation, Sustainability, Reuse of material and energy

*Corresponding author: Dr. Ahmad Farhad Talebi, Department of Microbial Biotechnology, Faculty of New Sciences and Technologies, Semnan University, 35131-19111, Semnan, Iran, e-mail: aftalebi@semnan.ac.ir

Dr. Sara Kabirnataj, Applied research and Seed production Center, Oilseeds Research and Development Company, Sari, Iran, e-mail: Kabirs.bio@gmail.com

Dr. Elham Soleimani, Genetic and Agricultural Biotechnology Institute of Tabarestan, Sari Agricultural Sciences and Natural Resources University, Sari, Iran, e-mail: elisoleimani86@gmail.com

https://doi.org/10.1515/9783110781267-022

22.1 Process integration and intensification through upstream strategies

22.1.1 Screening deposited native strains

More than 50,000 genetically variant microalgae strains are found in oceans, lakes, ponds, and rivers, but only 30,000 of them have been studied. Biomass from the studied strains has been subjected to extensive chemical processing, resulting in an array of high-value products such as polyunsaturated fatty acids (PUFA), carotenoids, phycobiliproteins, polysaccharides, and phycotoxins. It is worth mentioning that the limited number of algal species, namely, members of *Spirulina, Chlorella, Haematococcus, Dunaliella, Botryococcus, Phaeodactylum, Porphyridium, Chaetoceros, Crypthecodinium, Isochrysis, Nannochloris, Nitzschia, Schizochytrium, Tetraselmis, and Skeletonema* genus, have been commercially investigated (Sathasivam et al., 2019).

The selection of the most desired microalgae strain is the first vital step for traditionally and economically good algae-based outputs. Several novel microalgal strains have annually been deposited in microbial culture collection and described in the patents (Saxon et al., 2020). Furthermore, numerous data have been released in different databases about their valuable composition, cellular characteristics, field of application, specific growth condition, and their photosynthetic efficiency rate (Tabernero et al., 2013). These novel strains usually contain precious features that are helpful to increase productivity and also decrease energy waste in large-scale production, e.g., the capability to tolerate high temperatures, high CO_2 concentrations, and the ability to grow in wastewater. Some microalgae can cope with extreme environmental conditions such as variant temperature, pH, and hypotonic stresses (Abiusi et al., 2022).

Hereupon, there have been some patents introducing extremophilic strains such as *Scenedesmus* sp. (US20130164322A1), producing high levels of biomass in wastewater under adverse environmental conditions (unsuitable natural condition), which can save energy, particularly in large-scale production. The microalgal *Chlorella ohadii* are also introduced for producing biomass in extreme light and pollution (WO2015071908A1). *Chlorella* sp. strain ABC-001 (KR101855733B1) was introduced as a salt-resistant strain with elevated lipid productivity, which can grow under high CO_2 concentration condition and may overcome the issue of freshwater preparation for large-scale algal cultivation. Domestic wastewater, which is a rich source of nutrients, could also used as an alternative culture medium for growing *Scenedesmus bijuga* (CN104593265A). Since S. *bijuga* produces lipid in high yield, both economic and environmental benefits are achievable using the potential of heterotrophic cultivation. The *scenedesmus quadricauda* SDEC-13 (CN104593264A), which is a lipid-accumulating strain, could efficiently utilize the CO_2-saturated exhaust gases. So, it is possible to reduce culture costs by utilizing highly concentrated CO_2 in waste gases.

Therefore, proper strain selection could result in a reuse of elevated material and in decreasing emission. Cultivation of potent algal strains could not only enhance the energy efficiency during the tightly coupled integrated processes, but also represent the potential to achieve the goals of process intensification. A combination more than one specific unit process such as wastewater treatment, CO_2 sequestration, or production of value-added by-products under unified control operation could undoubtedly minimize the energy consumption and maximize the material recycle. Although many native strains have been patented with unique characteristics, nowadays, a lot of efforts are made to make interesting genetic modifications in desirable strains to increase their productivity. It is the subject of some patents that is further discussed.

22.1.2 Searching for newly designed strain

According to global market research, the global algae products market size is likely to grow at an annual growth rate of 4.88% from 2022 to 2031. Although major technical issues such as efficient biomass harvesting and dewatering, narrow metabolic diversity, and efficient biomolecule extraction threaten the future sustainability of this market, efforts toward overcoming these limitations have resulted in patents, research articles, and some specific databases. Recently, innovative microalgae-specific databases such as ALCOdb (http://alcodb.jp) (Aoki et al., 2016), OGDA (http://ogda.ytu.edu.cn/) (Liu et al., 2021), BioCyc (https://algae.biocyc.org/), and dEMBF (http://bbprof.immt.res.in/embf) (Misra et al., 2016) have provided huge information related to the microalgae genome sequence, gene expression, enzymes annotation, etc. However, these data are still limited to some model species. Progress in computational biology tools and bioinformatics will enrich these databases (Kumar et al., 2020).These platforms provide vigorous approaches to surmount the undesired features of promising native strains, and consequently improve the economic feasibility of the production procedure.

Genetic engineering approaches (GEA) design a host cell by downregulated and/or disrupted genes encoding enzymes involved in metabolic pathways that may compete for substrates, intermediates, and/or co-factors that influence the desired by-product. Modified metabolic characteristics could be obtained by engineering the activity or affinity of an endogenous host cell enzyme, or they might be achievable using introducing some heterologous genes into the host genome. There are several patents that try to make microalgae more productive through mutation, genetic engineering, and genome edition. Figure 22.1 shows the countries that are making the greatest scientific efforts in the field of recombinant microalgae and have obtained by-products.

Productivity enhancement through genetic alteration may happen by increasing the growth rate, improving carbon fixation, enhancing bioactive molecules production, and facilitating their extraction procedure. These alterations may be induced by

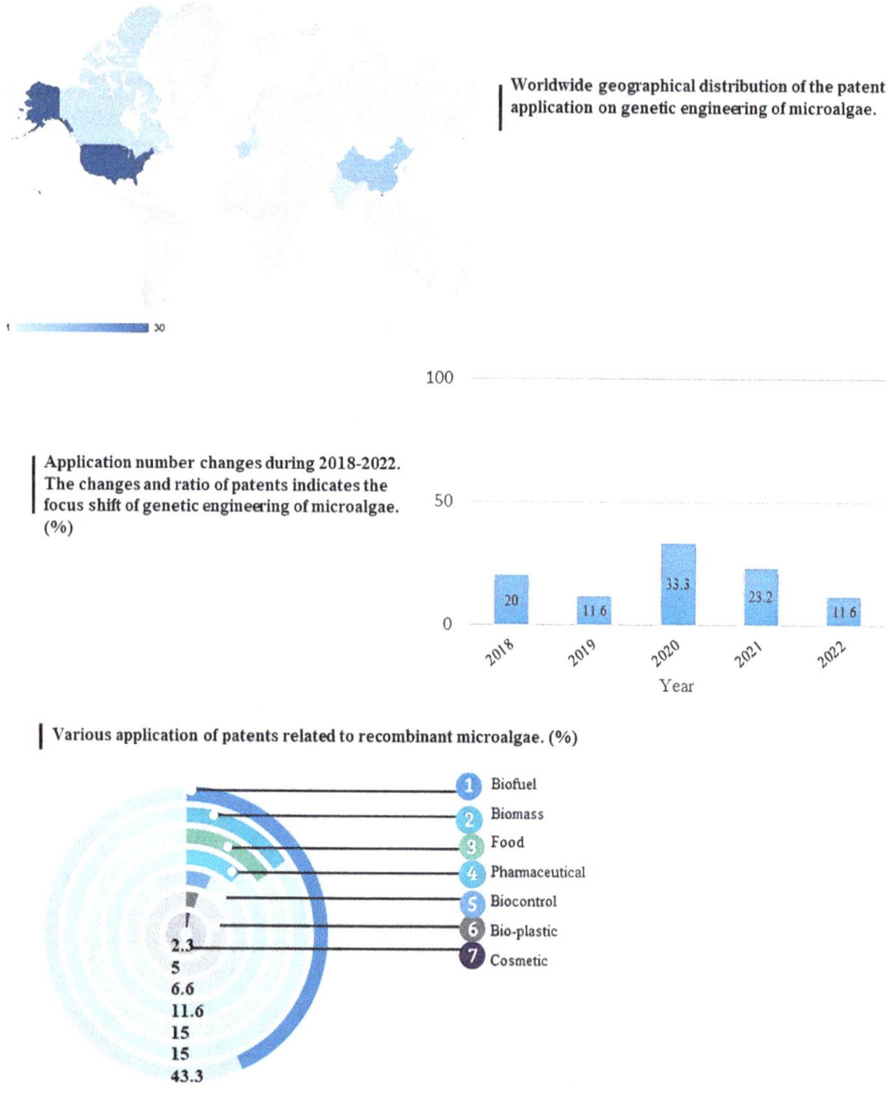

Figure 22.1: General overview of patent analysis for recombinant microalgae.

random mutation or precise gene insertion/edition methods. Several patents introduce the overproduction of biomolecules such as lipid. Patent US20220090002 has introduced a mutant with genetic modification to a nucleic acid sequence encoding, an RNA binding domain, which results in a 30% increase in lipid and/or biomass productivity. An engineered diatom has been patented (US10988773B2) that overproduces lipid during the exponential phase by enhancing the expression level or activity of a protein, which facilitates carbon assimilation.

Growth rate is also an essential parameter that increases the overall productivity. Some patents describe the master genes or their products, which have a vital role in the growth rate adjustment. Phototropin, a blue light receptor, intercedes several blue-light-elicited physiological processes in algae. It has been introduced as a cell growth suppressor. According to AU2022200067A1, the knockout mutants for phototropin represent a two-fold increase in growth and biomass accumulation, compared to the wild type. It also has been claimed that the overexpression of a combination of heterologous nucleotide sequences encoding bicarbonate transport proteins, carbonic anhydrase, light driven proton pump, cyclic electron flow regulators promotes the activity of cyclic electron transfer, increases the carbon concentration, and elevates the carbon fixation and biomass production in the algal cells (AU2021277727A1).

Several patents have introduced genetically engineered algae, producing numerous bioactive molecules, which finally save the energy consumption. Transgenic microalgae harboring heat-stable *cellulase*, has commercial application in biodegradation of cellulose or lignocellulose sources in large-scale production. Lignocellulose has a high potential for use in several areas such as the second-generation of bioethanol (Wi et al., 2015). The hurdle in its conversion into simple sugars limits its exploitation. Genetically engineered microalgae-producing plant cell wall-degrading enzymes will facilitate the biodegradation of cellulose or lignocellulose sources in the industrial field (WO2020089703A1).

Direct conversion of sunlight energy into carbohydrates makes the microalgal strains an attractive feedstock for bioethanol production (de Farias Silva and Bertucco, 2016). A patented ethanologenic cyanobacterial cells, with genetic modifications, showed improved ethanol productivity – 16 times higher than corn-based ethanol (US9650642B2). Moreover, the present invention reveals the capability of using infertile areas and saline water to address the problems associated with food-producing land and freshwater limitations. The subject of genetically modified photoautotrophic alcohols producing microorganisms have been numerously discussed in the patents published. At least 74 patents cover the "transgenic algae" keywords during the last two decades. In summary, an enhanced level of biosynthesis of acetaldehyde, pyruvate, acetyl-CoA, or other precursors in the transgenic microalgal strains provides an overexpressed carbohydrates and/or energy for the formation of ethanol (AU2014253543A1, US20150353961A1, US7973214B2).

Polyhydroxyalkanoates (PHA), the biodegradable, thermostable and elastomer bioplastics, are neutral lipids stored in several organisms, such as microalgal cells, with accumulation of up to 80% of its dry weight. In order to optimize PHA production in *Synechocystis* sp., cyanobacterial host cells have been genetically engineered to reduce or eliminate Slr1125 gene expression and manipulate the expression of some other genes such as NAD synthetase and/or NAD^+ kinase overexpression (US20150140622A1). This patent facilitates methods for cyanobacterial production of bioplastics and biomaterials in the industrial field.

The selection of a strain should be based on its advantages at the different phases of the process upstream as well as downstream and the steps to make it economically viable. The GEA is a valuable tool when the desired characteristics accumulate into an integrated process. The capability of simultaneously using infertile areas, wastewater treatment, and production of valuable recombinant by-products determine the final productivity and sustainability of the systems. The efficiency of the supply and utilization of energy, water, and other inputs as well as reduced pollutant emissions within the intensified process might be available through GEA. Moreover, selection of strains tolerant to ethanol, salt, temperature, light, above-neutral pH and mechanical stress guarantee the success of the process. Other determining factors will also be discussed in the following sections.

22.2 Process intensification for sustainable algal cultivation

22.2.1 Design of growing systems

The main objective in the upscale and intensification of the cultivation process is maximizing the production of high-quality biomass with desired characteristics during the mainstream processes. Overall improvement could be facilitated by the manipulation of key process variables. Light intensity, CO_2 concentration, aeration, and nutrient supply significantly influence the microalgae growth. The knowledge of specific lighting, temperature, and nutrient requirements of photoautotrophic organisms is essential for the successful application of different intensification strategies in cultivation for maximizing biomass and for the concentration of total biochemical products. Current trends in microalgal production are driven to explore low-cost, high-efficiency cultivation systems that consume less water and nutrients. The new design of novel equipment and processing methods should result in remarkable profits in energy efficiency, environmental footprint, and economic feasibility (Patil et al., 2020). A brief review of the recent innovations in the intensification of algal cultivation as mainstream processes is presented in this section.

The concept of ecological communities, which says that no species lives alone in nature (Maglie et al., 2021), has been utilized in several studies concerning the cocultivation of microalgae with different organisms, i.e., fungi, bacteria, and microalgae. The cocultivation of *Chlorella vulgaris* and Azotobacter *Mesorhizobium* sp. was patented as it led to 66.3%, 145.7%, and 215.0% increase in biomass, total lipid yield, and neutral lipid yield, respectively (CN109609382B). The addition of bacteria also had a positive effect on fatty acid profile; the content of unsaturated fatty acid, unlike saturated fatty acid, exhibits profound increase that could later result in improved quality of the properties of the prospective biodiesel. Cocultivation is a kind of peaceful and profitable symbiosis as microorganisms consume O_2 released by the algal photosynthesis apparatus. They also produce CO_2 that can be used by microalgae to reduce the effect of inhibition.

Cocultivation of algae and yeast strains is also a perspective from patent literature. The innovative systems gain the growth promoter and reduce energy consumption and production cost (Kumar et al., 2020). The mixed culture of two oleaginous microorganisms, *Chlorella pyrenoidosa* and *Rhodotorula glutinis*/*Candida tropicalis*, fully combines the mutual benefits of a symbiotic relationship between two oleaginous species. This alleviates the adverse effects of the pure culture process and facilitates the maintenance of stable culture conditions, specially pH and dissolved O_2 (CN106754383B). Therefore, an integrated process could survive and produce biomolecules more sustainably. During the cocultivation, the microalgal medium is supplemented with CO_2, which is supplied from yeast fermentation. It enhances the growth rate of photoautotrophic algal cells and also prevents CO_2 emission to the atmosphere.

Along with the physico-chemical protection of the cells present in the coculture, some potential for give and take in algae–microbe relationships supports the biochemical protection as well. Several vitamin-based symbiotic associations were also observed between microalgae and bacteria. Throughout the evolution of microalgae, biocompounds produced by bacteria act as vitamins in algal cells (Ray et al., 2022). Based on the patent number CN111440737B, *Alteromonas* sp., a gram-negative bacteria, can promote the growth and therefore yield dinoflagellates in coculturing conditions; a binary culture of microalgae increases in both the biomass and lipid yield, compared to the pure culture of microalgae.

Some microalgal strains can grow efficiently under mixotrophic conditions. They utilize organic carbon and CO_2 in light conditions (Benemann et al., 1987). Mixotrophic conditions tend to help *Chlorella* grow better than autotrophic ones. Therefore, mixotrophic cultivation through exogenous supplementation of organic carbons to open ponds or photobioreactors (PBRs) is suggested. Patent number CN108587920B introduced a mixotrophic culture condition in which an organic carbon source (acetic acid/sodium acetate) is added when the nitrogen source is consumed; consequently, the problem of fungus pollution caused in mixotrophic culture under non-sterile circumstances is also effectively solved. On the other hand, the yield of biomass, as well as secondary metabolites, is increased by more than 20%. An application of two separated cultivation phases, autotrophic culture and mixotrophic culture, is also claimed by CN113136339A and CN113136341A. To increase the energy recovery in such an integrated process, the discharged gas from the second culture unit is collected and transmitted to be used as a first-ventilation gas input. So, the efficiency of carbon use is expected to be close to 100%.

Reducing resource consumption, increasing the internal recycling, and optimization of energy and materials are among the targets of process intensification (RO135070A2). Projects of improved efficiencies can be very beneficial and also find opportunity to improve the public's perceptions. Further attempts in this field are discussed as follow.

22.2.2 Optimization of growing systems

There has been a steady increase in microalgal biomass productivity over the years due to innovations in equipment and technology. Steady state of the process operation transition, from the smaller to the larger scale, is considered to perform process intensification. Primary priorities for scaling up are environmental footprint and energy efficiency of the algal cultivation. However, cultivation of microalgae in large scale still requires strict control to maintain high productivity, even if they are easily cultured in a laboratory (Patil et al., 2020). The most common microalgae cultivation systems are open pond and closed photobioreactor. In addition to being easy to construct and maintain, open pond systems are ideal for large-scale cultivation, since they have low operating costs. Open systems, however, suffer from many disadvantages, including the requirement of large amounts of space, loss of water due to evaporation, very poor CO_2 absorption capabilities, CO_2 emissions, and high contamination risk by protozoa and bacteria, resulting in toxin and unusable products. Poor surface area-to-volume ratio is considered a dominant shortcoming of a traditional raceway, which directly impacts the system's ability to capture and release effectively.

Commercial-scale microalgae biotechnology relies heavily on open photobioreactors because of its low cost and simple construction (WO2013186626A1). About 99% of the biomass is produced in ponds and raceways (Olaizola and Grewe, 2019). It is important to note that mixed and circulated CO_2 and nutrients, cooling water, and refilling of evaporating water are some of the most energy-consuming parts of such cultivation systems (Chen et al., 2021). Since the principles of process intensification highlight reuse of energy and materials, to reduce resource consumption and increase internal recycling, several innovations have been recently published (CN107709537B, RO135070A2). Bioprocess engineers have made impressive and promising progress, and will be described later. There have been several patents claiming to reduce energy loss during cultivation in open ponds (for example, see US20190075743A1 and US20200367458A1). An important part of the operating cost of the raceway pond is the energy used in pumps that circulate the algae cultivation fluid or slurry. Keeping the cultivation media in a uniform manner would increase productivity by distributing nutrients evenly and preventing dead zones (Patil et al., 2020).

The limitations of an open system could be overcome by using a closed system. Energy, light, and other inputs are efficiently captured and utilized in a closed photobioreactor. Moreover, different environmental parameters could be controlled, and it also could be operated outdoors where sunlight is available. Therefore, it is expected that maximum production of algal biomass could solve the issue of cost investment, initial setup, and maintenance in a photobioreactor. However, other factors such as design, purity of the cultivation, user friendliness, space/land occupancy, and environmental footprint need to be optimized.

The cultivation of microalgae on an industrial scale is possible with only a few of these PBRs. Tubular PBRs, plastic bag PBRs, airlift column PBRs, and flat-panel PBRs belong to this group (Borowiak et al., 2021). By intensifying a photobioreactor, we can optimize the amount of light inside the reactor, improve gas exchange (supply of CO_2 and removal of oxygen), uniformly distribute nutrients within the cultivation medium, maximize reactor size for scale up, and increase biomass production (Patil et al., 2020).

All types of bioreactors have their own advantages and disadvantages. Hybrid types of bioreactors are considered as new innovation that could overcome the limitations of the developed cultivation systems. Integrated cultivation systems merge two different types of reactor to use their strengths in order to support their weaknesses. For example, one acts like a light harvesting unit using its high surface area-to-volume ratio and also controls the temperature of the culture. Another integrated reactor improves the gas exchange and also regulates the other culture variables. Exploiting such hybrid systems enables higher productivities and lower power consumption through better control over the culture variables (find more in: BR102019026874A2, CN114606103A, CN113136341A, AU2016285489B2, CN107709537B). A combination of the features of an open cultivation system and a photobioreactor to generate hybrid system has focused on two strategies: a) covered open pond to reduce the possibility of contamination, evaporative losses, and CO_2 emission, b) partially filled tube design that is widened and expanded to reduce costs (CN107709537B).

To achieve high-density cultures, many patents explored the optimization of environmental factors, including light delivery, CO_2 and O_2 gas transfer, medium supply, mixing, and temperature. For high productivity, the wavelength, intensity, and duration of the light are crucial factors in designing the PBRs (Brzychczyk et al., 2020). Solar energy is the most cost-effective source for microalgae production as it is free and abundant. There are, however, certain location-related disadvantages to this approach, such as day/night cycles, weather conditions, and seasonal changes (Blanken et al., 2013). Solar energy can be challenging to control and manage. The intense illumination, such as that in Southern Africa, can cause a temperature increase in a PBR, which will adversely affect cell growth and biomass yield (Sero et al., 2020). Controlling illumination is a relatively new method to manage nonphotosynthetically active radiation (PAR), which causes cell death. Phycologists have developed wavelength filtration systems to supply PAR to their cultures, while excluding nonPAR. Patent US20200229365 claimed that red light/far-red light in the life cycle management of green microalgae induces intense vegetative growth and enhances productivity. As a result, free natural light can be directly used in the aquaculture of *Haematococcus* without any concern about nonPAR radiation causing adverse effects. Therefore, by coupling this filter technology to PBR designs, biomass can be produced at a cost-effective rate using solar energy.

It is also possible to enhance the photosynthetic activity by artificial illumination, leading to a higher rate of biomass production and higher production of intracellular compounds. However, continuous artificial lighting will increase the electricity costs that will directly impact the final product prices. Industrial electricity prices, light source efficiency, and design of bioreactor that determines the light penetration depth are three factors that need to be taken into account when calculating the electricity cost.

So, different types of light source are commercially available, such as fluorescent tubes, high-intensity discharge lamps (HID), and light emitting diodes (LED). Recently patented photobioreactors (e.g. CN111411033A, CN208844056U, WO2019165689A1) are designed using LED light sources. LED has numerous benefits compared to other lighting sources. A light source's suitability will depend on its PAR spectrum, energy used, conversion efficiency, cost, and wavelength distribution (Nwoba et al., 2019). Based on the PAR efficiency of the lamps, HID and LED are the most convenient light sources that will better stimulate the photosynthesis process in algal cells. The PAR efficiency of HID is 1.87 μmol-phs^{-1} W^{-1} whereas commercially available LEDs exhibit a PAR efficiency of 1.91 μmol-phs^{-1} W^{-1} and it is anticipated to increase to 3 μmol-phs^{-1} W^{-1} over time; so the price of LED would decrease more in the near future. On the other hand, due to the lack of infrared radiation of LED, cooling the photobioreactor is more convenient and it reduces the cooling costs. Moreover, LEDs have other exciting characteristics such as the most extended lifespan and the ability to tolerate switching on and off effects (Darko et al., 2014). So, LEDs are the most promising artificial light source for constructing PBRs; however the issue of initial investment and maintenance cost of LEDs makes them unreasonable for commercial production of low-value products such as biofuel. Therefore, artificial lightening of high-value food additives, pharmaceutical products, and pigments such as carotenoids is more acceptable, since they are more economically valuable (CN108410939A, JP2020074723A). The irradiated wavelength of the LED light can also determine the quality and quantity of the desired biomolecule more simpler and faster, compared to the conventional methods (KR20190087938A). On the other hand, large-scale sunlight-illuminated microalgae cultivation is recommended to save production costs of biofuel (RO135070A2, US20140295448A1).

Biomass production would be much cheaper if sunlight and LED illumination are used together (JP2020074723A), particularly in a cultivation system that simultaneously generates electricity from the solar resource for operations. It is claimed by several patents that photovoltaic solar panels generate electricity to power the compressor or other energy-consuming instruments (e.g. US20210222111A1, US10093552B2). Installation of alternative energy sources, such as photovoltaic panels, enhances the sustainability of the process. 15% efficiency at converting light to electricity would be sufficient to meet typical electrical power needs. (Severo et al., 2020).

In summary, the production of a valuable biochemical requires further advancements in the engineering of a photobioreactor. It would enable the commercialization of noble algal products for special applications in the near future. Reduced environment footprint and low land use along with the large ratio of culture volume to surface area, optimized light penetration depth, efficient gas transfer to/from liquid phase, low energy requirement for pumps in recirculating the culture, with minimum associated shearing stress, and powered by renewable sources of energy are among the characteristics of promising future bioreactors.

22.2.3 Intensified biomass harvesting

The harvesting process involves separating microalgae cells from their cultivation medium in order to convert them into the desired product; it includes recovering, dewatering, and drying the biomass, sequentially. There is no doubt that the harvesting process is expensive, time/energy-consuming, and highly labor demanding, since algal cells only make up 0.1% and >1% of the total volume of an outdoor pond and PBR slurry, respectively. Hence, process intensification of downstream activities is crucial to save energy and reuse of material (Marrone et al., 2018). Microalgae harvesting involves mechanical, chemical, biological, and electricity-based methods. As a result, there are a number of ways to harvest algae, depending on their properties. The density of the resultant slurry, the quality of the product, and energy efficiency might vary under different conditions (Patil et al., 2020). The primary stage of harvesting involves forming a slurry containing 2–7% total suspended solids, followed by a secondary dewatering step, leading to an algal paste containing 15–25% total suspended solids. To obtain the desired microalgae concentration, performing a sequential dewatering process is suggested. Harvesting has a significant impact on the subsequent drying processes, hence it is crucial to the overall process (Uduman et al., 2010). Preconcentration, for example, by flocculation, filtration, or centrifugation, increases the efficiency significantly.

Flocculation is a phase separation process in which the algal biomass comes out of suspension in the form of floc, either spontaneously (Auto-flocculation) or after the addition of microorganisms (bio-flocculation). Patent number US10160989B2 is focused on using filamentous fungi *Cunninghamella echinulata* to pelletize single-cell microalgae *Chlorella vulgaris* and increase the yields using a low-cost separation process.

Electrocoagulation, as a physical concentration technique, forms aggregates by electrocoagulation of the cells of cyanobacteria and algae through the production of positively charged ions. Interaction between them and the negatively charged microalgae cells induces the coagulation of the biomass. Later, a sedimentation step could not only remove the excessive biomass, but also eliminate the released substances from the cells (Visigalli et al., 2021). In a patent from Czech (CZ32362U1), electrocoagulation has been introduced as a way to separate algal biomass cells from the culture medium. In this patent, more than 97% efficiency was verified for the electrocoagulation unit in a pilot plant for cyanobacterium *Microcystis aeruginosa* and green algae *Chlorella vulgaris*. CN105505780B patented an integrated algae harvesting process in which electricity shocks to microalgae cell cause a rapid loss of their activity. They claimed that algae with flagellum and long flocculation time could be harvested by reduced flocculants usage and flocculation time. In result, increasing the flocculating rate makes this method more economically efficient. Further details on electrocoagulation as a primary dewatering method for liquid/solid separation has been provided in the following patents (US9540258B2, CN101811757B, CN103043755B)

To reduce the burden of harvesting, microalgae cells can be immobilized. Solid-state cultivation process gains from several advantages, namely increased growth rate, improved utilization efficiency, higher nutrient adsorption, more contact area, elevated resistance to physical stresses, and, finally, more cost-effective cell harvesting and valueless substances removing (Xu et al., 2020). Unlike traditional liquid suspension cultures, immobilized culture involves embedding algae cells either directly in aligned polyurethane foam or adsorbing them to ready-made foam (CN103451140A, WO2014022734A1). Solid-state cultivation and biofilm cultivation are more cost-effective than suspension-based PBRs, since there is no homogeneous attachment between the biomass and the medium. *Chlorella vulgaris* (FACHB-31) has been applied in the biofilm culture method. It leads to simultaneous improvement in the yield of biomass and lipid (CN110684667B). A unique algal biofilm-based cultivation process was also introduced to produce a selenium-rich *Chlorella* by culturing in a selenium-rich medium used as a food/feed supplement (US20200022384A1). Therefore, due to the higher biomass production capacity of biofilm-attached culture, it could be a promising approach to reduce microalgae production costs in the near future.

A harvesting process would be completed if the harvested algal cells dewatered successfully. Therefore, for an efficient dewatering, membrane filtration, vacuum and pressure filtration, centrifugation, and spiral plate technology can be applied. Membrane filtration is effective in solutions of 5–7% dry matter. This method has been applied in several integrated patents such as WO2010120992A1, WO2010085619A1. Patent num. CN109876663A has introduced a kind of algae solution dehydration device, based on the positive infiltration technology at the dead end.

In order to make algae mass cultivation more economically viable, the growth medium can be reused after biomass harvesting and dewatering. A closed water loop reduces the water demand and the wastewater treatment costs. It also allows nutrients to be reused. However, various amounts of residual nutrients and algal exudates impede the spent medium recycling (Dourou et al., 2018; Fret et al., 2019). If the oxygen in water disappears, bacteria are easily mixed with water and consequently affect the microalgae growth. Patent num. CN109355190A has claimed that the recycling medium increases the dissolved oxygen. This leads to improved microalgae density and reduced bacterial growth as well as microalgae death rate. The integrated process improves microalgae quality and purity, and finally reduces the cost of culturing the microalgae. The patent number JP5305532B2 has also claimed that the recycling system through a filtering chamber in photobioreactors would decrease the production cost by removing the particles that can harm the growth of the culture, by filtering and reusing the spent growth medium.

Microalgae contain about 85% water content when harvested and dewatered. The water content should be decreased to 5% by drying processes (Fasaei et al., 2018). Compared to other processes, drying consumes the most energy to evaporate the water contents from microalgae biomass. The drying of microalgae relies heavily on heat transfer and water diffusion (Biz et al., 2019). Because of the excessive degree of latent heat of water vaporization, the drying procedures require massive quantities of energy, accounting for 75% of the entire value for the downstream process (Kim and Kim, 2022). Subsequently, the improvement of an integrated and effective drying process is an essential for the commercialization of microalgae-based systems. Bioscientists from China introduced a particularly microalgae drying system that is a kind of vapor recompression and heat exchange integrated system (CN206858559U). It is claimed that this technique has saved 41.7% energy, and the whole process substantially reduces the operating cost during an intensified process. Patent number KR101550992B1 has also introduced an automatic drying method by using hollow fiber membranes and near-hypotonic lines. This invention can reduce the water content through membrane filtration and expand the water content reduction effect by using near-infrared rays while harvesting and drying. Utilization of the close infrared rays is a relatively cheap heat source. The installation of a blower for wind speed could further develop the drying efficiency to 33% as compared with other techniques. Moreover, bacteria and contamination can be sterilized in the microalgae paste by using near-infrared rays.

22.3 Energy recovery during mainstream integrated processes

Several process integration options have been proposed to enhance the cost effectiveness of microalgae-based systems. Thus, three main types can be highlighted: mass, energy, and water integration. To reduce resource demand, it is necessary to maximize internal recycling, recovery logistics, reuse of material, and energy flows. Process integration improves the efficiency of these features using a more complex installation design. In brief, process integration could decrease production costs, process time, and material losses, to increase productivity, efficiency, and sustainability.

Box 1: What is process integration in algae innovations?

Process integration (PI) is an engineering approach toward design and operation for application in scale up of a microalgae cultivation process. It has been extended to the most varied areas of knowledge and also implemented in industrial concepts. PI aims to consolidate individual operations, envisioning synergy to increase final efficiency. The production chains of high-added value biochemicals obtained from algal cultivation harbor numerous technical, economic, environmental, and political changes to meet the challenges of global climate change, water/energy scarcity, food shortages, and land misuse. Research and development as an important step to increase the competitiveness and sustainability of microalgal cultivation is discussed in the patents. They aim to adopt alternative technologies improving the production chains and overcoming the technological bottlenecks. Biorefinery is considered an integrated solution for energy, water, and nutrient supply. Water integration is the most prevalent innovation claimed in different patents in algal cultivation. Maximized internal recycling, and recovery or reuse of water resources are the main objectives of PI technologies for coupled wastewater treatment and energy recovery using algal cultivation.

Integrated system for microalgae-based multitechnology coupling for mass cultivation of algae for biofuel production and bioremediation of environmental pollution are among the top innovations that might be exploited to enhance energy recovery during the process. A comprehensive list of progress made in process integration and intensification, applied to microalgae cultivation-coupled system, is summarized in Table 22.1.

Table 22.1: Overview of some recent progress in process integration and intensification applied to microalgae.

Applied area	Division	Mode of process integration/ intensification	Applied strain	Application number/ publication date
Bioremediation	Wastewater treatment	Nitrogen removal by an algal-sludge granule, retrieving chemical energy with minimal energy investment and reducing carbon footprint.	The cohabitation of microalgae (algae and cyanobacteria such as *Oscillatoria*, *Phormidium* and *Microcoleus*) and bacteria, and even protozoa, within the bio-granule	US10189732B2, 2019
		Treating oil-containing wastewater by applying algae to uptake the oil and grease, reducing energy costs, improving the properties of end products such as biodiesel.	*Ochromonas danica*	US11066316B2, 2021
		Degrading sewage estrogen (17β-estradiol) by algal cell secretion, removing its strong carcinogenicity and pathogenicity, with no secondary pollution.	*Chlorella* sp.	CN109305726A, 2019
Energy	Biodiesel	Oil overproduction by promoting sequential photoautotrophic and heterotrophic cultivation of algae to induce cell growth and oil production, respectively.	*Chlorella* sp, *Dunaliella* sp.	US20210071213A1, 2021

Table 22.1 (continued)

Applied area	Division	Mode of process integration/ intensification	Applied strain	Application number/ publication date
	Bioethanol	Providing a kind of construct of the high-performance biosynthesizing alcohol in cyanobacteria.	*Synechocystis* sp.	CN103361375B, 2018
	Biohydrogen	Improving hydrogen production in algae by coculturing algae with a nitrogen-fixing bacteria (*azotobacter chroococcum*).	*Chlamydomonas reinhardtii.*	CN106520897B, 2020
Agriculture	Fertilizer	A method for preparing a liquid composition, comprising living green microalgae to improve plant growth and decreasing the environmental pollution.	Selected from genus *Haematococcus, Pediastrum, Scenedesmus, Volvox*	US20210171413A1, 2021
	Fungicide/ bactericide	Cultivation of microalgae for its fungicidal and bactericidal activity of the cellular extract on fungi, oomycetes, and pathogenic bacteria.	*Amphidinium* sp.	US20190174768A1, 2019
	Crop protection	Mitigating crop losses in temporarily flooded areas using locally collected blue-green algae, capable of producing both dissolved oxygen and fixed nitrogen.	*Oscillatoria* sp.	US20210185956A1, 2021

Table 22.1 (continued)

Applied area	Division	Mode of process integration/ intensification	Applied strain	Application number/ publication date
Food and feed	Carboxylic acids	Using microalgae biomass as feedstock for lactic acid production, substitution of lignocellulose depleted algal biomass to reduce the inhibitory or toxic compounds produced through harsh pre-treatment of the lignocellulosic plant cell wall.	*Nannochloropsis salina* or *Haematococcus pluvialis*	US9920337B2, 2018
		Efficiently overproducing succinic acid as a raw material for the production of biodegradable biomass plastics, using cyanobacterial mutant.	*Synechocystis* sp.	JP6778870B2, 2020
	Pigment	Production of biomass with high content of lutein and low content of metals with proper antioxidant properties.	*Muriellopsis* sp.	WO2019071364A1, 2019
		Lutein extraction with high efficiency through shortening time, and decreasing toxicity and costs.	All blue-green algae	CN108586305A, 2018
		Overproduction of Zeaxanthin by mutant strain, and reducing the production cost using sea-water as a culture medium to cultivate a marine species.	*Dunaliella* sp.	US20190290712A1, 2019

Table 22.1 (continued)

Applied area	Division	Mode of process integration/ Intensification	Applied strain	Application number/ publication date
Pharmaceutical/ Cosmetic	Oil and Pigment	Inducing colored green algae to increase intracellular astaxanthin and oil accumulations in a fast-growth strains using plant hormones.	*Chromochloris zofingiensis*	WO2020103622A1, 2020
		Increasing the efficiency of astaxanthin production by optimizing light irradiation by having a ratio of photon flux density of the blue LED to the red LED to be 2:3 to 20:1.	*Haematococcus lacustris*	US20210010049A1, 2021
	Hair-care products	Caring hair and scalp by environmentally friendly algal extracts with complex composition.	*Chlorella* sp. and *Chondrus* sp.	EP3954356A2, 2022
	Photo-protective substances	Protecting skin and hair from ultraviolet rays by environmentally friendly algal extracts.	*Scenedesmus* sp.	KR101805899B1, 2018
	Biosensors	An affordable and simple method of electrochemical determination of vardenafil-active ingredient in biological media such as serum by using algae biomass as a carbon-based material, wherein hydrothermal carbonization is applied to algae biomass to be used for the preparation of the electrode material.	*Cystoseira* sp.	WO2021126095A1, 2021

Secured energy supply with biofuel poses great challenges for suppliers and producers. This young industry needs a good overview and the right strategic solutions. The goal itself is beyond question, but the way to get there raises many questions. Photoautotrophic organisms that act as highly efficient sunlight-driven cell factories are recently considered as a sustainable feedstock to produce the biofuels needed to address global energy needs, while mitigating the issues of carbon negative footprint.

The supply of biomass production with sufficient quality plays a system-critical requirement. How should the issues of freshwater and food supply be addressed? Is there any comprehensive energy outlook into the intensification of the process? Do 'clean up' strategies have potential for sustainable energy production? Or do integrated systems improve the environmental challenges?

To find answers to these questions and challenges like them, we take the opportunity to hear from leading experts from the algal industry about the strategies currently being pursued. The conversion of the eternal sun energy to alternative forms of energy entails many innovations. The production of algal biomass starts with energy and water. Although the energy is free and renewable, the demand on water is high. The production of feedstock from saline water could expand the places of cultivation and makes the process more flexible.

The choice of the appropriate biofuel generation is determined by the availability of the raw material, quality of water sources, and the required quantity and purity of prospective biofuel to meet the regulations. Other factors are the available strains and the flexibility of the technology and operation. Each factor is well discussed in the published patents. The retrieved patents introduced *C. vulgaris* as a most potent strain for various types of biofuel production. Carbohydrates can be exploited for the production of liquid fuel, bioethanol, and biogas (WO2014027871A1, WO2016097414A1). Moreover, accumulated lipid in the cells could be easily extracted and transesterified to produce biodiesel (EP3031932A1, CN103451101A). Further attempts are made to provide a more energy efficient process through process integration. Patent number WO2012038978A1 integrated jatropha biodiesel production by oil using *Chlorella* sp.; the waste organic compounds obtained from the first activity are utilized to promote lipid production in a mixotrophic culture condition. To maximize the material reuse, waste streams can be recycled back into mixotrophic microalgae cultivation or discharged as biofertilizer.

It has long been argued that oleaginous microalgae can be used for the development of integrated bioremediation and biofuel production. The issue of low-priced mass cultivation of microalgae as the most challenging step in microalgal biofuel production could be overcome when microalgal cultivation couples with wastewater treatment in open and closed systems. An interaction between the factors, e.g., potent microalgal strain, wastewater characteristics, and operation systems determines the final efficiency of the integrated process.

Microalgae can not only resist the toxic effects of different environmental pollutants, but these pollutants factors can also stimulate lipid accumulation in the algal cells. In summary, environmental stresses lead to oxidative stress in the algal cells. This leads to the formation of various reactive oxygen species (ROS) and it consequently induces different lipid induction mechanisms to increase the toxic effect resistance in cells (Nanda et al., 2021). The issue of harvesting algal biomass, and lipid extraction and biofuel generation are discussed in detail in the patent. Most valid patents that are cited more than 100 times have listed cultivation engineering, gene manipulation, and downstream biorefinery as the three main determining activities to promote algal biofuel production (for instance see: US6750048, US20080160593A1, US20080160591A1, US20160208243A1, US20110086386A1, US20080118964A1 and US20090011480A1). Patent number WO2014060973A1 highlighted the potential of microalgae belonging to the genus from the group comprising *Nannochloropsis*, *Chlorlla*, *Scenedesmus* and *Synechococciis* for bioenergy production. Accordingly, it could be concluded that the global driving force of microalgae biofuel has decreased after 2015. Development of other renewable energy sources, variation in oil price, increase in the popularity of the electric car, and changes in consumer choice are among the factors influencing this industry.

Due to the rapid development in the artificial modification of natural microalgal strains over the last two decades, a limited number of natural strains are suitable for biofuel production directly. Irradiated mutant microalgae stains could efficiently produce bioethanol or biodiesel with increased productivity of intracellular starch or lipid (KR101672407B1, CN114292838A). Moreover, the potential of genetically modified microalgae and diatom strains are also patented as methods for modifying the lipids quantity and/or quality in these oleaginous microorganisms (US10087453B2, US20160208243A1). Topics related to microalgae strain selection, processes of isolation, maintenance or propagation of cellular compositions as well as mutation or genetic engineering of microalgae are patented in class C12N, which form about 58% of the patents in the patents pool related to algal biofuel production (Li et al., 2020).

Direct power generation using microbial fuel cell (MFC)-based processes has developed as an emerging and promising technology for coupled wastewater treatment and energy recovery using algal cultivation. MFC is a device that converts chemical energy to electric current by producing electrons from a microbial oxidation process on the anode to oxidize a compound on the cathode. Since the cathodic reaction always limits the overall performance of the MFCs, bioscientists have focused to improve the cathodic reaction; the potential of the microalgae and their biocatalytic activities in oxygenation on the cathode through nonlimiting oxygen generation is investigated repeatedly (CN201877517U, TW202044658A).

The patented MFC technique, published by TW202043162A, requires the culturing of bacteria in the anode tank filled with nutrient source. Heterotrophic metabolism of bacteria generates electrons and protons. Immobilized algae are also inoculated in the cathode tank, which is filled with a specific nutrient source. The cathode tank is illuminated using a light source to enhance the microalgae's biocatalytic activities.

Process integration of wastewater treatment and MFCs poses several advantages including energy and material recirculating, wastewater treatment, microalgae biomass, byproducts preparation, and CO_2 utilization (Chiranjeevi and Patil, 2020). However, more research is required to improve the technical feasibility of the MFC-coupled wastewater treatment. In an interesting patent (WO2010088626A2), organic matter and dead algal cells convert to electrical energy within a coupled setup of a MFC and an algal cultivation for biodiesel production. More specifically, phototrophic algae sequestrate solar energy into biocompounds that may be converted into electrical energy later. This self-sustained power generator reduces the cost of biofuel production. In addition, it enhances the recirculating material and decreases the carbon footprint.

Synergistic action of the microalgae and the MFC forms a coupling system (CN109650548B). This integrated process of water pollution treatment and energy recovery could degrade 87.47 and 81.50% COD and ammonia nitrogen from domestic garbage leachate, respectively. Application of integrated processes for microalgae-based multi-technology coupling in the purification of high-salt water system is discussed in patent nu. WO2017101654A1; desalination chamber is located between the cathode chamber and the anode chamber. Extracellular electrogenerated bacteria in the anode chamber and photoautotrophic algae in cathode chamber generate a potential to desalt the wastewater, surrounded by anions and the cation exchange membrane within the desalination chamber. This design enables the energy of the whole system to be internally circulated; it effectively removes salt and other organic pollutants, and produces no secondary pollution. In conclusion, it should be noted that the choice of the appropriate wastewater treatment technology is determined by the required purity, available energy source, the variability of the raw water source, the adaptability of the required quantity, and the operational and theoretical limitations. This also applies to the integrated process of wastewater treatment and MFCs.

Recently, immobilized microalgae have been increasingly used for wastewater treatment. Since the biomass productivity in immobilized system is less than in other cultivation methods, several patents have been published to solve this problem for wastewater bioremediation systems. A patent (CN110776103A) provides an immobilized microcystis adsorption membrane system for purifying nitrogen and phosphorus in a water body, which not only enhances the contact area between the membrane and the wastewater but also enables the immobilized microalgae to acquire more strength; it solves the problems associated with the small contact area, low utilization efficiency, and enables microcystis to effectively remove salts and the partial organic matter from the water body. Biofilm-attached systems is also used for treating wastewater to minimize the harvest and recovery shortcomings in biomass harvesting as well as in removing pollutants with a desired species of microalgae.

Microalgae-based processes are too water-consuming. It might act as a strong constraint for large-scale cultivation of microalgae. Water expenditure in the cultivation and harvesting stages as well as in the transformation of the feedstock into products raises the need for better management and reuse of water in an integrated process. Water integration using seawater not only minimizes water requirements but also facilities mass integration (US8507254B1). Seawater is rich in nutrients such as carbonates, nitrates, phosphates, minerals, and other dissolved ions that promote cell growth in marine species. However, it should be noted that cell proliferation in seawater is extremely slow (JP2014113083A). In summary, bio-refinery of wastewater by the microalgae is an environmental friendly process with no secondary pollution. It presents numerous advantages, such as reduced cultivation cost, efficient recycling of nutrients, especially for phosphate residue, and biofixation of CO_2. However, the remaining obstacles, e.g., vulnerability to temperature and light variability and competition with the microflora need to be overcome during the future innovations.

Scaling up and mainstreaming might be restricted by narrow genetic pools, low culture cell densities, and inefficient harvesting/drying and other downstream processes. Further bioprocess performance improvements are explored by innovations in equipment and technologies. However, each technology has its own practical, operational, and theoretical limitations. It was found that mass cultivation, the overpriced and technically challenging step, exhibits a major impact on the final productivity of the process. Significant efforts are still required for the development and optimization of coupled systems to insure the environmental friendliness and the cost effectiveness of microalgae-based systems. Therefore, we reviewed the patent literature based on technological and environmental criteria during up- and mainstream processes. Coproduct fraction strategy is nominated in the top five key strategies to drive the intensification of microalgae processes. However, due to the health-related matters, there are few reports on the stable industrialization of bio-refinery, coupled with coproduction.

Finally, the integration of downstream processes, such as cell disruption, component extraction, catalytic conversion, and waste to bioproducts' production into a single step process, is an attractive area both from a techno-economic and intellectual property point of views. For establishing a zero-waste system, it is crucial to meet the regulation of process integration and this means that microalgae cultivation continues to hold potential as a low environment footprint industry.

References

Abiusi, F., Trompetter, E., Pollio, A., Wijffels, R. H., & Janssen, M. (2022). Acid tolerant and acidophilic microalgae: An underexplored world of biotechnological opportunities. *Frontiers in Microbiology*, *13*, 51. https://doi.org/10.3389/fmicb.2022.820907

Aoki, Y., Okamura, Y., Ohta, H., Kinoshita, K., & Obayashi, T. (2016). ALCOdb: Gene coexpression database for microalgae. *Plant and Cell Physiology*, *57*(1), e3–e3. https://doi.org/10.1093/pcp/pcv190

Benemann, J. R., Tillett, D. M., & Weissman, J. C. (1987). Microalgae biotechnology. *Trends in Biotechnology*, *5*(2). https://doi.org/10.1016/0167-7799(87)90037-0

Biz, A. P., Cardozo-Filho, L., & Zanoelo, E. F. (2019). Drying dynamics of microalgae (Chlorella pyrenoidosa) dispersion droplets. *Chemical Engineering and Processing – Process Intensification*, *138*, 41–48. https://doi.org/10.1016/j.cep.2019.03.007

Blanken, W., Cuaresma, M., Wijffels, R. H., & Janssen, M. (2013). Cultivation of microalgae on artificial light comes at a cost. *Algal Research*, *2*(4), 333–340. https://doi.org/10.1016/j.algal.2013.09.004

Borowiak, D., Lenartowicz, P., Grzebyk, M., Wiśniewski, M., Lipok, J., & Kafarski, P. (2021)). Novel, automated, semi-industrial modular photobioreactor system for cultivation of demanding microalgae that produce fine chemicals – The next story of H. pluvialis and astaxanthin. *Algal Research*, *53*(November 2020). https://doi.org/10.1016/j.algal.2020.102151

Brzychczyk, B., Hebda, T., & Pedryc, N. (2020). The influence of artificial lighting systems on the cultivation of algae: The example of chlorella vulgaris. *Energies*, *13*(22). https://doi.org/10.3390/en13225994

Chen, M., Chen, Y., & Zhang, Q. (2021). A review of energy consumption in the acquisition of bio-feedstock for microalgae biofuel production. *Sustainability (Switzerland)*, *13*(16). https://doi.org/10.3390/su13168873

Chiranjeevi, P., & Patil, S. A. (2020). Microbial Fuel Cell Coupled with Microalgae Cultivation for Wastewater Treatment and Energy Recovery. In: *Integrated Microbial Fuel Cells for Wastewater Treatment*. Elsevier, pp. 213–227. https://doi.org/10.1016/B978-0-12-817493-7.00010-2

Darko, E., Heydarizadeh, P., Schoefs, B., & Sabzalian, M. R. (2014). Photosynthesis under artificial light: The shift in primary and secondary metabolism. *Philosophical Transactions of the Royal Society B Biological Sciences*, *369*(1640). https://doi.org/10.1098/rstb.2013.0243

de Farias Silva, C. E., & Bertucco, A. (2016). Bioethanol from microalgae and cyanobacteria: A review and technological outlook. *Process Biochemistry*, *51*(11), 1833–1842. https://doi.org/10.1016/j.procbio.2016.02.016

Dourou, M., Tsolcha, O. N., Tekerlekopoulou, A. G., Bokas, D., & Aggelis, G. (2018). Fish farm effluents are suitable growth media for Nannochloropsis gaditana, a polyunsaturated fatty acid producing microalga. *Engineering in Life Sciences*, *18*(11), 851–860. https://doi.org/10.1002/elsc.201800064

Fasaei, F., Bitter, J. H., Slegers, P. M., & van Boxtel, A. J. B. (2018). Techno-economic evaluation of microalgae harvesting and dewatering systems. *Algal Research*, *31*, 347–362. https://doi.org/10.1016/j.algal.2017.11.038

Fret, J., Roef, L., Diels, L., Tavernier, S., Vyverman, W., & Michiels, M. (2019). Combining medium recirculation with alternating the microalga production strain: A laboratory and pilot scale cultivation test. https://doi.org/10.1016/j.algal.2019.101763

Kim, G. M., & Kim, Y. K. (2022). Drying techniques of microalgal biomass: A review. *Applied Chemistry for Engineering*, 33(2), 145–150. https://doi.org/10.14478/ace.2022.1007

Kumar, G., Shekh, A., Jakhu, S., Sharma, Y., Kapoor, R., & Sharma, T. R. (2020). Bioengineering of microalgae: Recent advances, a perspectives, and regulatory challenges for industrial application. *Frontiers in Bioengineering and Biotechnology*, *8*. https://doi.org/10.3389/fbioe.2020.00914

Li, D., Du, W., Fu, W., & Cao, X. (2020). A quick look back at the microalgal biofuel patents: Rise and fall. *Frontiers in Bioengineering and Biotechnology*, *8*, 1035. https://doi.org/10.3389/fbioe.2020.01035

Liu, T., Cui, Y., Jia, X., Zhang, J., Li, R., Yu, Y., Jia, S., Qu, J., & Wang, X. (2021). OGDA: A comprehensive organelle genome database for algae. *Database*, *2020*. https://doi.org/10.1093/DATABASE/BAAA097

Maglie, M., Baldisserotto, C., Guerrini, A., Sabia, A., Ferroni, L., & Pancaldi, S. (2021). A co-cultivation process of nannochloropsis oculata and tisochrysis lutea induces morpho-physiological and biochemical variations potentially useful for biotechnological purposes. *Journal of Applied Phycology*, 33(5), 2817–2832. https://doi.org/10.1007/s10811-021-02511-2

Marrone, B. L., Lacey, R. E., Anderson, D. B., Bonner, J., Coons, J., Dale, T., Downes, C. M., Fernando, S., Fuller, C., Goodall, B., Holladay, J. E., Kadam, K., Kalb, D., Liu, W., Mott, J. B., Nikolov, Z., Ogden, K. L., Sayre, R. T., Trewyn, B. G., & Olivares, J. A. (2018). Review of the harvesting and extraction program within the National Alliance for Advanced Biofuels and Bioproducts. *Algal Research*, 33, 470–485. https://doi.org/10.1016/j.algal.2017.07.015

Misra, N., Panda, P. K., Parida, B. K., & Mishra, B. K. (2016). DEMBF: A comprehensive database of enzymes of microalgal biofuel feedstock. *PLoS ONE*, *11*(1), e0146158. https://doi.org/10.1371/journal.pone.0146158

Nanda, M., Chand, B., Kharayat, S., Bisht, T., Nautiyal, N., Deshwal, S., & Kumar, V. (2021). Integration of Microalgal Bioremediation and Biofuel Production: A 'Clean Up' Strategy with Potential for Sustainable Energy Resources. In: Current Research in Green and Sustainable Chemistry. Elsevier, Vol. 4, p. 100128. https://doi.org/10.1016/j.crgsc.2021.100128

Nwoba, E. G., Parlevliet, D. A., Laird, D. W., Alameh, K., & Moheimani, N. R. (2019). Light management technologies for increasing algal photobioreactor efficiency. *Algal Research*, *39*(January). https://doi.org/10.1016/j.algal.2019.101433

Olaizola, M., & Grewe, C. (2019). Commercial Microalgal Cultivation Systems. *Grand Challenges in Algae Biotechnology*. http://www.springer.com/series/13485

Patil, R. A., Kausley, S. B., Joshi, S. M., & Pandit, A. B. (2020). Process Intensification Applied to Microalgae-Based Processes and Products. In: Handbook of Microalgae-Based Processes and Products: Fundamentals and Advances in Energy, Food, Feed, Fertilizer, and Bioactive Compounds. Elsevier Inc. https://doi.org/10.1016/B978-0-12-818536-0.00027-0

Ray, A., Nayak, M., & Ghosh, A. (2022). A review on co-culturing of microalgae: A greener strategy towards sustainable biofuels production. *Science of the Total Environment*, *802*, 149765. https://doi.org/10.1016/j.scitotenv.2021.149765

Sathasivam, R., Radhakrishnan, R., Hashem, A., & Abd_Allah, E. F. (2019). Microalgae metabolites: A rich source for food and medicine. *Saudi Journal of Biological Sciences*, *26*(4), 709–722. https://doi.org/10.1016/j.sjbs.2017.11.003

Saxon, R. J., Rad-Menéndez, C., & Campbell, C. N. (2020). Patent depositing of algal strains. *Applied Phycology*, *3*(1), 226–233, https://doi.org/10.1080/26388081.2020.1770124

Sero, E. T., Siziba, N., Bunhu, T., Shoko, R., & Jonathan, E. (2020). Biophotonics for improving algal photobioreactor performance: A review. *International Journal of Energy Research*, *44*(7), 5071–5092. https://doi.org/10.1002/er.5059

Severo, I. A., Deprá, M. C., & Jacob-Lopes, E. (2020). Process Integration Applied to Microalgae-based Systems. In: Handbook of Microalgae-based Processes and Products: Fundamentals and Advances in Energy, Food, Feed, Fertilizer, and Bioactive Compounds, pp. 709–735. https://doi.org/10.1016/B978-0-12-818536-0.00026-9

Tabernero, A., Martín Del Valle, E. A., & Galan, M. A. (2013). Microalgae technology: A patent survey. *International Journal of Chemical Reactor Engineering*, *11*(2), 733–763. https://doi.org/10.1515/ijcre-2012-0043

Uduman, N., Qi, Y., Danquah, M. K., Forde, G. M., & Hoadley, A. (2010). Dewatering of microalgal cultures: A major bottleneck to algae-based fuels. *Journal of Renewable and Sustainable Energy*, *2*(1), 012701. American Institute of PhysicsAIP. https://doi.org/10.1063/1.3294480

Visigalli, S., Barberis, M. G., Turolla, A., Canziani, R., Berden Zrimec, M., Reinhardt, R., & Ficara, E. (2021). Electrocoagulation–flotation (ECF) for microalgae harvesting – A review. *Separation and Purification Technology*, *271*, 118684. https://doi.org/10.1016/j.seppur.2021.118684

Wi, S. G., Cho, E. J., Lee, D. S., Lee, S. J., Lee, Y. J., & Bae, H. J. (2015). Lignocellulose conversion for biofuel: A new pretreatment greatly improves downstream biocatalytic hydrolysis of various lignocellulosic materials. *Biotechnology for Biofuels*, *8*(1), 228. https://doi.org/10.1186/s13068-015-0419-4

Xu, Z., Wang, H., Cheng, P., Chang, T., Chen, P., Zhou, C., & Ruan, R. (2020). Development of integrated culture systems and harvesting methods for improved algal biomass productivity and wastewater resource recovery – A review. *Science of the Total Environment*, *746*, 141039. https://doi.org/10.1016/j.scitotenv.2020.141039

Mehak Kaur, Hishita Peshwani, Mayurika Goel*

Chapter 23
A brief mapping of patents in microalgae-based systems

Abstract: Microalgae-based systems are one of the most sustainable substitutes for naturally occurring compounds, such as pigments, lipids, fatty acids, carotenoids, and proteins, which are well known to have various applications. In the past few decades, vigorous research has been conducted in the field of microalgae-based commercial applications. Global market trends have flourished tremendously with the increase in research and development of microalgae-based systems, focusing on the innovation of processes and development of products. Patents and publications are available on microalgae products that are present in the market along with their complete process. This chapter focuses on the available patents in the field of microalgae research from the past decades and their impact on the global market. Databases such as Google patents, espacenet, inPASS, and World Intellectual Property Organisation (WIPO) were surveyed with definite keywords. Trends on the growth of the microalgae industry as well as the fundamental challenges are elaborated. Innovation management for the conservation of intellectual property rights would also be discussed with respect to Indian and global initiatives in this area.

Keywords: microalgae, global market, patents, intellectual property rights

23.1 Introduction

Microalgae have been investigated extensively for their metabolic versatility and ecological importance. They are seen as potential sources of biofuels, cosmetics, food, feed, nanomaterials, and pharmaceuticals. They are planktons found in the sediments of both freshwater and marine systems; they are also referred to as macrophytes. These unicellular organisms can live alone, in communities, or in chains, and their size can vary from one micron to a hundred microns, depending on the species. They can tolerate a wide range of salinities, temperatures, and pH values (Khan et al., 2018). Current research efforts are primarily focused on isolating novel strains, characterizing the ecophysiology and metabolites of microalgac, modeling their productivity, controlling and

***Corresponding author: Mayurika Goel**, Sustainable Agriculture Program, The Energy and Resources Institute (TERI), Gurugram, Haryana, India, e-mails: mayurika.goel@teri.res.in, mayurikagoel@gmail.com
Mehak Kaur, Hishita Peshwani, Sustainable Agriculture Program, The Energy and Resources Institute (TERI), Gurugram, Haryana, India

https://doi.org/10.1515/9783110781267-023

improving the production of high-value chemicals, and developing novel and sustainable applications. From a commercial and market standpoint, microalgae production brings together significant number of researchers, biotech companies, and technological platforms to sustainably produce high-quality microalgae biomass. Over the last few decades, their industrial cultivation has increased dramatically and the global outlook has grown, diversified, and has been disseminated into various fields to utilize their full potential (Pulz and Gross, 2004; Rizwan et al., 2018). This book chapter provides insight into the patents available in the microalgae fields using different keywords, such as microalgae pigments, biofuels, food feed, pharmaceuticals, nutraceuticals, and cosmeceuticals. The global trend of research and development in microalgae research has also been discussed, focusing on the major research themes followed for the filing of patents.

23.2 Patent survey of microalgae commodities

The protection of products, processes, and technologies that have resulted from biotechnological interventions in microorganisms is done by the Patent Acts of legislations around the world. Although, patenting of living organisms that exist in nature is not possible, patenting of genetically modified novel phenotypes and strains, resulting from breeding, can be done. The intellectual property protection of microalgae-based technology has stimulated research and innovation in product formulations, and in utilization of novel techniques and strains for high yield, while attracting funding and investments. Therefore, it acts as a key component of innovation, bridging the gap between laboratories and factories (Murata et al., 2021). Analyzing patent databases is an effective way to track changes in a research field, particularly from a market perspective. It is also an important tool for predicting trends and for formulating research development strategies. One of the first patented microalgae-based inventions was on 'Fermentation process for producing alcoholic beverages from microalgae' in 1968 by Jorgen Jorgensen (Jorgen, 1968; source – WIPO database). For the current study, patents from 2002 to 2022 were surveyed from the World Intellectual Property Organisation (WIPO) database, Google Patents, espacenet (European Patent Office), and inPASS (Indian Patent Search Service) for the commodities biofuel, bio pigments, food and feed, cosmeceuticals, nutraceuticals, and pharmaceuticals with various keywords (results shown in Table 23.1). A general increase in the number of patents each year was observed, with most patents being related to bio-pigments, followed by food-feed (Figure 23.1).

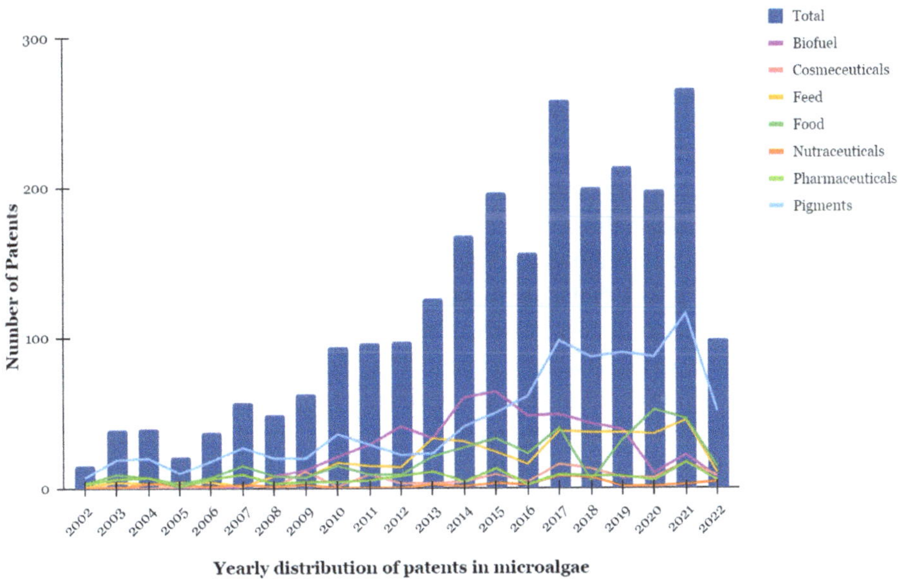

Figure 23.1: Evaluation of the distribution of patents among microalgae commodities across two decades (2002–2022); biofuel, bio-pigments, cosmeceuticals, feed, food, nutraceuticals, and pharmaceuticals.

23.2.1 Biofuel

Microalgae biofuel is considered the most promising solution for the global fossil fuel crisis. Microalgae offers numerous benefits over other sources: being a non-agricultural crop, it avoids the "food versus fuel" debate (Talebi et al., 2018) and has high reproducibility; natural resources required for biomass production are negligible and the oil yield per area is much higher compared to the best oilseed crops (Rocha et al., 2012). Considering this potential of microalgae for biofuel production, there has been a surge in publications as well as patents in the past decade, where the numbers have increased exponentially each year. Furthermore, the maximum number of patents was filed by developed countries such as the US, Australia, Japan, Korea, and by established organizations such as the WIPO, EPO, and others. A trend is evident as most of the applicants for US patents are companies, whereas in China, most patents are held by academies. Although India has significantly a smaller number of microalgae biofuel patents, most of them are held by universities and research institutes. A major chunk of the patents that were analyzed were utility patents that claimed multiple applications, while only a small fraction was of design patents. Most of those utility patents covered the various methods that can be used for producing biofuel from microalgae, but the search in the current chapter was for biomass cultivation, optimization processes, flocculation and enhanced lipid production, and oil extraction as well as processes applied for quality improvement.

Currently REG (Renewable Energy Group), US is the top biofuel producer in the world (Statista, 2018) along with Indonesia and China in the top five. This industry has overcome a lot of challenges in the past by the artificial modification of natural strains, with the development of gene engineering, by reducing the cost of biomass production and simultaneously increasing the lipid content through various modifications of bioreactors and fermenters (Li et al., 2020). Multinational platforms have also originated to work toward the promotion of alternatives to microalgae for destructive petroleum-based products. Biofuture Platform is a multistakeholder platform of 20 leading nations for policy dialogue to speed up the development and scale up the deployment of climate responsive, low carbon alternatives to fossil-based commodities. According to the recent 2022 amendment of the New Biofuel Policy (June 2018) by the Department of Biotechnology and Government of India, it is suggested that 20% ethanol and 5% biodiesel will be blended into gasoline and diesel, respectively, by 2025–26 after high performance was shown by microalgae biofuel groups across the country (DBT, 2018 and 2022; Biofuture Platform Facilitator and Clean Energy Ministerial, 2013).

23.2.2 Bio pigments

Pigments from microalgae sources such as carotenoids, phycobiliproteins, and chlorophylls are being used in the textile, cosmetic, food, feed, beverage, pharmaceutical, and other industries. *Haematococcus, Dunaliella, Chlorella,* and *Spirulina* are widely explored for the efficient production of microalgae pigments. Carotenoids are fat-soluble pigments containing oxygenated (xanthophylls) or hydrocarbon (carotenoid) C40 carbon skeletons that are naturally produced by plants and algae. The most common secondary xanthophyll is astaxanthin ($C_{40}H_{52}O_4$), which is a red-orange pigment with antioxidant, anticancer, anti-inflammatory, bio-remediation, neuroprotective, and immunity-enhancing properties (Kalra et al., 2020; Borowiak and Krzywonos, 2022). *Haematococcus* spp. is used commercially to produce astaxanthin at the industrial scale and is a major section of patents and publications per year in the field of microalgae pigments. Between 2002 and 2022, 560 patents were published on astaxanthin obtained from *Haematococcus* spp. The majority of these patents (64%) were application-based using astaxanthin extracted from *Haematococcus* as an antioxidant compound, or as cosmeceutical, nutraceutical, food, or feed additive, or as a red colorant. Methods for producing astaxanthin efficiently and to increase its accumulation in cells with the help of radiation, media optimization, genetic modifications, and others, followed the patent division. Efficient extraction of the compound from cells via various solvents and enzymes, and green extraction with CO_2 and others has also been patented. China is the leading country with 67% of patents. Microalgae *Dunaliella* is an approved natural source of beta-carotene and biofuels (Ambati et al., 2018; Novoveská et al., 2019; Kalra et al., 2020). Patents based on process optimization and recovery of beta-carotene constitutes significant number of patents (64) published from 2002–2022. China is the major contributor, with 76% of the patents under it. Chlorophyll ($C_{55}H_{72}MgN_4O_5$) is a green photosynthetic fat-soluble pigment that is produced by algae, cyanobacteria, and plants. It has antioxidant, anti-inflammatory, antimutagenic, and wound-healing properties, and is known to decrease the risk of colorectal cancer (Begum et al., 2016). *Chlorella* is commercially used for the industrial-scale production of chlorophyll for food, feed, cosmetics, nutraceuticals, and pharmaceutical applications. It was one of the earliest patented microalgae genera, with the first patent dating back to 1964. Chlorophyll obtained from *Chlorella* has 76 patents, from 2002 to 2022. China accounted for approximately 38% of patents, followed by the Republic of Korea (23%). Phycobiliproteins are light-harvesting water-soluble oligomeric proteins that are divided into phycocyanin (blue), phycoerythrin (red), allophycocyanin (bluish-green), and phycoerythrocyanin (orange). Phycocyanin from *Spirulina* is focused on in this chapter. It is commercially used as a blue pigment in the food, beverage, cosmetic, textile, fluorescent marker, and pharmaceutical industries. It has potent antioxidant, anticancer, anti-inflammatory, hepatoprotective, and neuroprotective effects (Begum et al., 2016). A total of 205 patents were published between 2002 and 2022 on phycocyanin extracted from *Spirulina,* with 60% of the patents contributed by China.

Table 23.1: Patent details of the various commodities of microalgae from 2002 to 2022.

Patent/ Application no.	Country	Granted to	Patent details	Year	Reference
CN1438240	China	The South-China-Sea Oceanography Inst., Chinese Academy of Sciences	Method for preparing phycocyanin of sea water	2005	(Wenzhou et al., 2005)
CN101003565	China	Shangdong Univ. of Tech	Method for preparing phycocyanin and allophycocyanin at the same time	2009	(Chunxiao and Zhengquan, 2009)
CN101607988	China	Weihai Grennan Bio-Engineering Co., Ltd.	Technique for extracting phycocyanin, chlorophyll and *Spirulina* polysaccharide in *Spirulina*	2009	(Daopeng et al., 2009)
CN102154330	China	South China University of Technology	Beta-carotene engineering bacteria based on *Dunaliella bardawil* metabolic pathway and the construction method thereof	2012	(Jianguo et al., 2012)

Table 23.1 (continued)

Patent/ Application no.	Country	Granted to	Patent details	Year	Reference
CN102168096	China	South China University of Technology	Beta-carotene engineering bacteria based on *Dunaliella salina* metabolic pathway and the construction method	2012	(Jianguo and Zhiwei, 2012)
CN101580857B	China	Xinao Science and Technology Development Co., Ltd.	Biomass energy prepared by the one-step method of microalgae	2012	(Lei and Minsheng, 2009)
CN101899102	China	South China University of Technology	Method for separating high purity phycocyanin from *Spirulina*	2013	(Zhang and Liao, 2013)
CN102617431	China	Tianjin Binhai suoerte Center for Biotechnology Co., Ltd.	Method for extracting beta-carotene from *Dunaliella* by inorganic base	2013	(Anjun et al., 2013)
CN101613618B	China	ENN Science and Technology Development Co Ltd	Method for preparing biodiesel by using microalgae lipid as raw material	2013	(Lei et al., 2009)
CN102618446B	China	Beijing Canfit Resource Recovery Technologies Inc.	Method for cultivating oleaginous microalgae by using fecal sewage	2013	(Dan et al., 2012)
CN104387468	China	Jiangnan University	Technology for separating phycocyanin in *Spirulina* by using polyvinyl ether alcohol amine through two-aqueous phase extraction	2015	(Song et al., 2015a)
CN104387469	China	Jiangnan University	Technology for separating phycocyanin in *Spirulina* by using polyvinyl ether phosphate through two-aqueous phase extraction	2015	(Song et al., 2015b)
CN102771836	China	Nikken Sohonsha Corporation	Method for producing beta-carotene-rich *Dunaliella* powder	2015	(Ami and Nobuo, 2015)

Table 23.1 (continued)

Patent/ Application no.	Country	Granted to	Patent details	Year	Reference
CN103554250	China	Yunnan Lanzuan Biology Science & Technology Co., Ltd.	Method for extracting phycocyanin	2015	(Wei et al., 2015)
CN104292326	China	Yantai Institute of Coastal Zone Research, Chinese Academy of Sciences	Industrially producing method of food-grade phycocyanin	2015	(Shao et al., 2015)
CN204509239U	China	SCUT- South China University of Technology	A kind of microalgae biodiesel preparation system of ultrasonic assistant shell calcium foundation stone ash	2015	(Xiaoxi et al., 2015)
CN102559374B	China	ENN Science and Technology Development Co Ltd.	Method for preparing biodiesel from microalgae through the one-step method by utilizing supercritical methanol.	2015	(Yanshan et al., 2012)
CN103642579A	China	Jinan Development Zone Xinghuo Science and Technology Research Institute	Method utilizing microalgae to produce biodiesel	2015	(Chunling and Changxiang, 2014)
CN103597069	China	Korean Ocean Research and Development Institute (KORDI)	Production plant for microalgae biofuel, bioreactor for producing biofuel, and method for producing microalgae biofuel	2015	(Do-Hyung et al., 2014)
CN103613661	China	Yunnan Lanzuan Biotechnology Co., Ltd.	Method of extracting high-purity phycocyanin	2016	(Wei et al., 2016)
CN103451101B	China	China Petroleum and Natural Gas Co Ltd	Production method of high-quality microalgae biodiesel	2016	(Hao et al., 2013)
CN103981017A	China	Institute of Oceanology of CAS	Method used for extracting oil from microalgae wet algae mud	2016	(Pengcheng et al., 2014)

Table 23.1 (continued)

Patent/ Application no.	Country	Granted to	Patent details	Year	Reference
CN104479008	China	Jiangnan University	Method for separating and purifying phycocyanin in *Spirulina* by using polyacrylic acid/ magnesium oxide hybrid microspheres	2017	(Song et al., 2017)
CN105646313	China	Beijing Boyuanxin Green Technology Co., Ltd.	Method for extracting beta-carotene from *Dunaliella salina* by utilizing subcritical fluid	2017	(Zhang et al., 2017)
CN104560409A	China	Guangzhou Institute of Energy Conservation of CAS	Method for direct preparation of biodiesel by utilizing microalgae ultrasonic-assisted ionic liquid composition	2017	(Zhen Hong et al., 2015)
CN104450849	China	Linyi University	Method for forcing *Dunaliella tertiolecta* to accumulate beta-carotene	2018	(Wang, 2018)
CN104762335A	China	Jiangsu Academy of Agricultural Sciences	Application of methyl jasmonate in the improvement of microalgae biodiesel quality	2018	(Pasa Emer et al., 2015)
CN109160947	China	Lin Yan	Preparation method and application of *Spirulina* phycocyanin	2020	(Huang, 2020)
CN110499272	China	South China University of Technology	Engineering bacteria with high beta-carotene yield based on *Dunaliella* metabolic pathway and the construction method, and the application of beta-carotene high yield engineering bacteria	2021	(Jianguo et al., 2021b)

Table 23.1 (continued)

Patent/ Application no.	Country	Granted to	Patent details	Year	Reference
CN110511919	China	South China University of Technology	Lycopene high-yield engineering bacterium based on *Dunaliella* metabolic pathway as well as the construction method and application thereof	2021	(Jianguo et al., 2021a)
CN113388026	China	Jiangxi University of Science and Technology, Pan Asia (Jiangmen) Institute of Biological Engineering and Health	Synchronous extraction method of Spirulina phycocyanin and algae oil	2022	(Pan et al., 2022)
202227002093	France	Total energies SE	Mutation of an acyl-co, a synthase for increased triacylglycerol production in microalgae	2022	(Magneschi et al., 2022)
WO2003027267	India	Parry Nutraceuticals Ltd.	Process to produce astaxanthin from *Haematococcus* biomass	2003	(Thomas et al., 2003)
IN3861/DELNP/ 2005	India	Tethys Research LLC	Biotransformation of compounds using non-prokaryotic microalgae	2007	(Kravit, 2007)
IN3008/CHENP/ 2007	India	–	Immunostimulatory preparations isolated from microalgae	2007	(NA, 2007)
221688	India	Applied Biotechnology Limited	A method of extraction and purification of naturally mixed carotenoid from microalgae	2008	(Venkatesh, 2008)
237338	India	Council Of Scientific and Industrial Research	A process for the preparation of food colorant from *Spirulina*.	2009	(Barhate et al., 2009)
8,051/DELNP/ 2008	India	Biotechmarine	Cosmetic active ingredient composed of arginine ferrulate and microalgae extract and its uses	2009	(Mekideche, 2009)

Table 23.1 (continued)

Patent/ Application no.	Country	Granted to	Patent details	Year	Reference
241826	India	Council Of Scientific and Industrial Research	Development of autotrophic bioreaction vessel for the growth process of *Haematococcus* for carotenoid production	2010	(Sarada et al., 2010)
IN9162/DELNP/ 2010	India	Ben-Gurion University of The Negev Research and Development Authority	Compositions comprising red microalgae polysaccharides and metals	2012	(Shosh, 2012)
251116	India	Council Of Scientific & Industrial Research	A process for the preparation of phycocyanin, a natural blue colorant, from *Spirulina* species	2012	(Ganapathi et al., 2012)
IN689/MUM/2010	India	North Maharashtra University	Natural blue pigment and process for producing the same	2012	(Puranik et al., 2012)
6,564/DELNP/ 2012	India	CSIR	Engine-worthy fatty acid methyl ester (biodiesel) from naturally occurring marine microalgae mats and marine microalgae cultured in open salt pans, together with the value addition of co-products	2012	(Mishra et al., 2015)
IN3120/DEL/2011	India	Council Of Scientific & Industrial Research	An improved medium composition and a process for enhanced lutein production from microalgae	2013	(N et al., 2013)
2,379/MUM/2010	India	Innovative Creations Business Modules Pvt. Ltd.	*Spirulina* biomass, with modified color characteristics, and process for preparing the same	2013	(Gajraj, 2013)

Table 23.1 (continued)

Patent/ Application no.	Country	Granted to	Patent details	Year	Reference
IN1787/DEL/2012	India	Council Of Scientific & Industrial Research, Oil and Natural Gas Corporation Ltd.	A composition and method for controlling rotifer in microalgae culture systems	2015	(Chauhan et al., 2015)
264507	India	Council Of Scientific and Industrial Research	A process for obtaining water-dispersible astaxanthin composition	2015	(Burde et al., 2015)
IN201641010123	India	Mohanraj Subramanian	Microalgae and methods for producing docosahexaenoic acid	2016	(Subramanian et al., 2016)
IN201747007941	India	Algobiotech	Method for inducing the synthesis of phycobiliproteins	2017	(Hatem and Ammar, 2017)
2,335/CHE/2006	India	Parry Nutraceuticals	Photoautotrophic growth of microalgae for omega-3-fatty acid production	2017	(Sebastian and Kumaravel, 2008)
IN201741018074	India	Dr. Kanapathy Gopalakrishnan	Scale up studies of microalgae *Chlorella rotunda* for biolipids (biodiesel) as third-generation biofuel, along with value-added product using indigenously developed photo bioreactor	2018	(Upendra et al., 2018a)
IN201741018095	India	Dr. Kanapathy Gopalakrishnan	Design of a novel hybrid photobioreactor for the cultivation of microalgae for biodiesel production	2018	(Upendra et al., 2018b)
IN201841006816	India	Achintya Ranjan	Regulated microalgae harvester	2019	(Ranjan, 2019)
IN201941044256	India	Raje Siddiraju Upendra	A hybrid photo bioreactor (PBR) for the cultivation of oleaginous microalgae for optimum lipid yield and subsequent biodiesel production	2019	(Lakshmi et al., 2019)

Table 23.1 (continued)

Patent/ Application no.	Country	Granted to	Patent details	Year	Reference
IN201811016783	India	Uttaranchal university	A process for microalgae harvesting and enhancing biomass productivity	2019	(Kumar et al., 2019a)
201811008547	India	Uttaranchal university	A method for the preparation of nutrient source from fermentation residues of food waste for the cultivation of microalgae	2019	(Kumar et al., 2019b)
IN201811050007	India	Tilak Raj Sharma	A simple method for extracellular production of high purity c-phycocyanin from *Spirulina* platensis	2020	(Yadav et al., 2020b)
IN202011045595	India	Lovely Professional University	An immunity boosting drink using microalgae	2020	(Rasane et al., 2020)
IN202021033216	India	IIT Bombay	A process for the extraction and recovery of high-quality protein from microalgae pool	2020	(Arora and Katiyar, 2022)
IN201811039752	India	Amity University	A method for flocculation and lipid extraction by microalgae using zirconium dioxide nanoparticles	2020	(Rai et al., 2020)
IN201811041629	India	ONGC (oil and natural gas corporation Ltd), IIT, Guwahati	Process for enhancing biomass productivity by high-density cultivation of microalgae	2020	(Goswami et al., 2020)
IN201611036660	India	CSIR	A process for sequestration of CO2 and traces of hydrocarbon from natural gas processing industry	2021	(Anguselvi et al., 2018)
IN202011010829	India	Amity University	A biopolymer-induced microalgae cell flocculation and enhanced lipid production for biodiesel application	2021	(Rai and Vasistha, 2021)

Table 23.1 (continued)

Patent/ Application no.	Country	Granted to	Patent details	Year	Reference
IN201811030379	India	DRDO	Novel composition of synthetic culture media for the growth of microalgae with biofuel potential	2021	(Yadav et al., 2020a)
IN201721042669	India	IITB- Monash research academy	Catalytic conversion of microalgae into methane-rich syngas using reactive flash volatilization	2022	(Gholkar et al., 2019)
IN201827009409	India	Algahealth (Ah) Ltd.	Improved process for producing fucoxanthin and/or polysaccharides from microalgae	2022	(Oran, 2022)
IN202231057904	India	Indian Institute of Technology Guwahati	An integrated process and system for harvesting and extraction of lipid from wet algal biomass	2022	(Goud and Suryawanshi, 2022)
LT5924	Lithuania	Simkus Almantas, Simkiene Aldona, Simkiene Aldona, Mazuras Linas	Method of isolating c-phycocyanin extract from cyanobacteria *Spirulina platensis* and other uses of biomass	2013	(Simkus et al., 2013)
KR1020020012351	Republic of Korea	Binex Co. Ltd., Cho, Man Gi, Kim, Won Suk	Method for producing natural beta-carotene using *Dunaliella salina*	2003	(Cho et al., 2003)
KR1020030077215	Republic of Korea	Esbiotech Co. Ltd.	Modified *Spirulina* and method for manufacturing the same	2004	(Han et al., 2004)
1020010101212	Republic of Korea	Kirin Brewery	Chlorophyll-rich and salt-resistant *Chlorella*	2005	(Koichi, 2005)
KR1020080040547	Republic of Korea	Gwangju Institute of Science and Technology	Method for preparing chlorophyll a by treating undisrupted intact *Chlorella* with ethanol and a method for preparing chlorin e6 at high yield from the obtained chlorophyll a	2008	(Kim et al., 2008a)

Table 23.1 (continued)

Patent/ Application no.	Country	Granted to	Patent details	Year	Reference
KR1020080055766	Republic of Korea	Gwangju Institute of Science and Technology	Simple and high-yield methods for preparing chlorophyll a and chlorin e6 by treating intact *Chlorella* with organic solvent	2008	(Kim et al., 2008b)
KR1020100081388	Republic of Korea	IUCF-HYU (Industry-university Cooperation Foundation Hanyang University)	Beta-carotene-producing strain that accumulates beta-carotene in low light	2012	(Jin and Park, 2012)
KR1020130035373	Republic of Korea	NBM Co. Ltd., Jeonju Biomaterials Institute	Medium composition for culturing a large amount of *Chlorella* with high chlorophyll content	2013	(Yim et al., 2013)
KR101389471	Republic of Korea	Hong Joo Heon, Kang Il Jun, No Hong Kyoon, Lee Dae Hoon, Park Hye Mi	Method for producing *Chlorella* powder with increased chlorophyll content and antioxidant activity	2014	(Hong et al., 2014)
KR1020160054239	Republic of Korea	Gangneungwonju University Industry-Academia Cooperation Group	Method for increasing chlorophyll a content in *Chlorella* ethanol extract	2017	(Byun, 2017)
KR102101988	Republic of Korea	Naturecoazen Farms Co. Ltd., Lee Heon	Production method of enzyme-treated *Spirulina* hydrolysate with improved heat stability and increased phycocyanin content	2020	(Suh et al., 2020)
RU2011143246A	Russia	Federal State Budgetary Educational Institution of Higher Professional Education "Moscow State University of Engineering Ecology" of the MSE of the Russian Federation (MGUIE) (RU)	Method for cultivation of microalgae for biofuel purpose	2013	(Григорьевич et al., 2013a)

Table 23.1 (continued)

Patent/ Application no.	Country	Granted to	Patent details	Year	Reference
RU0000131490	Russia	–	System of express remote control of aquatic environment parameters during the cultivation of microalgae for biofuel purpose	2013	(Григорьевич et al., 2013b)
KR100983023B1	South Korea	Korean Ocean Research and Development Institute (KORDI)	Method for manufacturing biodiesel using microalgae	2010	(Do Hyung et al., 2010)
KR101110068B1	South Korea	Korea Ocean Research and Development Institute	Manufacturing method of microalgae biofuel by culturing microalgae biofuel in large quantities	2012	(Do Hyung et al., 2012a)
KR101142358B1	South Korea	Korea Ocean Research and Development Institute	Biofuel production and a plant for manufacturing biofuel from microalgae	2012	(Do Hyung et al., 2012b)
KR101134294B1	South Korea	Korea Institute of Energy Research	Method for isolating oil from microalgae and converting it into bio diesel	2012	(Ji Yeon et al., 2011)
KR101264543B1	South Korea	Korea Institute of Energy Research	Method for extracting crude oil for biodiesel from microalgae by using a microwave pretreatment process and a manufacturing method of biodiesel using the same	2013	(Deog Keun et al., 2013)
KR101351281B1	South Korea	Korea Research Institute of Bioscience and Biotechnology, Korea Ocean Research and Development Institute, IAC	Microalgae *Chlamydomonas* strain with high-productivity of lipid, isolated from the Arctic ocean	2014	(Won Joong et al., 2011)
KR101547684B1	South Korea	STX Offshore Shipbuilding Co. LTD.	Offshore structure plant to manufacture microalgae biofuel	2015	(Tae Young et al., 2015)

Table 23.1 (continued)

Patent/ Application no.	Country	Granted to	Patent details	Year	Reference
US7838272B2	United States	Chevron USA Inc	Increased yield in gas-to-liquids processing via conversion of carbon dioxide to diesel via microalgae	2010	(Joseph, 2009)
US20090209014A1	United States	Washington State University Research Foundation	Heterotrophic algal high cell density production method and system	2011	(Zhanyou et al., 2009)
US8476060B2	United States	The University of Texas System	Process for separating lipids from a biomass	2013	(Russell et al., 2010)
US9359580B2	United States	Synaptic research LLC, Johns Hopkins University	Method for extraction and purification of oils from microalgae biomass using high-pressure CO2 as a solute	2016	(Donohue et al., 2014)
US9902977B2	United States	KAISAT and others	Process of producing bioenergy with low carbon dioxide emissions and zero-waste of biomass	2018	(Lim et al., 2015)
WO2009073816A1	WIPO (PCT)	Ohio State University Research Foundation	Optimization of biofuel production	2009	(Sayre and Richard, T., 2009)
WO2010045392A1	WIPO (PCT)	Kai Bioenergy Corporation	Hydrodynamic extraction of oils from photosynthetic cultures	2010	(Larach and Mario, C., 2010)

23.2.3 Food and feed

For centuries, microalgae have been exploited for food and feed. Microalgae-based functional food (dairy products, pastas, desserts, oil-derivatives, or supplements) and feed (for poultry, cattle, fish, and shellfish) are known to have favorable outcomes on human and animal health, including antimicrobial, antiviral, antioxidant, as well as anti-inflammatory effects. In food, there are two main categories of market products: dried algae that can directly be sold as dietary supplements as sources of proteins and carbohydrates, and high-value compounds that can be added to food/feed to enhance their nutritional value (Vigani et al., 2015). Microalgae are well suited for the food/ feed markets as they are able to synthesize the members of omega 6 as well as omega

3 fatty acids (GLA-gamma linolenic acid, ARA-arachidonic acid, and polyunsaturated fatty acids such as EPA- eicosapentaenoic acid and DHA- docosahexaenoic acid). Feed based on algal products has also been tested on animals (other than chicken and fish) and have proven to exert positive effects on animal physiology, immune response, gut function, and disease resistance as well as result in weight gain and increase in reproductive function (Camacho et al., 2019).

Commercial facilities for the production of microalgae are distributed worldwide; however, the actual commercialization is dominated by Asia and North America (Figure 23.2). The main reasons for this can be the different climatic conditions and the production systems used. China holds the highest number of patents in both food and feed sectors. The majority of the patents were related to production/manufacturing methods-based utility patents, various feed compositions, novel strains, and transgenic microalgae, and DHA-based oil and food items. A variety of food products such as fermented food, bio-oil, food seasoning, and puffed food have already been patented, but the most common category was food product enrichments.

23.2.4 Cosmeceuticals, nutraceuticals, and pharmaceuticals

High-value-added compounds (HVAC), synthesized by microalgae, have therapeutic value, in addition to commercial applications (Mehariya et al., 2021). Cosmeceuticals, nutraceuticals, and pharmaceuticals are used as HVAC microalgal compounds. Cosmeceuticals (cosmetics + pharmaceuticals) is the term used for cosmetics that have some additional health benefits such as skin whitening, antiaging, anticancer, antioxidant, anti-inflammation, antimicrobial, and UV protection among others. Using natural ingredients obtained from microalgae as gelling agents, stabilizers, thickeners, pigments, or as an active ingredient enhances the therapeutic properties of cosmetics while decreasing the exposure to harmful side effects caused by synthetic alternatives (Yarkent et al., 2020). Louis Vuitton Moët Hennessy, AGI Dermatics, Dermochlorella, Pentapharm, Greentech, and Givaudan are among the cosmetic companies that have launched microalgae-based cosmeceuticals worldwide (Mourelle et al., 2017).

Nutraceuticals (nutrition and pharmaceuticals) are compounds that enhance the medicinal benefits of food. Microalgae metabolites such as astaxanthin, lutein, amino acids, fatty acids, and others are used as functional foods and additives because of their antioxidant, antimicrobial, immunomodulatory, anticancer, and anti-inflammatory properties, and for their high nutritional value (Mehariya et al., 2021). *Chlorella, Dunaliella, Arthrospira, Haematococcus, Spirulina,* and *Isochrysis* are used as functional foods for bread, cookies, beverages, pasta, cakes, oil, and other consumables. The pharmaceutical activities of microalgae compounds include reduction in brain damage, cholesterol level, immune-stimulating, neuroprotective, anti-inflammatory, anticancer, antimicrobial, and antioxidant activities against wounds, peptic ulcers, and constipation. Patent and publication analyses showed that pharmaceuticals are the most explored application of bioactive compounds from microalgae, with approximately 136 patents from 2002 to 2022. It is followed by cosmeceuticals with 117 patents, and nutraceuticals with 42 patents. Companies such as Algahealth have the most patents in nutraceuticals (seven), and Biotechmarine has the most patents in cosmeceuticals (four). China is the leading country in patents, accounting for approximately 18% of patents.

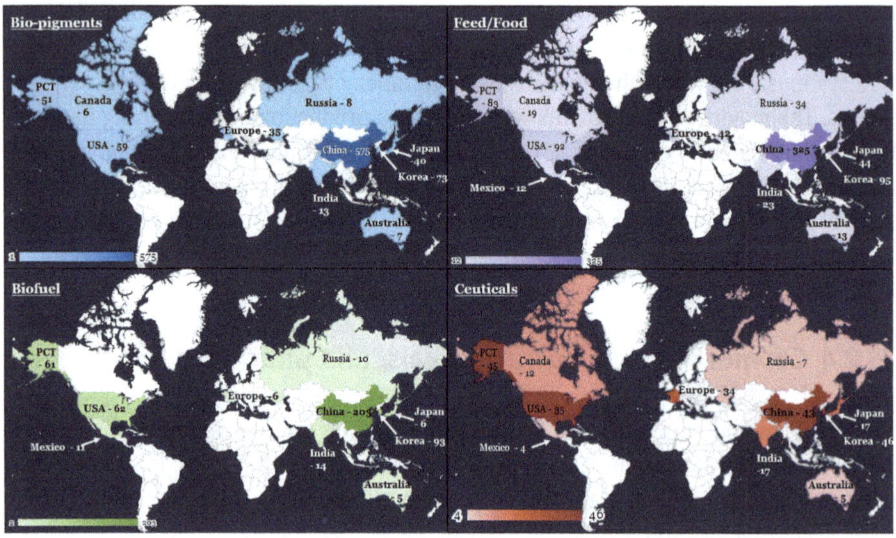

Figure 23.2: Country-wise distribution of patents in microalgae-related bio-pigments, biofuel, food/feed and cosmeceutical, nutraceutical and pharmaceutical commodities.

23.3 Global impact of microalgae market

The global microalgae market is expected to reach 18 billion by the year 2028, advancing at a heathy CAGR (Compound Annual Growth Rate) of approximately 8% between the forecast period of 2022 to 2028. The increasing number of research programs and businesses aiming to generate renewable energy and reduce carbon footprint will accelerate microalgae's current promising capabilities. Technological advancements, such as novel strains, synthetic biology, and automation as well as the genesis of new unions at the international and national levels to formulate regulatory standards, will be crucial to the global market's hike hereafter.

Recently, the COVID-19 pandemic affected various economies globally. The strategies adopted by the government to tackle COVID-19, such as quarantines and nationwide lockdowns, strict mandates on import/export, and the forced closures of corporations, impacted the microalgae industry adversely. The epidemic hampered the confectionery and food & beverage industry the most, as customers only bought necessities (Facts & Factors, 2022). The lower usage of functional foods and dietary supplements caused global supply chain disruptions, reducing the end-user consumption of microalgae biomass, impairing market growth. In the first quarter of 2020, a major drop in microalgae sales was seen due to obstructions in logistics and transportation. However, a strong demand for spirulina and chlorella was met through e-commerce platforms during the pandemic, owing to their immune boosting and nutritive properties. According to the FAO (Food and Agriculture Organization), China is one of the top producers and consumers of microalgae products, and its market was affected negatively. Furthermore, the epidemic caused undesirable repercussions across various nations including, India, the US, Brazil, Australia, and the EU-5. The rising demand for microscopic algae in the nutraceutical sector, along with the growing use of photobioreactors as a source of bulk proteins, is expected to boost demand over the projection period (Future Market Insights, 2022). Microalgae are a vital component in the generation of recombinant proteins; they have numerous health benefits, produce a wide array of bioactive compounds, and are GRAS (generally regarded as safe) certified.

However, the side effects related to microalgae products, along with the advancements required for cost reduction of down-streaming, the lack of commercialization of microalgae-based products, and the technological challenges, may hamper the growth of the global microalgae market. By the year 2028, food-grade microalgae are expected to dominate the global market as their application – food and beverages; dietary supplements' demand is proportionally increasing with consumer awareness, amid which, powder/dry form is anticipated to dominate as this form can be modified into tablets easily (Data Bridge Market Research, 2021).

23.4 Innovation management of microalgae-based systems in India

India is known as the 3rd largest ecosystem for startups in the world, with a consistent annual growth of 12–15%; biotechnology is an integral part of it and is also witnessing immense growth (Startup India). The government of India, with initiatives such as Make in India and Start-Up India (Department of Biotechnology), supports innovation and research based on various technologies. Funding agencies such as the Department of Biotechnology (DBT), Biotechnology Industry Research Assistance Council (BIRAC), India Science, Technology and Innovation, Centers of Excellence and Innovation in Biotechnology (CEIB), Sustainable Entrepreneurship and Enterprise Development (SEED) fund scheme, and others have launched various schemes for innovation research on microalgae biomass production, product development, and commercialization. The Crescent Innovation Incubation Council (CIIC) with BIRAC-BioNEST, the SEED fund

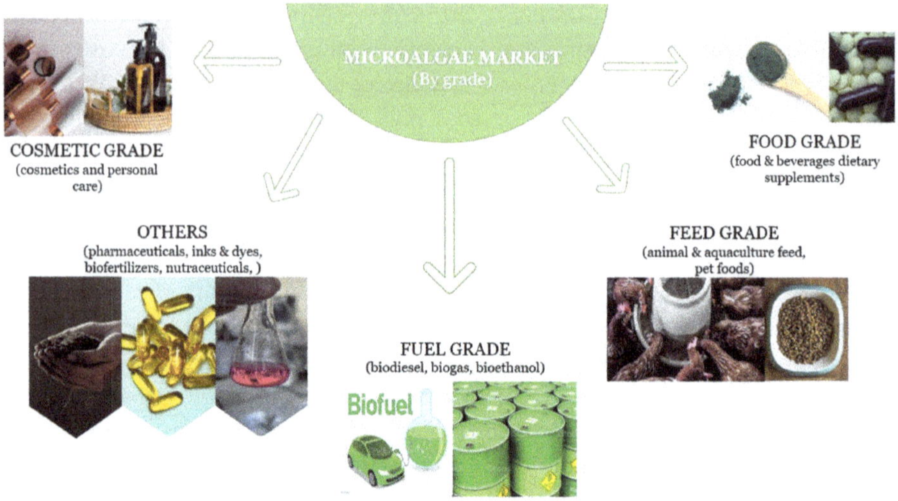

Figure 23.3: Grade-wise division of microalgae commodities in the global market – cosmetic-grade, feed-grade, food-grade, fuel-grade, and others such as biofertilizers, pigments, nutraceuticals, and pharmaceuticals.

scheme, and the Institution's Innovation Council established the Center for Microalgal and Nano-Biotechnology to promote sustainable utilization of microalgae biomass with nano-biological procedures and develop value-added products such as pigments, lipids, proteins, biodiesel, biochar, feed, fertilizers, low-cost high-quality biofuel, and carbon nanomaterials (Crescent Innovation Incubation Council). Projects funded by India Science, Technology, and Innovation include projects on waste management technology, usage of coal mine water for algal production, nutraceutical formulation from *Chlorella*, microalgae growth for biofertilizers, and high-value biofuel fractions (India Science, Technology and Innovation). The Bureau of Indian Standards had given specifications for the blending requirements, methods and testing of biodiesel, which is available in public domain under the designator identifier IS15607-2005 (Bureau of Indian Standards, 2005). Microalgae product-based startups such as Zaara Biotech (Zaara Biotech, 2019), Algallio Biotech (Algallio Biotech, 2016), and Sea6 Energy have also been provided monetary and appreciation support by Indian Central and State governments to promote use of natural products across the country.

23.5 Fundamental challenges in regulatory framework

Microalgae products do not just have to compete with chemical and plant sources but also with other microbiological sources such as fungi, yeast and, bacteria that could potentially be more profitable. The major challenge in this industry still remains with regard to upscaling, techniques for downstream processing, strict regulations, and protection rights. Improved or novel techniques are required to lower the production cost and solve the technical obstacles in extraction and refining, while favoring the stability and shelf-life extension of products. There are different regulations on microalgae-based food, feed, cosmetics, pharmaceuticals, nutraceuticals, cosmeceuticals, biofuels, and wastewater treatments in different geographical regions (Verdelho Vieira et al., 2022). Companies must adhere to these rules in order to market goods containing microalgae or their extracts around the world. This often involves applying to a regional authority for permission and submitting scientific data as well as health, safety evaluations, and records about minimal impact on the environment. The European legislation places more emphasis on the technology used to produce the final product than the American regulation, which pertains to the product, and determines whether it is safe or not. Protection of a technology, process, or commodity can be a boon but it can also hinder the research and availability of resources protected by it (Enzing et al., 2014).

For a non-GMO food-feed/nutraceuticals product, the European Food Safety Authority (EFSA) maintains a Qualified presumption of Safety (QSA) concept if the source is an already assessed microalgae source; otherwise, a framework provided by the European Community Regulation on Food Safety has to be followed (Vigani et al., 2015). The Novel Food Regulations are applicable to all phases of food and feed production, processing, and distribution, related to genetically modified (GM) foods (Spicer and Molnar, 2018). Adhering to these regulations requires years of time and rigorous quality assurance practices, which makes it harder for any novel commodity to even enter the market. The different restrictive regulatory requirements for food safety and approval in various countries, and consumer acceptance are the plausible reasons for the constrained pace of commercialization of microalgae products in the market.

The Food and Drug Administration (FDA) of USA regulates the Federal Food Drug and Cosmetic Act with the Dietary Supplement Health and Education Act; they are responsible for regulating food, feed, food additives, dietary ingredients, and supplements. It gives GRAS (generally recognized as safe) status to tested commodities and *Chlorella, Crypthccodinium cohnii, Porphyridium cruentum* and *Spirulina* have already acquired it (Mendes et al., 2008; Enzing et al., 2014). But it does not automatically mean that their extracts or active ingredients are also approved. They will have to enter the regulations cycle independently to get approval for human consumption, which again hinders further scientific breakthroughs. FDA and EFSA are also responsible for biosafety evaluation of GM microalgae. Commercial success of GM microalgae is still limited as its risk analysis, including strain identification, competition with wild kinds, vertical or horizontal gene transfer, safety considerations, and strain fitness needs to be assessed thoroughly (Spicer and Molnar, 2018; Henley et al., 2013).

23.6 Conclusion

Microalgae is a natural reservoir of various assets and can thus play a major role in the circular economy. It is suitable as food/feed, dietary supplements, pharmaceuticals, cosmetics, biofertilizers, bio stimulants, and their biomass can be used in wastewater treatment, carbon dioxide sequestration, and in biofuel production. Surging demand for microalgae in multiple industry verticals is augmenting the market's growth. Although the amounts produced and the market size of nutrients derived from microalgae are still considerably lower than those generated from crops, the industry has had a rapid expansion. With the growing consumer awareness and their preferences largely shifting toward natural and healthy products, microalgae seem to be an ideal candidate. Through the patent trend analysis in the current study, it is evident that this is the most appropriate time to shift toward a microalgae-based technology and encourage their commercialization. Advanced biotechnological interventions may help to overcome the fundamental challenges associated with the production and downstream processes. On the other hand, there can be fundamental changes in the market interests, impacting the trade value of patents as well. There is also a need for better inter-institutional cooperation between educational institutes and private sector along with more emphasis on better policy and regulations around commercialization of algal products.

References

Algallio Biotech. (2016). Algallio Biotech Private Limited, Viewed on October 17, 2022, at http://www.algalliobiotech.com/

Ambati, R. R., Gogisetty, D., Aswathanarayana, R. G., Ravi, S., Bikkina, P. N., Bo, L., & Yuepeng, S. (2018). Industrial potential of carotenoid pigments from microalgae: Current trends and future prospects. *Critical Reviews in Food Science and Nutrition*, *59*, 1880–1902.

Ami, B.-A., & Nobuo, M. (2015). Method for producing beta-carotene rich *Dunaliella* powder: CN102771836.

Anguselvi, V., Ram, L. C., Ebhinmasto, R., Chugh, P., Kashyap, R. K., & Sinha, R. (2018). A process for sequestration of CO2 and traces of hydrocarbon from natural gas processing industry using microalgae: IN201611036660.

Anila, N., Simon, D. P., Bhattacharaya, S., Sarada, R., & Ravishankar, G. A. (2013). An improved medium composition and a process for enhanced lutein production from microalgae: IN3120/DEL/2011.

Anjun, L., Jie, Z., Naihong, X., Yingfen, L., & Jiantao, L. (2013). Method for extracting beta-carotene from *Dunaliella* by inorganic base: CN102617431.
Arora, A. Y., & Katiyar, R. (2022). A process for the extraction and recovery of high-quality protein from microalgae pool: IN202021033216.
Barhate, R. S., Sampangi, C., & Raghavarao, K. S. M. S. (2009). A process for the preparation of food colourant from *Spirulina*: IN 237338.
Begum, H., Yusoff, F. M., Banerjee, S., Khatoon, H., & Shariff, M. (2016). Availability and utilization of pigments from microalgae. *Critical Reviews in Food Science and Nutrition*, *56*, 2209–2222.
Biofuture Platform Facilitator, and Clean Energy Ministerial. (2013). Biofuture Platform – Accelerate the Transition to a Global Bioeconomy, viewed on October 17, 2022, at https://biofutureplatform.org/
Borowiak, D., & Krzywonos, M. (2022). Bioenergy, biofuels, lipids and pigments – Research trends in the use of microalgae grown in photobioreactors. *Energies*, *15*, 5357.
Burde, S. K., Sarada, R., Raman, V., & Ravishankar, G. A. (2015). A process for obtaining water dispersible astaxanthin composition: 264507.
Bureau of Indian Standards. (2005). Bio-diesel (B 100) blend stock for diesel fuel: Petroleum, Coal, and Related Products.
Byun, H. (2017). Method for increasing chlorophyll a content in *Chlorella* ethanol extract: KR1020160054239.
Camacho, F., Macedo, A., & Malcata, F. (2019). Potential industrial applications and commercialization of microalgae in the functional food and feed industries: A short review. *Marine Drugs*, *17*, 312.
Chauhan, V. S., Kumudha, K., Swarnalatha, G. V., Kavitha, M. D., Vidyashankar, S., Sarada, R., Ravishankar, G. A., Jain, A. K., Bansal, A. K., Rathi, P., Kumar, J., & Pande, A. (2015). A composition and method for controlling rotifer in microalgal culture systems: IN1787/DEL/2012.
Cho, M. G., Kim, W. S., & Lee, B. C. (2003). Method for producing natural beta-carotene using *Dunaliella salina*: KR1020020012351.
Chunling, C., & Changxiang, X. (2014). Method utilizing microalgae to produce biodiesel: CN103642579A.
Chunxiao, M., & Zhengquan, G. (2009). Method for preparing phycocyanin and allophycocyanin at same time: CN101003565.
Crescent Innovation Incubation Council Center for Microalgal and Nano Biotechnology, viewed on October 17, 2022, at https://www.ciic.ventures/microalgal-and-nano-biotechnology/
Dan, T., Jingcheng, Y., Xuzhen, D., Guoliang, J., Xuemei, D., Liwei, S., Yiwei, T., & Jianhui, P. (2012). Method for cultivating oleaginous microalgae by using fecal sewage: CN102618446B.
Daopeng, Z., Jibin, D., & Zhiliang, L. (2009). Technique for extracting phycocyanin, chlorophyll and spirulina polysaccharide in *Spirulina*: CN101607988.
Data Bridge Market Research. (2021). Microalgae Market Size, Industry Report Analysis, Growth, Share, Application, Segmentation, Future Demands, Business Opportunities by Cellana, Corbion, Algarithm, Cyanotech, Algaecytes: Data Bridge Market Research, 350.
Deog Keun, K., You Kwan, O., Soon Chul, P., Jin Suk, L., Ji Yeon, P., & Jun Pyo, L. (2013). Method for extracting crude oil for biodiesel from microalgae by using a microwave pretreatment process, and a manufacturing method of the biodiesel using the same: KR101264543B1.
DBT. (2018). Department of Biotechnology Energy Science & Waste to Value, viewed on October 17, 2022, at https://dbtindia.gov.in/schemesprogrammes/research-development/energy-environment-and-bio-resource-based-applications-1
DBT. (2022). Department of Biotechnology Make in India & Start-up India | Department of Biotechnology, viewed on October 17, 2022, at https://dbtindia.gov.in/schemes-programmes/translational-industrial development-programmes/make-india-start-india
Do Hyung, K., Soo Jin, H., Cheol Hong, O., Se Jong, J., Affan, A., Dae Won, L., & Min, J. J. (2010). Method for manufacturing biodiesel using microalgae: KR100983023B1.

Do Hyung, K., Soo Jin, H., Chol Hong, O., Afaan, A., & Tae Ho, K. (2012a). Manufacturing method of microalgae biofuel by culturing microalgae biofuel in large quantities: KR101110068B1.

Do Hyung, K., Soo Jin, H., Chol Hong, O., & Dong Kyu, O. (2012b). Biofuel production and a plant for manufacturing biofuel from microalgae: KR101142358B1.

Do-Hyung, K., Soo-Jin, H., Chol-Hong, O., Tae-Ho, K., Heung-Sik, P., & Heung-Sik, A. (2014). Production plant for microalgae biofuel, bioreactor for producing biofuel, and method for producing microalgae biofuel: CN103597069.

Donohue, M. D., Betenbaugh, M. J., Oyler, G. A., & Rosenberg, J. N. (2014). Method for extraction and purification of oils from microalgal biomass using high-pressure CO2 as a solute: US9359580B2.

Enzing, C., Ploeg, M., Barbosa, M., & Sijtsma, L. (2014). Microalgae-based products for the food and feed sector: An outlook for Europe. *JRC Scientific and Policy Reports*, 49–60.

Facts and Factors. (2022). Demand for Global Microalgae Market Size & Share Worth USD 18.3 Billion by 2028, Exhibit a CAGR of 8.2% Growth | Microalgae Industry Trends, Value, Analysis & Forecast Exclusive Insight Report by Facts & Factors: Facts and Factors, 207.

Future Market Insights. (2022). Market Insights on Microalgae covering sales outlook, demand forecast & up-to-date key trends: Future Market Insights, 250.

Gajraj, R. S. (2013). *Spirulina* biomass with modified colour characteristic and process for preparing the same: 2379/MUM/2010.

Ganapathi, P., Sampangi, C., Narayan, A. V., Burde, K. S., Udayasankar, K., Ravishankar, G. A., Mallikarjuna, K. S., & Raghavarao, S. (2012). A process for preparation of natural blue colourant from *Spirulina* species: 251116.

Gholkar, P. V., Tanksale, A., & Shastri, Y. (2019). Catalytic conversion of microalgae into hydrogen rich syngas using reactive flash volatilzation: IN201721042669.

Goswami, G., Sinha, A., Kumar, R., Dutta, B. C., Singh, H., & Das, D. (2020). Process for enhancing biomass productivity by high density cultivation of microalgae: IN201811041629.

Goud, V. V., & Suryawanshi, P. G. (2022). An integrated process and system of harvesting and extraction of lipid from wet algal biomass: IN202231057904.

Григорьевич, С. В., Михайловна, И. Е., Григорьевич, Ч. В., Юрьевна, Р. М., & Петрович, П. С. (2013a). Method for cultivation of microalgae for biofuel purpose: RU2011143246A.

Григорьевич, С. В., Михайловна, И. Е., Игоревна, Я. А., Григорьевич, Ч. В., Петрович, П. С., & Владимирович, Ч. С. (2013b). System for express remote control of aquatic environment parameters during cultivation of microalgae for biofuel purpose: RU0000131490.

Han, B. H., Pyo, M. G., & Yang, S. J. (2004). Modified *Spirulina* and method for manufacturing the same: KR1020030077215.

Hao, H., Zhen, G., Jilong, Y., Xiaohong, T., Huang, H., Zipanlon, S., Ling, L., Xingguo, F., & And Xuteng, H. (2013). Production method of high-quality microalgae biodiesel: CN103451101B.

Hatem, B. O., & Ammar, J. (2017). Method for inducing the synthesis of phycobiliproteins: IN201747007941.

Henley, W. J., Litaker, R. W., Novoveská, L., Duke, C. S., Quemada, H. D., & Sayre, R. T. (2013). Initial risk assessment of genetically modified (GM) microalgae for commodity-scale biofuel cultivation. *Algal Research*, *2*, 66–77.

Hong, J. H., Kang, I. J., No, H. K., Lee, D. H., & Park, H. M. (2014). Method for producing *Chlorella* powder with increased chlorophyll content and antioxidant activity: KR101389471.

Huang, X. (2020). Preparation method and application of *Spirulina* phycocyanin: CN109160947.

India Science, Technology and Innovation Home | India Science, Technology & Innovation – ISTI Portal, viewed on October 17, 2022, at https://www.indiascienceandtechnology.gov.in/

Ji Yeon, P., You Kwan, O., Soon Chul, P., Jin Suk, L., Deok Keun, K., & Jun Pyo, L. (2011). Method for isolating oil from microalgae and converting into bio diesel: KR101134294B1.

Jianguo, J., Haohong, C., Fangchun, W., & And Hong, X. (2021a). Lycopene high-yield engineering bacterium based on dunaliella metabolic pathway as well as construction method and application thereof: CN110511919.

Jianguo, J., Haohong, C., Hong, X., & Zhicong, L. (2021b). Engineering bacteria with high beta-carotene yield based on dunaliella metabolic pathway and construction method and application of beta-carotene high yield engineering bacteria: CN110499272.

Jianguo, J., & Zhiwei, Y. (2012). Beta-carotene engineering bacteria based on *Dunaliella salina* metabolic pathway and construction method: CN102168096.

Jianguo, J., Zhiwei, Y., & Yongmin, L. (2012). Beta-carotene engineering bacteria based on *Dunaliella bardawil* metabolic pathway and construction method thereof: CN102154330.

Jin, E. S., & Park, S. H. (2012). Beta-carotene producing strain which accumulates beta-carotine at low light: KR1020100081388.

Jorgen, J. (1968). Fermentation process for producing alcoholic beverages from microalgae: US3389998A.

Joseph, M. S. (2009). Increased yield in gas-to-liquids processing via conversion of carbon dioxide to diesel via microalgae: US7838272B2.

Kalra, R., Gaur, S., & Goel, M. (2020). Microalgae bioremediation: A perspective towards wastewater treatment along with industrial carotenoids production. *Journal of Water Process Engineering*, *40*, 101794.

Kalra, R., Gaur, S., & Goel, M. (2022). Harnessing the potential of microalgal species Dunaliella: A biofuel and biocommodities perspective, in: Ahmad, A., Banat, F., Taher, H. *Algal Biotechnology. Integr. Algal Engin. Bioen., Bioremed., Biomed. Appl*, 259–279.

Khan, M. I., Shin, J. H., & Kim, J. D. (2018). The promising future of microalgae: current status, challenges, and optimization of a sustainable and renewable industry for biofuels, feed, and other products. *Microbial Cell Factories*, *17*, 0–21.

Kim, Y. C., Park, Z. Y., & Kim, H. J. (2008a). Method for preparing chlorophyll a by treating undisrupted intact *Chlorella* with ethanol and a method for preparing chlorin e6 at high yield from the obtained chlorophyll A: KR1020080040547.

Kim, Y. C., Park, Z. Y., & Kim, H. J. (2008b). Simple and high yield methods for preparing chlorophyll a and chlorin e6 by treating intact *Chlorella* with organic solvent: KR1020080055766.

Koichi, N. (2005). Chlorophyll-rich and salt-resistant *Chlorella*, 1020010101212.

Kravit, N. G. (2007). Biotransformation of compounds using non-prokaryoyic microalgae: IN3861/DELNP/2005.

Kumar, V., Nanda, M., Prasad, R., & Kumar, S. (2019a). A process for microalgae harvesting and enhancing biomass productivity: IN201811016783.

Kumar, Dr. V., Nanda, Dr. M., Verma, Dr. M., & Singh, Dr. A. (2019b). A method for preparation of nutrient source from fermentation residues of food waste for cultivation of microalgae: 201811008547.

Lakshmi, C., Hakeem, H. A., Muthappa, B. S., Ahmed, M. R., Yamanappa, V., & Upendra, R. S. (2019). A hybrid photo bioreactor (pbr) for the cultivation of oleaginous microalgae for optimum lipid yield and subsequent biodiesel production: IN201941044256.

Larach, & Mario, C. (2010). Hydrodynamic extraction of oils from photosynthetic cultures: WO2010045392A1.

Lei, S., Huimin, Z., & Minsheng, L. (2009). Method for preparing biodiesel by using microalgae lipid as raw material: CN101613618B.

Lei, S., & Minsheng, L. (2009). Biomass energy prepared by one-step method of microalgae: CN101580857B.

Li, D., Du, W., Fu, W., & Cao, X. (2020). A quick look back at the microalgal biofuel patents: Rise and fall. *Frontiers in Bioengineering and Biotechnology*, *8*.

Lim, K.-H., Lee, K., Um, Y.-S., & Lee, M.-W. (2015). Process of producing bioenergy with low carbon dioxide emissions and zero-waste of biomass: US9902977B2.

Magneschi, L., Billey, E., Jouhet, J., Marechal, E., & Collin, S. (2022). Mutation of an acyl-coA synthase for increased triacylglycerol production in microalgae: 202227002093.

Mehariya, S., Goswami, R. K., Karthikeysan, O. P., & Verma, P. (2021). Microalgae for high-value products: A way towards green nutraceutical and pharmaceutical compounds. *Chemosphere*, *280*, 130553.

Mekideche, N. (2009). Cosmetic active ingredient composed of arginine ferrulate and microalgae extract and its uses: 8051/DELNP/2008.

Mendes, A., Reis, A., Vasconcelos, R., Guerra, P., & Lopes da Silva, T. (2008). *Crypthecodinium cohnii* with emphasis on DHA production: A review. *Journal Applied Phycology*, *21*, 199–214.

Microalgae Market Share, Size, Trends, Industry Analysis Report, by Product Type (Dunaliella Salina, Spirulina, Chlorella, and Others), Microalgae Strain (Haematococcus Pluvialis, Phaeodactylum Tricornutum, Porphyridium Cruentum, Nannochloropsis, and Others), Grade (Food Grade, Feed Grade, Fuel Grade, Cosmetics Grade, and Others), Application, by Region; Segment Forecast, 2022–2030. (2022). Polaris Market Research, 110 p.

Mishra, S. C. P., Ghosh, P. K., Gandhi, M. R., Bhattacharya, S., Maiti, S., Upadhyay, S. C., Ghosh, A., Prasad, R. B. N., Kanjilal, S., Mishra, S. K., Shrivastav, A. V., Pancha, I., Paliwal, C., & Ghosh, T., and others. (2015). Engine worthy fatty acid methyl ester (biodiesel) from naturally occurring marine microalgal mats and marine microalgae cultured in open salt pans together with value addition of co products: 6564/DELNP/2012.

Mourelle, M., Gómez, C., & Legido, J. (2017). The potential use of marine microalgae and cyanobacteria in cosmetics and thalassotherapy. *Cosmetics*, *4*, 46.

Murata, M. M., Ito Morioka, L. R., Da Silva Marques, J. B., Bosso, A., & Suguimoto, H. H. (2021). What do patents tell us about microalgae in agriculture? *AMB Express*, *11*, 154.

NA. (2007). Immunostimulatory preparations isolated from microalgae: IN3008/CHENP/2007.

Novoveská, L., Ross, M. E., Stanley, M. S., Pradelles, R., Wasiolek, V., & Sassi, J.-F. (2019). Microalgal carotenoids: A review of production, current markets, regulations, and future direction. *Marine Drugs*, *17*, 640.

Oran, A. (2022). Improved process for producing fucoxanthin and/or polysaccharides from microalgae: IN201827009409.

Pan, T., Li, M., Wang, M., Zhang, J., Fu, G., Song, Q., & Dong, W. (2022). Synchronous extraction method of spirulina phycocyanin and algae oil: CN113388026.

Pasa Emer, A., Yifeng, C., & Zhilan, S. (2015). Application of methyl jasmonate in improvement of microalgae biodiesel quality: CN104762335A.

Pengcheng, L., Xiaolin, C., Ronge, X., Song, L., & Huahua, Y. (2014). A method for extracting oil from microalgae wet algae mud: CN103981017A.

Pulz, O., & Gross, W. (2004). Valuable products from biotechnology of microalgae. *Applied Microbiology and Biotechnology*, *65*, 635–648.

Puranik, P. R., Deshmukh, D. V., Pandav, P. V., Patel, V. E., Pawar, S. T., & Puranik, K. P. (2012). Natural blue pigment and process for producing the same: IN689/MUM/2010.

Rai, M. P., Khanra, A., & Vasistha, S. (2020). A method for flocculation and lipid extraction by microalgae using zirchonium di-oxide nanoparticles: IN201811039752.

Rai, M. P., & Vasistha, S. (2021). A biopolymer induced microalgae cell flocculation and enhanced lipid production for biodiesel application: IN202011010829.

Ranjan, A. (2019). Regulated microalgae harvester: IN201841006816.

Rasane, Dr. P., Mehta, Dr. C. M., Singh, Ms. J., Kaur, S., & Gupta, P. K. (2020). An immunity boosting drink using microalgae: IN202011045595.

Rizwan, M., Mujtaba, G., Memon, S. A., Lee, K., & Rashid, N. (2018). Exploring the potential of microalgae for new biotechnology applications and beyond: A Review:. *Renewable and Sustainable Energy Reviews*, *92*, 394–404.

Rocha, A. M., Sahoo, D., Ferrer, T., Quintella, C., and Torres, E. (2012). Biodiesel Production from Microalgae: A Mapping of Articles and Patents, in: Gordon, R., Seckbach, *J. Cellular Origin, Life in Extreme Habitats and Astrobiology, The Science of Algal Fuels*, 283–303.

Russell, C., Calvin, H., & Joaquin, R. (2010). Process for separating lipids from a biomass: US8476060B2.

Sarada, R., Raghavarao, K. S. M. S., & Ravishankar, G. A. (2010). A process for production of carotenoid from microalgae: 241826.

Sayre, & Richard, T. (2009). Optimization of biofuel production: WO2009073816A1.

Sebastian, T. S., & Kumaravel, S. (2008). Photoautotrophic growth of microalgae for omega-3 fatty acid production: 2335/CHE/2006.

Shao, M., Qin, S., & Gao, Z. (2015). Industrially producing method of food-grade phycocyanin: CN104292326.

Shosh, A. (2012). Compositions comprising red microalgae polysaccharides and metals: IN9162/DELNP/2010.

Simkus, A., Simkiene, A., Mazuras, C., & Mazuras, L. (2013). Method of isolating c-phicocyanin extract from cyanobacteria *Spirulina platensis* and other use of biomass: LT5924.

Song, J., Shi, M., Qin, X., Han, Z., & Wang, F. (2017). Method for separating and purifying phycocyanin in spirulina by using polyacrylic acid/magnesium oxide hybrid microspheres: CN104479008.

Song, J., Zhang, D., Shi, M., Tan, X., Han, Z., & Wang, F. (2015a). Technology for separating phycocyanin in *Spirulina* by using polyvinyl ether alcohol amine through two aqueous phase extraction: CN104387468.

Song, J., Zhang, D., Tian, X., Chen, Y., Yuan, Y., & Wang, F. (2015b). Technology for separating phycocyanin in *Spirulina* by using polyvinyl ether phosphate through two aqueous phase extraction: CN104387469.

Spicer, A., & Molnar, A. (2018). Gene editing of microalgae: Scientific progress and regulatory challenges in Europe. *Biology*, *7*, 21.

Startup India Indian Startup Ecosystem, viewed on October 17, 2022, at https://www.startupindia.gov.in/content/sih/en/international/go-to-market-guide/indian-startup-ecosystem.html

Statista. (2018). Biofuel production in leading countries, viewed on October 1, 2022, at <https://www.statista.com/statistics/274168/biofuel-production-in-leading-countries-in-oil-equivalent/>

Subramanian, M., Muthuirulappan, S., Marudhavanan, S., & Kumar, A. K. A. (2016). Microalgae and methods for producing docosahexaenoic acid: IN201641010123.

Suh, H. J., Ki, B. H., Gi, Y. B., & Lee, H. (2020). Production method of enzyme treated-*Spirulina* hydrolysate with improved heat stability and increased phycocyanin content: KR102101988.

Tae Young, P., Sung Kyun, A., Joon Bae, L., & Joong Kwan, L. (2015). Offshore structure plant to manufacture microalgae biofuel: KR101547684B1.

Talebi, A.F., Tabatabaei, M., & Aghbashlo, M. (2018). Recent Patents on Biofuels from Microalgae, in: Jacob-Lopes, E., Queiroz Zepka, L., & Queiroz, M. *Energy from Microalgae. Green Energy and Technology*, 291–306.

Thomas, S. S., Swaminathan, K., & Nagaraj, J. B. (2003). Process to produce astaxanthin from *Haematococcus* biomass: WO2003027267.

Upendra, R. S., Khandelwal, P., Shalini, R., Pallavi, B., Nayak, S., & Jain, S. (2018a). Scale up studies of microalgae *Chlorella rotunda* for biolipids (Biodiesel) as third generation biofuel along with value-added product using indigenously developed Photo bioreactor:IN201741018074.

Upendra, R. S., Khandelwal, P., Ruia, A., & Grace, A. (2018b). Design of a novel hybrid photobioreactor for cultivation of microalgae for biodiesel production: IN201741018095.

Venkatesh, N. S. (2008). A method of the extraction and purification of natural mixed carotenoid from microalgae: 221688.

Verdelho Vieira, V., Cadoret, J.-P., Acien, F. G., & Benemann, J. (2022). Clarification of most relevant concepts related to the microalgae production sector. *Processes*, *10*, 175.

Vigani, M., Parisi, C., Rodríguez-Cerezo, E., Barbosa, M. J., Sijtsma, L., Ploeg, M., & Enzing, C. (2015). Food and feed products from micro-algae: Market opportunities and challenges for the EU. *Trends in Food Science & Technology*, *42*, 81–92.

Wang, P. (2018). Method for forcing *Dunaliella tertiolecta* to accumulate beta-carotene: CN104450849.

Wei, X., Liu, H., & Xu, R. (2015). Method for extracting phycocyanin: CN103554250.

Wei, X., Liu, H., & Xu, R. (2016). Method of extracting high-purity phycocyanin: CN103613661.

Wenzhou, X., He, H., Lin, J., Dong, J., He, M., Wu, B., & Zeng, C. (2005). Method for preparing phycocyanin of sea water: CN1438240.

Won Joong, C., Youn Il, P., Han Gu, C., Sung Ho, K., Suk Weon, K., Su Hyun, S., & Jang Ryol, L. (2011). Microalgae *Chlamydomonas* strain with high productivity of lipid, isolated from arctic ocean: CN1438240.

Xiaoxi, M., Xiaowei, P., & Zhibin, X. (2015). A kind of microalgae biodiesel preparation system of ultrasonic assistant shell calcium foundation stone ash: CN204509239U.

Yadav, P. V., Negi, P. S., & Bala, M. (2020a). Novel composition of synthetic culture media for growth of microalgae with bio-fuel potential: IN201811030379.

Yadav, S. K., Purohit, A., & Kumar, V. (2020b). A simple method for extracellular production of high purity c-phycocyanin from *Spirulina platensis*: IN201811050007.

Yanshan, D., Minsheng, L., & Qiaoli, Y. (2012). Method for preparing biodiesel from microalgae through one-step method by utilizing supercritical methanol: CN102559374B.

Yarkent, Ç., Gürlek, C., & Oncel, S. S. (2020). Potential of microalgal compounds in trending natural cosmetics: A review. *Sustainable Chemistry and Pharmacy*, *17*, 100304.

Yim, S. K., Kwon, T. H., Doo, H. S., & Kwon, B. R. (2013). Medium composition for culturing a large amount of *Chlorella* with high chlorophyll content: KR1020130035373.

Zaara Biotech. (2019). Zaara Biotech – Feeding the nation forward, viewed on October 17, 2022, at https://zaarabiotech.com/

Zhang, X., & Liao, X. (2013). Method for separating high purity phycocyanin from *Spirulina*: CN101899102.

Zhang, X., Zhang, H., Zhao, Y., Zhang, Y., Li, H., & Wang, H. (2017). Method for extracting beta-carotene from *Dunaliella salina* by utilizing subcritical fluid: CN105646313.

Zhanyou, C., Zhiyou, W., Craig, F., & Shulin, C. (2009). Heterotrophic algal high cell density production method and system: US20090209014A1.

Zhenhong, Y., Changlin, M., Pengmei, L., rowan, Zhongming, W., Huiwen, L., Lingmei, Y., Xinshu, Z., Zhibing, L., & Shuna, L. (2015). Method for direct preparation of biodiesel by utilizing microalgae ultrasonic-assisted ionic liquid composition: CN104560409A.

Index

agricultural wastewater 168
algae 350, 430
algal biofuels 356
anaerobic digestate 199
anaerobic digestion 169
animal feed 4, 23, 90, 302
Arthrospira platensis 22
artificial intelligence 94
artificial neural networks 94
autotrophic 82
autotrophically 327

bioaccessibility 46
bioactive 10
bioactive compounds 56, 275
bioactive metabolites 108, 326
bioactive molecules 459
bioactivity 46
bioavailability 46
bio-based processes 134
bio-based products 121
biochar 301
biochemical engineering 49
biochemical processes 360
biochemicals 299
biocompounds 13
biodiesel 11, 25, 44, 149, 259, 277, 301, 394
bioeconomy 291
bioelectrochemical systems 425
bioenergy 58, 134, 214
bioethanol 11, 25, 257, 301, 390
biofertilizers 10, 25, 113
biofuel 25, 242, 472
biofuels 5, 20, 83, 108, 134, 184, 272, 299, 350, 378, 479
biogas 11, 30, 258, 397
biohydrogen 301
bioindustry 118
biomass 4, 20, 68, 82, 107, 134, 184, 214, 272, 300, 427, 456, 481
biomass pretreatments 76
biomaterials 90
biomethane 25
biomitigation 413
biomolecules 49, 69
bio-oils 399
biopesticides 10, 25
bioplastic 272
bioplastics 25
biopolymers 309
bioproducts 10, 134, 272
bioreactor 216
bioreactors 335
biorefinery 10, 42, 68, 83, 108, 148, 215, 291, 300, 378, 410
bioremediation 90, 108, 187, 216, 472
biostimulants 10
biosyngas 302
biotechnology 6, 217, 244, 291
Botryococcus braunii 7

carbon capture and storage 410
carbon capture and utilization 411
carotenoids 290
cell disruption 73, 255, 280
central carbon metabolism 413
chemical specialties 42
Chlorella 5, 61, 227
Chlorella vulgaris 22, 432
circular bio-economy 95
circular economy 95, 109, 272
climate change 43
closed photobioreactor 382
closed system 335
closed systems 246
CO_2 biofixation 285
commercial scale 232
cyanobacteria 82, 290, 300

dark fermentation 351
denitrification 221
dietary supplements 118, 302, 494
downstream 42, 56, 108, 168, 215, 243, 326, 378, 414
downstream bio-refinery 473
downstream processes 34, 93, 142, 475
downstream processing 281

eco-friendly 84, 245
energy efficiency 460
energy integration 134, 141, 416, 442
environmental footprint 460

https://doi.org/10.1515/9783110781267-024

environmental impact 165, 355
environmental stress 277
environmental sustainability 49
environmentally sustainable 442
eutectic solvents 68
exhaust gases 420

feedstuffs 24
fermenter 220
first generation biofuels 350
first-generation biomass 214
flat-plate reactors 225
food additive 47
food colorant 22
food ingredients 4, 12
food supplements 118
food supply 19
fourth biofuel generation 350
fourth-generation biomass 215
fuels 4

genetic engineering 49, 90, 215, 245, 457
green-fuels 275

Haematococcus 227
heat exchangers 156
heat integration 140
heat/mass transfer 378
heterotrophic 5, 82, 150, 381
heterotrophically 327
high-added-value products 48
high-value biomass 194
high-value chemicals 480
high-value compounds 83
high-value metabolites 194
high-value molecules 114
high-value products 20, 102, 137, 273, 456
high-value specialty molecules 242
human nutrition 90
hydrothermal carbonization 194
hydrothermal liquefaction 281, 351

industrial wastewater 168
integrated biorefinery 154
intellectual property protection 480
internet of things (IoT) 50

life cycle assessment 166, 448
Life Cycle Life Assessments (LCA) 57

marine organisms 243
mass integration 148, 154, 416, 442
mass transfer 246
mechanical cell disruption 74
medium-value microalgae 23
microalgae 4, 19, 41, 56, 67, 82, 108, 134, 148, 165, 183, 215, 241, 271, 299, 325, 378, 410, 456, 479
microalgae bioeconomy 48
microalgae biomass 48, 59, 223, 243, 497
microalgae biorefineries 148
microalgae biorefinery 68, 414
microalgae biotechnology 6, 42, 300, 462
microalgae harvesting 465
microalgae industry 57
microalgae market 47, 83, 300, 496
microalgae specialty chemicals 326
microalgae supply chain 108
microalgae systems 134
microalgae upstream processing 59
microalgae-based biofuels 48
microalgae-based bioproducts 414
microalgae-based processes 49
microalgae-based products 47, 114, 167, 243
microalgae-based systems 135, 165
microalgal biomass 148
microalgal biorefinery 447
microalgal bioremediation 193
microbial desalination cells 425
microbial electrolysis cells 425
microbial fuel cell 425, 473
microbial solar cells 425
midstream 56
mixotrophic 5, 82, 150, 381
mixotrophically 327
multiproduct biorefinery 102
municipal wastewater 168

natural food colorings 5
natural ingredients 495
nitrification 221
nonrenewable energy 44

open pond 382
open system 335
open systems 187, 246

patents 456, 480
photoautotrophic 5, 227, 277, 381
photo-autotrophic 150

photoautotrophic cultivation 60
photobioreactor 4, 216, 278
photobioreactors 60, 171, 335
photoheterotrophic 381
photoinhibition 228
physiochemical factors 191
phytoremediation 59
pigment-protein complex 71
plant-based meat 13
polyhydroxyalkanoate 304
primary metabolites 327
process integration 153, 173, 291, 317, 352, 410, 442, 467
process intensification 217, 243, 326, 355, 378, 442, 461
protein supplements 118

raceway ponds 219
reactors 156

second generation biofuels 350
secondary metabolites 33, 50, 236, 277
second-generation biomass 215
single-cell protein 184
Spirulina 5, 61
sustainability 283, 448, 457
sustainability metrics 175
sustainable 236
sustainable development 56
sustainable production 152

Tenochtitlan 22
third generation biofuels 350
third-generation biofuels 147
third-generation biomass 215
total chain integration 155
transcription factor engineering 49
transesterification 148
transgenic algae 459
tubular photobioreactors 224

ultrafiltration membrane 178
ultrasound 237
unicellular microalgae 6
unicellular protein 42
upstream 42, 56, 108, 142, 243, 378, 414, 459

value-added chemicals 326
value-added metabolites 45
value-added products 152, 179, 300, 415, 498

wastewater bioremediation 198
wastewater remediation 284
wastewater treatment 11, 109, 283
water footprint 174
water integration 165, 417, 442
water treatment 169

β-carotene 222

www.ingramcontent.com/pod-product-compliance
Lightning Source LLC
Chambersburg PA
CBHW061151240926
51846CB00006B/978